THE MELANOCORTIN SYSTEM

ANNALS OF THE NEW YORK ACADEMY OF SCIENCES
Volume 994

THE MELANOCORTIN SYSTEM

Edited by Roger D. Cone

The New York Academy of Sciences
New York, New York
2003

Library of Congress Cataloging-in-Publication Data

International Melanocortin Meeting (5th : 2002 : Sunriver, Or.)
  The melanocortin system / edited by Roger D.Cone.
      p. ; cm. -- (Annals of the New York Academy of Sciences ; v. 994)
  "Result of a conference entitled the Fifth International Melanocortin Meeting, held on August 25–28, 2002 in Sunriver, Oregon" -- Table of contents.
  Includes bibliographical references and indexes.
    ISBN 1-57331-442-0 (cloth : alk. paper) -- ISBN 1-57331-443-9(paper : alk. paper)
  1. MSH (Hormone)--Congresses. 2. Proopiomelanocortin--Receptors--Congresses.
    [DNLM: 1. Corticotropin--physiology--Congresses. 2. Corticotropin--agonists--Congresses. 3. Receptors, Corticotropin--Congresses. WK 515 I6036 2003] I. Cone, Roger D. II. Title. III. Series.
    Q11.N5 vol. 994
    [QP572.M75]
    500 s—dc21
    [612.4/   2003011820

GYAT/PCP
Printed in the United States of America
ISBN 1-57331-442-0(cloth)
ISBN 1-57331-443-9(paper)
ISSN 0077-8923

ANNALS OF THE NEW YORK ACADEMY OF SCIENCES

Volume 994
June 2003

THE MELANOCORTIN SYSTEM

Editor
ROGER D. CONE

Conference Organizers
GREGORY S. BARSH, ROGER D. CONE, AND LEX H.T. VAN DER PLOEG

This volume is the result of a meeting entitled the **Fifth International Melanocortin Meeting**, held on August 25-28, 2002 in Sunriver, Oregon.

CONTENTS

Part III. CNS Physiology

Part IV. The Pigmentary System

Financial assistance was received from:

- **AMGEN INC.**
- **MERCK & CO., INC**
- **MILLENNIUM PHARMACEUTICALS, INC.**
- **MINI MITTER CO., INC.**
- **NEUROCRINE BIOSCIENCES, INC.**
- **PALATIN TECHNOLOGIES**
- **PFIZER INC.**

Preface

ROGER D. CONE

Vollum Institute, Oregon Health & Science University, 3181 S.W. Sam Jackson Park Road, Portland, Oregon 97239-3098, USA

The Fifth International Melanocortin Meeting was held on August 25–28, 2002 in Sunriver, Oregon. The last major meeting covering the whole of melanocortin biology was held in Rouen, France almost eleven years ago and is recorded in Volume 680 of the *Annals* (*The Melanotropic Peptides* [1993]), edited by the organizers of that conference, Hubert Vaudry and Alex Eberle. The field has matured considerably since 1993, a time when melanocortin receptor genes had just been identified. Now scientists can describe the specific mechanisms of melanocortin action in diverse systems, including pigmentation, weight regulation, cortisol production, sexual behavior, and secretion from exocrine glands. These breakthroughs have been possible because of the remarkable synergy among research findings in chemistry, pharmacology, genetics, and neuroscience. Progress towards clinical application of melanocortin compounds has also increased significantly over the past decade, and tracking this progress was a salient new feature of the present symposium. Several pharmaceutical companies presented their work in areas as diverse as obesity and erectile dysfunction. In all, 62 speakers addressed four general areas of melanocortin biology: receptor structure/function and development of small molecule agonists/antagonists; peripheral non-pigmentary actions; central nervous system physiology; and the pigmentary system. Their findings are summarized in this volume.

The conference, attended by 193 scientists from 17 countries, was held at the Sunriver Resort near the town of Bend, in central Oregon, a well-known part of the northwestern United States owing to the beauty of the Cascade mountain range and the high desert ecosystem. Participants enjoyed a wide range of recreational activities, including white-water rafting, mountain biking, fly-fishing, hiking, golf, tennis, and stargazing, and the conference ended with a barbecue and country and western band in the Great Hall, an immense log lodge originally built as an officers' mess hall during World War II.

A number of people made a tremendous effort in organizing this meeting. Co-organizers Greg Barsh and Lex Van der Ploeg put in countless hours from beginning to end to ensure that the best science from around the world was represented and that the meeting ran smoothly. Generous sponsorship from Amgen, Merck & Co., Palatin Technologies, Neurocrine Biosciences, Millennium Pharmaceuticals, Mini Mitter, and Pfizer provided the funds necessary to cover a significant portion of the meeting expenses. The administration and staff from the convention services at Sunriver were extremely gracious. And, of course, Heather Takahashi, meeting coordinator, essentially did everything necessary to make the meeting happen from designing the website to arranging travel for participants; the meeting never would have happened

Ann. N.Y. Acad. Sci. 994: xi–xii (2003). © 2003 New York Academy of Sciences.

without her capable and tireless efforts. She also served as liaison to the editorial department of the New York Academy of Sciences and we must thank both for their skill and professionalism in seeing the book through the press.

We look forward to the next melanocortin meeting sometime soon and reporting on still further advances in this fascinating field of hormone research.

Molecular Mechanism of Agonism and Inverse Agonism in the Melanocortin Receptors

Zn^{2+} as a Structural and Functional Probe

BIRGITTE HOLST[a,b] AND THUE W. SCHWARTZ[a,b]

[a]*Laboratory for Molecular Pharmacology, The Panum Institute, University of Copenhagen, Denmark*

[b]*7TM Pharma A/S, Fremtidsvej 1, Hørsholm, Denmark*

ABSTRACT: Among the rhodopsin-like 7TM receptors, the MC receptors are functionally unique because their high constitutive signaling activity is regulated not only by endogenous peptide agonists—MSH peptides—but also by endogenous inverse agonists, namely, the proteins agouti and AGRP. Moreover, the metal-ion Zn^{2+} increases the signaling activity of at least the MC1 and MC4 receptors in three distinct ways: (1) by directly functioning as an agonist; (2) by potentiating the action of the endogenous agonist; and (3) by inhibiting the binding of the endogenous inverse agonist. Structurally the MC receptors are part of a small subset of 7TM receptors in which the main ligand-binding crevice, and especially extracellular loops 2 and 3, appear to be specially designed for easy ligand access and bias towards an active state of the receptor—i.e., constitutive activity. Thus, in the MC receptors extracellular loop 2 is ultrashort because TM-IV basically connects directly into TM-V, whereas extracellular loop 3 appears to be held in a particular, constrained conformation by a putative, internal disulfide bridge. The interaction mode for the small and well-defined zinc-ion between a third, free Cys residue in extracellular loop 3 and conceivably an Asp residue located at the inner face of TM-III gives important information concerning the activation mechanism for the MC receptors.

KEYWORDS: constitutive signaling; 7TM receptor; zinc; Zn^{2+}; metal-ion; agouti; AGRP; alpha-MSH; receptor activation; MC4

INTRODUCTION

The melanocortin receptor receptors constitute a highly interesting subfamily of 7TM, G-protein–coupled receptors both from a structural and a functional point of view. Functionally, at least the MC1, MC3, MC4, and MC5 receptors are characterized by a high degree of constitutive, that is, ligand-independent signaling.[1,2] This is a phenomenon observed in a number of 7TM receptors.[3,4] However, the MC receptors are at present unique among 7TM receptors because they are the only recep-

Address for correspondence: Birgitte Holst, Laboratory for Molecular Pharmacology, The Panum Institute, University of Copenhagen, Denmark. Voice: +45 3532 7602; fax: +45 3532 7610.

b.holst@molpharm.dk

Ann. N.Y. Acad. Sci. 994: 1–11 (2003). © 2003 New York Academy of Sciences.

tors where the high signaling activity is known to be regulated not only by endogenous agonists but also by endogenous inverse agonists.[1] Thus, the agouti protein functions as an inverse agonist for the MC1 receptor, whereas the homologous agouti-related protein (AGRP) is an inverse agonist for the MC-3 and MC4 receptors.[5,6] Another functionally interesting point is that the divalent metal-ion Zn^{2+} acts both as a partial agonist and as an enhancer or potentiator for the endogenous agonist α–MSH, at least in the MC1 and MC4 receptors.[7]

Structurally, the MC receptors clearly belong to the rhodopsin-like class of 7TM receptors. However, the MC receptors are among the few 7TM receptors that lack the characteristic long extracellular loop 2, which normally is connected through a disulfide bridge to the highly conserved CysIII:01 at the top of TM-III.[8] In contrast, extracellular loop 3 of the MC receptors appears to have evolved into a Cys- and Pro-rich, highly specialized structure. An attempt is made here to correlate the structural characteristics of the MC receptors with the functional characteristics.

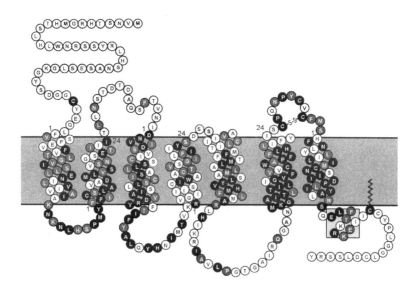

FIGURE 1. Serpentine model of the MC-4 receptor indicating residues that are identical among all MC receptors (white on black) or that are structurally conserved among three or more of the five receptors (white on grey). Note that conserved residues are located mainly in the transmembrane segments and in certain intracellular parts. The extracellular domains are in general poorly conserved as in 7TM receptors—except for extracellular loop 3. Here three Cys's and two Pro's are identical among all MC receptors. A putative disulfide bridge between two of the Cys residues is indicated (see text for details). A single Cys residue in the N-terminal extension is also totally conserved. Extracellular loop 2 is not conserved in sequence; however, in all MC receptors it is kept as an ultrashort loop.

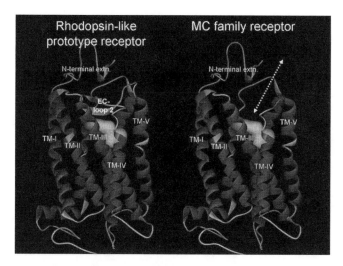

FIGURE 2. Molecular models of a prototype rhodopsin-like 7TM receptor (*left panel*) and a MC receptor (*right panel*) built over the X-ray structure of rhodopsin.[11] The receptors are seen from the lipid membrane with TM-IV in front. The surface of the retinal ligand of rhodopsin is shown as a white "cloud" in the middle of both models. Note that extracellular loop 2 is very short in the MC receptor and that the main ligand-binding crevice, i.e., basically the location of the retinal ligand, consequently is directly accessible for ligands or for extracellular loop 3 (see FIG. 5).

EXTRACELLULAR LOOP 2 IS MISSING AT "THE ENTRANCE" OF THE MAIN LIGAND BINDING CREVICE IN MC RECEPTORS

The MC receptors are highly homologous especially in the seven transmembrane helical segments and in the intracellular loops 1 and 2, as well as in helix eight, which connects the intracellular end of TM-VII with the conserved, presumably palmitylated, Cys in the C-terminal tail (FIG. 1). Within the transmembrane segments most of the rhodopsin-like conserved "fingerprint residues" are found in the MC receptors: AsnI:18, AspII:10, ArgIII:26 (of the so-called DRY motif), TrpIV:10, ProVI:15, and ProVII:17.[8] However, CysIII:01, which is one of the most conserved residues among 7TM receptors and which normally forms a disulfide bridge with a Cys residue located in the middle of extracellular loop 2 connecting TM-IV and TM-V, is missing in the MC receptors.[8] In all the MC receptors, CysIII:01 is exchanged with an Asp residue, which together with AspIII:05 located one helical turn further down in TM-III have been demonstrated to be crucially involved especially in the binding and function of agonists but also to some extent of inverse agonists[9,10] (FIG. 1).

Importantly, extracellular loop 2 is basically missing in the MC receptors. This loop is normally rather long in the 7TM receptors and in the X-ray structure of rhodopsin has been shown to form a β-sheet,[11] which is locked on top of the retinal ligand by the disulfide bridge to CysIII:01 (FIGS. 1 and 2). In all the MC receptors, extracellular loop 2 consists of only three to four not very conserved polar residues, which together with residues in extracellular loop 3 have been shown to have some importance for selective binding of the inverse agonists agouti and AGRP to the different MC receptors.[10,12,13]

The fact that the large extracellular loop 2 attached to TM-III is "missing" in the MC receptors has the consequence that the so-called main ligand-binding crevice, located between TM-III, -IV, -V, -VI, and -VII,[8,14] is in contrast to other 7TM receptors directly accessible for ligands from the extracellular space (arrow in FIG. 2, right panel). Moreover, in 7TM receptors part of the activation mechanism appears to be an inward movement of the extracellular end of TM-VI towards TM-III (see below), and such a movement in the MC receptors is not hindered by a large extracellular loop tethered to TM-III. This could be one of the explanations for the high degree of constitutive activity among the MC receptors.

THE CYS- AND PRO-RICH EXTRACELLULAR LOOP 3 IS STRUCTURALLY CONSERVED

Extracellular loop 3 is normally not conserved in respect of amino acid sequence among 7TM receptors, but it is generally kept as a relatively short loop of around a dozen residues.[8] As shown in FIGURE 1, among the MC receptors five of the residues in this loop are identical and another five are structurally conserved, which is in strong contrast to the otherwise poorly conserved extracellular domains of these receptors. Importantly, the five completely conserved residues are three Cys and two Pro residues, which, from a protein chemistry point of view, strongly indicates that in the MC receptors extracellular loop 3 is detained in a very specific, particular three-dimensional (3-D) structure or fold. It has been suggested that two of the Cys residues within this loop form an intraloop disulfide bridge as indicated in FIGURES 1 and 3.[7] Mutational analysis of the MC-1 and MC-2 receptors has shown that the two Cys residues are essential for expression and function of these receptors; furthermore, natural mutations which lead to malfunction of the MC-4 receptor and obesity in children expressing such mutations also affect one of these two Cys residues.[15,16] Because of the distance constraint that this disulfide bridge will impose upon the extracellular ends of TM-VI and -VII, it is very likely that these segments are kept closely together as a unit, which again could be part of the reason that the MC receptors show a high degree of constitutive activity (see below).

Interestingly, the third Cys residue in extracellular loop 3—namely, Cys^{271} in the MC-1 receptor—is apparently not involved in disulfide bridge formation, at least not in the resting state of the receptor[7,16] because it can be mutated without effect on α–MSH binding and action, and importantly because it is involved in the binding and function of Zn^{2+}, which requires a free thiol function.[7] Similarly, the Cys residue located in the N-terminal extension of all the MC receptors can be mutated without effect on α–MSH binding and function. Interestingly, mutational analysis with introduction of Cys residues in extracellular loop 3 and the N-terminal segment of the NK1 receptor would indicate that a disulfide bridge should form rather easily between Cys residues at these locations, as indicated in FIGURE 3.[17] That this is most likely *not* the case in the MC receptors could be a consequence of the presumed rather special 3-D structure of the Cys- and Pro-rich extracellular loop 3. However, the fact that a conserved Cys residue is found both in extracellular loop 3 and another equally well conserved Cys residue is found in a spatially close location in the N-

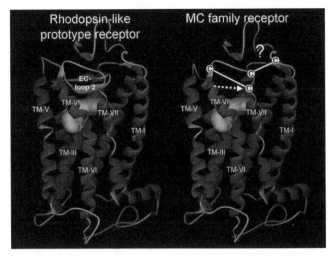

FIGURE 3. Molecular models of a prototype rhodopsin-like 7TM receptor (*left panel*) and a MC receptor (*right panel*) built over the X-ray structure of rhodopsin.[11] The receptors are seen from the lipid membrane—in this case with TM-VI in front—that is, an opposite view as shown in FIGURE 2. The surface of the retinal ligand of rhodopsin is shown as a white "cloud" in the middle of both models. The locations of the three totally conserved Cys residues in extracellular loop 3 and the single Cys residue in the N-terminal segment of the MC receptors are indicated. The distance between the two Cys residues that are supposed to be connected in a disulfide bridge is far too long in the structure of the extracellular loop 3 of rhodopsin, which indicates that the extracellular ends of TM-VI and VII will have to be much closer; this conceivably is achieved through an inward movement of the end of TM-VI towards TM-VII (and TM-III), as suggested by the *arrow*. The distance between the last Cys in extracellular loop 3 and the one in the N-terminal extension is also too far, and it is likely that at least under resting conditions these do not form a disulfide bridge (see text for experimental details).

terminal segment among all MC receptors would tend to suggest, that under some physiologically important circumstance these two Cys residues will form a disulfide bridge or be of some other kind of functional importance.

ZN^{2+} MAY BIND IN TWO DIFFERENT MODES WHEN ACTING AS AN AGONIST AND AS A POTENTIATOR

The function of the MC receptors is regulated not only by peptide agonists and protein inverse agonists, but it is also modulated by zinc-ions.[7] As shown in FIGURE 4, on the MC-1 receptor, Zn^{2+} acts as an almost 50% partial agonist with a reasonably high potency of 10 μM. Importantly, at slightly higher concentrations, Zn^{2+} is able to potentiate the action of the endogenous peptide agonist α–MSH by shifting the dose-response curve for the peptide approximately sixfold to the left (FIG. 4B). Similar effects of Zn^{2+} are observed in the MC-4 receptor although with a lower efficacy and a smaller shift in α-MSH potency.[7] At least the agonistic effect of the

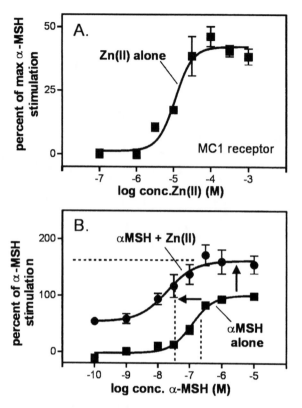

FIGURE 4. Zn^{2+} acting as an agonist (**panel A**) and as an agonist enhancer or potentiator (**panel B**) on the MC1 receptor. **Panel A:** cAMP accumulation in response to Zn^{2+} in COS-7 cells transiently transfected with the mouse MC1 receptor expressed as percentage of the maximal response to α-MSH. **Panel B:** α-MSH induced cAMP accumulation in the MC1 receptor in the presence (*filled circles*) and in the absence (*filled squares*) of Zn^{2+} at a concentration of 10^{-4} M. Zn^{2+} has an approximately sixfold potentiating effect on the α-MSH stimulation of signaling, and the agonistic effects of the metal-ion appear to be additive to that of the peptide agonist (Zn^{2+} has no effect in mock-transfected cells; for experimental detail see Ref. 7).

metal-ion is probably physiologically relevant as double or even triple digit μM concentrations of Zn^{2+} are observed in the synaptic cleft in certain areas of the brain upon neuronal stimulation.[7]

Zn^{2+} can also act as an agonist on the MC receptors in complex with small hydrophobic, aromatic metal-ion chelators. Importantly, the potency and efficacy of the metal-ion chelator complex depends highly upon the chemical structure of the chelator as such.[7] For example, among a small selection of 1,10-phenathroline analogues, some displayed almost full agonism in the MC-1 receptor. Such metal-ion chelators could be leads for a novel type of drugs, which act through a bridging Zn^{2+} ion either as agonists and or as potentiators for the endogenous agonist.

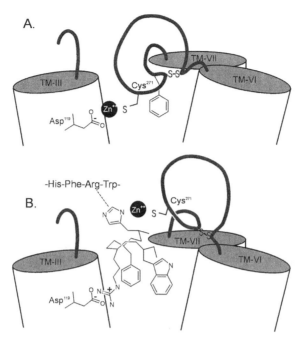

FIGURE 5. Schematic drawing of the putative molecular mechanism of action of Zn^{2+} as an agonist (**panel A**) and as a enhancer or positive modulator (**panel B**) in an MC receptor. For simplicity, only the extracellular ends of TM-III, -VI, and -VII, plus extracellular loop 3 that connects TM-VI and VII, are shown. **Panel A:** As an agonist, Zn^{2+} is envisioned to bind between Cys^{271} (mouse MC1 numbering) located in the disulfide-stabilized extracellular loop 3 and Asp^{119} (AspIII:05) located on the inner face of TM-III. Just a single other residue—a phenylalanine—of extracellular loop 3 is shown to indicate the possibility that such residues could be involved in the Zn^{2+} induced activation process by mimicking one of the important aromatic residues of the peptide ligand. **Panel B:** The putative binding of the core tetrapeptide sequence of MSH: -His-Phe-Arg-Trp- in the main ligand-binding crevice with the guanidinium moiety of the Arg interacting with Asp^{119} (AspIII:05) in TM-III and with Zn^{2+} acting as an enhancer by binding between Cys^{271} in extracellular loop 3 and the His residue of the ligand.

Mutational mapping of the Zn^{2+} binding site in the MC-1 receptor demonstrated that the metal-ion binds between the free Cys residue in extracellular loop 3 (described above) and conceivably AspIII:05, which also is a crucial interaction point for α-MSH (FIG. 5A). In general the inner face of TM-III is a major interaction face for agonists in 7TM receptors.[8] Thus, it is likely that Zn^{2+} induces receptor activation by holding the short extracellular loop 3—and thereby TM-VI and TM-VII—close towards TM-III. This fits very well with our view on the activation mechanism for 7TM receptors in general (see below). However, it is likely that the enhancing, potentiating effect of Zn^{2+} is mediated through another mechanism. The reason is that α-MSH peptide itself binds with high affinity to AspIII:05 and probably will exclude the metal-ion from binding here. In this connection it is interesting that the functionally important, core sequence of MSH peptides consists of the tetrapeptide:

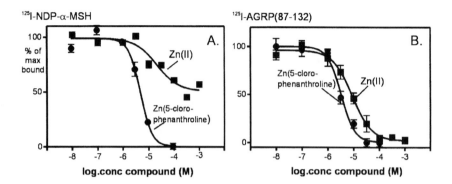

FIGURE 6. Different effects of Zn^{2+} on the binding of radiolabeled agonist (**panel A**) and radiolabeled inverse agonist (**panel B**). **Panel A:** Whole-cell competition binding experiments performed with ^{125}I-NDP-α-MSH in transiently transfected COS-7 cells expressing the human MC4 receptor. Note that the free Zn^{2+} can only partially displace the radioactive agonist in accordance with the fact that at higher concentrations the metal-ion functions as a functional enhancer and conceivably helps instead of inhibits binding of the agonist (for details see Ref. 7). **Panel B:** Whole-cell competition binding experiments performed with ^{125}I-AGRP(87-132) (Amersham Pharmacia Biotech, TRK 311) in transiently transfected COS-7 cells expressing the human MC4 receptor (binding experiments performed as described for ^{125}I-NDP-α-MSH in Ref. 7). Note that both the free Zn^{2+} as well as the metal-ion in complex with a metal-ion chelator, 5-chloro-1,10-phenanthroline, displaces the inverse agonist AGRP totally and with high affinity.

-His-Phe-Arg-Trp- of which it is the guanidinium function of the Arg, which is believed to bind to AspIII:05 in the receptor.[9,10] In FIGURE 5B it is suggested that Zn^{2+} may act as an enhancer for α-MSH function through binding in between the His residue of the core tetrapeptide ligand and the free Cys residue in extracellular loop 3. This hypothesis can readily be tested by use of peptide analogues having variable ability to bind metal-ions at the position of the His residue.[18] It should be noted that a potential metal-ion binding Asp residue located at position III:01 also could be involved in the potentiating effect of Zn^{2+}.

ZN^{2+} ALSO ACTS AS PURE BLOCKER OF AGRP BINDING IN THE MC4 RECEPTOR

As shown in FIGURE 6B, Zn^{2+} blocks the binding of the inverse agonist AGRP to the MC4 receptor with a potency similar to the one at which it acts as an agonist at this receptor. It should be noted that the free zinc-ion, which at higher concentrations acts as an enhancer for the peptide agonist and consequently does not displace the ^{125}I-α-MSH totally (FIG. 6A), in fact does displace the AGRP inverse agonist completely (FIG. 6B). Inhibition of the binding of an inverse agonist or antagonist is obviously one way of achieving agonism provided that there is a tonus of the inverse agonist to act on. Thus, Zn^{2+} has the interesting functional property of being able to modulate the melanocortin receptors positively towards increased signaling through three different mechanisms: (1) direct agonism, (2) enhancement of the endogenous agonist function, and (3) inhibition of the natural inverse agonist. A compound having such properties would be extremely interesting as a drug candidate, for example, on the MC4 receptor for the treatment of obesity or erectile dysfunction. Importantly, although the presumed metal-ion binding site is conserved among the MC receptors, metal-ion chelator complexes have different potency and efficacies in, for example, the MC1 and the MC4 receptors.[7] Thus, it is likely that metal-ion binding compounds can be developed to selectively target different MC receptors.

MOLECULAR MECHANISM OF AGONISM AND INVERSE AGONISM IN MC RECEPTORS

There is general agreement in the field that a major, important part of the activation mechanism for 7TM receptors is an outward movement of the intracellular end of TM-VI.[19,20] However, it should be noted that the initial triggering event in the activation mechanism occurs at the ligand binding site. This is perhaps most clearly demonstrated in rhodopsin, where the key, primary activating event is the light-induced isomerization of the 11-*cis* to all-*trans* form of the retinal ligand located in the main ligand-binding crevice. Photo-affinity labeling studies have demonstrated that the ring of the retinal molecule that labels the inner face of TM-VI on the dark, inactive state of rhodopsin during light activation shifts position away from TM-VI to instead be positioned close to the inner face of TM-IV.[21] Our studies with engineered, activating metal-ion sites located between residues on the inner faces of TM-III and residues in TMs VI and VII, respectively, demonstrate that the spatial gap between these three helices closes during activation. This is interpreted as being an inward movement especially of TM-VI but also of TM-VII towards TM-III[22,23] (unpublished observations). The shift in the position of the retinal away from the inner face of TM-VI would obviously also allow for such an inward movement of TM-VI to occur.

It is clear that the MC receptors basically are set for being in the active state. The large extracellular loop 2, which in other 7TM receptors occupies the space between TM-III and TM-VI, is absent in the MC receptors, allowing for free movement inwards of TM-VI (FIGS. 2 and 3). Moreover, TM-VI is probably held closely in towards TM-VII by the short extracellular loop 3 generated by the suggested, highly conserved disulfide bridge which efficiently constrains and shortens this loop (FIG. 3). Finally, the extracellular end of TM-V is conceivably held over towards TM-IV, that is, away from TM-VI, by the ultrashort extracellular loop 2. Thus, it is suggested that the high constitutive activity of the MC receptors is based on the design of a highly specialized main ligand-binding crevice, which in these receptors allows for free movement of the extracellular end of TM-VI in towards TM-III. Agonists such as the peptide α-MSH or Zn^{2+} will stabilize such a conformation where TM-VI and -VII and in the "active inwards conformation" at the extracellular ends (FIG. 5). It can be envisioned that the inverse agonists agouti and AGRP function by holding TM-VI and -VII away from TM-III. In this connection, it has been shown that the action of agouti/AGRP mainly is dependent on residues in extracellular loops 2 and 3.[10]

It should be noted that in the large family of rhodopsin-like 7TM receptors, a subset of receptors structurally cluster around the MC receptors. For example, the EDG receptors, the cannabinoid receptors, the mas oncogene, and some orphan receptors all lack the disulfide attachment of extracellular loop 2 to the top of TM-III, and they all have a short, often interloop, disulfide-stabilized extracellular loop 3. Among these receptors a large number have been demonstrated to display high constitutive activity. We would suggest that also in these cases the high constitutive activity is to a large degree determined by the special design of their extracellular loops 2 and 3. It remains to be demonstrated to what degree the signaling activity of these receptors is regulated not only by agonists but also by inverse agonists such as in the case of the MC receptors.

REFERENCES

1. NIJENHUIS, W.A., J. OOSTEROM & R.A. ADAN. 2001. AgRP(83-132) acts as an inverse agonist on the human-melanocortin-4 receptor. Mol. Endocrinol. **15:** 164–171.
2. EBERLE, A.N., J. BODI, G. OROSZ, et al. 2001. Antagonist and agonist activities of the mouse agouti protein fragment (91-131) at the melanocortin-1 receptor. J. Recept. Signal. Transduct. Res. **21:** 25–45.
3. SEIFERT, R. & K. WENZEL-SEIFERT. 2002. Constitutive activity of G-protein-coupled receptors: cause of disease and common property of wild-type receptors. Naunyn Schmiedebergs Arch. Pharmacol. **366:** 381–416.
4. PARNOT, C., S. MISEREY-LENKEI, S. BARDIN, et al. 2002. Lessons from constitutively active mutants of G protein-coupled receptors. Trends Endocrinol. Metab. **13:** 336–343.
5. DINULESCU, D.M. & R.D. CONE. 2000. Agouti and agouti-related protein: analogies and contrasts. J. Biol. Chem. **275:** 6695–6698.
6. OLLMANN, M.M., B.D. WILSON, Y.K. YANG, et al. 1997. Antagonism of central melanocortin receptors in vitro and in vivo by agouti-related protein. Science **278:** 135–138.
7. HOLST, B., C.E. ELLING & T.W. SCHWARTZ. 2002. Metal-ion mediated agonism and agonist-enhancement in the melanocortin MC1 and MC4 receptors. J. Biol. Chem. **277:** 47662–47670.
8. SCHWARTZ, T.W. & B. HOLST. 2002. Molecular structure and function of 7TM/G-protein coupled receptors. *In* Textbook of Receptor Pharmacology. J.C. Forman & T. Johansen, Eds.: 65–84. CRC Press. Boca Raton, FL.
9. YANG, Y., C. DICKINSON, C. HASKELL-LUEVANO, et al. 1997. Molecular basis for the interaction of [Nle4,D-Phe7]melanocyte stimulating hormone with the human melanocortin-1 receptor. J. Biol. Chem. **272:** 23000–23010.
10. YANG, Y.K., T.M. FONG, C.J. DICKINSON, et al. 2000. Molecular determinants of ligand binding to the human melanocortin-4 receptor. Biochemistry **39:** 14900–14911.
11. PALCZEWSKI, K., T. KUMASAKA, T. HORI, et al. 2000. Crystal structure of rhodopsin: a G protein-coupled receptor. Science **289:** 739–745.
12. HASKELL-LUEVANO, C., R.D. CONE, E.K. MONCK, et al. 2001. Structure activity studies of the melanocortin-4 receptor by in vitro mutagenesis: identification of agouti-related protein (AGRP), melanocortin agonist and synthetic peptide antagonist interaction determinants. Biochemistry **40:** 6164–6179.
13. MCNULTY, J.C., D.A. THOMPSON, K.A. BOLIN, et al. 2001. High-resolution NMR structure of the chemically-synthesized melanocortin receptor binding domain AGRP(87-132) of the agouti-related protein. Biochemistry **40:** 15520–15527.
14. SCHWARTZ, T.W. 1994. Locating ligand-binding sites in 7TM receptors by protein engineering. Curr. Opin. Biotechnol. **5:** 434–444.
15. FAROOQI, I.S., G.S. YEO, J.M. KEOGH, et al. 2000. Dominant and recessive inheritance of morbid obesity associated with melanocortin 4 receptor deficiency. J. Clin. Invest. **106:** 271–279.

16. DOUFEXIS, M. & S. KAPAS. 2002. Importance of cysteine residues in ligand binding and signalling by melanocortin 2 receptors [abstract]. 5th International Melanocortin Meeting. Sunriver, OR.

17. ELLING, C.E., U. RAFFETSEDER, S.M. NIELSEN, et al. 2000. Disulfide bridge engineering in the tachykinin NK1 receptor. Biochemistry **39:** 667–675.

18. HOLDER, J.R., R.M. BAUZO, Z. XIANG, et al. 2002. Structure-activity relationships of the melanocortin tetrapeptide Ac-His-DPhe-Arg-Trp-NH(2) at the mouse melanocortin receptors. 1. Modifications at the His position. J. Med. Chem. **45:** 2801–2810.

19. FARRENS, D.L., C. ALTENBACH, K. YANG, et al. 1996. Requirement of rigid-body motion of transmembrane helices for light activation of rhodopsin. Science **274:** 768–770.

20. GETHER, U., S. LIN, P. GHANOUNI, et al. 1997. Agonist induced conformational changes in transmembrane domains III and VI of the beta2 adrenoceptor. EMBO J. **16:** 6737–6747.

21. BORHAN, B., M.L. SOUTO, H. IMAI, et al. 2000. Movement of retinal along the visual transduction path. Science **288:** 2209–2212.

22. ELLING, C.E., K. THIRSTRUP, B. HOLST, et al. 1999. Conversion of agonist site to metal-ion chelator site in the beta(2)-adrenergic receptor. Proc. Natl. Acad. Sci. USA **96:** 12322–12327.

23. HOLST, B., C.E. ELLING & T.W. SCHWARTZ. 2000. Partial agonism through a zinc-ion switch constructed between transmembrane domains III and VII in the tachykinin NK(1) receptor. Mol. Pharmacol. **58:** 263–270.

Exploring the Stereostructural Requirements of Peptide Ligands for the Melanocortin Receptors

VICTOR J. HRUBY, MINYING CAI, PAOLO GRIECO, GUOXIA HAN, MALCOLM KAVARANA, AND DEV TRIVEDI

Department of Chemistry, University of Arizona, Tucson, Arizona 85721, USA

ABSTRACT: The melanotropin peptides α-MSH, γ-MSH, and β-MSH are believed to be the natural ligands for the four melanocortin receptors, MC1R, MC3R, MC4R, and MC5R. However, these peptides generally have low selectivity for these receptors. We report on some approaches to the development of selective agonists and antagonists peptide ligands for these receptors.

KEYWORDS: α-melanocyte stimulating hormone (α-MSH); γ-melanocyte-stimulating hormone (γ-MSH); melanocortin receptors; agonists; antagonists; melanotropin peptides; conformational constraint; receptor selective melanotropins; cyclic melanotropins

INTRODUCTION

The discovery and structural elucidation of α-melanocyte-stimulating hormone (α-MSH), α-melanotropin, Ac-Ser-Tyr-Ser-Met-Glu-His-Phe-Arg-Trp-Gly-Lys-Pro-Val-NH$_2$, by Harris and Lerner in 1957,[1] led immediately to structure-activity studies from several prominent peptide laboratories worldwide (as reviewed, for example, in Ref. 2). In a New York Academy of Sciences Meeting on ACTH and Related Peptides in 1976, Robert Schwyzer, referring to α-MSH, suggested that "...we have reached the limits of insight that can be reasonably provided by structure-activity studies ...,"[3] although he recognized that additional analogues might be needed to examine stimulus → effect coupling. However, from our studies on the mechanism controlling the release of α-MSH,[4] our perception was that there still were a number of missing gaps, including the lack of a good antagonist, and unexplained novel biological effects ascribed to MSH. To explore these, we prepared the Nle^4, D-Phe^7-substituted analogue, [Nle^4,D-Phe^7]α-MSH,[5] and found it had highly prolonged biological activity[6] that persisted for hours, days, or weeks, and the cyclic disulfide constrained analogue of α-MSH c[Cys^4,Cys^{10}]α-MSH that had superagonist activity even though it contained no D-Phe^7 residue.[8] Thus, α-MSH has been an excellent ligand for de novo ligand-based design of peptide hormones and neurotransmitters.[7] Furthermore, evidence continued to accumulate that α-MSH was a

Address for correspondence: Dr. Victor J. Hruby, Department of Chemistry, Univeristy of Arizona, Tucson, Arizona 85721. Voice: 520-621-6332; fax: 520-621-8407.

hruby@u.arizona.edu

Ann. N.Y. Acad. Sci. 994: 12–20 (2003). © 2003 New York Academy of Sciences.

TABLE 1. Suggested potential physiological and medical applications of melanotropic peptides

Function	Medical application
Pigmentation	Enhanced pigmentation—UV protection
Pigmentation	Cancer prevention—skin cancer
Learning	Memory retention, attention
Pain	Pain modulation
Immune response	Anti-inflammatory effects
Injury	Repair of nerve damage
Feeding behavior	Obesity
Feeding behavior	Anorexia
Sexual response	Erectile function
Addiction	Treatment of addictions
Kidney function	Natriuresis
Temperature control	Antipyretic
Cardiovascular function	Heart function
Stress response	Stress-related disorders

hormone that evolved very early in the evolution of animals, and was of central importance for the modulation of many critical biological effects (TABLE 1). This steady progress, however, was largely ignored by drug companies.

This changed suddenly and dramatically in the 1990s with the discovery and cloning of the melanocortin receptors (reviewed in Refs. 8 and 9). Of particular interest was the finding that in addition to the two expected receptors, the melanocortin-1 receptor (MC1R) for pigmentation and the melanocortin-2 receptor (MC2R) for response to stress (adrenocorticotropic hormone [ACTH] receptor), three other closely related receptors were found—the MC3R (found primarily in the brain), the MC4R (found primarily in the brain), and the MC5R (found primarily in the periphery). The discovery of these new melanocortin receptors, especially those found in the central nervous system (CNS), provided a new impetus to reexamine the central importance of α-MSH as a neuropeptide.

By the early 1980s our laboratory was fortunate to have developed [Nle^4,D-Phe^7]α-MSH (NDP-α-MSH),[5] which had been shown to be considerably more stable than α-MSH in *in vitro* assays and animals.[10] This compound was [^{125}I]-labeled in several laboratories and has been used widely as the ligand of choice for binding and other bioassays. In addition, we had just developed a super potent, highly efficacious and even more stable cyclic lactam-containing heptapeptide analogue of α-MSH, Ac-Nle-c[Asp-His-D-Phe-Arg-Trp-Lys]α-MSH(4-10)-NH$_2$, (Ac-Nle^4-c[Asp^5,D-Phe^7,Lys^{10}]α-MSH(4-10)-NH$_2$, MT-II).[11,12] This analogue has been widely used because of its high potency *in vivo* and its ability to cross the blood–brain barrier (BBB).[13] At about the same time, we also developed the first highly potent α-MSH receptor antagonist for the MC3R and MC4R, Ac-Nle[4]-c[Asp[5]-His-D-Nal(2′)-Arg-Trp-Lys]α-MSH(4-10)-NH$_2$,　　(Ac-Nle[4]-c[Asp[5],D-Nal(2′)[7],Lys[10]]-α-MSH(4-10)-NH$_2$, SHU-9119),[14] which also has been an invaluable tool (this volume). In collaboration with numerous scientists, we have used these analogues for *in vivo* studies related to examination of feeding behavior,[13] cardiovascular function,[15] erectile function,[16] and several others. Despite the successful use of these compounds, it was found that none of them is very selective for any of the melanocortin receptors, and thus we set a goal of obtaining selective analogues for the four melanocortin receptors.

RESULTS AND DISCUSSION

Selectivity Considerations

The early observations from several laboratories that α-MSH and NDP-α-MSH and MT-II have only modest selectivities for the MC1 vs. MC3 vs. MC4 vs. MC5 receptors raised the critical question of how α-MSH might be modified to obtain more selective ligands for the melanocortin receptor, and the level of selectivity. Many scientists consider ligands that are 10–20-fold selective for a subtype of receptor as sufficient. However, we have maintained for some time that this generally is not adequate.[7,17] In the case of agonists, problems associated with efficacy, tissue differences in transduction efficiency, etc., and in the case of antagonists,[18] the situation *in vivo* can be even more confounding, depending, for example, upon on/off rates in comparison with the endogenous ligand. Hence, selectivities of 100 to 1000 or more often are needed to have confidence that the effect observed is a consequence of the ligand's interaction with the receptor. A more perplexing question is whether we can obtain melanotropins that are selective for the same receptor in different tissue (for example, CNS tissue vs. peripheral tissue for the same receptor), and we would like to address this issue as well.

Design Consideration

A robust ligand-based design strategy[7,18,19] for the design of peptide hormone and neurotransmitter agonists and antagonists has been developed (TABLE 2). Some of these strategies, especially from sections I and II in TABLE 2, are illustrated by NDP-α-MSH, by our cyclic disulfide-containing ligands, by our cyclic lactam ligands, for example, MT-II, by our bicyclic analogues with disulfide and lactam structures in a single molecule, by our use of proline-6 analogues[20,21] and related peptides,[21] and the use of topographically constrained aromatic amino acids (constrained in chi-space, χ^1 and χ^2).[22,23] We also showed that the minimum active sequence for frog and lizard[2] (His-Phe-Arg-Trp-) also appear to be the case for the more recently discovered MC3R, MC4R, and MC5R,[24] that antagonists for the MC5R could be formed based on SHU-9119, but with modifications of the His residue, and that a potent antagonist for the human MC1R could be obtained based on a cyclic chimeric deltorphin II (a delta opioid ligand)-melanotropin analogue.[25]

TABLE 2. Design strategies to obtain selective peptide and peptidomimetic agonist and antagonist ligands for the melanocortin receptors

I. Examination of ligands based on POMC structures

   A. Alanine and D-amino acid scans

   B. Conformational constraints

      1. Side chain–side chain cyclization

      2. Side chain–backbone cyclization

      3. Backbone–backbone cyclization

   C. Peptidomimetics/topically constrained amino acids

II. Examination of previously discovered templates for bioactive conformations

   A. Cyclic disulfide—various sized rings

   B. Cyclic lactams—various sized rings

   C. Bicyclic analogues

   D. D-Phe, D-Trp and topographically designed amino acid substitutions

III. New templates

   A. Somatostatin-designed melanotropins

   B. Deltorphin II-related peptides

   C. Non-peptide scaffolds

IV. Novel libraries—β-turn mimetics

TABLE 3. Most potent ligands for the hMC3R

1. NDP-α-MSH-EC_{50} = 0.4 nM at hMC3R
                   = 0.1 nM at mMC3R

2. SHU-9119—IC_{50} = 0.1 nM at hMC3R

3. [D-Trp^8]-γ-MSH—EC_{50} = 0.33 nM

4. H-Tyr-Val-Nle-Gly-Pro-D-Nal(2′)-Arg-Trp-Asp-Arg-Phe-Gly-NH$_2$ - IC_{50} = 0.21 nM at hMC3R

5. Ac-Nle-c[Asp-Pro-D-Nal(2′)-Arg-Trp-Lys]-NH$_2$ - IC_{50} = 0.19 nM

DESIGNING PEPTIDE LIGANDS THAT ARE POTENT AND/OR SELECTIVE FOR MELANOCORTIN RECEPTORS

Potency

Efforts to increase potency relative to α-MSH and MT-II will only be briefly discussed here (TABLE 3). As shown in TABLE 3, the potent agonist NDP-α-MSH has subnanomolar potency at the human and mouse MC3 receptor, as does the antagonist analogue SHU-9119. Most interesting, as part of a D-amino acid scan of γ-MSH, we found[26] that substitution of L-Trp with D-Trp in γ-MSH led to a highly potent ligand for the MC3R (**3**, TABLE 3). Interestingly, we found that a Nle^3,Pro^5,D-Nal(2′)-analogue **4** of γ-MSH (**4**, TABLE 3) also was a very potent ligand for the MC3R. We

1. Minimum Active Sequence*

      MC1R - His-Phe-Arg-Trp
      MC3R - H-Gly-His-Phe-Arg-Trp-Asp-Arg-Phe-Gly-OH
      MC4R - H-Gly-His-Phe-Arg-Trp-Asp-Arg-Phe-Gly-OH
      MC5R - H-Met-Gly-His-Phe-Arg-Trp-Asp-Arg-Phe-Gly-OH

2. Key Pharmacophore Residues

      MC1R - His-Phe-Arg-Trp
      MC3R - His-Phe-Arg-Trp
      MC4R - His-Phe-Arg-Trp-Asp-Arg-Phe
      MC5R - Tyr-Ala-His-Phe-Arg-Trp-Asp-Arg-Phe

*Full agonist activity as measured by 2nd messenger assays in stably transfected cell lines.

FIGURE 1. Aspects of structure-activity relationships of γ-MSH at human melacortin receptors.

further found that modification of MT-II by introducing Pro^6 followed by cyclization, gave an analogue **5** (TABLE 3) with pmolar binding at the hMC3R. In general, we found that rational ligand-based design principles can readily be used to provide highly potent ligand at all four receptors that are targets for MSH. There is a need for a much more careful examination of structure-activity relationships for γ-MSH (H-Tyr-Val-Met-Gly-His-Phe-Arg-Trp-Asp-Arg-Phe-Gly-OH). It is very clear from our D-amino acid scan, L-Ala scan, and truncation scan (manuscript in preparation) that γ-MSH—though it has the same core sequence, His-Phe-Arg-Trp- as α-MSH— has unique structure-activity relationships at the melanocortin receptors MC1, MC3, MC4, and MC5 (FIG. 1) as illustrated, for example, by the high potency *and* selectivity of [D-Trp^8]γ-MSH at the hMC3R[26] (TABLE 3).

Selectivity

To obtain potent and *highly* selective ligands for the human melanocortin receptors MC1R, MC3R, MC4R, and MC5R has not been easy, although it now is clear that these four receptors do have uniquely different ways in which they interact with melanotropin peptide ligands. In TABLE 4 we list some of the more selective peptide ligands obtained thus far for the human melanocortin receptors. Many of the compounds are from our laboratory; for those that are not, references are provided in the discussion below. For comparison, the selectivity for the human MC1R by MT-II is given in TABLE 4.[27,31] As can be seen, MT-II is modestly selective for the human MC1R vs. the other human melanotropin receptors [vs. MC3R (50×); vs. MC4R (10×); vs. MC5R (69×)]. Probably the most potent and selective human MC1R agonist is the linear MSH analogues of Szardenings *et al.*[28] with a 1000-fold selectivity for the hMC1R vs. the hMC3R, and greater than 100,000× selectivity vs. the hMC4R and hMC5R. The most selective agonist peptide ligand for the hMC3R is [D-Trp^8]γ-MSH (**3**, TABLE 4).[26] The compound has subnanomolar agonist potency and is about 300-fold selective for the hMC3R vs. the hMC4R and hMC5R. There are a few modestly selective agonists for the hMC4R. For example, the N-terminal to side-

TABLE 4. Selective agonists for human melanocortin receptor

Receptor	Compound	K_i^a/EC$_{50}^a$/Selectivity^b
hMC1R	**1** MT-II	K_i, MC1R = 0.67 nM; MC3R, 50×; MC4R, 10×; MC5R, 69×
hMC1R	**2** H-Ser-Ser-Ile-Ile-Ser-His-Phe-Arg-Gly-Lys-Pro-Val-NH$_2$	K_i, MC1R = 0.87 nM; MC3R, 1250×; MC4R, >100,000×; MC5R, >100,000×
hMC3R	**3** H-Tyr-Val-Met-Gly-His-Phe-Arg-D-Trp-Asp-Arg-Phe-Gly-OH	EC$_{50}$, MC3R = 0.33 nM; MC4R, 300×; MC5R, 250×
hMC4R	**4** (O)C-(CH2)x-C(O)-c[His-D-Phe-Arg-Trp-Lys]-NH$_2$	EC$_{50}$, MC4R = 1.6 nM; MC3R, 45×; MC5R, 350×
	5 c[NH-CH$_2$)$_2$-(O)-His-D-Phe-Arg-Trp-Glu]-NH$_2$	EC$_{50}$, MC4R = 0.56 nM; MC3R, 90×; MC5R, 3400×
hMC5R	**6** Ac-Nle-c[Asp-Pro-D-Nal(2')-Arg-Trp-Lys]-NH$_2$	EC$_{50}$, MC5R = 0.072 nM; MC3R, >10,000×; MC4R, > 10,000×

^{a}Binding affinity or bioactivity (nM) at designated receptor.
^{b}Selectivity *vs.* designated receptors measured.

TABLE 5. Selective antagonists for human melanocortin receptors

Receptor	Compound	IC$_{50}^a$/pA$_2^a$/Selectivity^b
MC1R	**1** c[Gly-Cpg-D-Nal(2')-Arg-Trp-Glu]Val-Val-Gly-NH$_2$	IC$_{50}$, MC1R = 12 nM; MC3R, 4×; MC4R, 100×
MC3R	**2** Ac-Nle-c[Asp-Acpc-D-Nal(2')-Arg-Trp-Lys]-NH$_2$	IC$_{50}$, MC3R = 2.5 nM; MC4R, 100×; pA$_2$ = 7.6
MC4R	**3** Ac-Nle-c[Asp-Cpe-D-Nal(2')-Arg-Trp-Lys]-NH$_2$	IC$_{50}$, MC4R = 0.51 nM; MC3R, 200×; pA$_2$ = 8.0
MC5R	**4** H-Tyr-Val-Nle-Gly-Pro-D-Nal(2')-Arg-Trp-Asp-Arg-Phe-Gly-NH$_2$	IC$_{50}$, MC5R = 15 nM

^{a}Binding affinities or antagonist activities at designated receptor.
^{b}Selectivity *vs.* designated receptors measured.

chain cyclic analogue[29] (TABLE 4) is 45-fold selective vs. the hMC3R and 350-fold selective vs. the hMC5R. The cyclic N-terminal to side-chain cyclic compound[30] (**5**, TABLE 4) has slightly improved binding selectivities of about 90-fold vs. the hMC3R and 3,400-fold vs. the hMC5R. Finally, for selectivity at the hMC5R, we have found a Pro[6] substituted analogue[21] (compound **6**), which is a highly potent agonist at the hMC5R, and has good selectivity vs. the hMC3R and hMC4R of about >10000× in each case. In TABLE 5 we list some of the most selective peptide antagonists for the various human melanocortin receptors. It has been difficult to obtain highly selective peptide antagonist analogues for the hMC3R, hMC4R, and hMC5R receptors. For the human MC1R, we recently discovered that a chimeric peptide of melanotropin and deltorphin II is a potent and selective ligand for the human MC1R (**1**, TABLE 5.)[25] For the human MC3R, Ac-Nle-c[Asp-Acpc-D-Nal(2′)-Arg-Trp-Lys]-NH$_2$ (**2**, TABLE 5) has fairly high binding affinity for the hMC3R (2.5 nM) and no agonist activity with modest selectivity for the hMC3R vs. hMC4R of about 100 (TABLE 5).[31] The situation for the hMC4R is a little better with the MT-II modified ligand **3** (TABLE 5) having a high binding affinity for the hMC4R, no agonist activity, and good selectivity vs. the hMC3R of about 200.[31] It is interesting to note that a change in the amino acid at position 6 of α-MSH in an MT-II analogue can lead to large changes in potency at the human MC3R vs. MC4R, and suggests that modulation of structure at the N-terminal position of α-MSH and γ-MSH may be an excellent approach to obtaining selective antagonist analogues for the hMC3R and hMC4R. As for the hMC5R, there still is not a highly selective antagonist ligand reported. However, listed in TABLE 5 is a linear analogue of γ-MSH, H-Tyr-Val-Nle-Gly-Pro-D-Nal(2′)-Arg-Trp-Asp-Arg-Phe-Gly-NH$_2$ that has modest binding affinity for the human MC5R (15 nM) with no bioefficacy at all, but it is a full agonist at hMC3R and hMC4R (Balse-Srinivasan, Hruby *et al.*, in preparation).

CONCLUDING REMARKS

The many critical biological effects that are modulated and controlled through the melanotropin peptides related to α-MSH and γ-MSH (TABLE 1) present a scientifically intriguing and challenging set of problems. It seems to this investigator that before good drugs can be obtained that have minimal or no toxicity, and ligands that clearly differentiate the physiological roles of the various melanocortin receptors, it will be necessary to discover much more selective melanotropin peptides or peptide mimetics. In this regard, it now seems likely that each of the melanocortin receptors serves more than one physiological function. If one of these functions is peripheral and the other central, it may be possible to target one or the other receptor with the same ligand with different delivery systems. On the other hand, it may be possible to obtain subtype melanotropin selectivity which would be a more ideal solution. Yet another consideration, especially if there is cross talk among the melanocortin receptors, is the design of melanotropin ligands that can act as balanced agonist at two of the receptors with little or no activity at the other two receptors. Alternatively, one may need a ligand that acts as an antagonist at one melanocortin receptor (e.g., hMC3R), but as an agonist at another melanocortin receptor (e.g., hMC4R). Clearly, there is still much to be learned about melanotropin peptide and nonpeptide structure-activity relationships, about the similarities and differences in the binding pockets of the various melanocortin receptors and how this might be exploited for drug design, and about the nature of the transduction chemistry at the various melanocortin receptors (e.g., one or more G proteins involved for each receptor). However, in view of the critical biological, medical, and behavior effects that are modulated, controlled, and expressed via these receptors, the payoff both in terms of scientific insight and understanding, and in terms of the treatment of disease, is great.

ACKNOWLEDGMENTS

The support of this research from the U.S. Public Health Service (Grant DK 17420) is greatly appreciated. The views expressed are those of the authors and not necessarily those of the USPHS.

REFERENCES

1. HARRIS, J.I. & R.M. LERNER. 1957. Amino acid sequence of the α-melanocyte stimulating hormone. Nature **179:** 1346–1347.
2. HRUBY, V.J., B.C. WILKES, W.L. CODY, *et al.* 1984. Melanotropins: structural, conformational and biological considerations in the development of superpotent and superprolonged analogs. Pept. Protein Rev. **3:** 1–64, and references therein.
3. SCHWYZER, R. 1977. ACTH: A short introductory review. Ann. N. Y. Acad. Sci. **297:** 3–26.
4. HADLEY, M.E., V.J. HRUBY & A. BOWER. 1975. Cellular mechanisms controlling melanophore stimulating hormone (MSH) release. Gen. Comp. Endocrinol. **26:** 24–35.
5. SAWYER, T.K., P.J. SANFILIPPO, V.J. HRUBY, *et al.* 1980. [Nle4,DPhe7]α-Melanocyte stimulating hormone: a highly potent α-melanotropin with ultralong biological activity. Proc. Natl. Acad. Sci. USA **77:** 5754–5758.
6. SAWYER, T.K., V.J. HRUBY, P.S. DARMAN & M.E. HADLEY. 1982. [4-Half-cystine, 10-half-cystine]-α-melanocyte stimulating hormone: a cyclic α-melanotropin exhibiting superagonist Biological activity. Proc. Natl. Acad. Sci. USA **79:** 1751–1755.
7. HRUBY, V.J. 1982. Conformational restrictions of biologically active peptides via amino acid side chain groups. Life Sci. **31:** 189–199.
8. CONE, R.D., Ed. 2000. The Melanocortin Receptors. Humana Press. Totowa, NJ.
9. VAUDRY, H. & A.N. EBERLE, Eds. 1993. The Melanotropic Peptides. 1993. N.Y. Acad. Sci. **690:** 1–687.
10. DE L. CASTRUCCI, A.-M., M.E. HADLEY, E.I. YORULMAZOGLU, *et al.* 1985. Synthesis and studies of superpotent melanotropins resistant to enzyme degradation. Pigm. Cell **1985:** 145–151.
11. AL-OBEIDI, F., M.E. HADLEY, B.M. PETTITT & V.J. HRUBY. 1989. Design of a new class of superpotent cyclic α-melanotropins based on quenched dynamic simulations. J. Am. Chem. Soc. **111:** 3413–3416 and references therein.
12. AL-OBEIDI, F., A.-M. DE LAURO CASTRUCCI, M.E. HADLEY & V.J. HRUBY. 1989. Potent and prolonged acting cyclic lactam analogues of α-melanotropin: design based on molecular dynamics. J. Med. Chem. **32:** 2555–2561.
13. FAN, W., B.A. BOSTON, R.A. KESTERSON, *et al.* 1997. Role of melanocortinergic neurons in feeding and the agouti obesity syndrome. Nature **385:**165–168.
14. HRUBY, V.J., D. LU, S.D. SHARMA, *et al.* 1995. Cyclic lactam α-melanotropin analogues of Ac-Nle^4-c[Asp^5,D-Phe^7,Lys^{10}]α-MSH(4-10)-NH$_2$ with bulky aromatic amino acids at position 7 show high antagonist potency and selectivity at specific melanocortin receptors. J. Med. Chem. **38:** 3454–3461.
15. LI, S-J., K. VARGA, P. ARCHER, *et al.* 1996. Melanocortin antagonists define two distinct pathways of cardiovascular control by α- and γ-melanocyte-stimulating hormones. J. Neurosci. **16:** 5182–5188.

16. WESSELLS, H., K. FUCIARELLI, J. HANSEN, *et al.* 1998. Synthetic melanotropic peptide initiates erections in men with psychogenic erectile dysfunction: double-blind, placebo controlled crossover study. J. Urol. **160:** 389–393.

17. HRUBY, V.J. 1993. Conformational and topographical considerations in the design of biologically active peptides. Biopolymers **33:** 1073–1082.

18. HRUBY, V.J. 1992. Strategies in the development of peptide antagonists. *In* Progress in Brain Research, Vol. 92. J. Joosse, R.M. Buijs & F.J.H. Tilders, Eds.: 215–224. Elsevier Science Publ.

19. HRUBY, V.J. 2002. Designing peptide receptor agonists and antagonists. Nat. Rev. Drug Discovery **1:** 847–858.

20. GRIECO, P., G. HAN & V.J. HRUBY. 2000. New Dimensions in the Design of Potent and Receptor Selective Melanotropin Analogues. Peptides for the New Millenium. G.B. Fields, J.P. Tam & G. Barany, Eds.: 541–542. Kluwer Academic Publ., Dordrecht, the Netherlands.

21. GRIECO, P., G. HAN, D. WEINBERG, *et al.* 2002. Design and synthesis of highly potent and selective melanotropin analogues of SHU9119 modified at position 6. Biochem. Biophys. Res. Commun. **292:** 1075–1080.

22. HASKELL-LUEVANO, C., L.W. BOTEJU, H. MIWA, *et al.* 1995. Topographical modification of melanotropin peptide analogues with β-methyltryptophan isomers at position 9 leads to differential potencies and prolonged biological activities. J. Med. Chem. **38:** 4720–4729.

23. HASKELL-LUEVANO, C., K. TOTH, L. BOTEJU, *et al.* 1997. β-Methylation of the Phe[7] and Trp[9] melanotropin side chain pharmacophores affects ligand-receptor interactions and prolonged biological activity. J. Med. Chem. **40:** 2740–2749.

24. HASKELL-LUEVANO, C., S. HENDRATA, C. NORTH, *et al.* 1997. Discovery of prototype peptidomimetic agonists at the human melanocortin receptors MC1R and MC4R. J. Med. Chem. **40:** 2133–2139.

25. HAN, G., J.M. QUILLAN, K. CARLSON, *et al.* 2003. Design of novel chimeric melanotropin-deltorphin analogues. Discovery of the first potent human melanocortin 1 receptor antagonist. J. Med. Chem. **46:** 810–819.

26. GRIECO, P., P.M. BALSE, D. WEINBERG, *et al.* 2000. D-Amino acid scan of γ-melanocyte-stimulating hormone: importance of Trp[8] on human MC3 receptor selectivity. J. Med. Chem. **43:** 4998–5002.

27. SCHIÖTH, H.B., F. MUTULIS, R. MUCENIECE, *et al.* 1998. Discovery of novel melanocortin[4] receptor selective MSH analogues. Br. J. Pharmacol. **124:** 75–82.

28. SZARDENINGS, M., R. MUCENIECE, I. MUTULE, *et al.* 2000. New highly specific agonistic peptides for human melanocortin MC1 receptor. Peptides **21:** 239–243.

29. KAVARANA, M.J., D. TRIVEDI, M. CAI, *et al.* 2002. Novel cyclic templates of alpha-MSH give highly selective and potent antagonists/agonists for human melanocortin-3/4 receptors. J. Med. Chem. **45:** 2644–2650.

30. BEDNAREK, M.A., T. MACNEIL, R. TANG, *et al.* 2001. Potent and selective peptide agonists of α-melanotropin action at human melanocortin receptor 4: their synthesis and biological evaluation in vitro. Biochem. Biophys. Res. Commun. **286:** 641–645.

31. GRIECO, P., A. LAVECCHIA, M. CAI, *et al.* 2002. Structure-activity studies of the melanocortin peptides: discovery of potent and selective antagonists at MC3 and MC4 receptors. J. Med. Chem. **45:** 5287–5294.

Melanocortin Receptors: Ligands and Proteochemometrics Modeling

J.E.S. WIKBERG, F. MUTULIS, I. MUTULE, S. VEIKSINA, M. LAPINSH, R. PETROVSKA, AND P. PRUSIS

Department of Pharmaceutical Biosciences, Uppsala University, Box 591, SE-751 24 Uppsala, Sweden

ABSTRACT: The melanocortin receptors exist in five subtypes, $MC_{1-5}R$. These receptors participate in important regulations of the immune system, central behavior, and endocrine and exocrine glands. Here we provide a short review on MCR subtype selective peptides and organic compounds with activity on the MCRs, developed in our laboratory. Also provided is an overview of our new proteochemometric modeling technology, which has been applied to model the interaction of MSH peptides with the MCRs.

KEYWORDS: melanocortin receptors; MSH; melanocyte-stimulating hormone; proteochemometrics; selective ligands

INTRODUCTION

In 1992 the MC_1 and MC_2 receptors ($MC_{1-2}Rs$) were cloned from DNA libraries from melanoma cell lines and human genomic DNA,[1,2] and sequences for the MC_3 and MC_5Rs were identified.[3] During the subsequent year, the MC_3 and MC_5Rs were cloned in their full lengths, as was the MC_4R (see Ref. 4). While the MC_2R was the ACTH receptor, the other MCRs are melanocyte-stimulating hormone (MSH) binding receptors.

Since then knowledge on the MCRs has been rapidly expanding. Their locations to different regions in the body were mapped out, and their physiological roles have begun to be understood. While the role for the MC_1R as regulator of melanin pigment formation in the skin was early appreciated, more recent results indicate that the MC_1R has important regulatory roles in the immune system. Not only has the receptor been found in immune-presenting cells, such as macrophages and monocytes, where it has a distinctive role in suppressing cell activity,[4,5] but it is also present on the antigen-presenting B cells, on natural killer cells, and on a subpopulation of the cytotoxic T cells.[6] The role of the MC_2 receptor as regulator of corticosteroid production of the adrenals is established, but the receptor may have other roles as it is present in the skin where the ACTH peptide itself is also produced.[7] Much interest has been devoted to the roles of the MC_3 and MC_4Rs in the central nervous system.

Address for correspondence: J.E.S. Wikberg, Department of Pharmaceutical Biosciences, Uppsala University, Box 591, SE-75124 Uppsala, Sweden. Voice: +46-18-471 42 38; fax: +46-18-559718.

Jarl.Wikberg@farmbio.uu.se

Ann. N.Y. Acad. Sci. 994: 21–26 (2003). © 2003 New York Academy of Sciences.

These receptors are involved in regulation of complex feeding and sexual behaviors. At least part of this regulation may be related to control of central reward[8] and the central dopaminergic system.[9] The MC_5R has an established role to control sebum secretion from sebaceous glands.[7]

Because of these roles of the MCRs, there is a substantial interest to develop selective agonists and antagonists. It is hoped that MC_1R selective agonists will be useful in the treatment of disorders in the immune system and that MC_3 and MC_4R agonists will be useful in the treatment of sexual dysfunctions and overweight conditions. Moreover, MC_4R antagonists might be useful in the treatment of anorexia and MC_5R antagonists to treat acne and seborrheic dermatis.[7] Development of such agonists and antagonists is now well under way, and promising low molecular weight compounds of several chemical classes exist.[7,10] However, peptides are still the most selective high-affinity compounds for the MCRs.

Drug development is a costly and quite empirical process relying on iterative chemical synthesis and biological testing. While computer-assisted structure-based methods have been developed, they are of limited value when accurate 3-D structures of receptors are not available. We will briefly summarize here our development of some ligands for the MCRs, and our new proteochemometrics technology, which is useful to gain insight into receptor–ligand interactions and which may potentially find use in drug development.

LIGANDS TO MC RECEPTORS

Using a combination of phage-display and empirical approaches we developed the MS05 and MS09 peptides.[11] Both are potent full agonists on the MC_1R; the MS05 peptide being the most MC_1R-selective agonist known with better than 1000-fold selectivity over the other MCRs. The MS09 peptide is even more potent but shows lower selectivity, its K_d being about 0.2 nM for the MC_1R. Both of these compounds show potent anti-inflammatory actions.

In our laboratory we also developed a number of MC_4R selective cyclic antagonists, among them the HS014, HS024, and HS131 peptides (see Ref. 7). These show from 0.3–12 nM K_d for the MC_4R. HS014 and HS131 are the most selective, with about 10–20-fold selectivity over the MC_3R and higher for the other MCRs. They have been used to map various physiological functions of the MCRs.

More recently we prepared a series of compounds containing presumed pharmacophoric groups of the melanocortic peptides (naphtalene, amino or guanidine, and indole moieties).[12,13] The most active compounds among this series reached to less than 1 μM K_d for the MCRs, some of them showing moderate selectivity for the MC_1R. Compounds among this series appear to be antagonist or partial agonists.

PROTEOCHEMOMETRICS

An outline of proteochemometrics is shown in FIGURE 1. Proteochemometrics is based on the modeling of data of sets of receptors interacting with sets of ligands. To create a proteochemometric model, one needs first to measure the activities for the interactions of a number of receptors showing sequence variation with a number of

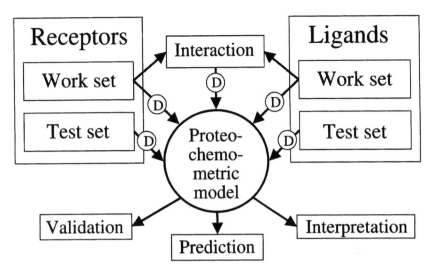

FIGURE 1. Principles of proteochemometrics. In proteochemometrics, the interactions of a set of receptors with a set of ligands are first evaluated experimentally creating "work sets." The work sets are then described using descriptors ("D"), and so-called receptor–ligand, receptor–receptor, and ligand–ligand cross-descriptors. These descriptors are then correlated to the interaction activities using a suited mathematical approach, thereby resulting in the proteochemometric model. Before acceptance as a model, a descriptor needs to be validated for reliability, accuracy, and resolution, after which it can be used for interpretations and predictions. Predictions can be applied onto the "test sets," thus making it possible to perform *a priori* computations of the interaction activity of receptors and/or ligands.

different ligands. The interaction data are then correlated to descriptors of the receptors and ligands, thereby creating the proteochemometric model.

Descriptors of receptors and ligands can be derived in different ways. They fall into two different broad categories: physicochemical descriptors and structural descriptors. Physicochemical descriptors are derived from measured or computed physicochemical properties of the receptors and ligands. Structural descriptors are derived from the structures of the entities under study in one, two, three, or even more dimensions. Moreover, so-called binary descriptors can also be used. Such descriptors may represent the absence or presence of a particular structural feature in a molecule. A binary descriptor could also be used to characterize the values of a physicochemical property over or below certain limit.

In the development of proteochemometrics we have used physicochemical, structural, and binary descriptors to represent receptors and ligands. In our modeling of MCR and MSH-peptide interactions we used binary descriptors,[14–16] while we used both physicochemical and structural descriptors to model the interaction of amine G-protein coupled receptors with organic amines.[17,18]

The above-described ligand and receptor descriptors are termed ordinary descriptors. In addition, cross-descriptions are created by multiplying any two ordinary descriptors. Three types of cross-descriptors exist: ligand–ligand, receptor–receptor, and ligand–receptor. In proteochemometric modeling, these cross-descriptions have special meaning and will be discussed further below.

DISCOVERY OF PROTEOCHEMOMETRIC MODELING

We discovered proteochemometrics when we analyzed the binding of α-MSH and MS05, and chimeric peptides created from these, to wild-type MC_1 and MC_3Rs receptors, and chimeric MC_1/MC_3Rs.[14] The "manual" interpretation of the complex data obtained proved difficult, and we wanted to develop a mathematical approach to analyze it. We reasoned the data could be coded using binary numbers. Thus, a unity number would represent a sequence stretch taken from one of the sequence possibilities, while a zero would represent the other. From a chemical point of view, a binary number represents the structural and/or physicochemical properties of one of the two possible sequence stretches at a particular place in the receptor or MSH peptide. Since the chimeric receptors were constructed by dividing the receptors into four parts, each receptor was represented by a vector with four numbers; e.g., the MC_1 receptor by [0,0,0,0], the MC_3R by [1,1,1,1], and one of the chimeras by [[0,1,1,0], etc. Because the chimeric MSH peptides were divided into two parts, two binary numbers were used to code them. Six binary numbers, constituting ordinary descriptors, would thus describe any one of the possible receptor ligand combinations.

The data were then subjected to modeling by applying the partial least squares projection to latent structures (PLS) algorithm, and reasonable models could be created relating the ordinary descriptors to the logarithm of ligand–receptor affinities ($R^2 = 0.71$ and $Q^2 = 0.62$). (PLS models are characterized by the R^2 and Q^2, which represent fractions for modeled and predicted Y-variations, respectively). However, after cross-descriptors were computed and added to the descriptors, substantial improvements of the PLS model ($R^2 = 0.93$ and $Q^2 = 0.75$) resulted.

Interpretation of a proteochemometric model can be performed by assessing the signs and values of the coefficients of the obtained regression equation, or by using the PLS modeling variable importance in projection (VIP) measures (see Ref. 14). For example, a coefficient of an ordinary term of ligands describes the influence of an aspect of the ligands on their interactions with non-varied part(s) of the receptors. Ligand–receptor cross-terms, on the other hand, relate to the ability of the corresponding varied portions of both the ligands and receptors to form complimentary interactions.

Interpreting the proteochemometric model showed that the MSH peptides bind to a binding pocket located between TM1, TM2/3, and TM6/7. TM4/5 is, however, located at a remote place from this pocket. It is noteworthy that, despite the low number of descriptors, this proteochemometric model placed the binding pocket in the region where it was independently placed by molecular modeling and ligand docking.[19]

FURTHER DEVELOPMENTS IN PROTEOCHEMOMETRICS

In further studies[15] we applied proteochemometrics to model the interaction of a wider set of cyclic and linear MSH peptides to the above wild-type and chimeric

MC_1/MC_3 receptors. In these studies binary coding was used to describe the richer variation of structural properties of the MSH peptides used, while the receptor descriptions were the same as above. Further, in this study very good models could be created when we applied state-of-the-art PLS modeling ($R^2 = 0.97$; $Q^2 = 0.91$). We also used the very sturdy approach of external prediction to validate the correctness of the proteochemometric model. This was done by excluding half of the experimental data and creating a model from the remaining, and thereafter predicting the binding activities of the excluded receptor–peptide combinations. The analysis showed that excellent external prediction could be achieved for the data that had never been part of the model building, the measure for external Y-variation, eQ^2, being as high as 0.85.

Interpretation of the final full proteochemometric model[15] indicated the placement of the MSH peptides in the receptor to exactly the same place as in our first study.[14] However, one further very interesting use of proteochemometric modeling came from the interpretation of the sizes and signs of the regression coefficients for ligand–ligand cross-descriptors. The analysis indicated that certain structural features in the MSH peptides group them into similar conformational spaces, which may explain their similar way of interacting with the MC receptors.[15]

FUTURE PERSPECTIVES OF PROTEOCHEMOMETRICS

We have applied proteochemometrics in two further studies. One was performed on a data set of 18 wild-type, cassette and point mutated α_1-adrenoceptors interacting with 12 structurally closely related low molecular weight α_1-adrenoceptor blocking compounds.[17] The more detailed variation of amino acids in the receptor sequences allowed the receptors to be described in larger detail, which we did by using amino acid z-scales.[20] By use of these z-scales, which represent the principal components of measured and computed physicochemical properties of amino acids, excellent proteochemometric models resulted. Moreover, the z-scales allowed identification of the amino acids in the receptor sequences involved in the interactions with the α_1-adrenoceptor blocking compounds, and a detailed interpretation of the physicochemical properties of importance for the ligand binding.[17]

The other study performed proteochemometric modeling onto a set of 21 amine G-protein coupled receptors (GPCRs) interacting with 23 organic amines. The receptor set included wild-type GPCRs only, namely subtypes of serotonin, dopamine, adrenergic, and histaminergic receptors. The organic amines showed wide diversity and represented antipsychotics, alkaloids, and experimental compounds. Receptor sequences were described using amino acid z-scales, while the organic compounds were described using a new approach: GRINDs based on molecular interaction fields (MIFs) see Ref. 18). Also, in this case surprisingly good proteochemometric models that validated well could be created. A number of important and detailed conclusions could be drawn on the mode of interaction of the organic amines with the amine GPCRs.[18]

Based on our studies, proteochemometric modeling can be used to gain detailed insights into molecular recognition processes. Proteochemometrics might be a useful tool in ligand design as well. Future studies will show whether this can be substantiated, for example, for design of MCR subtype selective compounds.

ACKNOWLEDGMENTS

Experimental portions of this research were supported by the Swedish Research Council (04X-05957; 230-2000-291) and Melacure Therapeutics.

REFERENCES

1. CHHAJLANI, V. & J.E.S. WIKBERG. 1992. Molecular cloning and expression of the human melanocyte stimulating hormone receptor cDNA. FEBS Lett. **309:** 417–420.
2. MOUNTJOY, KG., LS. ROBBINS, MT. MORTRUD, *et al.* 1992. The cloning of a family of genes that encode the melanocortin receptors. Science **257:** 1248–1251.
3. CHHAJLANI, V. & J.E.S. WIKBERG, INVENTORS; WAPHARM, ASSIGNEE. 1992. A polypeptide, DNA coding for the polypeptide, and uses thereof. Danish patent 104,692. Date of application: August 21.
4. WIKBERG, J.E.S., R. MUCENIECE, I. MANDRIKA, *et al.* 2000. New aspects on the melanocortins and their receptors. Pharmacol. Res. **42:** 393–420.
5. MANDRIKA, I., R. MUCENIECE & J.E.S. WIKBERG. 2001. Effect of melanocortin peptides on lipopolysaccharide/interferon-γ induced NF-κB DNA-binding and nitric oxide production in macrophage-like RAW 264.7 cells: evidence for dual mechanisms of action. Biochem. Pharmacol. **61:** 613–621.
6. NEUMAN-ANDERSEN, G., O. NAGAEVA, I. MANDRIKA, *et al.* 2001. MC_1 receptors are constitutively expressed on leucocyte subpopulations with antigen presenting and cytotoxic functions. Clin. Exp. Immunol. **126:** 441–446.
7. WIKBERG, J.E.S. 2001. Melanocortin receptors: new opportunities in drug discovery. Exp. Opin. Ther. Patents **11:** 61–76.
8. LINDBLOM, J. 2002. The role of the melanocortin system in linking energy homeostasis with reward mechanisms. Ph.D. thesis, Uppsala University, Uppsala.
9. LINDBLOM, J., B. OPMANE, F. MUTULIS, *et al.* 2001. The MC_4 receptor mediates α-MSH induced release of nucleus accumbens dopamine. Neurochemistry **12:** 2155–2158.
10. ANDERSSON, PM., A. BOMAN, E. SEIFERT, *et al.* 2001. Ligands to the melanocortin receptors. Expert Opinion Ther. Patents **11:** 1583–1592.
11. SZARDENINGS, M., R. MUCENIECE, I. MUTULE, *et al.* 2000. New highly specific peptides agonistic for human melanocortin MC_1 receptor. Peptides **21:** 239–243.
12. MUTULIS, F., I. MUTULE, M. LAPINSH, *et al.* 2002. Reductive amination products containing naphthalene and indole moieties bind to melanocortin receptors. Bioorg. Med. Chem. Lett. **12:** 1035–1038.
13. MUTULIS, F., I. MUTULE & J.E.S. WIKBERG. 2002. N-Alkylaminoacids and their derivatives interact with melanocortin receptors. Bioorg. Med. Chem. Lett. **12:** 1039–1042.
14. PRUSIS, P., R. MUCENIECE, P. ANDERSSON, *et al.* 2001. PLS modeling of chimeric MS04/MSH-peptide and MC_1/MC_3-receptor interactions reveals a novel method for analysis of ligand receptor interactions. Biochem. Biophys. Acta **1544:** 350–357.
15. PRUSIS, P., T. LUNDSTEDT & J.E.S. WIKBERG. 2002. Proteo-chemometrics analysis of MSH peptide binding to melanocortin receptors. Protein Eng. **15:** 305–311.
16. PRUSIS, P., R. MUCENIECE, T. LUNDSTEDT, *et al.* 2000. Identification of the binding pocket for the TRH peptide in the melanocortin 1 receptor. Lett. Peptide Sci. **7:** 225–228.
17. LAPINSH, M., P. PRUSIS, A. GUTCAITS, *et al.* 2001. Development of proteo-chemometrics: a novel technology for the analysis of drug-receptor interactions. Biochem. Biophys. Acta **1525:** 180–190.
18. LAPINSH, M., P. PRUSIS, T. LUNDSTEDT, *et al.* 2002. Proteo-chemometrics modeling of the interaction of amine G-protein coupled receptors with a diverse set of ligands. Mol. Pharmacol. **61:** 1465–1475.
19. PRUSIS, P., H.B. SCHIÖTH, R. MUCENIECE, *et al.* 1997. Modeling of the three-dimensional structure of the human melanocortin-1 receptor using an automated method and docking of a rigid cyclic MSH core peptide. J. Mol. Graphics Modelling **15:** 307–317.
20. SANDBERG, M., L. ERIKSSON, J. JONSSON, *et al.* 1998. New chemical descriptors relevant for the design of biologically active peptides. A multivariate characterization of 87 amino acids. J. Med. Chem. **41:** 2481–2491.

Loops and Links: Structural Insights into the Remarkable Function of the Agouti-Related Protein

GLENN L. MILLHAUSER,[a] JOE C. McNULTY,[a] PILGRIM J. JACKSON,[a]
DARREN A. THOMPSON,[a] GREGORY S. BARSH,[b] AND IRA GANTZ[c]

[a]Department of Chemistry and Biochemistry, University of California, Santa Cruz,
California 95064, USA

[b]Howard Hughes Medical Institute and the Department of Pediatrics and Genetics,
Stanford University Medical Center, Stanford, Califorrnia 94305, USA

[c]Department of Surgery, University of Michigan Medical Center, Ann Arbor,
Michigan 48109, USA

ABSTRACT: The agouti-related protein (AGRP) is an endogenous antagonist of
the melanocortin receptors MC3R and MC4R found in the hypothalamus and
exhibits potent orexigenic activity. The cysteine-rich C-terminal domain of this
protein, corresponding to AGRP(87-132), exhibits receptor binding affinity
and antagonism equivalent to that of the full-length protein. We recently determined the NMR structure of AGRP(87-132) and demonstrated that a portion of the domain adopts the inhibitor cystine-knot fold. Remarkably, this is
the first identification of a mammalian protein with this specific architecture.
Further analysis of the structure suggests that melanocortin receptor contacts
are made primarily by two loops presented within the cystine knot.[10] To test
this hypothesis we designed a 34-residue AGRP analogue corresponding to
only the cystine knot. We found that this designed miniprotein folds to a homogeneous product, retains the desired cystine-knot architecture, functions as
a potent antagonist, and maintains the melanocortin receptor pharmacological
profile of AGRP(87-132).[26] The AGRP-like activity of this molecule supports
the hypothesis that indeed the cystine-knot region possesses the melanocortin
receptor contacts. Based on these design and structure studies, we propose that
the N-terminal loop of AGRP(87-132) makes contact with a receptor exoloop
and helps confer AGRP's selectivity for the central MCRs.

KEYWORDS: melanocortin receptor; agouti-related protein; nuclear magnetic
resonance

INTRODUCTION

The agouti-related protein (AGRP) plays a key role in the regulation of feeding
behavior and energy homeostasis in mammals.[1-3] This orexigenic, paracrine signaling molecule is produced in the hypothalamus and is a potent antagonist of α-MSH

Address for correspondence: Glenn L. Millhauser, Department of Chemistry and Biochemistry, University of California, Santa Cruz, CA 95064. Voice: 831-459-2176; fax: 831-459-2935.
glennm@hydrogen.ucsc.edu

Ann. N.Y. Acad. Sci. 994: 27–35 (2003). © 2003 New York Academy of Sciences.

Agouti-related protein (AGRP)

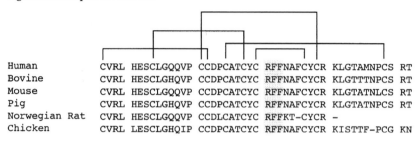

Human	CVRL HESCLGQQVP CCDPCATCYC RFFNAFCYCR KLGTAMNPCS RT
Bovine	CVRL HESCLGHQVP CCDPCATCYC RFFNAFCYCR KLGTTTNPCS RT
Mouse	CVRL HESCLGQQVP CCDPCATCYC RFFNAFCYCR KLGTATNLCS RT
Pig	CVRL HESCLGHQVP CCDPCATCYC RFFNAFCYCR KLGTATNPCS RT
Norwegian Rat	CVRL HESCLGQQVP CCDLCATCYC RFFKT-CYCR -
Chicken	CVRL LESCLGHQIP CCDPCATCYC RFFNAFCYCR KISTTF-PCG KN

Agouti protein

Human	CVAT RNSCKPPAPA CCDPCASCQC RFFRSACSCR VLSLNC
Bovine	CVAT RDSCKPPAPA CCDPCAFCQC RFFRSACSCR VLNPTC
Mouse	CVAT RDSCKPPAPA CCDPCASCQC RFFGSACTCR VLNPNC
Pig	CVAN RDSCKPPALA CCDPCAFCQC RFFRSACSCR VLNPTC
Norwegian Rat	CVAT RDSCKPPAPA CCNPCASCQC RFFGSACTCR VLNPNC
Red Fox	CVAT RNSCKSPAPA CCDPCASCQC RFFRSACTCR VLSPSC

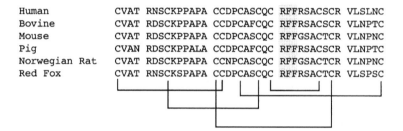

FIGURE 1. Sequences of the agouti and AGRP cysteine-rich, C-terminal domains for various species. Disulfide bonds between cysteines are indicated. Residues of the RFF triplet, which dock at a receptor transmembrane pocket, are shaded.

activity at melanocortin receptors (MC3R and MC4R) of the central nervous system. The melanocortin receptor family consists of five subtypes (MC1R–MC5R) and belongs to the superfamily of G-protein coupled receptors (GPCRs) which activate the adenylate cyclase signal transduction pathway.[4,5] AGRP was discovered by its homology with the agouti protein, which participates in the control of coat pigmentation in rodents by antagonizing MC1 receptors.[1,2] As shown in FIGURE 1, both AGRP and the agouti protein possess a cysteine-rich C-terminal domain, and these domains alone are sufficient for high-affinity MCR binding, selectivity, and potent antagonism.[6,7] Between AGRP and the agouti protein, the C-terminal domains are approximately 40% identical, yet they bind to distinct sets of MCR subtypes; AGRP binds with high affinity to MC3R and MC4R whereas the agouti protein binds to MC1R and MC4R.[1,8] The molecular basis for this receptor subtype selectivity is currently unknown.

AGRP is the focus of current attention for several reasons. First, AGRP and the agouti protein exhibit unique biochemical function, as they are the only known endogenous competitive antagonists of GPCRs.[1,9] Second, 6 of the 10 cysteines in AGRP's C-terminal domain participate in a network of disulfide cross-links with spatial positioning that is more reminiscent of an invertebrate toxin than a mammalian protein.[10] Third, controlling the interaction between AGRP and its receptors in the brain may open up new avenues for treating consumptive disorders. Indeed, obesity and other diseases of energy balance such as diabetes are approaching epidemic proportions in the United States.[11] MC3R and MC4R, the two receptors for which AGRP exhibits high affinity, may be ideal pharmacological targets for treating these disorders.[3,12] In addition, the wasting condition known as cachexia is one of the major contributors to declining health in AIDS and cancer. It has been suggested that molecules with AGRP-like activity may be useful in treating this condition.[13]

STRUCTURE OF AGRP(87-132) AND DESIGN OF A MINI-AGRP

Shortly after the discovery of AGRP, we sought to determine the three-dimensional structure of this novel protein. Given the unique function of AGRP as an endogenous antagonist and its emerging importance in diabetes and obesity research, structural studies were clearly warranted. Disulfide connectivities had been determined for both agouti[7] and AGRP.[14] The Cys spacing within the C-terminal domains and the disulfide maps are similar to that found in a number of invertebrate toxins. Homology models indeed suggested that agouti and AGRP would adopt a specific toxin-like fold called an inhibitor cystine-knot (ICK) motif.[15,16] Although such a prediction seems reasonable by the criteria of homology models, suggesting an ICK fold for AGRP or agouti is remarkable because this type of structure had not been previously observed in mammalian proteins. In addition, members of the ICK motif class differ widely in their detailed structures, and the pseudoknot nature of this fold allows for non-ICK arrangements of the polypeptide.[17-19] Thus, we concluded that structural modeling alone would not provide a robust structural characterization.

Active AGRP(87-132) was prepared by total chemical synthesis.[20] Nuclear magnetic resonance (NMR) data were obtained at 800 MHz from a sample at 15°C in pH 5.0 phosphate buffer.[10] A suite of homonuclear experiments yielded 799 distance restraints and 92 torsion restraints. Using these restraints along with known disulfide connectivities, we calculated a family of 40 low-energy structures as shown in FIGURE 2. The backbone fold is well defined as indicated by the overall backbone RMSD of 0.535 Å. The first 34 residues corresponding to AGRP(87-120) show an even greater degree of order with a backbone RMSD of 0.360 Å. In contrast, the final 12 residues show significant conformational variability that, in turn, reflects the near absence of long-range NOEs linking this region of the protein to the first 34 residues.

A ribbon diagram of the three-dimensional structure of AGRP(87-132) is shown in FIGURE 2.[10] The protein is characterized by a three-strand anti-parallel β-sheet stabilized by a network of disulfide bonds. We refer to the short loop presenting residues 111–113 as the active loop, and the flanking loops as the N-terminal and C-terminal loops, respectively. A defining feature of the ICK motif is that the disulfide emerging from the third cysteine threads through a topological circle formed by two other disulfides and their intervening peptide segments. Examination of AGRP(87-132) indeed demonstrates that the disulfide involving the third Cys passes from the front of the protein to the back through the appropriate circle. Thus, AGRP(87-132) adopts the ICK fold and is the first mammalian protein assigned to this structure class.

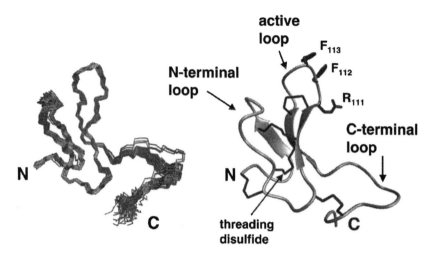

FIGURE 2. NMR structure of the AGRP C-terminal domain, AGRP(87-132). (*Left*) Family of 40 low-energy structures. The structures are aligned from residues 87–120 to highlight the observed ordering of this region and the relative disorder of residues 121–132. Only backbone atoms are shown. (*Right*) Ribbon diagram of AGRP(87-132). Backbone *arrows* identify a three-strand β-sheet; the Cys side chains involved in disulfide bonds are in *black*. The threading disulfide assigns this protein to the ICK family. Relevant loops and residues of the RFF triplet are indicated.

The triplet of residues at positions 111–113, highlighted for both agouti and AGRP sequences in FIGURE 1, is essential for MC4R antagonism. An alanine scanning experiment on agouti showed that replacement of any residue within the RFF triplet caused at least a 40-fold loss of affinity for MC4R binding.[15] We have shown that replacement of R with A in the triplet leads to a complete loss of AGRP antagonism.[21] FIGURE 2 shows that the RFF triplet resides within a short octapeptide loop —the active loop—closed by a single disulfide bond.

Tota *et al.* reasoned that the active loop forms a primary contact point for the ligand–receptor interaction.[16] They tested their hypothesis by screening short cyclic peptides derived from the RFF-containing region of agouti and AGRP for MC3R and MC4R antagonism. Indeed, they demonstrated inhibition of NDP-MSH binding and inhibition of α-MSH–stimulated cAMP production. However, the AGRP-derived cyclic peptide exhibited a greater than 10-fold lower affinity than that of the full AGRP C-terminal domain at MC4R. Also, their cyclic peptides showed almost no affinity for MC3R regardless of whether the peptide sequences were derived from agouti or AGRP. Thus, other structural determinants must play a role in selectivity and tight binding at central MCRs.

Mutagenesis studies suggest that the RFF triplet interacts with its cognate receptors by docking to a cluster of negatively charged residues presented by two adjacent transmembrane helices.[22,23] Both agonists and antagonists appear to share this common transmembrane MCR docking location. However, there are additional regions in the receptors that influence only the binding of AGRP and the agouti protein. Cassette mutagenesis studies have shown that the second and third extracellular loops

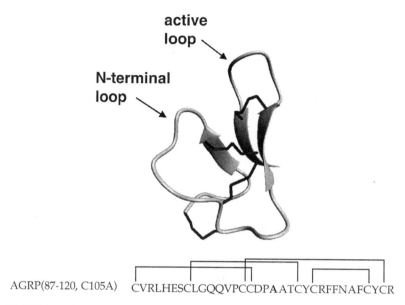

FIGURE 3. Structure and sequence of AGRP(87-120, C105A), mini-AGRP. Structural features follow FIGURE 2 legend.

(exoloops) connecting transmembrane helices are particularly important.[24,25] Although the MC1R does not bind AGRP, a chimeric MC1R containing the second and third exoloops from MC4R results in a receptor that binds AGRP with high affinity.[24] In addition, chimeric MC4R with MC1R exoloops loses its high affinity for AGRP.

Based on these mutagenesis studies and the AGRP(87-132) structure, we hypothesized that the ordered region of AGRP's C-terminal domain possessed the determinants that control its MCR function. To test this hypothesis, we developed a miniprotein designed to present only the ordered region.[26] The sequence of this "mini-AGRP," designated AGRP(87-120, C105A), is shown in FIGURE 3. The rationale of this sequence is as follows. First, the ICK motif is found for numerous toxins with widely varying sequences and thus should fold as an independent domain.[17] Second, while the 33-residue segment corresponding to AGRP(87-119) contains all of the disulfide bonds involved in the ICK structure, the NMR structure demonstrates that Arg120 participates directly in the hydrogen bonded β-sheet and thus was included in the sequence.[10] To avoid having a free cysteine that in turn might lead to non-native disulfides, we eliminated Cys105 by replacing it with alanine. Finally, AGRP(87-120, C105A) was synthesized with an acetyl group at the N-terminus and an amide ($-NH_2$) at the C-terminus to avoid non-native electrostatic charges at these backbone positions.

Side-by-side receptor activity assays were performed on AGRP(87-120, C105A) and AGRP (87-132).[26] AGRP(87-120, C105A) functions as an antagonist and displaces NDP-MSH with IC_{50} values of 7.5 nM at MC3R and 6.1 nM at MC4R. These values are within experimental error of those determined for AGRP(87-132) (5.2 nM and 11.0 nM). Considering the full characterization over the family of MCRs, the function of mini-AGRP is indistinguishable from that of AGRP(87-132). We also determined the NMR structure of this designed protein following the protocols used for AGRP(87-132).[26] The structure family revealed a low RMSD of 0.41 Å and thus a well-defined backbone fold. The ribbon diagram, shown in FIGURE 3, shows that the β-hairpin region from residues 106–120, which includes the active loop, is structured the same as in AGRP(87-132). On the other hand, it appears that the N-terminal loop and the β-hairpin are further separated from each other in AGRP(87-120, C105A) than in AGRP(87-132). Interestingly, hydrogen/deuterium exchange experiments identify only one marginally protected hydrogen bond between the β-strand from residues 92–94 and the β-hairpin in AGRP(87-120, C105A), whereas in AGRP(87-132) two hydrogen bonds are identified with each showing significant protection from exchange.[26]

INSIGHTS INTO POSSIBLE RECEPTOR CONTACTS

The above findings identify clearly the core region of AGRP responsible for MCR selectivity and antagonism. We may now begin to identify regions of AGRP that interact with MC receptors. There are no structures available for any member of the melanocortin receptor family. However, the MCRs are homologous to rhodopsin, another member of the GPCR family for which a high-quality X-ray structure has been determined.[27] Homology models of MC1R[28,29] and MC4R[23] have been developed and locate the transmembrane region responsible for AGRP active loop recognition to helices 2 and 3 near the receptor's extracellular surface.[22,23] FIGURE 4 shows the rhodopsin structure, representative of MC4R, along with AGRP's C-terminal domain. AGRP(87-132) is oriented such that its active loop presenting the RFF triplet points towards the transmembrane binding region, and its N-terminal loop is directed to the receptor's third exoloop (between helices 6 and 7). The C-terminal loop, which our miniprotein shows is not needed for MCR recognition, is pointed away from the receptor surface. The bars in the figure are drawn to establish the distance between the relevant interaction sites. In the receptor, the bar reaches from the transmembrane site (left) to the third exoloop (right), whereas in AGRP the bar reaches from the active loop to a distal point on the N-terminal loop. In each case, the bar is approximately 24 Å long and thus suggests that the relevant regions of AGRP can indeed span the portions of the receptor that influence AGRP binding. Moreover, the specific interaction between the AGRP(87-132) N-terminal loop and the receptor exoloop may be responsible for AGRP's ability to bind selectively to central MCRs.

Also highlighted in FIGURE 4 is a collection of basic residues in AGRP(87-132) that contribute to a patch of positive charge on the protein surface. One of these residues is from the RFF triplet. However, the remaining four are all within, or near, the C-terminal loop. Although no function has yet been ascribed to this loop, its conserved sequence and participation in surface charge is suggestive. Recent studies suggest that syndecans may play a role in potentiating the effects of AGRP. Syndecans are a class of heparin sulfate proteoglycans (HSPGs), which are found ubiquitously on cell surfaces. Syndecan-3 is found on neurons in the hypothalamic feeding centers; feeding state regulates its level of expression, and syndecan-3 knockout

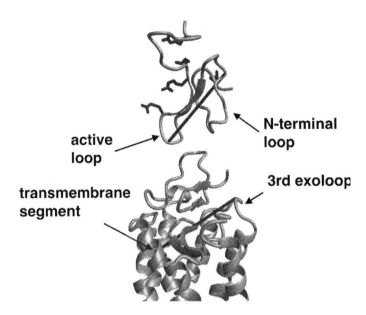

FIGURE 4. Model of AGRP(87-132) oriented for receptor docking. AGRP is positioned above the extracellular surface of rhodopsin, a GPCR with homology to MCRs. The bars are 24 Å and show that the distance between AGRP's active loop and N-terminal loop is approximately matched to the distance between the receptor transmembrane segment and the third exoloop. The Arg and Lys residues on AGRP(87-132) shown in black (positions 89, 111, 120, 121 and 131, FIG. 1) contribute to a patch of positive charge on the AGRP surface. Arg111, in the active loop, docks to the transmembrane segment. The remaining positively charged residues point away from the receptor surface and may facilitate interactions with syndecans.

mice show reduced reflex hyperphagia in response to food deprivation. Pharmacological studies by Reizes *et al.* suggest that syndecan-3 potentiates AGRP function at MC4R.[30] Although it is currently thought that the AGRP N-terminus mediates this interaction, the finding of a positive patch on a surface region of AGRP(87-132) that points away from the receptor suggests that the C-terminal domain may participate as well.

By combining protein design with structure determination, we have identified the core region responsible for AGRP function at MCRs. In light of our studies, and those of others, it seems very plausible that the active loop and flanking region of the N-terminal loop comprise the receptor-binding interface, with the C-terminal loop perhaps a contributor to AGRP's interaction with syndecans. The next step is to directly test these hypotheses. An unexpected benefit of our investigations is the finding that the synthesis of mini-AGRP is straightforward. Thus, mini-AGRP may serve as a scaffold into which one may engineer biophysical probes for exploring receptor interactions. Such studies, now ongoing in our laboratory, should greatly advance the understanding of how AGRP uses its unique fold to exert its antagonist function at melanocortin receptors.

ACKNOWLEDGMENTS

This work was supported by National Institutes of Health grants DK58606 (G.L.M), DK64265 (G.L.M.), DK48506 (G.S.B.), and DK54032 (I.G.) and by the University of Michigan Gastrointestinal Peptide Research Center (P30DK34933).

REFERENCES

1. OLLMANN, M.M. *et al.* 1997. Antagonism of central melanocortin receptors in vitro and in vivo by agouti-related protein. Science **278:** 135–138.
2. SHUTTER, J.R. *et al.* 1997. Hypothalamic expression of ART, a novel gene related to agouti, is up-regulated in obese and diabetic mutant mice. Genes Dev. **11:** 593–602.
3. WILSON, B.D., M.M. OLLMANN & G.S. BARSH. 1999. The role of agouti-related protein in regulating body weight. Mol. Med. Today **5:** 250–256.
4. GANTZ, I. *et al.* 1993. Molecular cloning, expression, and gene localization of a 4th melanocortin receptor. J. Biol. Chem. **268:** 15174–15179.
5. ABDEL-MALEK, Z.A. 2001. Melanocortin receptors: their functions and regulation by physiological agonists and antagonists. Cell. Mol. Life Sci. **58:** 434–441.
6. YANG, Y.K. *et al.* 1999. Characterization of agouti-related protein binding to melano-cortin receptors. Mol. Endocrinol. **13:** 148–155.
7. WILLARD, D.H. *et al.* 1995. Agouti structure and function: characterization of a potent alpha-melanocyte stimulating hormone receptor antagonist. Biochemistry **34:** 12341–12346.
8. YANG, Y.K. *et al.* 1997. Effects of recombinant agouti-signaling protein on melanocortin action. Mol. Endocrinol. **11:** 274–280.
9. LU, D.S. *et al.* 1994. Agouti protein is an antagonist of the melanocyte-stimulating-hormone receptor. Nature **371:** 799–802.
10. MCNULTY, J.C. *et al.* 2001. High resolution NMR structure of the chemically-synthe-sized melanocortin receptor binding domain AGRP(87-132) of the agouti-related protein. Biochemistry **40:** 15520–15527.
11. FRIEDMAN, J.M. 2000. Obesity in the new millennium. Nature **404:** 632–634.
12. CHEN, A.S. *et al.* 2000. Inactivation of the mouse melanocortin-3 receptor results in increased fat mass and reduced lean body mass. Nat. Genet. **26:** 97–102.
13. MARKS, D.L., N. LING & R.D. CONE. 2001. Role of the central melanocortin system in cachexia. Cancer Res. **61:** 1432–1438.
14. BURES, E.J. *et al.* 1998. Determination of disulfide structure in agouti-related protein (AGRP) by stepwise reduction and alkylation. Biochemistry **37:** 12172–12177.
15. KIEFER, L.L. *et al.* 1998. Melanocortin receptor binding determinants in the agouti protein. Biochemistry **37:** 991–997.
16. TOTA, M.R. *et al.* 1999. Molecular interaction of agouti protein and agouti-related protein with human melanocortin receptors. Biochemistry **38:** 897–904.
17. CRAIK, D.J., N.L. DALY & C. WAINE. 2001. The cystine knot motif in toxins and impli-cations for drug design. Toxicon **39:** 43–60.
18. PALLAGHY, P.K. *et al.* 1994. A common structural motif incorporating a cystine knot and a triple-stranded beta-sheet in toxic and inhibitory polypeptides. Protein Sci. **3:** 1833–1839.
19. NORTON, R.S. & P.K. PALLAGHY. 1998. The cystine knot structure of ion channel toxins and related polypeptides. Toxicon **36:** 1573–1583.
20. YANG, Y.K. *et al.* 1999. Characterization of aouti-related protein binding to melanocor-tin receptors. Mol. Endocrinol. **13:** 148–155.
21. BOLIN, K.A. *et al.* 1999. NMR structure of a minimized human agouti related protein prepared by total chemical synthesis. FEBS Lett. **451:** 125–131.
22. YANG, Y.K. *et al.* 2000. Molecular determinants of ligand binding to the human mel-anocortin-4 receptor. Biochemistry **39:** 14900–14911.
23. HASKELL-LUEVANO, C. *et al.* 2001. Structure activity studies of the melanocortin-4 receptor by in vitro mutagenesis: identification of agouti-related protein (AGRP),

melanocortin agonist and synthetic peptide antagonist interaction determinants. Biochemistry **40:** 6164–6179.

24. YANG, Y.K. *et al.* 1999. Contribution of melanocortin receptor exoloops to agouti-related protein binding. J. Biol. Chem. **274:** 14100–14106.

25. OOSTEROM, J. *et al.* 2001. Common requirements for melanocortin-4 receptor selectivity of structurally unrelated melanocortin agonist and endogenous antagonist, agouti protein. J. Biol. Chem. **276:** 931–936.

26. JACKSON, P.J. *et al.* 2002. Design, pharmacology and NMR structure of a minimized cystine knot with agouti-related protein activity. Biochemistry **41:** 7565–7572.

27. PALCZEWSKI, K. *et al.* 2000. Crystal structure of rhodopsin: A G protein-coupled receptor. Science **289:** 739–745.

28. PRUSIS, P. *et al.* 2001. PLS modeling of chimeric MS04/MSH-peptide and MC1/MC3-receptor interactions reveals a novel method for the analysis of ligand-receptor interactions. Biochim. Biophys. Acta **1544:** 350–357.

29. PRUSIS, P. *et al.* 1997. Modeling of the three-dimensional structure of the human melanocortin 1 receptor, using an automated method and docking of a rigid cyclic melanocyte-stimulating hormone core peptide. J. Mol. Graphics Modelling **15:** 307–317.

30. REIZES, O. *et al.* 2001. Transgenic expression of syndecan-1 uncovers a physiological control of feeding behavior by syndecan-3. Cell **106:** 105–116.

Melanocortin Tetrapeptides Modified at the N-Terminus, His, Phe, Arg, and Trp Positions

JERRY RYAN HOLDER AND CARRIE HASKELL-LUEVANO

University of Florida, Department of Medicinal Chemistry, Gainesville, Florida 32610, USA

ABSTRACT: The endogenous melanocortin agonists all contain the conserved His-Phe-Arg-Trp sequence proposed to be important for melanocortin receptor selectivity and stimulation. We have generated peptide libraries consisting of over 100 peptides modified at the N-terminus and at each of the four amino acid positions. These peptides were characterized at the mouse melanocortin MC1, MC3, MC4, and MC5 receptors for agonist or antagonist functional activity. The results from these studies include the identification of a nM MC4 versus MC3 receptor selective (>4700-fold) agonist (JRH 420-12), a nM MC4 receptor agonist that is a nM MC3 receptor antagonist (JRH 322-18), a nM MC5 receptor selective (>100-fold) agonist versus the MC1, MC3, and MC4 receptors (FFM 1-60), and side-chain substitutions that may be utilized for non-peptide design considerations.

KEYWORDS: structure-activity relationship; melanocortin receptor; MTII; SHU9119; NDP-MSH; peptide modification; receptor pharmacology

INTRODUCTION

Truncation studies have been performed on NDP-MSH,[1] a highly potent and enzymatically stable analogue of α–MSH (FIG. 1), which identified the DPhe-Arg-Trp tripeptidyl sequence as the minimal sequence required for biological activity in the classical frog (*Rana pipiens*) and lizard (*Anolis carolinensis*) skin bioassays.[2–4] Recently we have carried out similar truncation studies of NDP-MSH at the cloned central and peripheral mouse melanocortin (MC) receptors.[5] The Ac-His-DPhe-Arg-Trp-NH$_2$ tetrapeptide was identified as possessing an EC$_{50}$ value of 10 nM at the murine MC4 receptor, which is similar to the 8 nM results published by Yang *et al.* for the tetrapeptide at the human MC4 receptor.[6] This tetrapeptide exhibits full agonist efficacy at all the mouse MC receptors and is equipotent to the endogenous hormone α-MSH (within experimental error), and only 30-fold less potent than NDP-MSH at the mouse MC4R. Because of the small size and potency of this peptide, relative to the tridecapeptide α-MSH (TABLE 1), Ac-His-DPhe-Arg-Trp-NH$_2$ was used as a starting point in a study aimed at improving the potency and receptor selectivity of the ligand at the melanocortin receptors. In a series of extensive structure-activity re-

Address for correspondence: Carrie Haskell-Luevano, University of Florida, Department of Medicinal Chemistry, Gainesville, FL 32610. Voice: 352-846-2722; fax: 352-392-8182.
Carrie@cop.ufl.edu

Ann. N.Y. Acad. Sci. 994: 36–48 (2003). © 2003 New York Academy of Sciences.

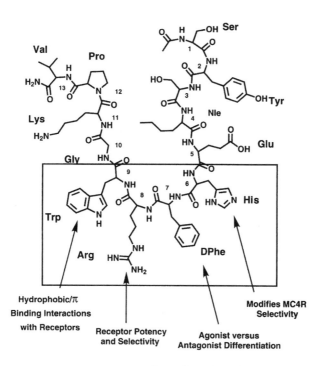

FIGURE 1. Chemical structure of [Nle^4,DPhe^7]-MSH (NDP-MSH). This peptide was the lead compound in a truncation study which identified the DPhe-Arg-Trp tripeptidyl sequence as the minimal active fragment needed for melanocortin receptor activity. Amino acid abbreviations and sequence number are indicated for reference. Important features of each of the core residues are emphasized.

lationship (SAR) studies, modifications were systematically made at each of the four amino acid residues as well as the N-terminus of the tetrapeptide, with subsequent pharmacological evaluation at the mouse MC1R, MC3R–MC5R.[7–11] This chapter briefly summarizes key results of various modifications of the melanocortin tetrapeptide library.

METHODS

Synthesis

Peptide synthesis was performed as previously published.[7–11] Briefly, peptide syntheses were performed using standard Fmoc methodology[12] on an automated synthesizer (Advanced ChemTech 440MOS, Louisville, KY). Peptides were purified to homogeneity using high performance liquid chromatography and were >95% pure as determined by analytical HPLC in two diverse solvent systems. Structural integrity was assessed using 1D-^{1}H NMR and the peptides had the correct molecular mass as verified by mass spectrometry.

Table 1. Amino Acid Sequences of α-MSH and MSH Analogues

Compound	Amino Acid Sequence
α-MSH	Ac-Ser^1-Tyr^2-Ser^3-Met^4-Glu^5-His^6-Phe^7-Arg^8-Trp^9-Gly^{10}-Lys^{11}-Pro^{12}-Val^{13}-NH$_2$
NDP-MSH	Ac-Ser^1-Tyr^2-Ser^3-**Nle**^{4}-Glu^5-His^6-**DPhe**^{7}-Arg^8-Trp^9-Gly^{10}-Lys^{11}-Pro^{12}-Val^{13}-NH$_2$
MTII	Ac-Nle^4-cyclo[Asp^5-His^6-**DPhe**^{7}-Arg^8-Trp^9-Lys^{10}]-NH$_2$
SHU 9119	Ac-Nle^4-cyclo[Asp^5-His^6-**DNal(2')**^{7}-Arg^8-Trp^9-Lys^{10}]-NH$_2$
SHU 8914	Ac-Nle^4-cyclo[Asp^5-His^6-**(pI)DPhe**^{7}-Arg^8-Trp^9-Lys^{10}]-NH$_2$
NDP-MSH(6-9)	Ac-His^6-**DPhe**^{7}-Arg^8-Trp^9-NH$_2$
JRH 322-18	Ac-His^6-**(pI)DPhe**^{7}-Arg^8-Trp^9-NH$_2$
JRH 420-12	Ac-**Anc**-DPhe^7-Arg^8-Trp^9-NH$_2$

Functional Bioassay

HEK-293 cells stably expressing the mouse melanocortin receptors were transfected with 4 μg CRE/β-galactosidase reporter gene as previously described.[7–11,13] EC$_{50}$ and pA$_2$ values were determined using dose-response curves (10^{-4} to 10^{-12} M). The sample absorbance, OD$_{405}$, was measured using a 96-well plate reader (Molecular Devices). Data points were normalized both to the relative protein content and non-receptor–dependent forskolin stimulation. The antagonistic properties of these compounds were evaluated by the ability of these ligands to competitively displace the MTII agonist (Bachem) in a dose-dependent manner, at up to 10 μM concentrations. The pA$_2$ values were generated using the Schild analysis method.[14]

Data Analysis

EC$_{50}$ and pA$_2$ values represent the mean of duplicate wells performed in quadruplet or more independent experiments. EC$_{50}$ and pA$_2$ estimates, and their associated standard errors, were determined by fitting the data to a nonlinear least-squares analysis using the PRISM program (v3.0, GraphPad Inc.).

RESULTS AND DISCUSSION

Modification of the N-Terminus with Acyl "Capping" Groups

Both α-MSH and NDP-MSH (TABLE 1) have been the lead compounds in several SAR studies; only a few studies, however, have reported "capping" modifications of the N-terminus and the corresponding structure-activity relationships. The addition of fatty acid conjugates,[15,16] biotin,[17] and chlorotriazinylaminofluorescein[18] to the N-terminus of α-MSH analogues have been made, and resulted in enhanced or decreased potencies in the classical lizard and frog skin bioassays and melanoma cell tyrosinase assay, depending on the type of modification made. In a more recent study, analogues of MTII modified at the N-terminus were evaluated at the human MC3–MC5 receptors.[19] Several of the MTII analogues were found to have improved hMC4 receptor selectivity, with respect to the hMC3 and hMC5 receptors. This por-

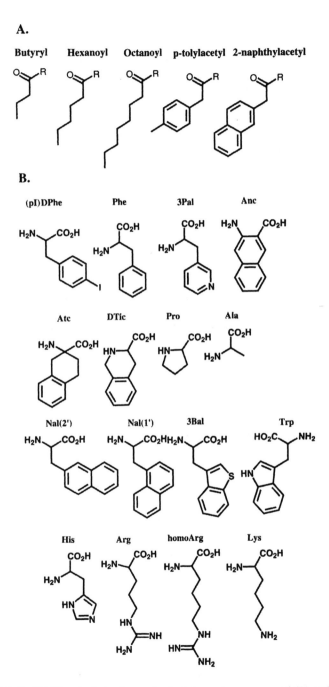

FIGURE 2. (A) Chemical structures of selected capping groups and (B) amino acids used in the SAR studies of the melanocortin Ac-His-DPhe-Arg-Trp-NH$_2$ tetrapeptide.

tion of our research was undertaken to determine if the addition of various linear, cyclic, or aromatic hydrophobic moieties to the N-terminus of the His-DPhe-Arg-Trp-NH$_2$ tetrapeptide would enhance ligand potency and/or selectivity at the melanocortin receptors.[11] During the progress of this study, Cheung *et al.* have reported the Bu-His-DPhe-Arg-Trp-Gly-NH$_2$ pentapeptide, which contains an N-terminal butyryl moiety that is structurally reminiscent to the norluecine residue of MTII, to possess potent agonist activity at the hMC4 and hMC1 receptors, with no detectable hMC3 and hMC5 receptor activity (no agonist activity up to 50 mM).[20]

We observed that by increasing the chain length of linear alkyl groups (FIG. 2A) the potency was increased at the melanocortin receptors, especially the MC1R, MC3R, and MC4R. Specifically, at the MC1R there was a 4-fold increase when the butyryl moiety was added to the peptide, a 45-fold increase when the hexanoyl was added, and a 110-fold increase when the octanoyl group was added. The most noteworthy alkyl addition to the N-terminus of the tetrapeptide is the octanoyl moiety. The octanoyl peptide is equipotent to NDP-MSH and 14-fold more potent than α-MSH at the MC4R, although the peptide is only four amino acid residues in length (NDP-MSH and α-MSH contain 13 amino acid residues). The increased potencies at the melanocortin receptors may be the result of increased hydrophobic ligand–receptor interactions, or possibly the result of increased lipid–peptide interactions that may enhance the interaction of the ligand with the receptor within the lipid bilayer.[21]

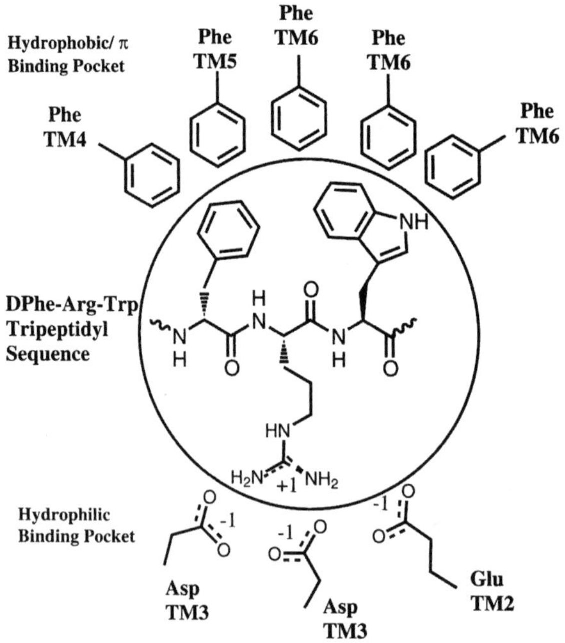

FIGURE 3. Illustration of the melanocortin agonist NDP-MSH amino acids DPhe-Arg-Trp in the putative melanocortin receptor binding pocket consisting of an aromatic hydrophobic receptor domain and an hydrophilic electrostatic receptor domain.

Receptor mutagenesis studies and homology molecular modeling of the MC1R and MC4R have indicated that the two receptors contain similar putative ligand binding pockets consisting of a hydrophobic-aromatic network of Phe receptor residues (FIG. 3), which are proposed to interact with the aromatic Phe and Trp residues of the melanocortin ligands.[6,22–26] Several N-terminal aromatic moieties, consisting of mono- and multi-ring systems, as well as alkyl-substituted aromatic ring systems, were introduced to the N-terminus of the tetrapeptide. The addition of aryl moieties generally resulted in peptides with increased potencies, or at least were equipotent at the melanocortin receptors, and only one of the sixteen aromatic modified tetrapeptides resulted in a decrease in potency. The addition of N-terminal p-tolylacetyl and 2-naphthylacetyl groups (FIG. 2A) resulted in a 30-fold increased potency at the MC4R, with similar results at the MC1R, MC3R, and MC5R. These data would suggest that incorporation of aromatic moieties in additional melanocortin ligands may improve ligand potency. These data also provide additional experimental support for the hypothesis that the MC1R and MC4R contain a similar hydrophobic-aromatic ligand binding pocket that interacts with ligand aromatic residues. Interestingly, the only aromatic modification that resulted in a loss of potency also led to the identifi-

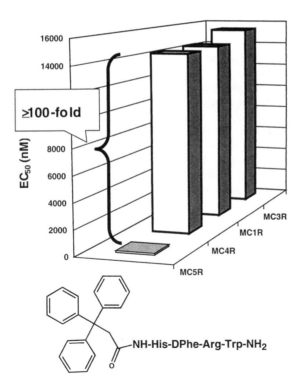

FIGURE 4. Illustration demonstrating the MC5R selectivity versus the remaining melanocortin receptors of the His-DPhe-Arg-Trp-NH$_2$ tetrapeptide modified at the N-terminus with the 3,3,3-triphenylpropionyl moiety.

Table 2. Central Melanocortin Receptor Selective Agonist Peptide Comparisons

Structure	mMC3R EC$_{50}$ (nM)	mMC4R EC$_{50}$ (nM)	Selectivity Ratio mMC3R EC$_{50}$/mMC4R EC$_{50}$
Ac-**Anc**-DPhe-Arg-Trp-NH$_2$	>100,000	21.1	>4700
Ac-**rac(Atc)**-DPhe-Arg-Trp-NH$_2$	>100,000	714	>140
Ac-**3PAL**-DPhe-Arg-Trp-NH$_2$	>100,000	887	>113
Ac-**Phe**-DPhe-Arg-Trp-NH$_2$	11,900	70.6	169
Ac-**3Bal**-DPhe-Arg-Trp-NH$_2$	19,800	211	94
Ac-**DNal(2')**-DPhe-Arg-Trp-NH$_2$	>100,000	1,500	>67
Ac-**DPhe**-DPhe-Arg-Trp-NH$_2$	>100,000	2,170	>46
Ac-**DTrp**-DPhe-Arg-Trp-NH$_2$	>100,000	6,200	>16
Ac-**DTic**-DPhe-Arg-Trp-NH$_2$	>100,000	11,700	>9

cation of a peptide with useful pharmacological properties. The 3,3,3-triphenylpropionyl-tetrapeptide (FFM 1-60) possessed an EC$_{50}$ value of 140 nM at the MC5R and was 100-fold selective versus the remaining melanocortin receptors (FIG. 4). This MC5R-selective tetrapeptide may provide a valuable *in vivo* tool for further physiological characterization of this melanocortin receptor.

Core Residue Modifications

The benchmark melanocortin peptides (α-MSH, NDP-MSH, MTII, and SHU9119) have been the lead compounds in several SAR studies aimed at providing a better understanding of the interactions that occur between melanocortin ligands and the melanocortin receptors, as well as improving ligand potency and receptor selectivity. These SAR studies include alanine scanning experiments,[27,28] D-amino acid scans,[29] truncation studies,[2,5,30,31] and introduction of conformational restrictions, such as global (cyclization strategies)[32,33] and topographical[34] (β–methyl amino acids)[35,36] constraints; however, the role that each amino acid residue in the core His-Phe-Arg-Trp sequence plays in receptor recognition and activation has not been examined in great detail using the tetrapeptide template. The following sections review the results of the recent SAR studies of the melanocortin Ac-His-DPhe-Arg-Trp-NH$_2$ tetrapeptide modified at each amino acid residue.

Previous reports have indicated that the His[6] position may be utilized to improve MC4R selectivity in regard to one or more of the remaining melanocortin receptors.[19,37,38] Following initial modifications at the His[6] position in the Ac-His-DPhe-Arg-Trp-NH$_2$ tetrapeptide, it was observed that agonist activity at the MC3R could be eliminated or significantly lowered by substituting large and/or constrained aromatic groups (FIG. 2B) for the relatively small imidazole ring of histidine, without complete loss of activity at the MC4R (TABLE 2). Replacement of histidine with the topographically constrained D-1,2,3,4-tetrahydroisoquinoline-3-carboxylic acid (DTic), the bulky D-naphthylalanine (DNal), the relatively large D-tryptophan (DTrp), 3-pyridinylalanine (3Pal), or D-phenylalanine (DPhe) amino acids resulted in tetrapeptides that were unable to generate a maximal functional response at the MC3R at concentrations up to 100 μM; the peptides, however, were full agonists at the MC4R. These initial observations, combined with the observations made by Danho *et al.*,[37] prompted us to search for additional bulky and constrained aromatic amino acids to replace histidine in the tetrapeptide template. Two conformationally constrained aromatic amino acids, racemic aminotetrahydro-2-naphthyl-carboxylic acid (Atc) and amino-2-naphthylcarboxylic acid (Anc), when used in the His[6] position, resulted in peptides with nanomolar potencies at the MC4R and with only slight stimulatory activity at the MC3R at concentrations up to 100 μM. The most potent and selective (MC4R versus MC3R) tetrapeptide, Ac-Anc-DPhe-Arg-Trp-NH$_2$ (JRH 420-12), has a MC4R EC$_{50}$ value of 21 nM, which is equipotent, within experimental error, to the endogenous agonist α-MSH at the MC4R. This peptide (JRH 420-12) is >4,700-fold selective for the centrally located MC4R versus the MC3R, and is currently being used as an *in vivo* characterization tool to further correlate the physiological functions of the MC4R in animal models. Danho *et al.* have previously used racemic Atc and additional constrained amino acids at the His[6] position in cyclic analogues of MTII, which resulted in some of the most MC4R selective peptides disclosed at that time.[37] These data lead to the hypothesis that the His[6] position is the most critical position for increasing MC4R versus MC3R selectivity.[37] Interestingly, simple replacement of the imidazole ring with a phenyl ring resulted in a potent MC4R agonist (EC$_{50}$ = 70 nM), 170-fold selective against the MC3R (EC$_{50}$ = 11,900 nM). This data would indicate that aromatic moieties, such as the ones used herein, may be used as a replacement of the imidazole ring in the design of small molecule melanocortin ligands, which may simplify the synthetic strategy (the imidazole ring generally requires protection during synthesis, whereas the naphthyl and benzyl rings do not). The results of our SAR studies of melanocortin tetrapeptides modified at the His[6] position support the hypothesis that this position is a critical position responsible for MC4R selectivity, and they are in agreement with the observations of Danho *et al.*[37]

Previous SAR studies of cyclic melanocortin analogues modified at the Phe[7] position with bulky aromatic amino acids resulted in the discovery of two highly potent MC3R and MC4R antagonists, SHU8914 (Ac-Nle-cyclo[Asp-His-(pI)DPhe-Arg-Trp-Lys]-NH$_2$) and SHU9119 (Ac-Nle-cyclo[Asp-His-DNal(2′)-Arg-Trp-Lys]-NH$_2$).[33] Since the discovery of SHU9119 by Hruby *et al.*, SHU9119 has been used extensively to aid in characterization of the physiological roles of the central MC3R and MC4R.[22,9–43] To test the hypothesis that large stereoelectronic modifications of the Phe[7] position can provide MC3R and MC4R antagonists, various modifications were made at this position using the linear tetrapeptide template Ac-His-DPhe-Arg-Trp-NH$_2$.[9] The Nal(1′) and Nal(2′) analogues were neither agonists nor antagonists at both the MC3R and MC4R (at concentrations up to 100 μM); the DNal(1′) analogue was an agonist at the MC3R (EC$_{50}$ = 4100 nM) and MC4R(EC$_{50}$ = 303 nM), and the DNal(2′) analogue was a partial agonist/antagonist (pA$_2$ = 6.5) at the MC3R and potent antagonist at the MC4R (pA$_2$ = 7.8). These data support the hypothesis that not only is a bulky aromatic ring needed for antagonism at the MC4R, but the correct stereochemistry (D versus L) and orientation of the naphthyl ring (1′ versus 2′) are additional requirements for MC4R antagonism.[22] When (pI)DPhe was used to replace DPhe[7] in the cyclic MTII template[33] as well as the linear NDP-MSH template,[22] this resulted in potent MC3R and MC4R antagonists. When the same substitution was made in the linear tetramer, the result was a melanocortin peptide with novel pharmacology at the cloned murine receptors.[9] As predicted, Ac-His-(pI)DPhe-Arg-Trp-NH$_2$ (JRH 322-18) is a potent antagonist at the MC3R (pA$_2$ = 7.25); however and quite unexpectedly, the peptide is a full agonist at the MC4R

(EC_{50} = 25 nM). To determine whether the observed pharmacology was species specific, JRH 322-18 was also tested at the human MC4R and similar agonist pharmacology was observed (EC_{50} = 5 nM). Compounds with interesting and novel pharmacology at the melanocortin receptors, such as JRH 322-18 and JRH 420-12, may provide useful tools to aid in the physiological characterization of the melanocortin receptors where knockout animals are not viable options, and we are currently pursuing *in vivo* studies with these peptides in our animal models.

Several studies have emphasized the important role that the Arg^8 residue has in melanocortin receptor recognition and stimulation.[3,4,44–46] It has been suggested that the basic Arg residue interacts with acidic residues in the transmembrane (TM) domains 2 and 3 of the MC1 and MC4 receptors[6,22,23,26,47,48] (FIG. 3), and this interaction may be responsible for placing the aromatic Phe and Trp residues in the correct orientation needed to allow the receptor conformational changes required for signal transduction to occur.[49] Recently, it has been suggested that it is the two terminal gaunidino NH groups, and not the positive charge of Arg, that are essential for efficient interactions of melanocortin ligands with the receptors.[20] To further correlate the role of the Arg^8 residue, several modifications were made using basic and acidic moieties as well as alanine and proline substitutions.[7] Replacement of Arg with either Ala, Pro, or acidic groups resulted in tetrapeptides with greatly reduced potencies, as compared with Ac-His-DPhe-Arg-Trp-NH$_2$, whereas substitutions with basic residues affected potency to a lesser extent. Interestingly, two amino acid substitutions were found to increase the selectivity of the MC4R in respect to the MC3R (FIG. 5). Replacement of Arg with Lys, which substitutes the gaunidino functionality for that of a free amino terminus, results in a tetrapeptide that is 31-fold selective for the MC4R versus the MC3R. When the side-chain length is increased by one methylene group (homoArg), nanomolar potency is retained at the MC4R (EC_{50} =

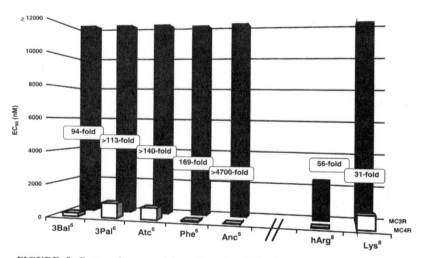

FIGURE 5. Bar graph summarizing selected modifications that enhance MC4R versus MC3R agonist selectivity. The superscript numbers indicate the amino acid residue in which the modification was made in the Ac-His^6-DPhe^7-Arg^8-Trp^9-NH$_2$ peptide template.

43 nM), but the potency at the MC3R is reduced to 2.4 µM (56-fold selectivity). These data indicate that further manipulation of the Arg[8] residue may be an effective means of increasing receptor selectivity in the design of future melanocortin ligands.

The aromatic Trp[9] residue is proposed to interact with a hydrophobic-aromatic binding pocket formed by aromatic residues in the MC1 and MC4 receptors,[22,23,47] and the critical role that the Trp[9] residue plays in receptor activity has been determined experimentally.[28,45,46] To further examine the role of the Trp[9] residue in melanocortin receptor activity and selectivity, alanine replacement and various aromatic substitutions were made in the Ac-His-DPhe-Arg-Trp-NH$_2$ tetrapeptide template.[8] All of the aromatic substitutions made at the Trp[9] position were tolerated without complete loss of activity at the MC1R and MC4R, supporting the hypothesis that this residue interacts with an aromatic receptor binding pocket. We also observed that several modifications could be made that resulted in equipotent or only slight reductions in potencies at the melanocortin receptors, as compared with the Trp-containing tetrapeptide. The most potent of these modifications, the Nal(2') substitution, resulted in a peptide equipotent to the parent tetrapeptide at the MC1R, MC4R, and MC5R, with only a 5-fold decrease in potency at the MC3R. Other modifications that negligibly effected potency are Nal(1'), DNal(2'), and 3-benzothienylalanine (3Bal) (FIG. 2B). These modifications may be considered for the design of the melanocortin small molecule, non-peptide compounds because the indole ring is chemically reactive and may be prone to oxidation and/or reduction under certain synthetic conditions.

CONCLUSIONS

We have synthesized, purified to homogeneity, analytically and pharmacologically characterized over 100 modified melanocortin tetrapeptides to evaluate structure-activity trends and to identify peptides with novel and useful pharmacology. Two peptides have been discovered with very interesting pharmacological profiles, JRH 322-18 and JRH 420-12, which are currently being characterized for their *in vivo* properties in our animal models. We have identified modifications that increase potency and receptor selectivity (summarized in FIG. 5), as well as identified chemically non-reactive moieties that can be used to replace reactive indole and imidazole groups in the design of future melanocortin ligands.

ACKNOWLEDGMENTS

This work is supported by National Institutes of Health grant DK57080. C.H.-L. is a recipient of a Burroughs Wellcome fund Career Award in the Biomedical Sciences.

REFERENCES

1. SAWYER, T.K., P.J. SANFILIPPO, V.J. HRUBY, *et al.* 1980. 4-Norleucine, 7-D-phenylalanine-α-melanocyte-stimulating hormone—a highly potent α-melanotropin with ultralong biological-activity. Proc. Natl. Acad. Sci. USA **77:** 5754–5758.

2. SAWYER, T.K., A.M. CASTRUCCI, D.J. STAPLES, *et al.* 1993. Structure-activity relationships of [Nle4, D-Phe7]α-MSH. Discovery of a tripeptidyl agonist exhibiting sustained bioactivity. Ann. N. Y. Acad. Sci. **680:** 597–599.
3. HASKELL-LUEVANO, C., T.K. SAWYER, S. HENDRATA, *et al.* 1996. Truncation studies of α-melanotropin peptides identify tripeptide analogues exhibiting prolonged agonist bioactivity. Peptides **17:** 995–1002.
4. EBERLE, A.N. 1988. The Melanotropins: Chemistry, Physiology, and Mechanism of Action. Karger. Basel.
5. HASKELL-LUEVANO, C., J.R. HOLDER, E.K. MONCK & R.M. BAUZO. 2001. Characterization of melanocortin NDP-MSH agonist peptide fragments at the mouse central and peripheral melanocortin receptors. J. Med. Chem. **44:** 2247–2252.
6. YANG, Y.K., T.M. FONG, C.J. DICKINSON, *et al.* 2000. Molecular determinants of ligand binding to the human melanocortin-4 receptor. Biochemistry **39:** 14900–14911.
7. HOLDER, J.R., R.M. BAUZO, Z. XIANG, *et al.* 2003. Structure-activity relationships of the melanocortin tetrapeptide Ac-His-DPhe-Arg-Trp-NH2 at the mouse melanocortin teceptors: part 3, modification at the Arg position. Peptides **24:** 73–82.
8. HOLDER, J.R., R.M. BAUZO, Z. XIANG, *et al.* 2002. Structure-activity relationships of the melanocortin tetrapeptide Ac-His-DPhe-Arg-Trp-NH2 at the mouse melanocortin receptors: part 4, modification at the Trp position. J. Med. Chem. **45:** 5736–5744.
9. HOLDER, J.R., R.M. BAUZO, Z. XIANG, *et al.* 2003. Structure-activity relationships of the melanocortin tetrapeptide Ac-His-DPhe-Arg-Trp-NH2 at the mouse melanocortin receptors: part 2, modification at the Phe position. J. Med. Chem. **45:** 3073–3081.
10. HOLDER, J.R., R.M. BAUZO, Z. XIANG, *et al.* 2002. Structure-activity relationships of the melanocortin tetrapeptide Ac-His-DPhe-Arg-Trp-NH2 at the mouse melanocortin receptors: part 1, modification at the His position. J. Med. Chem. **45:** 2801–2810.
11. HOLDER, J.R., F.F. MARQUES, R.M. BAUZO, *et al.* 2003. Characterization of aliphatic, cyclic, and aromatic N-terminally "capped" His-Phe-Arg-Trp-NH2 tetrapeptides at the melanocortin receptors. Eur. J. Pharmacol. **462:** 41–52.
12. CARPINO, L.A. & G.Y. HAN. 1972. 9-Fluorenylmethoxycarbonyl amino-protecting group. J. Org. Chem. **37:** 3404–3405.
13. CHEN, W., T.S. SHIELDS, P.J. STORK & R.D. CONE. 1995. A colorimetric assay for measuring activation of Gs- and Gq-coupled signaling pathways. Anal. Biochem. **226:** 349–354.
14. SCHILD, H.O. 1947. PA, a new scale for the measurement of drug antagonism. Br. J. Pharmacol. **2:** 189–206.
15. HADLEY, M.E., F. AL-OBEIDI, V.J. HRUBY, *et al.* 1991. Biological activities of melanotropic peptide fatty acid conjugates. Pigm. Cell Res. **4:** 180–185.
16. AL-OBEIDI, F., V.J. HRUBY, N. YAGHOUBI, *et al.* 1992. Synthesis and Biological Activities of Fatty Acid Conjugates of a Cyclic Lactam alpha-Melanotropin. J. Med. Chem. **35:** 118–123.
17. CHATURVEDI, D.N., J.J. KNITTEL, V.J. HRUBY, *et al.* 1984. Synthesis and biological actions of highly potent and prolonged acting biotin-labeled melanotropins. J. Med. Chem. **27:** 1406–1410.
18. CHATURVEDI, D.N., V.J. HRUBY, A.M. CASTRUCCI, *et al.* 1985. Synthesis and biological evaluation of the superagonist [N alpha-chlorotriazinylaminofluorescein-Ser1,Nle4,D-Phe7]-alpha-MSH. J. Pharm. Sci. **74:** 237–240.
19. BEDNAREK, M.A., T. MACNEIL, R.N. KALYANI, *et al.* 1999. Analogs of MTII, lactam derivatives of alpha-melanotropin, modified at the N-terminus, and their selectivity at human melanocortin receptors 3, 4, and 5. Biochem. Biophys. Res. Commun. **261:** 209–213.
20. CHEUNG, A., W. DANHO, J. SWISTOK, *et al.* 2002. Structure-activity relationship of linear peptide Bu-His-DPhe-Arg-Trp-Gly-NH(2) at the human melanocortin-1 and -4 receptors: arginine substitution. Bioorg. Med. Chem. Lett. **12:** 2407–2410.
21. SARGENT, D.F. & R. SCHWYZER. 1986. Membrane lipid phase as catalyst for peptide-receptor interactions. Proc. Natl. Acad. Sci. USA **83:** 5774–5778.
22. HASKELL-LUEVANO, C., R.D. CONE, E.K. MONCK, *et al.* 2001. Structure activity studies of the melanocortin-4 receptor by in vitro mutagenesis: identification of agouti-related protein (AGRP), melanocortin agonist and synthetic peptide antagonist interaction determinants. Biochemistry **40:** 6164–6179.

23. HASKELL-LUEVANO, C., T.K. SAWYER, S. TRUMPP-KALLMEYER, et al. 1996. Three-dimensional molecular models of the hMC1R melanocortin receptor: complexes with melanotropin peptide agonists. Drug Design Discov. 14: 197–211.

24. LU, D., C. HASKELL-LUEVANO, D.I. VÄGE, et al. 1997. Functional variants of the MSH receptor (MC1-R), agouti, and their effects on mammalian pigmentation. In G Proteins, Receptors, and Disease. A.M. Spiegel, Ed.: 231–259. The Humana Press Inc. Clifton, NJ.

25. LU, D., D.I. VÄGE & R.D. CONE. 1998. A ligand-mimetic model for constitutive activation of the melanocortin-1 receptor. Mol. Endocrinol. 12: 592–604.

26. YANG, Y., C. DICKINSON, C. HASKELL-LUEVANO, et al. 1997. Molecular basis for the interaction of [Nle4,D-Phe7]melanocyte stimulating hormone with the human melanocortin-1 receptor. J. Biol. Chem. 272: 23000–23010.

27. SAHM, U.G., G.W.J. OLIVIER, S.K. BRANCH, et al. 1994. Influence of alpha-MSH terminal amino-acids on binding-affinity and biological-activity in melanoma-cells. Peptides 15: 441–446.

28. GRIECO, P., P. BALSE-SRINIVASAN, G. HAN, et al. 2002. Synthesis and biological evaluation on hMC(3), hMC(4) and hMC(5) receptors of gamma-MSH analogs substituted with L-alanine. J. Pept. Res. 59: 203–210.

29. GRIECO, P., P.M. BALSE, D. WEINBERG, et al. 2000. D-Amino acid scan of gamma-melanocyte-stimulating hormone: importance of Trp(8) on human MC3 receptor selectivity. J. Med. Chem. 43: 4998–5002.

30. CASTRUCCI, A.M.L., M.E. HADLEY, T.K. SAWYER, et al. 1989. alpha-Melanotropin—the minimal active sequence in the lizard skin bioassay. Gen. Comp. Endocrinol. 73: 157–163.

31. HRUBY, V.J., B.C. WILKES, M.E. HADLEY, et al. 1987. alpha-Melanotropin—the minimal active sequence in the frog skin bioassay. J. Med. Chem. 30: 2126–2130.

32. HASKELL-LUEVANO, C., M.D. SHENDEROVICH, S.D. SHARMA, et al. 1995. Design, synthesis, biology, and conformations of bicyclic alpha-melanotropin analogues. J. Med. Chem. 38: 1736–1750.

33. HRUBY, V.J., D. LU, S.D. SHARMA, et al. 1995. Cyclic lactam alpha-melanotropin analogues of Ac-Nle4-cyclo[Asp5, D-Phe7,Lys10] alpha-melanocyte-stimulating hormone-(4-10)-NH2 with bulky aromatic amino acids at position 7 show high antagonist potency and selectivity at specific melanocortin receptors. J. Med. Chem. 38: 3454–3461.

34. HRUBY, V.J., G. LI, C. HASKELL-LUEVANO, et al. 1997. Design of peptides, proteins, and peptidomimetics in chi space. Biopolymers 43: 219–266.

35. HASKELL-LUEVANO, C., L.W. BOTEJU, H. MIWA, et al. 1995. Topographical modification of melanotropin peptide analogues with beta-methyltryptophan isomers at position 9 leads to differential potencies and prolonged biological activities. J. Med. Chem. 38: 4720–4729.

36. HASKELL-LUEVANO, C., K. TOTH, L. BOTEJU, et al. 1997. beta-Methylation of the Phe7 and Trp9 melanotropin side chain pharmacophores affects ligand-receptor interactions and prolonged biological activity. J. Med. Chem. 40: 2740–2749.

37. DANHO, W., J. SWISTOK, A. CHEUNG, et al. 2001. Highly selective cyclic peptides for human melanocortin-4 receptor: design, synthesis, bioactive conformation, and pharmacological evaluation as an anti-obesity agent. In Peptides: The Wave of the Future: Proceedings of the Second International and the Seventeenth American Peptide Symposium. M. Lebl & R.A. Houghten, Eds.: 701–704. Kluwer Academic Publishers. Norwell, MA.

38. GRIECO, P., G.X. HAN, D. WEINBERG, et al. 2002. Design and synthesis of highly potent and selective melanotropin analogues of SHU9119 modified at position 6. Biochem. Biophys. Res. Commun. 292: 1075–1080.

39. FAN, W., B.A. BOSTON, R.A. KESTERSON, et al. 1997. Role of melanocortinergic neurons in feeding and the agouti obesity syndrome. Nature 385: 165–168.

40. ADAN, R.A., A.W. SZKLARCZYK, J. OOSTEROM, et al. 1999. Characterization of melanocortin receptor ligands on cloned brain melanocortin receptors and on grooming behavior in the rat. Eur. J. Pharmacol. 378: 249–258.

41. LI, S.J., K. VARGA, P. ARCHER, et al. 1996. Melanocortin antagonists define two distinct pathways of cardiovascular control by alpha- and gamma-melanocyte-stimulating hormones. J. Neurosci. 16: 5182–5188.

42. RAPOSINHO, P.D., E. CASTILLO, V. D'ALLEVES, et al. 2000. Chronic blockade of the melanocortin 4 receptor subtype leads to obesity independently of neuropeptide Y action, with no adverse effects on the gonadotropic and somatotropic axes. Endocrinology **141:** 4419–4427.

43. VRINTEN, D.H., W.H. GISPEN, G.J. GROEN, et al. 2000. Antagonism of the melanocortin system reduces cold and mechanical allodynia in mononeuropathic rats. J. Neurosci. **20:** 8131–8137.

44. BEDNAREK, M.A., T. MACNEIL, R.N. KALYANI, et al. 2000. Analogs of lactam derivatives of alpha-melanotropin with basic and acidic residues. Biochem. Biophys. Res. Commun. **272:** 23–28.

45. BEDNAREK, M.A., M.V. SILVA, B. ARISON, et al. 1999. Structure-function studies on the cyclic peptide MT-II, lactam derivative of alpha-melanotropin. Peptides **20:** 401–409.

46. SAHM, U.G., G.W.J. OLIVIER, S.K. BRANCH, et al. 1994. Synthesis and biological evaluation of alpha-MSH analogs substituted with alanine. Peptides **15:** 1297–1302.

47. HASKELL-LUEVANO, C. 2000. In vitro mutagenesis studies of melanocortin receptor coupling and ligand binding. In The Melanocortin Receptors. R.D. Cone, Ed.: 263–306. The Humana Press Inc., Clifton, NJ.

48. YANG, Y., M. CHEN, Y. LAI, et al. 2002. Molecular determinants of human melanocortin-4 receptor responsible for antagonist SHU9119 selective activity. J. Biol. Chem. **277:** 20328–20335.

49. HASKELL-LUEVANO, C., G. NIKIFOROVICH, S.D. SHARMA, et al. 1997. Biological and conformational examination of stereochemical modifications using the template melanotropin peptide, Ac-Nle-c[Asp-His- Phe-Arg-Trp-Ala-Lys]-NH2, on human melanocortin receptors. J. Med. Chem. **40:** 1738–1748.

Molecular Genetics of Human Obesity-Associated MC4R Mutations

CECILE LUBRANO-BERTHELIER,[a] MARTHA CAVAZOS,[a]
BEATRICE DUBERN,[a,b] ASTRID SHAPIRO,[a] CATHERINE LE STUNFF,[c]
SUMEI ZHANG,[a] FRANCK PICART,[a] CEDRIC GOVAERTS,[d]
PHILIPPE FROGUEL,[e,f] PIERRE BOUGNÈRES,[c] KARINE CLEMENT,[g]
AND CHRISTIAN VAISSE[a]

[a]Diabetes Center and Department of Medicine, University of California, San Francisco,
San Francisco, California 94143, USA

[b]Pediatric Gastroenterology and Nutrition Department, Armand-Trousseau Hospital,
75012 Paris, France

[c]Department of Pediatric Endocrinology, Saint Vincent de Paul Hospital,
University René Descartes, INSERM U561, 75014 Paris, France

[d]Department of Cellular and Molecular Pharmacology, University of California,
San Francisco, San Francisco, California 94143, USA

[e]CNRS 80 90-Institut de Biologie de Lille, Institut Pasteur,
59000 Lille, France

[f]Barts & The London Genome Center, Queen Mary School of Medicine,
London, United Kingdom

[g]Medicine and Nutrition Department, Hotel-Dieu Hospital,
75004 Paris, France

ABSTRACT: Heterozygous coding mutations in the melanocortin 4 receptor (MC4R) are implicated in 1 to 6% of early onset or severe adult obesity cases. To better address the problem of the genotype:phenotype relationship within this specific form of obesity, we systematically studied the functional characteristics of 50 different obesity-associated MC4R mutations. Structure modeling of MC4R indicates that obesity-associated MC4R mutations are not localized in a single domain of the protein. We developed a flow cytometry–based assay to compare cell membrane expression of obesity-associated MC4R mutants. Using this assay, we demonstrate that over 54% of the obesity-associated MC4R mutations impair the membrane expression of MC4R. All other mutations impair the basal constitutive activity and/or the EC_{50} for the physiological agonist α-MSH as measured in a cAMP- dependent luciferase assay. The extent of the alterations in receptor activity ranges from a total suppression of MC4R activation in response to α-MSH to a mild alteration of the basal constitutive activity of the receptor. Since most patients are heterozygous for

Address for correspondence: Christian Vaisse, Diabetes Center and Department of Medicine, University of California, San Francisco, CA 94143. Voice: 415-514-0530; fax: 415-564-5813.
vaisse@medicine.ucsf.edu

Ann. N.Y. Acad. Sci. 994: 49–57 (2003). © 2003 New York Academy of Sciences.

MC4R mutations, these data indicate that a small decrease in overall MC4R activity can cause obesity, strongly supporting the hypothesis that the MC4R is a critical component of the adipostat in humans.

KEYWORDS: obesity; genetics; melanocortin receptors

THE MC4R IS IMPLICATED IN THE REGULATION OF ENERGY METABOLISM

MC4R transduces signal by coupling to the heterotrimeric Gs protein and activating adenylate cyclase.[1] The expression of MC4R is restricted to the nervous system and is found in hypothalamic nuclei involved in food intake regulation.[2] MC4R regulates food intake by integrating an agonist (satiety) signal provided by α-melanocyte-stimulating hormone (α-MSH)[3] and an antagonist (orexigenic) signal provided by the agouti-related protein (AGRP).[4] In addition, recent data suggest that MC4R exhibits a constitutive, food intake inhibiting activity on which AGRP acts as an invert agonist.[5,6] The critical role of MC4R in the long-term regulation of body weight has been well demonstrated in mice.[7] Mice lacking both alleles of MC4R (MC4R–/– mice) develop a severe obesity syndrome while mice heterozygous for an MC4R deletion show an average weight that is intermediate between that of wild type and MC4R–/– mice.

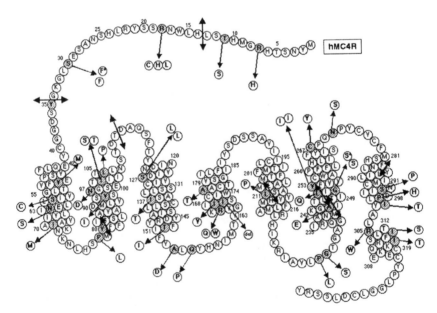

FIGURE 1. Schematic representation of MC4R and the sequence variants detected in human obesity. The positions of the sequence variants are indicated on the secondary structure of MC4R. Amino acids are indicated as *circles in single-letter code.* Mutated amino acids are indicated in bold in a *hatched circle.* S30F*/G252S* and Y35STOP**/D37V** are double mutants.

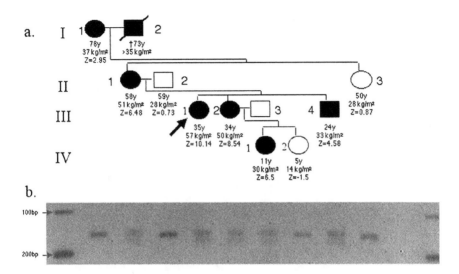

FIGURE 2. Example of pedigree, mutation screening, and phenotypes in the available family members of the carrier of 732-733insGATT MC4R mutation. (**a**) The upper part of the figure represents the pedigree structure of the family of the carrier. Age (years), BMI (kg/m²), and Z score (standard deviation of BMI) are indicated under each individual. *Arrows* indicate the proband. *Filled symbols* denote obesity. (**b**) The *lower portion* of the figure represents the result of PCR-RFLP genotyping of each available family member.

MC4R MUTATIONS ARE ASSOCIATED WITH HUMAN OBESITY

The human melanocortin receptor MC4R is a 332 amino acid protein encoded by a single exon gene localized on chromosome 18q22.[1,8] We and others have reported that multiple rare mutations in MC4R cause a common nonsyndromic form of obesity.[9–15] By screening a total of 1,787 adults with morbid obesity or children with early onset obesity from 7 cohorts, we detected a total of 35 different MC4R mutations (5 frameshift or non-sense, 30 missense) in 36 patients representing 2% of the screened population. The localization of these mutations as well as that of the other mutations described in the literature are presented on FIGURE 1.

The implications of these rare heterozygous mutations in the obesity of the probands is based on four different types of arguments. First, the *in vivo* animal study demonstrating an altered body weight regulation in heterozygous knock-out animals. Second, the presence of rare variants is associated with obesity, because systematic screening of non-obese controls does not detect any rare variants in MC4R.[13] Third, the fact that MC4R mutations co-segregate with obesity in the family of the proband as shown, for example in FIGURE 2, for mutation 732-733 ins-GATT. Finally, as most of the obesity-associated MC4R mutations are missense mutations, confirmation of their pathogenic role requires functional characterization of the mutant receptor. We systematically investigated three aspects of the functional alterations caused by 50 obesity-associated MC4R mutations.

Membrane Localization

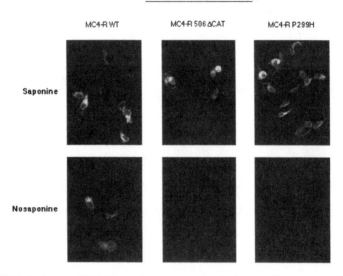

FIGURE 3. Assay of MC4R membrane expression by immunofluorescence. We quali-tatively assessed the membrane expression of WT and variants of MC4R. HEK 293 cells are transiently transfected with the MC4R wild type or mutants on which was added an N-ter-minal Flag epitope. Twenty-four hours after transfection, cells are split and grown on cov-erslips for another 24 h before immunofluorescence. Cells are rinsed, fixed in 3.7% paraformaldehyde (PFA), stained with a monoclonal anti-Flag antibody M2, and FITC-con-jugated anti-mouse IgG in the presence or absence of a membrane-permeabilizing agent (Sa-ponine). Cells are mounted on slides and observed by fluorescent microscopy. In the absence of a membrane-permeabilizing agent, only cell surface–expressed receptors are stained. This figure shows the result of immunofluorescence of two sequence variants leading to complete intracellular retention of the receptor, one frameshift mutation, 506ΔCAT and one missense mutation P299H.

A LARGE NUMBER OF MC4R MUTATIONS ALTER THE CELL SURFACE EXPRESSION OF THE RECEPTOR

Defective intracellular protein transport is an increasingly recognized anomaly of hereditary disease-causing mutations. Examples include the failure of proper trans-port of the low-density lipoprotein (LDL) receptor in some types of familial hyper-cholesterolemia,[16] and the cystic fibrosis transmembrane conductance regulator (CFTR) in cystic fibrosis.[17] Of particular relevance is the implication of rhodopsin mutations in retinitis pigmentosa (RP). Rhodopsin is a G protein–coupled photore-ceptor expressed in specialized neuronal cells. As for MC4R, mutations in rhodopsin can cause either a dominant or recessive form of RP.[18–21] The vast majority of the 150 dominant RP-causing rhodopsin mutations described to date are so-called class II mutations that are intracellularly retained. Transgenic expression in rodents of such mutants has subsequently demonstrated their intracellular accumulation and role in photoreceptor degeneration.[20,22] We evaluated the intracellular retention of all obesity-associated MC4R mutations. Major alterations of cell surface expression of MC4R can easily be demonstrated by immunocytochemistry analysis as shown in FIGURE 3 for mutations P299H and 506delCAT, and was demonstrated, using this method, for all frameshift and non-sense mutations.[23] However, this method provides only qualitative information on the presence or absence of the receptor at the cell surface. To allow for the rapid evaluation of cell surface expression of MC4R relative to total expression of the receptor in individual transiently transfected cells, we developed a method based on immunostaining and fluorescence detection by flow cytometry. We constructed a chimeric receptor containing a C-terminal intrac-

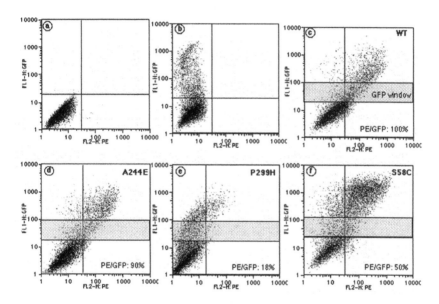

FIGURE 4. Flow cytometry analysis of MC4R cell surface expression. To evaluate MC4R cell surface expression relative to total expression of the receptor, we developed a method based on immunostaining and fluorescence detection by flow cytometry. We constructed a chimeric receptor by adding an extracellular N-terminal Flag epitope and an intracellular C-terminal green fluorescent protein (GFP). HEK 293 cells were transiently transfected with the chimeric wild type and mutated MC4R. Twenty-four hours after transfection HEK 293 cells were immunostained at 4°C with a mouse anti-Flag antibody (M2) and a PE-conjugated antimouse in the absence of detergent. GFP and PE emissions are analyzed in each individual cell by flow cytometry with a FACS® Calibur Beckton-Dickinson and results analyzed using the software CellQuest are shown as dot plot. (**a**) untransfected HEK 293 cells; (**b**) HEK 293 cells transfected with WT chimeric MC4R and immunostained without anti-Flag antibody; (**c**) HEK 293 cells transfected with WT chimeric MC4R and immunostained with both antibodies. GFP emission represents MC4R total expression and PE emission represents cell surface MC4R expression. To limit artifacts caused by receptor overexpression, we restrict our analysis to cells expressing low levels of the receptor as defined by the GFP window shown. Representative dot plots are shown for three missense mutations: (**d**) A244E, (**e**) P299H, and (**f**) S58C. The ratio PE/GFP, representing relative membrane expression, is calculated for each individual cell using the FlowJo software and is expressed as a percentage of the value obtained in the same experiment for the wild-type MC4R. This ratio is indicated on the *right bottom* corner of dot plot graphs.

ellular green fluorescence protein (GFP) and an N-terminal extracellular Flag epitope. This chimeric construct remains responsive to the natural agonist α-MSH with the same EC_{50} as the native receptor. The GFP fluorescence emitted by chimera expressing cells can be detected by FACS allowing for the detection and analysis of cells expressing a similar level of receptor for further analysis. Immunostaining of the transfected cells with a mouse anti-Flag primary antibody and a phycoerythrin (PE)-conjugated anti-mouse secondary antibody in the absence of detergent allows for the detection of cell surface expression of the transfected receptor. Concurrent FACS analysis of both GFP and PE fluorescence for each individual cell, therefore, allows quantification of cell surface expression relative to total expression for each selected cell. FIGURE 4 shows the results obtained with mutations P299H, S58C, and A244E. Using this method, we demonstrated that 54% of obesity-associated MC4R mutations are partially or totally intracellularly retained.

Although further studies will be necessary to demonstrate a defective processing of MC4R mutations as well as neuronal degeneration upon transgenic expression of MC4R mutants, our finding that obesity-associated MC4R mutations have a decreased membrane expression in a heterologous expression system is the first suggestion of a dominant negative neurodegenerative disease mechanism for such mutations.

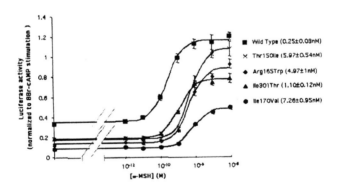

FIGURE 5. α-MSH activation of obesity-associated MC4R mutants. Activity of the receptors is assayed by analyzing their ability to activate the expression of a cAMP-induced luciferase reporter gene. Cells stably expressing each receptor and transiently expressing the fusion construct were stimulated for 6 h with medium alone, increasing amount of α-MSH or 8Br-cAMP, after which luciferase was measured. Data points represent the mean of five determinations divided by maximal level of luciferase activity achieved by 1 mM of 8Br-cAMP. *Error bars* indicate SD. The $EC_{50} \pm SD$ is indicated in parentheses for the WT MC4R and each variant receptor. We indicate here the ratio PE/GFP representing membrane expression of MC4R mutant. PE/GFP is expressed as a percentage of the value obtained for the wild type in the same experiment. The following values are the MEAN ± SEM of at least three independent experiments. I301T: 37 ± 3%; T150I: 113 ± 12%; I170V: 84 ± 3%; R165W: 26 ± 6%.

MOST OBESITY-ASSOCIATED MC4R MUTATIONS ALTER THE AGONIST ACTIVATION OF THE RECEPTOR

MC4R decreases food intake in response to α-MSH activation of the receptor. In patients carrying MC4R mutations, obesity could be caused by an altered response to α-MSH. We compared the α-MSH induction of cAMP production by the different mutated MC4Rs to that of the wild-type receptor. Activity of the receptors is assayed by analyzing their ability to activate the expression of a cAMP-induced reporter luciferase gene (pCRE-luciferase) in the presence of α-MSH. HEK293 cells were transfected with the wild type and all mutated receptors together with pCRE-luciferase. Thirty-six hours after transfection, cells were stimulated for 6 h at 37°C with medium alone, increasing amounts of α-MSH, after which luciferase activity, proportional to the amount of cAMP produced in response of agonist stimulation, is assessed. Results are normalized to the maximal luciferase activity obtained in the presence of 8Br-cAMP (1 mM). In this assay, we observed that 56% of obesity-associated MC4R missense mutations presented with an alteration of the EC_{50} showing a defect in agonist activation of the receptor. FIGURE 5 shows α-MSH activation curves for the variants I301T, T150I, I170V, and R165W. A decrease in cell surface expression of the receptor could be responsible for the alteration of the receptor activity in response to α-MSH. However, the impairment of receptor activity is not always associated with a diminution of membrane expression and, in this case, could be due to a decreased affinity of the receptor for α-MSH, to a decrease in signal transduction, or a combination of these defects.

THE CONSTITUTIVE ACTIVITY OF MC4R IS ALTERED BY OBESITY-ASSOCIATED MUTATIONS

Recent *in vitro* data have demonstrated that MC4R exhibits a constitutive activity on which AGRP acts as an invert agonist.[5,6] These data suggest that, in the absence of ligand, MC4R can exert a food intake inhibiting activity but the *in vivo* relevance of this constitutive activity is unknown. We systematically assessed the basal activity of the obesity-associated MC4R variants that display normal membrane expression. HEK 293 cells were transfected with the wild-type or mutated receptors together with pCRE-luciferase reporter gene. Thirty-six hours after transfection, cells were incubated 6 h at 37°C in stimulating medium alone or in the presence of 8Br-cAMP (1mM). Results were normalized to the maximal luciferase activity obtained in the presence of 1 mM 8Br-cAMP and expressed as the percentage of the value obtained for the wild type in the same experiment. We found 55% of the tested MC4R mutations had a significantly decreased constitutive activity. For five of the obesity-associated MC4R mutants, this decreased constitutive activity was the only functional defect detected. This strongly supports a physiological role for the constitutive activity of MC4R. Interestingly, most mutations exhibiting decreased constitutive activity as the sole functional defect are localized in the extracellular N-terminal domain of MC4R, suggesting a role for this domain in the maintenance of the basal activity of the receptor.

TOWARD A FUNCTIONAL CLASSIFICATION OF OBESITY-ASSOCIATED MC4R MUTATIONS

The prevalence of obesity, a multifactorial disease caused by the interaction of genetic and environmental factors,[24,25] is increasing at an alarming rate in industrialized countries. Because obesity is resistant to currently available treatments, the elucidation of the pathophysiologic mechanisms underlying it is a necessary prerequisite to the identification of new therapeutic strategies for this condition. The recent description of at least six different monogenic forms of childhood obesity[9–15,26–31] has confirmed the implication of genetic factors in the development of human obesity and suggests that this genetic predisposition is extremely heterogeneous. Together our studies confirm that all obesity-associated MC4R mutations alter the function of the protein, providing an additional argument for their pathogenic role. In addition, our studies indicate that the genetic heterogeneity of obesity extends to the allelic level and that multiple pathogenic mechanisms may underlie the obesity phenotype of MC4R mutation carriers. A better understanding of these different disease mechanisms will be a prerequisite for clarifying the genotype-phenotype relationship of this condition and for eventually adapting the preventive and pharmacologic interventions to the specific molecular defects.

ACKNOWLEDGMENTS

This work was supported by National Institutes of Health grant RO1 DK60540 and an American Diabetes Association Career Development Award to C.V. We thank Leslie Spector for editorial work.

REFERENCES

1. GANTZ, I., H. MIWA, Y. KONDA, et al. 1993. Molecular cloning, expression, and gene localization of a fourth melanocortin receptor. J. Biol. Chem. **268:** 15174–15179.
2. MOUNTJOY, K.G., M.T. MORTRUD, M.J. LOW, et al. 1994. Localization of the melanocortin-4 receptor (MC4-R) in neuroendocrine and autonomic control circuits in the brain. Mol, Endocrinol, **8:** 1298–1308.
3. FAN, W., B.A. BOSTON, R.A. KESTERSON, et al. 1997. Role of melanocortinergic neurons in feeding and the agouti obesity syndrome. Nature **385:** 165–168.
4. OLLMANN, M.M., B.D. WILSON, Y.K. YANG, et al. 1997. Antagonism of central melanocortin receptors in vitro and in vivo by agouti-related protein. Science **278:** 135–138.
5. NIJENHUIS, W.A., J. OOSTEROM & R.A. ADAN. 2001. AgRP(83-132) acts as an inverse agonist on the human-melanocortin-4 receptor. Mol. Endocrinol. **15:** 164–171.
6. HASKELL-LUEVANO, C. & E.K. MONCK. 2001. Agouti-related protein functions as an inverse agonist at a constitutively active brain melanocortin-4 receptor. Regul. Pept. **99:** 1–7.
7. HUSZAR, D., C.A. LYNCH, V. FAIRCHILD-HUNTRESS, et al. 1997. Targeted disruption of the melanocortin-4 receptor results in obesity in mice. Cell **88:** 131–141.
8. SUNDARAMURTHY, D., D.A. CAMPBELL, J.P. LEEK, et al. 1998. Assignment of the melanocortin 4 receptor (MC4R) gene to human chromosome band 18q22 by in situ hybridisation and radiation hybrid mapping. Cytogenet. Cell Genet. **82:** 97–98.
9. VAISSE, C., K. CLEMENT, B. GUY-GRAND, et al. 1998. A frameshift mutation in human MC4R is associated with a dominant form of obesity. Nat. Genet. **20:** 113–114.
10. YEO, G.S., I.S. FAROOQI, S. AMINIAN, et al. 1998. A frameshift mutation in MC4R associated with dominantly inherited human obesity. Nat. Genet. **20:**111–112.

11. HINNEY, A., A. SCHMIDT, K. NOTTEBOM, *et al.* 1999. Several mutations in the melano-cortin-4 receptor gene including a nonsense and a frameshift mutation associated with dominantly inherited obesity in humans. J. Clin. Endocrinol. Metab. **84:**1483–1486.
12. GU, W., Z. TU, P.W. KLEYN, *et al.* 1999. Identification and functional analysis of novel human melanocortin-4 receptor variants. Diabetes **48:** 635–639.
13. VAISSE, C., K. CLEMENT, E. DURAND, *et al.* 2000. Melanocortin-4 receptor mutations are a frequent and heterogeneous cause of morbid obesity. J. Clin. Invest. **106:** 253–262.
14. DUBERN, B., K. CLEMENT, V. PELLOUX, *et al.* 2001. Mutational analysis of melanocor-tin-4 receptor, agouti-related protein, and alpha-melanocyte-stimulating hormone genes in severely obese children. J. Pediatr. **139:** 204–209.
15. MERGEN M., H. MERGEN, M. OZATA, *et al.* 2001. A novel melanocortin 4 receptor (MC4R) gene mutation associated with morbid obesity. JCEM **86:** 3448.
16. HOBBS, H.H., D.W. RUSSEL, M.B. BROWN, *et al.* 1990. The LDL receptor locus in familial hypercholesterolemia: mutational analysis of a membrane protein. Annu. Rev. Genet. **24:**133–170.
17. MORELLO, J., U. PETÄJÄ-REPO, D. BICHET, *et al.* 2000. Pharmacological chaperones: a new twist on receptor folding. TIPS **21:** 466–469.
18. SUNG, C. H., C.M. DAVENPORT & J. NATHANS. 1993. Rhodopsin mutations responsible for autosomal dominant retinitis pigmentosa. Clustering of functional classes along the polypeptide chain. J. Biol. Chem.**268:** 26645–26649.
19. SUNG, C. H. & A.W. TAI. 2000. Rhodopsin trafficking and its role in retinal dystro-phies. Int. Rev. Cytol. **195:** 215–267.
20. KAUSHAL, S. & H.G. KHORANA. 1994. Structure and function in rhodopsin. 7. Point mutations associated with autosomal dominant retinitis pigmentosa. Biochemistry **33:** 6121–6128.
21. FARRAR, G.J., P.F. KENNA & P. HUMPHRIES. 2002. On the genetics of retinitis pigmen-tosa and on mutation-independent approaches to therapeutic intervention. EMBO J. **21:** 857–864.
22. ROLLAND-CACHERA, M.F., T.J. COLE, M. SEMPE, *et al.* 1991. Body mass index varia-tions: centiles from birth to 87 years. Eur. J. Clin. Nutr. **45:**13–21.
23. HO, G. & R.G. MACKENZIE. 1999. Functional characterization of mutations in melano-cortin-4 receptor associated with human obesity. J. Biol. Chem. **274:** 35816–35822.
24. COMUZZIE, A.G. & D.B. ALLISON. 1998. The search for human obesity genes. Science **280:** 1374–1377.
25. HILL, J.O. & J.C. PETERS. 1998. Environmental contributions to the obesity epidemic. Science **280:** 1371–1374.
26. MONTAGUE, C.T., I.S. FAROOQI, J.P. WHITEHEAD, *et al.* 1997. Congenital leptin defi-ciency is associated with severe early-onset obesity in humans. Nature **387:** 903–908.
27. STROBEL, A., T. ISSAD, L. CAMOIN, *et al.* 1998. A leptin missense mutation associated with hypogonadism and morbid obesity. Nat. Gen. **18:** 213–215.
28. CLEMENT, K., C. VAISSE, N. LAHLOU, *et al.* 1998. A mutation in the human leptin receptor gene causes obesity and pituitary dysfunction. Nature **392:** 398–401.
29. JACKSON, R.S., J.W. CREEMERS, S. OHAGI, *et al.* 1997. Obesity and impaired prohor-mone processing associated with mutations in the human prohormone convertase 1 gene. Nat. Genet. **16:** 303–306.
30. KRUDE, H., H. BIEBERMANN, W. LUCK, *et al.* 1998. Severe early-onset obesity, adrenal insufficiency and red hair pigmentation caused by POMC mutations in humans. Nat. Genet. **19:** 155–157.
31. HOLDER, J.L.J., N.F. BUTTE & A.R. ZINN. 2000. Profound obesity associated with a balanced translocation that disrupts the SIM1 gene. Hum. Mol. Genet. **9:** 101–108.

α-MSH and Desacetyl-α-MSH Signaling through Melanocortin Receptors

KATHLEEN G. MOUNTJOY,[a,b] CHIA-SHAN JENNY WU,[b] JILLIAN CORNISH,[c] AND KAREN E. CALLON[c]

Departments of [a]Physiology, [b]Molecular Medicine, and [c]Medicine, Faculty of Medical and Health Sciences, University of Auckland, Auckland 1, New Zealand

ABSTRACT: The functional significance of N-terminal acetylation of ACTH[1-13]NH$_2$ is unknown. N-terminal acetylation of ACTH[1-13]NH$_2$ (known as desacetyl-α-MSH) to produce α-MSH enhances some activities of ACTH[1-13]NH$_2$ and virtually eliminates others. To determine whether α-MSH and desacetyl-α-MSH diverge in their coupling to melanocortin receptors *in vitro*, we measured the sensitivity of MC1, MC3, MC4, and MC5 receptors stably expressed in HEK293 cells to these peptides, functionally coupling them to adenylyl cyclase and a calcium signaling pathway. α-MSH and desacetyl-α-MSH similarly coupled these overexpressed receptors to both signaling pathways. In contrast, we discovered that α-MSH significantly increased primary rat osteoblast proliferation while for desacetyl-α-MSH there was only a trend to do the same. Osteoblast cells expressing very low levels of endogenous melanocortin receptors, in contrast with transfected HEK293 cells overexpressing a single melanocortin receptor, may provide an *in vitro* model for differentiating between α-MSH and desacetyl-α-MSH signaling.

KEYWORDS: α-MSH; desacetyl-α-MSH; melanocortin receptor; signaling; osteoblast

INTRODUCTION

The physiological responses to pro-opiomelanocortin (POMC)-derived peptides include pigmentation, adrenal development and corticosteroid synthesis, food intake and feed efficiency, metabolism, body weight, immune and cardiovascular regulation. POMC, produced most abundantly in the pituitary and brain, is also expressed in several peripheral tissues including skin, pancreas, and testis. POMC is processed through a coordinated, tissue-specific, series of proteolytic cleavages and post-translational modifications which influence the activity of the peptide products. For example, in the corticotropic cells of the anterior lobe of the pituitary, the major end product is the 39 amino acid, ACTH[1-39] (FIG. 1). In the melanotrophs of the intermediate lobe of the pituitary, ACTH[1-39] is the precursor of ACTH[1-13] and

Address for correspondence: Kathleen G. Mountjoy, Department of Physiology, Faculty of Medical and Health Sciences, University of Auckland, Auckland 1, New Zealand. Voice: 64-9-373 7599 ext. 86447; fax: 64-9-373 7499.
kmountjoy@auckland.ac.nz

Ann. N.Y. Acad. Sci. 994: 58–65 (2003). © 2003 New York Academy of Sciences.

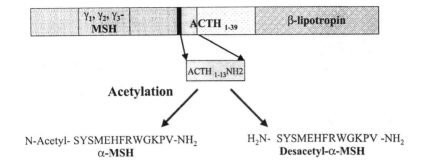

FIGURE 1. POMC processing to produce desacetyl-α-MSH and α-MSH.

corticotropin-like intermediate lobe peptide (CLIP = ACTH[18-39]). ACTH[1-13] is amidated at the carboxy terminus to form ACTH[1-13]NH_2 known as desacetyl-α-melanocyte-stimulating hormone (MSH). Desacetyl-α-MSH can be N-terminal acetylated to form α-MSH in secretory granules prior to exocytosis.[1] The major fraction of ACTH[1-13]NH_2 produced by pituitary melanotrophs is acetylated at the amino terminus, while most of brain-derived ACTH[1-13]NH_2 is not. In rodent circulation, α-MSH is the major circulating form of ACTH[1-13]NH_2, but desacetyl-α-MSH is more abundant than α-MSH in human circulation. Desacetyl-α-MSH also predominates over α-MSH in human amniotic fluid as well as in human and rodent fetuses.

The functional significance of N-terminal acetylation of ACTH[1-13]NH_2 is unknown, although it appears to enhance some activities of ACTH[1-13]NH_2 and virtually eliminate others. α-MSH injected daily to rats is 10–100-fold more effective than desacetyl-α-MSH at increasing pigmentation,[2] arousal, memory, attention, and excessive grooming.[3] Desacetyl-α-MSH, however, is more effective than α-MSH at blocking opiate analgesia and opiate receptor binding *in vivo*.[3] α-MSH and desacetyl-α-MSH also differentially affect feeding and weight gain. Weight gain of yellow *agouti* obese mice is increased by subcutaneouly administered desacetyl-α-MSH, as is food intake and fat pad weight, but α-MSH injections do not significantly increase food intake or body weight.[4]

Desacetyl-α-MSH Is a Weak Agonist for Inducing Pigmentation

Viable yellow mice (A^{vy}/a) proved useful for studying pigmentation following Geshwind's discovery in 1966 that α-MSH induced pigmentation most markedly in A^{vy}/a mice but not in *e/e* mice.[5] Indeed, data collected from mouse genetic studies over many decades greatly facilitated the discovery that agouti protein is a natural antagonist of α-MSH–induced pigmentation through the MC1-R.[6,7] Viable yellow (A^{vy}/a) are yellow mice in which the *a* gene is ectopically overexpressed in all tissues.[8]

In 1989, Shimizu discovered that a 10-fold higher concentration of desacetyl-α-MSH compared with α-MSH was required to induce pigmentation in viable yellow

TABLE 1. EC_{50} values for melanocortin peptides coupling mouse melanocortin receptors stably expressed in HEK293 cells to adenylyl cyclase (AC) or mobilization of $[Ca^{2+}]_i$

	EC_{50} (nM)			
	MC1-R	MC3-R	MC4-R	MC5-R
α-MSH (AC)	0.2 ± 0.1 (n = 3)	0.9 ± 0.4 (n = 3)	1.1 ± 0.5 (n = 3)	1.3 ±0.5 (n = 3)
desacetyl-α-MSH (AC)	0.1 ± 0.0 (n = 3)	1.0 ± 0.5 (n = 3)	0.5 ± 0.0 (n = 3)	0.8 ± 0.2 (n = 3)
α-MSH ($[Ca^{2+}]_i$	4.3 ± 1.8 (n = 5)	1.3 ± 0.6 (n = 4)	0.5 ± 0.3 (n = 15)	1.3 ± 0.7 (n = 4)
desacetyl-α-MSH ($[Ca^{2+}]_i$	2.7 ± 1.9 (n = 6)	1.3 ± 0.3 (n = 4)	0.3 ± 0.1 (n = 9)	1.6 ± 0.8 (n = 4)

Mean ± SD, n = number of dose-response curves. (Modified from Mountjoy et al.,[18,19] with permission.

mice.[4] This was not the first time that α-MSH had been identified as the primary form of ACTH[1-13]NH$_2$ stimulating pigmentation. Guttmann and Boissonnas in 1961 showed that α-MSH was the primary stimulus for melanophore activity.[9] Since the cloning of the melanocortin receptors,[10] the mechanism by which α-MSH is deemed more potent than desacetyl-α-MSH at inducing pigmentation has not been determined.

Desacetyl-α-MSH Compared to α-MSH Is a Weak Agonist at Inhibiting Food Intake

Intracerebroventricular (i.c.v.)-administered α-MSH and ACTH[1-24] inhibit food intake in food deprived[11] and fed[12] rats, whereas i.c.v. injections of desacetyl-α-MSH are either without any effect, or an increased dose of desacetyl-α-MSH compared to α-MSH is required to inhibit feeding.[13] Both MC3-R and MC4-R are expressed in brain and both have distinct roles in regulating energy homeostasis. Agouti protein was discovered to be an antagonist of α-MSH at the MC4-R in addition to the MC1-R[14], accounting for the obese phenotype in dominant yellow obese agouti mice. The MC4-R knockout mice develop a phenotype remarkably similar to the A^{vy}/a mouse except they do not have a yellow coat color.[15] By homology with agouti protein, agouti gene-related peptide (AGRP) was discovered as a natural antagonist of α-MSH at both MC3-R and MC4-R in the brain.[16] The MC3-R knockout mice develop a milder form of obesity compared with the MC4-R knockout mice.[17] The obesity phenotype in the MC4-R knockout mouse results from hyperphagia and defects in metabolism, while the obesity in the MC3-R knockout mouse is a result of increased feed efficiency with preferential partitioning of energy stores toward adipose tissue, and reduced energy expenditure. The effects of melanocortin peptides inhibiting food intake are therefore mediated through the MC4-R in the brain and the mechanism for α-MSH being more potent than desacetyl-α-MSH at inhibiting food intake has not been determined.

α-MSH and Desacetyl-α-MSH Couple Melanocortin Receptors in a Similar Manner to Adenylyl Cyclase

To determine whether α-MSH and desacetyl-α-MSH diverge in their functional coupling to melanocortin receptors *in vitro*, we have measured the sensitivity of these peptides functionally coupling four melanocortin receptors stably expressed in HEK293 cells to adenylyl cyclase.[18] α-MSH and desacetyl-α-MSH activate mouse MC1-R, MC3-R, MC4-R, and MC5-R coupling to increased adenylyl cyclase activity with similar EC_{50}'s (TABLE 1). Desacetyl-α-MSH is slightly more potent than α-MSH at the mouse MC4-R.

α-MSH and Desacetyl-α-MSH Couple Melanocortin Receptors in a Similar Manner to Mobilization of Intracellular Free Calcium

Melanocortin peptides activate a second intracellular signaling pathway in HEK293 cells that results in mobilization of intracellular calcium without an increase in inositol triphosphate.[19] α-MSH and desacetyl-α-MSH couple mouse MC1-R, MC3-R, MC4-R, and MC5-R expressed in HEK293 cells similarly to mobilization of intracellular calcium (TABLE 1).

Agouti Protein Is a Competitive Antagonist of α-MSH and Desacetyl-α-MSH Coupling Mouse MC1-R to Adenylyl Cyclase and Mobilization of Intracellular Calcium

Mouse agouti protein (10–100 nM) shifted the EC_{50} for mouse MC1-R coupling to adenylyl cyclase in a dose-dependent manner from 1.74×10^{-10} M to 1.02×10^{-9} M (5.9-fold) and from 1.26×10^{-10} M to 8.67×10^{-10} M (5.4-fold) for α-MSH and desacetyl-α-MSH, respectively. The maximum response appeared to be decreased by 1.2- and 1.3-fold for α-MSH and desacetyl-α-MSH, respectively, but these changes were not significant.[18] Mouse agouti protein (55 nM) significantly decreased the sensitivity of α-MSH and desacetyl-α-MSH coupling the mouse MC1-R to mobilization of intracellular calcium 6- and 8-fold, respectively. The maximum responses were not significantly changed. Mouse agouti protein is therefore a competitive antagonist of both peptides at the MC1-R coupling to adenylyl cyclase and mobilization of intracellular calcium signaling pathways in HEK293 cells.[18,19]

Agouti Protein Is More Effective at Antagonizing Desacetyl-α-MSH than α-MSH Coupling Mouse MC4-R to Adenylyl Cyclase

Mouse agouti protein potently antagonized desacetyl-α-MSH coupling mouse MC4-R to adenylyl cyclase in a non-competitive manner. While 100 nM agouti protein increased the EC_{50} 4.8-fold, as little as 10 nM agouti protein significantly reduced the maximum response 1.4-fold (FIG. 2C and D). Although agouti protein appeared to increase the EC_{50} and decrease the maximum response for α-MSH coupling the mouse MC4-R, these changes were not significant.[18]

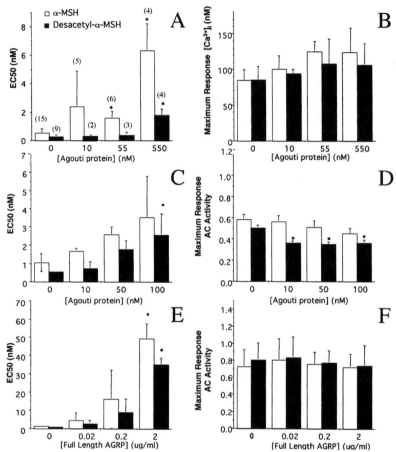

FIGURE 2. Antagonism of α–MSH and desacetyl-α-MSH coupling mouse MC4-R to [Ca^{2+}]$_i$ [**A,B**] and adenylyl cyclase (AC) [**C–F**]. Means ± SD are shown. The number of experiments for each data point in **A** and **B** are shown in parentheses. Data points shown in **C, D, E,** and **F** are derived from three separate experiments. *Significant difference 2. determined using ANOVA between absence of antagonist or presence of antagonist. (*P* <0.05). (Modified from Mountjoy et al.,[18,19] with permission.)

Agouti Protein Is More Effective at Antagonizing α-MSH than Desacetyl-α-MSH Coupling Mouse MC4-R to Mobilization of Intracellular Calcium

Agouti protein (55 nM) significantly increased the EC$_{50}$ for α-MSH (3-fold) and 550 nM agouti protein significantly increased the EC$_{50}$ for desacetyl-α-MSH (4-fold) coupling the mouse MC4-R to a rise in [Ca2+]i[19] (FIG. 2A and B). Although mouse agouti protein decreases the sensitivity of melanocortin peptides coupling the MC4-R to the calcium signaling pathway, there is a dose-dependent trend for it to increase maximum responsiveness. It is unclear from our data, therefore, whether agouti protein is a competitive antagonist, or enhancer, of the MC4-R coupling to the calcium signaling pathway.

AGRP Similarly Antagonizes α-MSH and Desacetyl-α-MSH Coupling Mouse MC4-R to Adenylyl Cyclase

Almost full-length recombinant human AGRP (2 mg/mL) significantly increased the EC_{50}'s for α-MSH and desacetyl-α-MSH coupling mouse MC4-R to adenylyl cyclase 40–42-fold (FIG. 2E and F). AGRP did not significantly change the maximum responsiveness and therefore it is a competitive antagonist and its mechanism of action at the mouse MC4-R appears to differ from agouti protein.

α-MSH, but not Desacetyl-α-MSH Stimulates Proliferation of Primary Rat Osteoblasts

In our search for cells that express endogenous MC4-R we discovered that MC4-R mRNA is expressed in low abundance in rat primary osteoblasts. Primary rat osteoblasts were isolated from 20 day fetal rat calvariae[20] and grown in DMEM supplemented with 10% FCS, 50 U/mL penicillin and 50 μg/mL streptomycin 24-well plates for proliferation assays. Subconfluent osteoblasts were growth arrested and then stimulated with MSH peptides. α-MSH ($10^{-9}–10^{-7}$ M) significantly increased thymidine incorporation and cell number into growth arrested primary rat osteo-

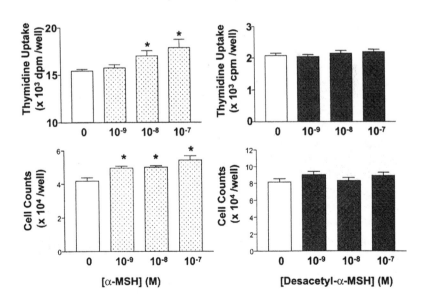

FIGURE 3. Thymidine uptake and cell number in growth-arrested primary cultures of fetal rat osteoblasts treated for 24 h with either α-MSH or desacetyl-α-MSH. There were six wells in each group and each experiment was repeated 3 or 4 times. Data are mean ± SEM. The significance of differences between groups was determined using Student's *t* tests for unpaired data and a 5% significance level. *Significant differences from untreated cells. ($P < 0.04$).

blasts, while there was only a trend for desacetyl-α-MSH to do the same (FIG. 3). This pattern of response is remarkably similar to some of the *in vivo* differences in biological activities reported for these peptides.

CONCLUSIONS

In contrast to the differences in biological responses documented for α-MSH and desacetyl-α-MSH *in vivo*, we have been unable to detect clear differences between these two peptides coupling to either mouse MC1-R or mouse MC4-R overexpressed in HEK293 cells. Agouti protein, however, is more effective at antagonizing desacetyl-α-MSH than α-MSH coupling mouse MC4-R to adenylyl cyclase, but more effective at antagonizing α-MSH than desacetyl-α-MSH coupling MC4-R to mobilization of intracellular calcium. While these differences in agouti antagonism of the MSH peptides may contribute to the yellow agouti obese mouse phenotype, they could not account for the different *in vivo* biological actions of α-MSH and desacetyl-α-MSH. AGRP is the natural antagonist of MC4-R in the brain, and full-length AGRP similarly antagonized α-MSH and desacetyl-α-MSH coupling the mouse MC4-R to adenylyl cyclase. A novel *in vitro* model for furthering our understanding of differences between α-MSH and desacetyl-α-MSH signaling may be rat primary osteoblasts. In contrast to overexpressing a single melanocortin receptor subtype in HEK293 cells, MC4-R mRNA is expressed in very low abundance in rat primary osteoblasts; this, together with expression of other melanocortin receptor subtypes in osteoblasts, may be important in distinguishing activity between α-MSH and desacetyl-α-MSH.

ACKNOWLEDGMENTS

This work was funded by The Wellcome Trust, Health Research Council of New Zealand, Auckland Medical Research Foundation, and New Zealand Lottery Health. We thank D.H. Willard and W.O. Wilkison (Glaxo-Wellcome Research Institute, Research Triangle Park, North Carolina, USA) for supplying mouse agouti protein and K.L. Stark (Amgen Inc., Thousand Oaks, California, USA) for supplying AGRP.

REFERENCES

1. DORES, R.M. 1988. Pituitary melanotropin biosynthesis. *In* The Melanotropic Peptides. A.N. Eberle, Ed.: 25–38.
2. CANGEMI, L. *et al.* 1995. N-Acetyltransferase mechanism for α-melanocyte stimulating hormone regulation in rat ageing. Neurosci. Lett. **201:** 65–68.
3. O'DONOHUE, T.L. *et al.* 1982. N-Acetylation regulates the behavioral activity of α-melanotropin in a multineurotransmitter neuron. Science **215:** 1125–1127.
4. SHIMIZU, H. *et al.* 1989. Effects of MSH on food intake, body weight and coat color of the yellow obese mouse. Life Sci. **45:** 543–552.
5. GESCHWIND, I.I. 1966. Change in hair color in mice induced by injection of α-MSH. Endocrinology **79:** 1165–1167.
6. CONE, R.D. & K.G. MOUNTJOY. 1993. Molecular genetics of the ACTH and melanocyte-stimulating hormone receptors. Trends Endocrinol. Metab. **4:** 242–247.

7. ROBBINS, L.S. *et al.* 1993. Pigmentation phenotypes of variant extension locus alleles result from point mutations that alter MSH receptor function. Cell **72:** 827–834.
8. BULTMAN, S.J., E.J. MICHAUD & R.P. WOYCHIK. 1992. Molecular characterization of the mouse agouti locus. Cell **71:** 1195–1204.
9. GUTTMANN, S. & R. BOISSONNAS. 1961. Influence of the structure of the N-terminal extremity of a-MSH on the melanophore stimulating activity of this hormone. Experientia **17:** 265–267.
10. MOUNTJOY, K.G. *et al.* 1992. The cloning of a family of genes that encode the melanocortin receptors. Science **257:** 543–546.
11. TSUJII, S. & G.A. BRAY. 1989. Acetylation alters the feeding response to MSH and beta-endorphin. Brain Res. Bull. **23:** 165–169.
12. POGGIOLI, R., A.V. VERGONI & A. BERTOLINI. 1986. ACTH-(1-24) and alpha-MSH antagonize feeding behavior stimulated by kappaopiate agonists. Peptides **7:** 843–848.
13. ABBOTT, C.R. *et al.* 2000. Investigation of the melanocyte stimulating hormones on food intake. Lack of evidence to support a role for the melanocortin-3-receptor. Brain Res. **869:** 203–210.
14. LU, D. *et al.* 1994. Agouti protein is an antagonist of the melanocyte-stimulating-hormone receptor. Nature **371:** 799–802.
15. HUSZAR, D. *et al.* 1997. Targeted disruption of the melanocortin-4 receptor results in obesity in mice. Cell **88:** 131–141.
16. OLLMANN, M.M. *et al.* 1997. Antagonism of central melanocortin receptors in vitro and in vivo by agouti-related protein. Science **278:** 135–138.
17. BUTLER, A.A. *et al.* 2000. A unique metabolic syndrome causes obesity in the melanocortin-3 receptor-deficient mouse. Endocrinology **141:** 3518–3521.
18. MOUNTJOY, K.G., D.H. WILLARD & W.O. WILKISON. 1999. Agouti antagonism of melanocortin-4 receptor: greater effect with desacetyl-α-MSH than with α-MSH. Endocrinology **140:** 2167–2172.
19. MOUNTJOY, K.G. *et al.* 2001. Melanocortin receptor-mediated mobilization of intracellular free calcium in HEK293 cells. Physiol. Genomics **5**(1): 11–19.
20. CORNISH, J. *et al.* 1999. Trifluoroacetate, a contaminant purified proteins, inhibits proliferation of osteoblasts and chondrocytes. Am. J. Physiol. Endocrinol. Metab. **126:** 1416–1420.

Syndecan-3 Modulates Food Intake by Interacting with the Melanocortin/AgRP Pathway

OFER REIZES,[a] STEPHEN C. BENOIT,[b] APRIL D. STRADER,[b]
DEBORAH J. CLEGG,[b] SHAILAJA AKUNURU,[a] AND RANDY J. SEELEY[b]

[a]*Procter & Gamble Pharmaceuticals, Inc., Health Care Research Center,
Mason, Ohio 45040, USA*

[b]*Department of Psychiatry, University of Cincinnati, Cincinnati, Ohio 45267, USA*

ABSTRACT: Syndecan-3, expressed in the developing nervous system and adult
brain, alters feeding behavior through its interaction with the CNS melanocortin system, which provides critical tonic inhibition of both food intake and
body adipose stores. A variety of both *in vitro* and transgenic data supports the
hypothesis that syndecan-3 modulates melanocortin activity via syndecan-3
facilitation of agouti-related protein (AgRP), a competitive antagonist of α-
melanocyte-stimulating hormone (α-MSH) at the melanocortin-3 and -4
receptors. Consistent with this hypothesis, mice lacking syndecan-3, which
therefore would be predicted to have less effective AgRP, are more sensitive to
inhibition of food intake by the melanocortin agonist MTII. Additionally, we
took advantage of the fact that syndecan-3 facilitation of AgRP is limited to
when it is bound to the cell membrane. Pharmacologic inhibition of the enzyme
that cleaves syndecan-3 from the cell membrane leads to increased food intake
in fasted rats, which have elevated levels of AgRP. Furthermore, the shedding
process appears to be regulated under physiologic conditions, because a putative inhibitor of the shedding process, tissue inhibitor of metalloprotease-3
(TIMP-3), is increased by food deprivation. These observations contribute to
the hypothesis that syndecan-3 regulation of melanocortin signaling contributes to the normal control of energy balance. Collectively, the data suggest that
the modulation of melanocortin regulation of energy balance by syndecan-3 is
modulated by the action of a TIMP-3–sensitive metalloprotease.

KEYWORDS: syndecan; heparan sulfate proteoglycan; metalloprotease inhibitor

INTRODUCTION

More than 50% of Americans are considered overweight and a third are considered obese despite increased public awareness about obesity's deleterious effects,
which include hypertension, diabetes, and some forms of cancer.[1] Yet, body weight
is known to be a regulated variable. Most animals including humans can precisely

Address for correspondence: Dr. Ofer Reizes, Procter & Gamble Pharmaceuticals, Inc., Health
Care Research Center, Mason, Ohio 45040. Voice: 513-622-4462; fax: 513-622-1195.
reizes.o@pg.com

Ann. N.Y. Acad. Sci. 994: 66–73 (2003). © 2003 New York Academy of Sciences.

match caloric intake with caloric expenditure, resulting in relatively stable energy stores as adipose tissue.[2-4] Obesity or increased body fat occurs when the body's mechanisms for regulating weight (i.e., energy balance) are disrupted by increased feeding, reduced metabolism, reduced physical activity, or a combination of these processes.[5] An important and pressing question for the medical research community is, "what processes have been disrupted that might have brought about the rapid increase in the prevalence of obesity?" The central nervous system plays a critical role in maintaining energy balance. The hypothalamus, in particular, is pivotal as it integrates a number of peripheral signals that reflect the status of peripheral energy stores and ongoing utilization to influence both food intake and energy expenditure.[4,6]

Ingestive behavior is an important component of the body weight regulation equation. This behavior is influenced by a variety of short- and long-term regulatory signals that control meal initiation, termination, and size.[7,8] Myriad potential candidates for this regulation have been investigated, including peripheral hormones and hypothalamic neuropeptides, but much remains to be elucidated.[9,10]

Syndecans are a family of highly abundant cell surface heparan sulfate proteoglycans (HSPGs; proteins with covalently attached highly acidic sugar chains) that are unique in their ability to bind many extracellular peptides, such as hormones and growth factors.[11,12] They act as co-receptors by modulating and facilitating interactions of ligands with their receptors.[13,14] Importantly, most syndecans can also be released or shed from the cell surface in a highly regulated way, reducing the levels of cell surface syndecan. The shed extracellular domain of syndecan competes with the cell surface syndecan for hormone or growth factor binding and therefore may act as an inhibitor. These molecules have been studied extensively and shown to interact with multiple ligands *in vitro*; yet, there are only a few examples of their physiological function.[15]

Transgenic expression of cell surface syndecan-1 (which is insensitive to ectodomain shedding) in the hypothalamus leads to hyperphagia and obesity in mice.[16] These mice recapitulate the obese phenotype seen with hypothalamic melanocortin deficiency. In fact, *in vitro* cAMP analyses demonstrated that syndecans modulate the activity of the melanocortin-4 receptor (MC-4R) by potentiating the activity of agouti-related protein (AgRP), a competitive antagonist of this receptor.[16,17,18] Subsequent studies indicated that the endogenous neural syndecan, syndecan-3, is the physiological regulator again by facilitation of the antagonist properties of AgRP. In normal mice, cell surface syndecan-3 levels increase as a result of food deprivation, and syndecan-3 null mice have a blunted food intake response to food deprivation. Because syndecan-3 can be shed from the cell surface, we hypothesized that shedding of the syndecan-3 extracellular domain via a matrix metalloprotease contributes to the termination of feeding.

We have extended this original observation by demonstrating that modulation of syndecan-3–regulated feeding behavior occurs via a TIMP-3 inhibitable metalloprotease. First, we determined that TIMP-3 is induced by fasting in a manner similar to syndecan-3 and neuropeptides known to regulate feeding. Second, in fasted rats, a pan-selective metalloprotease inhibitor increases food intake to a greater extent than does the normal hyperphagic response to food deprivation. Finally, the administration of the melanocortin agonist MTII resulted in a greater reduction in food intake in syndecan-3 null mice than in wild-type mice. The data support a role for

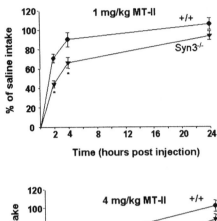

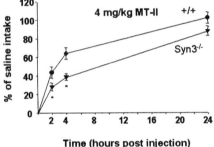

FIGURE 1. MTII reduces food intake to a greater extent in syndecan-3 null than in wild-type mice. Syndecan-3 null mice ($n = 14$) and wild-type mice ($n = 15$) were given intraperitoneal injections of saline (vehicle), 1 mg/kg MTII, and 4 mg/kg MTII. All mice received all three doses in random order. Each injection day was separated by one "wash-out" day to allow the mice to recover. On the injection days, food was removed 2.5 h prior to lights-off and weighed. Immediately before lights-off, mice were injected and food was replaced. Food weights were measured at 2, 4, and 24 h after injection. *$P < 0.05$.

syndecan-3 as a physiological modulator of feeding behavior and suggest that pro-teases, which regulate its presence at the cell surface, regulate syndecan and melano-cortin function to control food intake.

RESULTS

Syndecan-3 Null Mice Are More Sensitive to the Melanocortin Agonist MTII

Syncecan-3 alters feeding behavior by enhancing the action of AgRP to antago-nize the MC-3 and MC-4R. In the absence of syndecan-3, mice show reduced reflex hyperphagia after an overnight (16-h) fast, suggesting that AgRP's ability to reduce melanocortin tone is disrupted. We predicted that the action of melanocortins or their exogenous pharmacologic mimetics should be more pronounced in syndecan-3 null mice. Indeed, syndecan-3 null mice are more sensitive to the inhibitory action of MTII, a melanocortin agonist injected peripherally (FIG. 1).[19] In wild-type mice, 4 mg/kg MTII reduces food intake by 40% over a 4-h period; by contrast, 1 mg/kg

MTII reduces food intake to the same extent in syndecan-3 null mice. Thus, the syndecan-3 null mice are more than 4 times more sensitive to the action of MTII than are wild-type mice. At 4 mg/kg MTII, food intake in syndecan-3 null mice is reduced by 65% over the same 4-h period. These data support the hypothesis that syndecan-3 alters feeding behavior by modulating the interaction of AgRP with the MC-3 and MC-4R. That is, as in the *in vitro* results presented previously, melanocortin agonists, such as α–melanocyte-stimulating hormone (α-MSH) or MTII, are more potent in the absence of syndecan-3.

Pan-Selective Matrix Metalloprotease Inhibitor Increases Food Consumption in Hyperphagic Rats

Shedding of syndecan extracellular domains is a regulated process that is induced rapidly and is often detectable *in vitro* within 15–30 min.[20] Expression of cell surface syndecan-3 is induced by food deprivation and reduced rapidly on refeeding.[16] Therefore, if shedding of the extracellular domain of syndecan-3 terminates its activity, inhibition of shedding would maintain a state of hunger, resulting in greater food intake. Because the sheddase/protease is unknown, a pan-selective MMPI was used for these experiments. This compound inhibits MMP-1, -2, -3, -7, 8, -9, and -13 as well as several sheddases including tumor necrosis factor-α (TNF-α) convertase and aggrecanase. As a test of the hypothesis, rats were fasted overnight to induce expression of endogenous AgRP and cell surface syndecan-3. The shedding inhibitor was then introduced into the third ventricle (I3VT) of the brain, and food intake was measured (FIG. 2). FIGURE 2A represents 4-h food intake data using 10 nmol of MMPI, and FIGURE 2B shows cumulative intake data at 5 and 10 nmol of MMPI. As depicted in the figure, MMPI increased food intake only in fasted but not in ad libitum fed rats (FIG. 2A). Similar increases in food intake are seen with I3VT administration of the carboxyl terminal domain of AgRP.

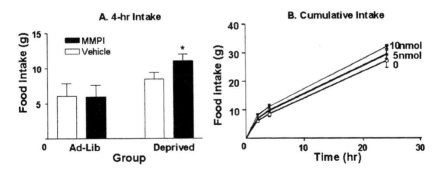

FIGURE 2. MMP inhibitor enhances food intake in fasted, but not in ad libitum fed rats. (**A**) Ad libitum fed rats (*left bars*) and fasted rats (*right bars*) were I3VT infused with either 10 nmol MMPI or vehicle (1% DMSO). Food intake was measured at 4 h. (**B**) Fasted rats were injected with 0, 5, or 10 nmol of the MMPI. Food intake was measured at 2, 4, and 24 h. Injections were given in a volume of 2 μL at the onset of dark. *$P < 0.05$ ($n = 6$ rats/condition).

TABLE 1. Nutritional status affects expression of hypothalamic genes

	Fed ad libitum	24-hour Food Deprivation	48-hour Food Deprivation
AgRP	1×	3×	4×
MCH1R	1×	2×	2×
TIMP-3	1×	1.8×	1.7×

NOTE: Expression normalized to GAPDH.

Tissue Inhibitor Metalloprotease-3 and Food Deprivation

The matrix metalloproteases consist of a large family including over 20 MMPs, 30 ADAMs (a disintegrin and metalloprotease), and 4 TIMPs (tissue inhibitor of metalloproteases).[21,22] Several of these have been studied during nervous system development, although little is known of their function in the adult nervous system. We have evaluated the differential expression of hypothalamic genes during periods of food deprivation using affymetrix chips to perform a genome-wide screen. Using the U34A rat affymetrix chip, 4,532 genes were expressed in the hypothalamus from a total of 8,799 genes present on the chip. Of the expressed genes, 120 showed statistically significant increases or decreases in expression in response to fasting. Importantly, one of the genes that increased in response to a 72-hour fast is the TIMP-3.[23,24] To confirm this observation, rats were fasted for 24 and 48 h, hypothalamus was collected, and RT-PCR was performed to evaluate expression of TIMP-3 as well as other peptides including AgRP, MCH, and the MCH receptor. Indeed, we confirmed the differential expression observed in the affymetrix experiment and determined that mRNA expression of TIMP-3 as well as AgRP, MCH, and the MCH receptor was increased during this fast (TABLE 1). The increased TIMP-3 expression in fasted rats supports the hypothesis that shedding modulates feeding behavior. These data provide information on the nature of the metalloprotease that results in shedding of the syndecan and potential modulation of feeding behavior. TIMP-3 is known to inhibit MMP-1, MMP-2, MMP-3, and MMP-9 as well as several ADAMs. Therefore, finding that TIMP-3 is regulated by energy balance offers clues into the specific MMP or ADAM that may cleave syndecan-3 and inhibit food intake.

CONCLUSIONS

Melanocortin receptors (MC-Rs) regulate multiple physiological processes including feeding behavior, energy metabolism, stress, pigmentation, and immune response. Feeding behavior and skin pigmentation are modulated by an interplay between the MC-R ligands, α-MSH, a stimulatory ligand, and its competitive antagonists, agouti-related protein (AgRP) or agouti-signaling protein (ASP), respectively. AgRP and its homologue, ASP, are 132 amino acid proteins that compete with α-MSH for binding to the MC-4R. These proteins have conserved C-terminal domains that are critical for their specific binding to the MC-Rs and basic N-terminal domains

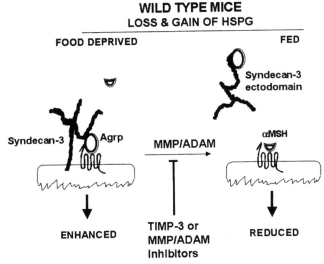

FIGURE 3. Model of mechanism for modulation of syndecan-3 at the cell surface and regulation of feeding behavior. Food deprivation leads to increased syndecan-3 expression at the cell surface, whereas feeding reduces cell surface expression. Cell surface expression of syndecan-3 would, in turn, modulate the activity of AgRP at the melanocortin-3 or melanocortin-4 receptors. During food deprivation, TIMP-3 is induced, resulting in inhibition of a sheddase or matrix metalloprotease, leading to an increase in cell surface expression of syndecan-3. Exogenous matrix metalloprotease inhibitor or increased TIMP-3 expression results in increased syndecan-3 expression and increased food intake.

that interact with the HS chains of syndecan. Whereas the C-terminal domain of the agouti proteins determines specificity of interaction with the MC-Rs, the N-terminal domain has unknown function, but it likely is involved in AgRP regulation.

Syndecan-3 appears to be a co-receptor for AgRP action at the melanocortin receptors. Our model indicates that regulation of syndecan-3 at the cell surface modulates the action of AgRP (FIG. 3). In the absence of syndecan-3, AgRP signaling is disrupted; therefore, melanocortin agonists are apparently more potent *in vitro* and *in vivo*. Indeed, the melanocortin agonist MTII is more potent in mice lacking syndecan-3, suggesting decreased antagonism by AgRP. Furthermore, a careful balance between an unidentified metalloprotease and an endogenous inhibitor, TIMP-3, likely regulates cell surface syndecan-3 expression. In fasted rats, a pan-selective metalloprotease inhibitor increases feeding behavior in a dose-dependent manner, and fasting increases the expression of the metalloprotease inhibitor TIMP-3.

Syndecan-3 interacts with the amino terminal domain of AgRP. This interaction likely modulates the activity of the AgRP with the melanocortin receptor. The C-terminal domain of AgRP interacts with the melanocortin receptor and inhibits the activity of α-MSH. Interestingly, hypothalamic injection of the C-terminal domain of AgRP leads to a prolonged increase in feeding behavior that can last 6 days.[25,26] We propose that the intact full-length AgRP would not cause the prolonged feeding effect. Because the amino terminal domain of AgRP interacts with syndecan, a full-length AgRP would be regulated, leading to a much shorter induction in feeding behavior. Therefore, modulation of cell surface expression of syndecan-3 is a regulator of feeding behavior, providing a novel target of regulating melanocortin signaling and energy homeostasis.

ACKNOWLEDGMENTS

We would like to thank Mike Janusz for the pan-selective matrix metalloprotease inhibitor used in the study and Kihmberly A. Wilmer for assistance with animal breeding. We would also like to thank Chuck McOsker and Russell Sheldon for input and critical reading of the manuscript.

REFERENCES

1. CROWLEY, V.E., G.S. YEO & S. O'RAHILLY. 2002. Obesity therapy: altering the energy intake-and-expenditure balance sheet. Nat. Rev. Drug Discovery 1: 276–286.
2. PORTE, D., JR., R.J. SEELEY, S.C. WOODS, et al. 1998. Obesity, diabetes and the central nervous system. Diabetologia 41: 863–881.
3. SCHWARTZ, M.W., S.C. WOODS, D. PORTE, JR., et al. 2000. Central nervous system control of food intake. Nature 404: 661–671.
4. SEELEY, R.J. & M.W. SCHWARTZ. 1999. Neuroendocrine regulation of food intake. Acta Paediatr. Suppl. 88: 58–61.
5. WEIGLE, D.S. 1994. Appetite and the regulation of body composition. FASEB J. 8: 302–310.
6. SCHWARTZ, M.W., M.F. DALLMAN & S.C. WOODS. 1995. Hypothalamic response to starvation: implications for the study of wasting disorders. Am. J. Physiol. 269: 949–957.
7. HALFORD, J.C. & J.E. BLUNDELL. 2000. Pharmacology of appetite suppression. Prog. Drug Res. 54: 25–58.
8. HAVEL, P.J. 2001. Peripheral signals conveying metabolic information to the brain: short-term and long-term regulation of food intake and energy homeostasis. Exp. Biol. Med. (Maywood) 226: 963–977.
9. FLIER, J.S. & E. MARATOS-FLIER. 1998. Obesity and the hypothalamus: novel peptides for new pathways. Cell 92: 437–440.
10. SPIEGELMAN, B.M. & J.S. FLIER. 2001. Obesity and the regulation of energy balance. Cell 104: 531–543.
11. BERNFIELD, M., M. GÖTTE, P.W. PARK, et al. 1999. Functions of cell surface heparan sulfate proteoglycans. Annu. Rev. Biochem. 68: 729–777.
12. PARK, P.W., O. REIZES & M. BERNFIELD. 2000. Cell surface heparan sulfate proteoglycans: selective regulators of ligand-receptor encounters. J. Biol. Chem. 275: 29923–29926.
13. GALLAGHER, J.T. & J.E. TURNBULL. 1992. Heparan sulfate in the binding and activation of basic fibroblast growth factor. Glycobiology 2: 523–528.
14. RAPRAEGER, A.C. 1995. In the clutches of proteoglycans: how does heparan sulfate regulate FGF binding? Chem. & Biol. 2: 645–649.
15. CONRAD, H.E. 1998. Heparin-Binding Proteins. Academic Press. San Diego.
16. REIZES, O., J. LINCECUM, Z. WANG, et al. 2001. Transgenic expression of syndecan-1 uncovers a physiological control of feeding behavior by syndecan-3. Cell 106: 105–116.
17. OLLMANN, M.M., B.D. WILSON, Y.-K. YANG, et al. 1997. Antagonism of central melanocortin receptors in vitro and in vivo by agouti-related protein. Science 278: 135–138.
18. SHUTTER, J.R., M. GRAHAM, A.C. KINSEY, et al. 1997. Hypothalamic expression of ART, a novel gene related to agouti, is up-regulated in obese and diabetic mutant mice. Genes & Dev. 11: 593–602.

19. FAN, W., B.A. BOSTON, R.A. KESTERSON, *et al.* 1997. Role for melanocortinergic neurons in feeding and the *agouti* obesity syndrome. Nature **385:** 165–168.
20. FITZGERALD, M.L., Z. WANG, P.W. PARK, *et al.* 2000. Shedding of syndecan-1 and -4 ectodomains is regulated by multiple signaling pathways and mediated by a TIMP-3-sensitive metalloproteinase. J. Cell Biol. **148:** 811–824.
21. BLOBEL, C.P. 2000. Remarkable roles of proteolysis on and beyond the cell surface. Curr. Opin. Cell Biol. **12:** 606–612.
22. SCHLONDORFF, J. & C.P. BLOBEL. 1999. Metalloprotease-disintegrins: modular proteins capable of promoting cell-cell interactions and triggering signals by protein-ectodomain shedding. J. Cell Sci. **112:** 3603–3617.
23. BREW, K., D. DINAKARPANDIAN & H. NAGASE. 2000. Tissue inhibitors of metalloproteinases: evolution, structure and function. Biochim. Biophys. Acta **1477:** 267–283.
24. NAGASE, H. & K. BREW. 2002. Engineering of tissue inhibitor of metalloproteinases mutants as potential therapeutics. Arthritis Res. **4:** S51–S61.
25. HAGAN, M.M., P.A. RUSHING, L.M. PRITCHARD, *et al.* 2000. Long-term orexigenic effects of AgRP-(83–132) involve mechanisms other than melanocortin receptor blockade. Am. J. Physiol. Regul. Integr. Comp. Physiol. **279:** 47–52.
26. HAGAN, M.M., S.C. BENOIT, P.A. RUSHING, *et al.* 2001. Immediate and prolonged patterns of agouti-related peptide-(83–132)-induced c-Fos activation in hypothalamic and extrahypothalamic sites. Endocrinology **142:** 1050–1056.

Functional Role, Structure, and Evolution of the Melanocortin-4 Receptor

HELGI B. SCHIÖTH,[a] MALIN C. LAGERSTRÖM,[a] HAJIME WATANOBE,[b] LOGI JONSSON,[c] ANNA VALERIA VERGONI,[d] ANETA RINGHOLM,[a] JON O. SKARPHEDINSSON,[c] GUDRUN V. SKULADOTTIR,[c] JANIS KLOVINS,[a] AND ROBERT FREDRIKSSON[a,e]

[a]Department of Neuroscience, Uppsala University, BMC, 751 24, Uppsala, Sweden

[b]Division of Internal Medicine, Clinical Research Center, International University of Health and Welfare, Tochigi, Japan

[c]Department of Physiology, University of Iceland, Reykjavik, Iceland

[d]Department of Biomedical Sciences, Section of Pharmacology, University of Modena, Modena, Italy

[e]Department of Animal Breeding and Genetics, Swedish University of Agricultural Sciences, 751 24, Uppsala, Sweden

ABSTRACT: The melanocortin (MC)-4 receptor participates in regulating body weight homeostasis. We demonstrated early that acute blockage of the MC-4 receptor increases food intake and relieves anorexic conditions in rats. Our recent studies show that 4-week chronic blockage of the MC-4 receptor leads to robust increases in food intake and development of obesity, whereas stimulation of the receptor leads to anorexia. Interestingly, the food conversion ratio was clearly increased by MC-4 receptor blockage, whereas it was decreased in agonist-treated rats in a transient manner. Chronic infusion of an agonist caused a transient increase in oxygen consumption. Our studies also show that the MC-4 receptor plays a role in luteinizing hormone and prolactin surges in female rats. The MC-4 receptor has a role in mediating the effects of leptin on these surges. The phylogenetic relation of the MC-4 receptor to other GPCRs in the human genome was determined. The three-dimensional structure of the protein was studied by construction of a high-affinity zinc binding site between the helices, using two histidine residues facing each other. We also cloned the MC-4 receptor from evolutionary important species and showed by chromosomal mapping a conserved synteny between humans and zebrafish. The MC-4 receptor has been remarkably conserved in structure and pharmacology for more than 400 million years, implying that the receptor participated in vital physiological functions early in vertebrate evolution.

KEYWORDS: melanocyte-stimulating hormone (MSH); agouti-related protein (AgRP); melanocortin receptor; food intake; reproduction; evolution

Address for correspondence: Helgi B. Schiöth, Department of Neuroscience, Uppsala University, BMC, Box 593, 751 24, Uppsala, Sweden. Voice: +46 18 471 41 60; fax: + 46 18 51 15 40.
helgis@bmc.uu.se

Ann. N.Y. Acad. Sci. 994: 74–83 (2003). © 2003 New York Academy of Sciences.

INTRODUCTION

The melanocortin (MC)-4 receptor was originally cloned by Gantz and colleagues.[1] It is expressed in the central nervous system and is found in multiple sites in almost every brain region, including the cortex, thalamus, hypothalamus, brain stem, and spinal cord.[1,2] The different natural melanocyte-stimulating hormone (MSH) peptides bind the MC-4 receptor with a distinct pharmacologic profile. β-MSH has the highest affinity to this receptor of the melanocortins, whereas γ-MSH peptides have very low affinity.[3] Agouti-related protein (AgRP) is a potent endogenous antagonist for the MC-4 receptor.[4] This receptor has an important role in the regulation of food intake and energy balance.[5–7] In this overview, we address the physiological role of the MC-4 receptor in the long-term regulation of food intake and its putative role in reproduction. We also give insights into the structural characterization of the MC-4 receptor protein and studies on the evolution of the MC-4 receptor genes.

ROLE OF THE MC-4 RECEPTOR IN LONG-TERM REGULATION OF FOOD INTAKE

Acute administration of the nonselective MC-3/MC-4 receptor agonists MTII, α-MSH, and β-MSH is very effective in reducing food intake.[8,9] Nonselective MC-3/MC-4 receptor antagonists, such as SHU9119, and selective MC-4 antagonists, such as HS014 and HS024, and AgRP, are also very effective in increasing food intake after acute injections.[10–13] We showed early that blockage of the MC-4 receptor relieved stress-induced anorexia.[14] This study provided the first indication that MC-4 receptor antagonists may be helpful in the treatment of anorexia. However, little was known about how long-term injections of MC receptor antagonists or agonist affect food intake, body weight, or energy expenditure. We therefore studied how chronic infusion of the MC-4 receptor antagonist for two weeks affected food intake. Studies showed that blockage of the MC-4 receptor produced a robust increase in

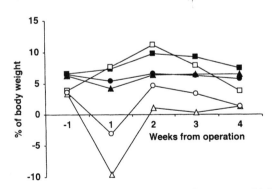

FIGURE 1. Daily food consumption after 4 weeks of treatment with MTII (▲), HS024 (■), and saline (●) and weekly weight changes after 4 weeks of treatment with MTII (△), HS024 (□), and saline (○). Results are redrawn from Ref. 19.

food intake and body weight without tachyphylaxis.[15–17] To further investigate the long-term effects, we performed a study with four weeks of chronic intracerebroventricular (i.c.v.) infusion of the selective MC-4 receptor antagonist HS024 and the nonselective MC receptor agonist MT-II and monitored food intake and body weight homeostasis.[18] Some of the main results of this study (redrawn and shown in FIG. 1) show that agonist (MT-II)-treated rats ate less and lost considerably more weight than did control rats during the first week of treatment. However, during the second and third weeks, they gained weight, and by the end of the treatment period their weight gain was similar to that of the control rats. Antagonist (HS024) treatment caused hyperphagia and development of obesity during the entire period. Large accumulations of fat and large increases in leptin levels were observed. Interestingly, the food conversion ratio, defined as body weight increase relative to the weight of ingested food, was clearly increased in the antagonist-treated rats, whereas it was decreased in the agonist-treated rats. Importantly, the effect on the food conversion ratio was transient, being greatest for both experimental groups during the first week and then attenuated to reach the control levels at the end of the study. We also investigated the effects of 8-day chronic i.c.v. infusion of both antagonist (HS024) and agonist (MTII) on oxygen consumption.[18] We observed a significant increase in oxygen consumption 2 days after the start of agonist infusion. This increase disappeared at day 4 and later in the study. No difference was observed in oxygen consumption after injection of the antagonist. The immediate effect of the MC receptor agonist on both food intake and metabolism was thus transient, even though weight loss was maintained. These studies provide insight into the effects that can be expected after therapeutic treatment with exogeneous MC-4 receptor agonists and antagonists.

ROLE OF THE MC-4 RECEPTOR IN REPRODUCTION

Obesity, anorexia, and general body weight fluctuations cause a variety of effects on the reproductive system. The neurobiologic mechanisms of the connections between body weight and the reproductive axis, however, are not well understood. Leptin could play a key role in connecting energy balance and reproduction through its action on both luteinizing hormone (LH) and prolactin (PRL) release. The melanocortin peptides and MC-4 receptors are suggested to participate in possible downstream effects of leptin on food intake.[20,21] Other studies show that α-MSH may be involved in PRL release, that α-MSH is released after a suckling stimulus, and that immunoneutralization of α-MSH leads to a clear decrease in a suckling-induced PRL release.[22] We showed that MC receptor antagonists decrease the magnitude of the LH and PRL surges in normally fed steroid-primed ovariectomized female rats.[23] Moreover, the antagonists also block the leptin stimulation of these hormonal surges in starved rats.[24] The results suggest that as with food intake, the effects of leptin on steroid-induced LH and PRL surges in female rats are mediated, at least in part, through the MC-4 receptor. AgRP also completely prevents both LH and PRL surges in female rats.[25] Injections of anti-AgRP serum in fasted rats, which are devoid of LH and PRL surges, also reinstated both surges.[25] These results indicated that endogenous AgRP may be involved in LH and PRL surges during starvation and suggest that the MC receptor has a role in the alleged surge center in the

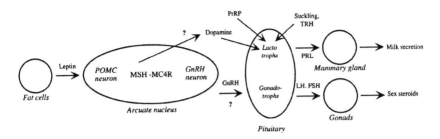

FIGURE 2. Putative role of the MC-4 receptor in integrating the hypothalamic regulation of food intake with the HPG axis. Redrawn from Ref. 23.

endocrine hypothalamus that entrains both LH and PRL surges. Injection of MTII and γ-MSH (MC-3 receptor agonist) on LH and PRL surges in starved steroid-primed ovariectomized female rats showed that the agonist significantly recovered the PRL surge, whereas γ-MSH was without effect.[26] This provides further evidence that it is indeed the MC-4 receptor, and not the MC-3 receptor, that mediates the pre-ovulatory PRL surge in female rats. This link between the MC-4 receptor and the GnRH-LH system was further supported by recent results showing that immortalized mouse GnRH neurons specifically express functional MC-4 receptors that are coupled to GnRH release.[27] FIGURE 2 shows a schematic representation of the putative role of the MC-4 receptor in integrating the HPG axis with the hypothalamic regulatory mechanism of food intake.[22]

STRUCTURAL ANALYSIS OF THE MC-4 RECEPTOR

Site-directed mutagenesis has played an important role in determining the putative interaction of a ligand to a single amino acid within G-protein–coupled receptors (GPCRs). Such interaction, however, may not always be informative about the orientation of the helix bundles. Moreover, results of inactivation (alanine scanning) studies, in most cases, cannot discriminate between specific ligand-receptor interactions and changes that cause unspecific conformational alterations that perturb ligand binding. This is particularly evident when the ligand is a flexible molecule such as a peptide or a protein. One way of studying the three-dimensional structures of GPCRs is construction of a high affinity zinc binding site between the helices, using two His-residues facing each other[28] (see FIG. 3). Coordination of the metal ion binding sites is well characterized in numerous X-ray structures of soluble proteins, and distances from the chelating atoms to the metal ion are known. Only a few mutagenesis studies on the human MC-4 receptor are available.[29,30] We constructed a molecular model of the human MC-4 receptor based on the crystal structure of rhodopsin.[31] We also inserted His residues into several positions in the receptor to generate metal ion binding sites. We were able to create artificial high-affinity zinc sites between specific TM regions. The metal ion induced increases in second-messenger levels predicting the interaction of the TM region when the agonist activates the receptor (Lagerström *et al.*, unpublished data). We also identified a natural metal

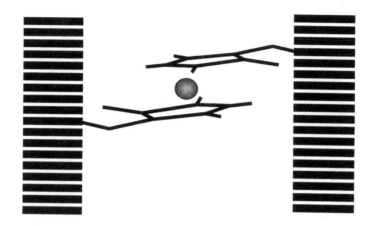

FIGURE 3. Schematic representation of how metal ion can form a highly defined bond between two histidine residues that can determine the orientation of helices within a receptor.

ion site in the receptor. Results were subsequently used to construct a refined model of the human MC-4 receptor protein. Reliable molecular models of the MC-4 receptor may be useful in designing and refining potent and selective substances binding to this receptor.

EVOLUTION OF THE MC-4 RECEPTOR

Very little is known, however, about the molecular mechanisms of central regulation of energy balance in nonmammalian species despite the fact that the growth rate of animals from several taxa, including fish, has important global ecological and economical impact. Novel genes arise by both individual and/or block duplication. The former often leaves the new gene in close proximity to the parent gene, whereas the latter contributes to duplication of whole or part of chromosomal regions. Large-scale duplications, including polyploidizations, are believed to be an important mechanism in the evolution of vertebrate genomes.[32] We have studied the relation of the MC-4 receptor to other GPCRs within the human genome by phylogenetic analysis also considering paralogous chromosomal regions and "fingerprint" sequences found in this family of genes. The receptors that are phylogenetically most closely related to the MC receptors are shown in FIGURE 4. To obtain further insight into the mechanism of gene family evolution and evolution of the central regulation of food intake, we are currently undertaking several studies on MC receptor family. The teleost zebrafish *(Danio rerio)* is an important experimental animal in genomic and developmental research. It has a compact genome with a similar number of genes as humans, but it is only one third in size, while still sharing large blocks of conserved synteny.[33] We recently reported the cloning, genome mapping, functional expression, pharmacology, and anatomic distribution of MC-4 receptor from

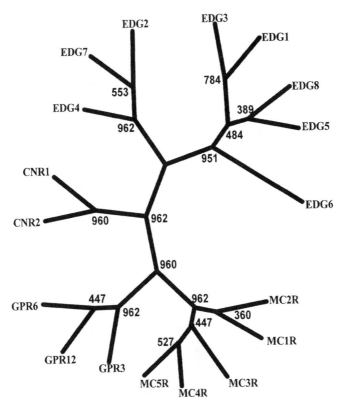

FIGURE 4. Phylogenetic tree showing the relation of MC receptors to other GPCRs in the human genome. Large numbers of human GPCRs were assembled, and the 13 most related are shown in the figure. Phylogenetic trees were built using the Win32 version 2.1 of the MEGA2 software, using the maximum parsimony method with the Max-Mini Branch and Bound Tree search algorithm. Stability of the nodes was assessed using the bootstrapping method with 1,000 replicas, and all sites were treated as informative. Trees were plotted in Tree-view. Numbers above the nodes indicate number of bootstrap replicates.

zebrafish $(z)^{34}$ (see FIG. 5). Chromosomal mapping of the receptor genes showed conserved synteny between regions containing zMC-4 and human MC-4 receptors. The zMC-4 receptor shares 71% overall amino acid identity with the respective human orthologues and over 90% in the TM regions that are believed to be most important for ligand binding. The zMC-4 receptor also showed pharmacologic properties remarkably similar to those of their human orthologue, with similar affinities and the same potency order, with α-MSH also causing a dose-dependent increase in intracellular cAMP levels. The zMC-4 receptor was expressed in the brain, eye, ovaries, and gastrointestinal tract. The studies represented the first characterization of MC receptors in a non-amniote species and indicate that the MC receptor subtypes arose very early in vertebrate evolution. Important pharmacologic and functional properties as well as gene structure and syntenic relationships have been

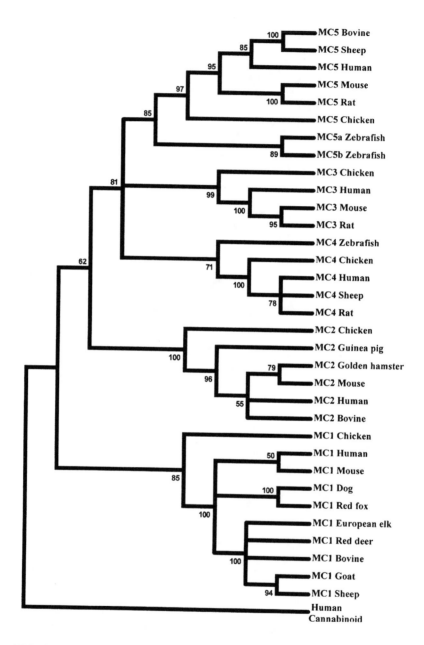

FIGURE 5. Phylogenetic tree of the family of MC receptors in different species. The tree with the full-length amino acid sequences of the MC receptors was created in a similar way as the tree in FIGURE 4, but with 100 replicas. Trees were plotted in Tree-explorer. The human cannabinoid 1 receptor sequence was used to root the tree. The numbers above the nodes indicate the number of bootstrap replicates.

highly conserved over a period of more than 400 million years, implying that these receptors participate in vital physiological functions.

CONCLUDING REMARKS

It is likely that drugs affecting the MC-4 receptor and many other hypothalamic receptors will be developed within the next few years. It is conceivable that MC receptor active substances and other substances acting on central GPCRs may find therapeutic value in the treatment of obesity. There is, however, a range of other diseases that are getting less attention, but for which such a compound may become useful. Pathologic conditions such as eating disorders, including geriatric anorexia, anorexia nervosa, bulimia, and neurologic food craving disorders, do not have effective pharmacotherapies. Many neuropeptides, such as MSH, AgRP, neuropeptide Y, leptin, and their receptors are, to a greater or lesser extent, involved in reproductive processes. Pharmacologic agents affecting obesity and body weight homeostasis may also have an inherent potential to affect the reproductive system. It is possible that substances intended to affect body weight may find a role in curing reproductive disorders. Such conditions may include malfunctioning of LH and PRL secretion or different conditions of infertility caused by anorexia, hypothalamic amenorrhea, and other types of idiopathic amenorrhea, all of which resist current medical treatment. It is also important to be aware that drugs affecting either body weight or reproduction may have an inherent capacity to cause adverse effects on the other systems. Further understanding of the role of the genes, their origin, structure, and function, will increase the likelihood of the development of new successful treatment of pathologic conditions.

ACKNOWLEDGMENTS

The studies were supported by the Swedish Research Council (VR, medicine), Swedish Society for Medical Research (SSMF), Åke Wibergs Stiftelse, Petrus och Augusta Hedlunds Stiftelse, and Melacure Therapeutics AB, Uppsala, Sweden.

REFERENCES

1. GANTZ, I., H. MIWA, Y. KONDA, *et al.* 1993. Molecular cloning, expression, and gene localization of a fourth melanocortin receptor. J. Biol. Chem. **268:** 15174–15179.
2. MOUNTJOY, K.G., M.T. MORTRUD, M.J. LOW, *et al.* 1994. Localization of the melanocortin-4 receptor (MC4-R) in neuroendocrine and autonomic control circuits in the brain. Mol. Endocrinol. **8:** 1298–1308.
3. SCHIÖTH, H.B., R. MUCENIECE & J.E.S. WIKBERG. 1996. Characterisation of melanocortin 4 receptor by radioligand binding analysis. Pharmacol. Toxicol. **79:** 161–165.
4. BARSH, G.S., M.M. OLLMANN, B.D. WILSON, *et al.* 1999. Molecular pharmacology of Agouti protein *in vitro* and *in vivo*. Ann. N.Y. Acad. Sci. **885:** 143–152.
5. HUSZAR, D., C.A. LYNCH, V. FAIRCHILD-HUNTRESS, *et al.* 1997. Targeted disruption of the melanocortin-4 receptor results in obesity in mice. Cell **88:** 131–141.
6. CONE, R.D. 1999. The central melanocortin system and energy homeostasis. Trends Endocrinol. Metab. **10:** 211–216.

7. SCHIOTH, H.B. 2001. The physiological role of melanocortin receptors. Vitam. Horm. **63:** 195–232.
8. FAN, W., B.A. BOSTON, R.A. KESTERSON, *et al.* 1997. Role of melanocortinergic neurons in feeding and the agouti obesity syndrome. Nature **385:** 165–168.
9. KASK, A., L. RAGO, J.E. WIKBERG, *et al.* 2000. Differential effects of melanocortin peptides on ingestive behaviour in rats: evidence against the involvement of MC(3) receptor in the regulation of food intake. Neurosci. Lett. **283:** 1–4.
10. KASK, A., L. RAGO, F. MUTULIS, *et al.* 1998. Selective antagonist for the melanocortin 4 receptor (HS014) increases food intake in free-feeding rats. Biochem. Biophys. Res. Commun. **245:** 90–93.
11. KASK, A., F. MUTULIS, R. MUCENIECE, *et al.* 1998 Discovery of a novel superpotent and selective melanocortin-4 receptor antagonist (HS024): evaluation *in vitro* and *in vivo*. Endocrinology **139:** 5006–5014.
12. ROSSI, M., M.S. KIM, D.G. MORGAN, *et al.* 1998. A C-terminal fragment of Agouti-related protein increases feeding and antagonizes the effect of alpha-melanocyte stimulating hormone in vivo. Endocrinology **139:** 4428–4431.
13. VERGONI, A.V., A. BERTOLINI, J.E. WIKBERG, *et al.* 1998. Differential influence of a selective melanocortin MC4 receptor antagonist on melanocortin-induced behavioral effects in rats. Eur. J. Pharmacol. **362:** 95–101.
14. VERGONI, A.V., A. BERTOLINI, J.E. WIKBERG, *et al.* 1999. Selective melanocortin MC4 receptor blockage reduces immobilization stress-induced anorexia in rats. Eur. J. Pharmacol. **369:** 11–15.
15. SKULADOTTIR, G.V., L. JONSSON, J.O. SKARPHEDINSSON, *et al.* 1999. Long term orexigenic effects of a novel selective MC4 receptor antagonist. Br. J. Pharmacol. **126:** 27–34.
16. KASK, A., R. PAHKLA, A. IRS, *et al.* 1999. Long-tem administration of MC4 receptor antagonist HS014 causes hyperphagia and obesity in rats. Neuroreport **10:** 707–711.
17. VERGONI, A.V., A. BERTOLINI, G. GUIDETTI, *et al.* 2000. Chronic melanocortin 4 receptor blockage causes obesity without influencing sexual behavior in male rats. J. Endocrinol. **166:** 419–426.
18. JONSSON, L., J.O. SKARPHEDINSSON, G.V. SKULADOTTIR, *et al.* 2002. Food conversion is transiently affected during 4-week chronic administration of melanocortin agonist and antagonist in rats. J. Endocrinol. **173:** 517–523.
19. JONSSON, L., J.O. SKARPHEDINSSON, G.V. SKULADOTTIR, *et al.* 2001. Melanocortin receptor agonist transiently increases oxygen consumption in rats. Neuroreport **12:** 3703–3708.
20. SEELEY, R.J., K.A. YAGALOFF, S.L. FISHER, *et al.* 1997. Melanocortin receptors in leptin effects. Nature **390:** 349.
21. KASK, A., L. RÄGO, J.E. WIKBERG & H.B. SCHIÖTH. 1998. Evidence for involvement of the melanocortin 4 receptor in the effects of leptin on food intake. Eur. J. Pharmacol. **360:** 15–19.
22. SCHIÖTH, H.B. & H. WATANOBE. 2002. Melanocortins and reproduction. Brain Res. Brain Res. Rev. **38:** 340–350.
23. WATANOBE, H., H.B. SCHIOTH, J.E. WIKBERG, *et al.* 1999. The melanocortin 4 receptor mediates leptin stimulation of luteinizing hormone and prolactin surges in steroid-primed ovariectomized rats. Biochem. Biophys. Res. Commun. **257:** 860–864.
24. WATANOBE, H., T. SUDA, J.E. WIKBERG, *et al.* 1999. Evidence that physiological levels of circulating leptin exert a stimulatory effect on luteinizing hormone and prolactin surges in rats. Biochem. Biophys. Res. Commun. **263:** 162–165.
25. SCHIÖTH, H.B., Y. KAKIZAKI, A. KOHSAKA, *et al.* 2001. Agouti-related peptide prevents steroid-induced luteinizing hormone and prolactin surges in female rats. Neuroreport **12:** 687–690.
26. WATANOBE, H., M. YONEDA, Y. KAKIZAKI, *et al.* 2001. Further evidence for a significant participation of the melanocortin 4 receptor in the preovulatory prolactin surge in the rat. Brain Res. Bull. **54:** 521–525.
27. KHONG, K., S.E. KURTZ, R.L. SYKES, *et al.* 2001. Expression of functional melanocortin-4 receptor in the hypothalamic gt1-1 cell line. Neuroendocrinology **74:** 193–201.

28. ELLING, C.E., K. THIRSTRUP, S.M. NIELSEN, *et al.* 1997. Metal-ion sites as structural and functional probes of helix-helix interactions in 7TM receptors. Ann. N.Y. Acad. Sci. **814:** 142–151.

29. YANG, Y.K., T.M. FONG, C.J. DICKINSON, *et al.* 2000. Molecular determinants of ligand binding to the human melanocortin-4 receptor. Biochemistry **39:** 14900–14911.

30. YANG, Y., M. CHEN, Y. LAI, *et al.* 2002. Molecular determinants of human melano-cortin-4 receptor responsible for antagonist SHU9119 selective activity. J. Biol. Chem. **277:** 20328–20335.

31. PALCZEWSKI, K., R. KUMASAKA, T. HORI, *et al.* 2000. Crystal structure of rhodopsin: a G protein-coupled receptor. Science **289:** 739–745.

32. LUNDIN, L.G. 1993. Evolution of the vertebrate genome as reflected in paralogous chromosomal regions in man and the house mouse. Genomics **16:** 1–19.

33. POSTLETHWAIT, J.H., Y.L. YAN, M.A. GATES, *et al.* 1998. Vertebrate genome evolution and the zebrafish gene map. Nature Genet. **18:** 345–349.

34. RINGHOLM, A., R. FREDRIKSSON, N. POLIAKOVA, *et al.* 2002. One melanocortin 4 and two melanocortin 5 receptors from zebrafish show remarkable conservation in structure and pharmacology. J. Neurochem. **82:** 6–18.

Anti-Inflammatory Potential of Melanocortin Receptor-Directed Drugs

A. SKOTTNER,[a,b] C. POST,[a,c] A. OCKLIND,[a] E. SEIFERT,[a] E. LIUTKEVICIUS,[d] R. MESKYS,[d] A. PILINKIENE,[d] G. BIZIULEVICIENE,[d] AND T. LUNDSTEDT[a,e]

[a]Melacure Therapeutics AB, Ulleråkersvägen 38, SE-756 43 Uppsala, Sweden

[b]Department of Biological Research on Dependence, Uppsala Biomedicinska Centrum (BMC), Husargatan 3, SE-751 23 Uppsala, Sweden

[c]Karolinska Institutet, Neurotec Department, Section of Experimental Geriatrics Novum, KFC, SE-141 86 Huddinge, Sweden

[d]Institute of Immunology, Moletu Pl 29, Vilnius 2021, Lithuania

[e]Department of Medicinal Chemistry, Uppsala Biomedicinska Centrum (BMC), Husargatan 3, SE-751 23 Uppsala, Sweden

ABSTRACT: Melanocortin receptor–based drug discovery is particularly active in the field of neuroendocrine systems and is mostly related to food intake and novel obesity therapies. The immunomodulatory and anti-inflammatory effects of nonpeptidic, low molecular weight compounds activating the melanocortin-1 receptor (MC1R) provide a new principle for treating various types of inflammation, such as dermal, joint, and gastrointestinal, probably by virtue of the effects acting through modulation of proinflammatory and anti-inflammatory cytokines. Several reports demonstrate that α-MSH, for example, has anti-inflammatory effects in different models. The aim of our study was to design, synthesize, and characterize compounds that bind to and activate the MC1R *in vitro*. The binding affinities are submicromolar to this receptor, and activation of the receptor (cAMP assay) varies from full agonists to partial agonists as well as antagonists. *In vivo*, the compounds exert prominent anti-inflammatory effects, with efficacy in the same range as that of dexamethasone, for example. The potential advantages of MC1R-based anti-inflammatory effects versus glucocorticosteroids, for example, are that the latter, albeit exerting prominent anti-inflammatory effects, also have many side effects that most likely will not characterize an MC1R-based anti-nflammatory drug.

KEYWORDS: MC1R; agonist; anti-inflammation; edema; low molecular weight compounds

INTRODUCTION

The melanocortin 1 receptor (MC1R) was first demonstrated in cell lines from melanoma tumors,[1] but it was later shown to be present in many different organs, both centrally and peripherally. Its presence on immunologically important cells,

Address for correspondence: Dr. A. Skottner, Melacure Therapeutics AB, Ulleråkersvägen 38, SE-756 43 Uppsala, Sweden. Voice: +46 18 567246; fax: +46 18 567201.
anna.skottner@melacure.com

Ann. N.Y. Acad. Sci. 994: 84–89 (2003). © 2003 New York Academy of Sciences.

such as macrophages, monocytes, dendritic cells, and neutrophils,[2–5] suggests that it is involved in the anti-inflammatory effects of α-MSH. The MC1 receptor expression has been shown to be up-regulated in melanocytes, keratinocytes, and monocytes in response to external stimuli, such as UVB-irradiation and retinoic acid,[6] as well as in response to α-MSH. Further support for the anti-inflammatory effects of α-MSH is found in results of studies in a model of contact hypersensitivity (CHS) and the induction of hapten-specific tolerance[7] in models measuring dinitrofluorobenzene (DNFB)-induced edema in mouse paws or ears. Significant anti-inflammatory actions were found on both variables by the administration of α-MSH.[8] The immunomodulatory action of α-MSH includes both immunostimulatory and immunosuppressive effects. Several studies have shown that α-MSH antagonizes the effects of proinflammatory cytokines such as interleukin (IL)-1α, IL-1β, IL-6, and tumor necrosis factor-α (TNF-α) and induces the production of the anti-inflammatory cytokine IL-10.[9,10]

There is also evidence that α-MSH acts as an antagonist to IL-1 and inhibits binding of IL-1β to IL-1 receptor type I. Rajora *et al.*[11] demonstrated a direct effect of α-MSH on brain cells to inhibit TNF-α after the induction of acute brain inflammation, probably by binding to MC1 receptors. α-MSH is produced in the skin, the highest concentration being found in the epidermis,[12] and it has the well-known effect in melanocytes of stimulating the formation of melanin in the skin. Available data demonstrate intense staining for MSHs in lesions of inflammatory skin diseases and tumors,[13] indicating local action of the endogenous peptide.

The goal of the current study was to evaluate low molecular weight (lmw) compounds with affinity for the MC1R for their anti-inflammatory potential and possible use in dermatitis.

MATERIAL AND METHODS

Cell Cultures

Mouse melanoma cells (B16-F1, European Collection of Cell Culture, Wiltshire, UK) endogenously expressing the MC1 receptor in abundance were maintained in equal volumes of Dulbecco's modified Eagle medium (DMEM) and Ham's F-12, supplemented with 10% fetal bovine serum (FBS) and 1% antibiotic antimycotic solution (AAS) (all from Invitrogen). Three cell lines of chinese hamster ovary cells stably transfected with mouse MC3, human MC4, and mouse MC5 receptors (Euroscreen, Belgium) were maintained in Ham's F-12 supplemented with 10% FBS, 1% AAS, and 400 μg/mL geneticin (Invitrogen).

Ligand Binding

Cells in 96-well plates were used one day postseeding. The cells were washed, incubated with a fixed concentration of tracer and different concentrations of the test compound for 2 h at room temperature, and washed. Radioactivity was counted in a Microbeta scintillation gamma-counter (Wallac Oy, Finland), and the K_i values were determined by GraphPad Prism v 3.0 (GraphPad Software, San Diego, California, USA).

cAMP Stimulation

Cells were cultured overnight in 96-well plates, washed, and stimulated with 1, 3, 10, and 30 μM test compound or 1 pM to 10 μM α-MSH (Neosystems) for 20 min at 37°C. The concentration of intracellular cAMP was determined on cell lysates by a commercial enzyme immunoassay (Amersham Biosciences, Uppsala, Sweden).

Animals

Female BALB/c mice (weighing 20–22 g), obtained from the Laboratory Animal Center, Institute of Immunology, Vilnius, Lithuania, were used for all experiments. Animals were maintained in an environment of controlled temperature (22 ± 2°C). Food and water were provided *ad libitum*. All procedures were carried out in accordance with guidelines of the European Union and were approved by the Ethics Committee on Animal Experimentation, Institute of Immunology, Vilnius, Lithuania.

Contact Hypersensitivity Sensitivity Experiment

Mice were sensitized by application of 30 μL of 0.5% 2, 4-dinitrofluorobenzene (DNFB, Sigma) on the shaved abdomen. After 5–7 days, animals were challenged with an injection of 10 μL of 0.3% DNFB into the paw. In the topical experiments, mice were challenged on the fourth day after the first sensitization in the same manner. The DNFB solution was prepared just prior to application of the acetone-olive oil mixture (4:1). Unchallenged mice paws served as the control. Twenty-four hours after the last challenge, differences in paw weight were evaluated as an indicator of the contact sensitivity response (paw edema). At the same time, blood samples were collected by heart puncture for evaluation of blood variables (hematology), IL-10, TNF-α, and corticosterone concentration. Test compounds (EL-13, α-MSH, and other lmw compounds) were given 2 h prior to sensitization and then daily for 4–6 days. Experiments were also performed in which the compounds were administered 2 h before the last challenge. When the single compound was tested, doses ranged from 0.05 to 6 mg/kg for EL13 or lmw compounds, 0.5 mg/kg for α-MSH. The reference groups were treated intraperitoneally with prednisolone (Sigma, 20 mg/kg in sterile water). Nontreated mice with edema and healthy mice were included as reference groups.

Mice were topically administered with 200 μg, 20 μg, and 2 μg of EL13 dissolved in 50 μL of vehicle (2% ethanolamine, 20% ethanol, and 78% sterile water) 2 h before sensitization. After sensitization, the aforementioned doses of the EL-13 were topically administered daily for 4 days. On the fourth day, mice were challenged with DNFB, as just mentioned. Dexamethasone (2 μg/50 μL of vehicle), α-MSH (20 μg/50 μL of vehicle), and vehicle alone were administered topically as a control. Hematologic analysis of mouse blood was performed using MASCOT™ Multi-species Hematology System, HEMAVET® 850 (CDC Technologies Inc., Oxford, CT, USA). EDTA-2Na, 20 μL of 7.5%, prepared in 0.9% NaCl, were added per milliliter of blood before analysis. Concentrations of cytokines IL-10 and TNF-α were quantitatively determined in mice sera using ELISA kits for IL-10 and TNF-α assays (BIOTRAK, Amersham Pharmacia Biotech), according to the manufacturer's instructions. The corticosterone concentration was quantitatively determined in mouse serum samples by a competitive enzyme immunoassay kit (Immunodiagnostic Systems Ltd, UK) according to the manufacturer's instructions.

Multivariate Evaluation

The multivariate method used was Principal Component Analysis (PCA).[14] Briefly, PCA is a method of making projections of the observations (in this case, different tested compounds) down to a plane of a lower dimension to give an overview of large data tables. This overview is very useful in showing groups, trends, and outliers.

RESULTS AND DISCUSSION

Affinities (K_i values) of the test compound for the MC receptors were obtained from competition binding assays. The affinity of the test compound for the mouse (m)MC1, mMC3, human (h)MC4, and mMC5 receptor was 0.4, 16, 18, and 5 µM, respectively. For comparison, the affinity of α-MSH was 0.004, 0.007, 0.065, and 0.02 µM for the respective mMC1, mMC3, hMC4, and mMC5 receptors. The functional activity of the test compound was tested as cAMP-stimulating capacity. Results were expressed as % maximum stimulation of the maximum α-MSH-induced response. Stimulation was 30, 6, 118, and 6% for the mMC1, mMC3, hMC4, and mMC5 receptors, respectively.

Intraperitoneal administration of EL13 starting before sensitization resulted in significant and dose-dependent effects in the mouse model of DNFB-induced paw edema. Similar results were obtained in mice models of ear edema when the compound was given prior to challenge, that is, before the second dose of DNFB, a more therapeutic approach. A concomitant effect was observed on hematologic variables, and decreases in the same range as that with prednisolone were seen on circulating levels of egg white blood cells. Several lmw compounds have been tested in the model for their effects on paw edema, and the results vary from significantly effective to noneffective, but are clearly correlated to the affinity to and functional effect on MC1R expressing cells *in vitro*. By using multivariate analysis of all of our lmw compounds tested, we could clearly show that several compounds exerted anti-inflammatory effects in the mouse model and that some of these were as effective as the reference compounds, that is, prednisolone or dexamethasone. The data set analyzed contained healthy controls, edema controls, reference compounds, and lmw compounds. FIGURE 1 shows a score plot of a PCA model of the tested compounds. This model explains 65% of the variance by two components. Different groups of compounds are encircled. The edema and the healthy controls are lying as antipoles in the plot. Members of the dexamethasone group are situated mainly within the healthy group. The prednisolone group is situated in the middle, indicating a partial recovery. Our own lmw compounds are showing a diversity of anti-inflammatory effects. One very interesting group of compounds is situated adjacent to the healthy group, thus showing promising results. Significant results were further obtained when one of our most promising compounds, EL13, was administered topically on the paw at three different doses (FIG. 2). The edema was significantly decreased with all doses, and the effect was in the same range as that seen with

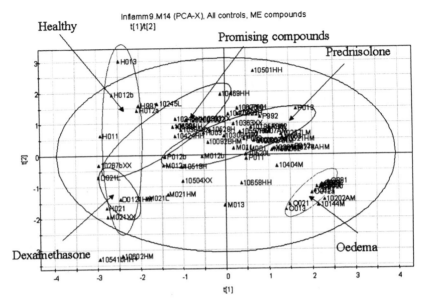

FIGURE 1. Score plot of a PCA model containing all tested compounds; controls, references, and our own lmw compounds.

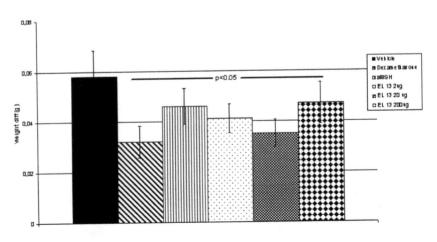

FIGURE 2. Effects on paw edema in mice after topical administration of different doses of EL13.

dexamethasone given topically, and which served as a reference, in the same experiment. Serum concentrations of selected cytokines measured after treatment indicated an increase in the anti-inflammatory cytokine IL-10 (vehicle 52.9, dexamethasone 157.3, α-MSH 259.5, and EL13 92.1–192.0 pg/mL, respectively) and a slight decrease in the proinflammatory cytokine TNF-α (vehicle 159.9, dexamethasone 111.0, α-MSH 200.0, and EL13 74.2–222.2 pg/mL, respectively), which approached the levels observed with α-MSH.

In conclusion, we have identified several lmw compounds with interesting anti-inflammatory properties, and one of the most interesting compounds, EL13, exhibiting exciting *in vivo* effects is now being further evaluated for use in dermatitis.

REFERENCES

1. CHHAJLANI, V. & J.E.S. WIKBERG. 1992. Molecular cloning and expression of the human melanocyte stimulating hormone receptor cDNA. FEBS Lett. **309:** 417–420.
2. STAR, R.A., N. RAJORA, *et al.* 1995. Evidence of autocrine modulation of macrophage nitric oxide synthase by α-melanocyte stimulating hormone. Proc. Natl. Acad. Sci. USA **92:** 8016–8020.
3. HARTMEYER, M., T. SCHOLZEN, *et al.* 1997. Human dermal microvascular endothelial cells express the melanocortin receptor type i and produce levels of IL-8 upon stimulation with alpha-melanocyte stimulating hormone. J. Immunol. **159:** 1930–1937.
4. CATANIA, A., A. RAJORA, *et al.* 1996. The neuropeptide αMSH has specific receptors on neutrophils and reduces chemotaxis *in vitro*. Peptides **17:** 675–679.
5. BECHER, E., K. MAHNKE, *et al.* 1999. Human peripheral blood derived dendritic cells express functional melanocortin receptor MC1R. Ann. N.Y. Acad. Sci. **885:** 188–195.
6. LUGER, T.A., T. SCHOLZEN, *et al.* 1997. The role of α-melanocyte stimulating hormone in cutaneous biology. J. Invest. Dermatol. Symp. Proc. **2:** 87–93.
7. GRABBE, S., R.S. BHARDWAJ, *et al.* 1996. α-Melanocyte-stimulating-hormone induces hapten-specific tolerance in mice. J. Immunol. **156:** 473–478.
8. HILTZ, M.E. & J.M. LIPTON. 1990. Alpha-MSH peptides inhibit acute inflammation and contact sensitivity. Peptides **11:** 979–982.
9. CATANIA, A. & J.W. LIPTON. 1993. α-Melanocyte stimulating hormone in the modulation of host reactions. Endocr. Rev. **14:** 564–576.
10. WIKBERG, J.E.S., R. MUCENIECE, *et al.* 2000. New aspects on the melanocortins and their receptors. Pharmacol. Res. **42:** 393–420.
11. RAJORA, N., G. BOCCOLI, *et al.* 1997. Alpha-MSH modulates experimental inflammatory bowel disease. Peptides **18:** 381–385.
12. THODY, A.J., K. RIDLEY, *et al.* 1983. MSH peptides are present in mammalian skin. Peptides **4:** 813–816.
13. SLOMINSKI, A., J. WORTSMAN, *et al.* 1993. Detection of proopiomelanocortin-derived antigens in normal and pathologic human skin. J. Lab. Clin. Med. **122:** 658–666.
14. ALBINSSON, A., G. ANDERSSON, *et al.* 1993. Use of neuroendocrine effects in discrimination between CNS-active drugs in the rat. Drug Dev. Res. **29:** 227–234.

MT-II Induces Penile Erection via Brain and Spinal Mechanisms

HUNTER WESSELLS,[a] VICTOR J. HRUBY,[b] JOHN HACKETT,[c] GUOXIA HAN,[b] PREETI BALSE-SRINIVASAN,[b] AND TODD W. VANDERAH[c]

[a]Department of Urology, University of Washington School of Medicine, Seattle, Washington 98195, USA

[b]Department of Chemistry, University of Arizona, Tucson, Arizona 85721, USA

[c]Departments of Anesthesiology and Pharmacology, University of Arizona College of Medicine, Tucson, Arizona 85719, USA

ABSTRACT: α-Melanocyte–stimulating hormone induces penile erection via melanocortin (MC) receptors in areas surrounding the third ventricle, but spinal and peripheral mechanisms have not been demonstrated. We used pharmacological strategies to localize the site of the proerectile action of the melanocortin receptor agonist MT-II. We administered MT-II intracerebroventribularly (i.c.v.), intrathecally (i.th.), and intravenously (i.v.) and scored penile erection and yawning for 90 min in awake male rats. In some animals i.c.v. or i.th. SHU-9119 was injected 10 minutes before i.c.v. and i.th. MT-II to confirm the MC receptor action of the agonist and to distinguish spinal from supraspinal effects. To exclude a site of action in the penis, we recorded intracorporal pressure responses to intracavernosal injection of MT-II in the anesthetized rat. MT-II induced penile erections in a dose-dependent fashion, with optimal response at 1 µg for both i.c.v. and i.th. routes. Supraspinal MT-II-induced erections were completely suppressed by 1 µg SHU-9119 i.c.v. Yawning was observed with i.c.v. and i.v. MT-II, whereas spinal injection did not produce this behavior. SHU-9119 blocked the erectile responses to i.th. MT-II when injected i.th. but not i.c.v. Intracavernosal MT-II neither increased intracorporal pressure nor augmented neurostimulated erectile responses. The lumbosacral spinal cord contains MC receptors that can initiate penile erection independent of higher centers. We confirmed the supraspinal proerectile action of MT-II. These results provide insight into the central melanocortinergic pathways that mediate penile erection and may allow for more efficacious melanotropin-based therapy for erectile dysfunction.

KEYWORDS: MSH; SHU-9119; spinal cord; rats; sex behavior

INTRODUCTION

Neuropeptides have been implicated in the control of penile erection and sex behavior in animals. Although oxytocin has been identified as a key transmitter in erec-

Address for correspondence: Hunter Wessells, Department of Urology, University of Washington School of Medicine, Seattle, WA 98195. Voice: 206-731-3205; fax: 206-731-4709.
wessells@u.washington.edu

Ann. N.Y. Acad. Sci. 994: 90–95 (2003). © 2003 New York Academy of Sciences.

tile signaling,[1] neurokinins, proopiomelanocortin (POMC), and melanocortins also play a role.[2–4] α-Melanocyte–stimulating hormone (α-MSH), adrenocorticotrophic hormone (ACTH), and melanocortin (MC) receptors mediate penile erection, sexual motivation, and grooming behavior.[5–8] Administration of α-MSH and ACTH into the central nervous system of rats induces penile erection because of action in the paraventricular nucleus, dorsomedial nucleus, and other hypothalamic nuclei.[9] Dopamine and oxytocin have proerectile activity at the level of the lumbosacral spinal cord,[10–12] and POMC and MC_4 receptor mRNA has been detected in the spinal cord,[13–15] leading us to hypothesize that melanocortins may induce erection via a spinal site of action. To test this hypothesis, we devised complementary pharmacological strategies to determine whether MT-II, a superpotent agonist of MC_3 and MC_4 receptors, induces erection via brain, spinal cord, or peripheral mechanisms.[16]

METHODS

We determined the behavioral effects of MT-II in awake animals by intracerebroventricular (i.c.v.), intrathecal (i.th.), and intravenous (i.v.) administration of vehicle and MT-II using methods described by Bernabe *et al.*[17] Adult male Long Evans rats were housed in a reverse light-dark cycle, and behavioral studies were performed between 1300 and 1800 h Each rat was used for a single injection of drug or placebo, and the Institutional Animal Care and Use Committees of our institutions approved all experimental procedures. Intracerebroventricular experiments used stereotactically placed 22-gauge stainless steel single guide cannulae in the lateral ventricle (1.5 mm lateral to the midline and 1.0 mm caudal to bregma to a depth of 3.5 mm). Intrathecal injections used an 8-cm length of PE-10 tubing inserted through an incision made in the atlantooccipital membrane and advanced to the level of the lumbar enlargement of the rat.[18] For intravenous injections, an acute tail vein method was used. Vehicle or MT-II (0.3, 1, and 3 μg per rat i.c.v. and i.th.; 0.3, 1, and 3 mg/kg i.v.) was injected in awake freely moving rats after a 5-min period of exploration; penile erection, yawning, and genital grooming episodes were recorded for 90 min. An erection was scored when the rat stood up on the extremity of its hindlimbs, bent its head toward the genital region, and the erect penis was visible.[17] At the end of each experiment, we verified proper drug delivery by injection of toluidine blue dye i.c.v. or i.th. under ether anesthesia, after which the rat was killed and decapitated. Brains and or spinal cords were removed, grossly sectioned, and inspected for the presence of blue dye. The initial stereotactic i.c.v. study of MT-II–induced erection contained 19 to 26 animals per group. Subsequent experiments used four to six rats per group to minimize the number of animals.

Once we established an optimal erectogenic dose of i.c.v. and i.th. MT-II, simultaneously placed i.c.v. cannulae and i.th. catheters were used to determine whether the action of i.th. MT-II was truly spinal versus supraspinal. The nonselective MC-R antagonist SHU-9119, in concentrations known to inhibit MT-II–induced erection,[16] was injected either i.c.v. or i.th. 10 min before i.th. injection of MT-II. Penile erection and yawning were scored as described below.

To determine whether local mechanism in the penis could contribute to the erectogenic effect of MT-II, in anesthetized animals, we recorded changes in intracorporal pressure (ICP) and blood pressure (BP) in response to intracavernosal injection of vehicle (0.9% NaCl) and MT-II (0.1, 1, 10, and 100 µg in 50 µL). This entailed cannulating the carotid artery, attaching it to a pressure transducer, exposing the penis and inserting a 25-gauge needle into the right corpus cavernosum.[17]

All data are expressed as mean ± SEM for each group. Differences in mean number of erections for the best dose of MT-II i.c.v., i.th., and i.v. were compared with one-way ANOVA. Mean maximum ICP and ICP/BP in response to vehicle and escalating doses of MT-II were compared with repeated measures one-way ANOVA. The level of significance was chosen as $P = 0.05$.

RESULTS

Penile erections were induced in a dose-dependent fashion with supraspinal, spinal, and parenteral administration of MT-II. The mean number of erections (± SEM) with vehicle was 0.8 ± 0.19, 0 ± 0, and 0.5 ± 0.5 for i.c.v., i.th., and i.v., respectively. Both i.c.v. and i.th. administration produced an optimal erectile effect at 1 µg per rat. Intravenous MT-II had a peak erectile effect at 1 mg/kg. The number of erections obtained at the best dose of MT-II (FIG. 1) was significantly greater than vehicle for each of the routes of administration ($P < 0.05$). The mean number of erections with intravenous MT-II was significantly lower than with i.c.v. or i.th. administration (2.5 vs. 4.5 and 8.33, respectively, $P < 0.05$). No statistically significant increase in the mean number of yawning episodes was observed with i.th. administration, whereas i.c.v. and i.v. MT-II induced significant yawning behavior (data not shown). Intracerebroventricular SHU-9119, at doses sufficient to inhibit supraspinal

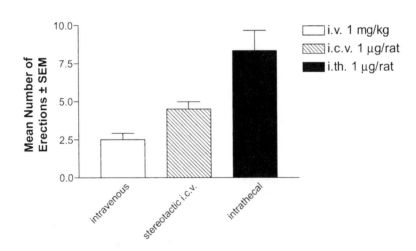

Route of administration

FIGURE 1. Number of penile erections with optimal doses of MT-II for each route of delivery. Values are presented as mean ± SEM ($n = 4-26$ animals/group). Each dose was significantly different than vehicle. Adapted from Wessells *et al.*[16]

MT-II induced erections, had no effect when administered 10 min before intrathecal MT-II (FIG. 2A). In contrast, the same dose of SHU-9119, when administered i.th., almost completely abolished penile erection induced by spinal MT-II (FIG. 2B). Intracavernosal MT-II had no effect on ICP when compared with vehicle in the anesthetized rat (TABLE 1). Intracavernosal MT-II had no significant effect on BP or heart rate in this model (data not shown).

DISCUSSION

α-MSH and ACTH have long been known to induce erection and yawning when injected into the lateral ventricle or specific hypothalamic nuclei. We used MT-II and SHU-9119 to demonstrate that melanocortins induce penile erection via a novel ac-

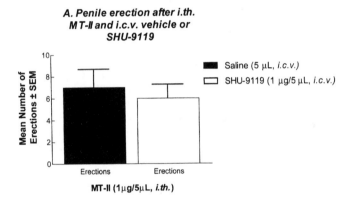

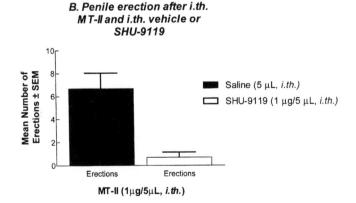

FIGURE 2. Number of penile erections with injection of vehicle and SHU-9119 i.c.v. or i.th. 10 min before 1 µg i.th. MT-II. Values are presented as mean ± SEM (*n* = 6 animals/ group). *$P < 0.05$ vs. vehicle. Adapted from Wessells *et al.*[16]

TABLE 1. Maximum intracorporal pressure (mean ± SEM) in response to vehicle or MT-II

MTII (μg)	0	0.1	1	10	100
Maximum ICP	22.32	24.59	27.13	25.96	29.64
SEM		4.88	4.38	3.63	4.75	6.41

NOTE: No significant difference was found between vehicle and MT-II (ANOVA repeated measures). (Adapted from Wessells et al.[16])

tion in the spinal cord. Intrathecal MT-II induced nearly twice the number of erections as did i.c.v. administration, without a significant increase in yawning. The greater number of erections with i.th. compared with i.c.v. administration in our awake rat model may be the result of phasic increases in ICP, and thus more easily visible erectile events.[10] Melanocortins previously have not been thought to induce erection via an action in the spinal cord. We saw an effect with i.th. MT-II, whereas Mizusawa et al.[12] did not with α-MSH. This discrepancy may relate to differences in the potency and stability of the two peptides.[19,20] Intrathecal SHU-9119 blocked the action of i.th. MT-II, whereas i.c.v. injection did not, demonstrating that the proerectile effect of i.th. MT-II is not the result of drug reaching the hypothalamus from the i.th. space. We did not determine whether suprapinal melanocortinergic pathways depend on spinal MC-R activation to induce erections.

A recent study localized MC-R to the corpus cavernosum,[15] but we did not find any proerectile effects of MT-II either via a direct action on ICP or by modulating neurogenic signals (data not shown).[16] This is in agreement with a study in which MT-II had no effect on electrical field stimulation–evoked relaxation of rabbit corpus cavernosal strips in vitro.[21]

The finding of a spinal site of action for MT-II combined with other evidence of spinal melanocortin signaling[22] leads us to propose a complete proerectile melanocortinergic pathway. We previously showed that MT-II initiates penile erection in men with erectile dysfunction.[23–25] Receptor specificity or other methods that allow targeting of spinal MC-R may reduce side effects that may limit the clinical applications of MT-II and thus enhance efficacy of melanocortin based therapy for erectile dysfunction.

ACKNOWLEDGMENTS

This work was supported by Grant IRG 110T from the American Cancer Society (H.W.) and the U.S. Public Health Service DK 17420 (V.J.H.).

REFERENCES

1. ARGIOLAS, A., M.R. MELIS & G.L. GESSA. 1988. Yawning and penile erection: central dopamine-oxytocin-adrenocorticotropin connection. Ann. N. Y. Acad. Sci. **525:** 330–337.
2. DORNAN, W.A. et al. 1993. Site-specific effects of intracerebral injections of three neurokinins (neurokinin A, neurokinin K, and neurokinin gamma) on the expression of male rat sexual behavior. Physiol. Behav. **54:** 249–258.

3. HUGHES, A.M., B.J. EVERITT & J. HERBERT. 1988. The effects of simultaneous or separate infusions of some pro-opiomelanocortin-derived peptides (beta-endorphin, melanocyte stimulating hormone, and corticotrophin-like intermediate polypeptide) and their acetylated derivatives upon sexual and ingestive behaviour of male rats. Neuroscience **27:** 689–698.

4. MEDINA, F. *et al.* 1998. The inter-relationship between gonadal steroids and POMC peptides, beta-endorphin and alpha-MSH, in the control of sexual behavior in the female rat. Peptides **19:** 1309–1316.

5. MEYERSON, B.J. & B. BOHUS. 1976. Effect of ACTH 4-10 on copulatory behavior and on the response for socio-sexual motivation in the female rat. Pharmacol. Biochem. Behav. **5:** 539–545.

6. VERGONI, A.V. *et al.* 1998. Differential influence of a selective melanocortin MC4 receptor antagonist (HS014) on melanocortin-induced behavioral effects in rats. Eur. J. Pharmacol. **362:** 95–101.

7. WIKBERG, J.E. *et al.* 2000. New aspects on the melanocortins and their receptors. Pharmacol. Res. **42:** 393–420.

8. MACNEIL, D.J. *et al.* 2002. The role of melanocortins in body weight regulation: opportunities for the treatment of obesity. Eur. J. Pharmacol. **440:** 141–157.

9. ARGIOLAS, A. *et al.* 2000. ACTH- and alpha-MSH–induced grooming, stretching, yawning and penile erection in male rats: site of action in the brain and role of melanocortin receptors. Brain Res. Bull. **51:** 425–431.

10. GIULIANO, F. *et al.* 2001. Spinal proerectile effect of oxytocin in anesthetized rats. Am. J. Physiol. Regul. Integr. Comp. Physiol. **280:** R1870–R1877.

11. GIULIANO, F. *et al.* 2002. Pro-erectile effect of systemic apomorphine: existence of a spinal site of action. J. Urol. **167:** 402–406.

12. MIZUSAWA, H., P. HEDLUND & K.E. ANDERSSON. 2002. alpha-Melanocyte stimulating hormone and oxytocin induced penile erections, and intracavernous pressure increases in the rat. J. Urol. **167:** 757–760.

13. VAN DER KRAAN, M. *et al.* 1999. Expression of melanocortin receptors and pro-opiomelanocortin in the rat spinal cord in relation to neurotrophic effects of melanocortins. Brain Res. Mol. Brain Res. **63:** 276–286.

14. ELMQUIST, J.K. 2001. Hypothalamic pathways underlying the endocrine, autonomic, and behavioral effects of leptin. Physiol. Behav. **74:** 703–708.

15. VAN DER PLOEG, L.H. *et al.* 2002. A role for the melanocortin 4 receptor in sexual function. Proc. Natl. Acad. Sci. USA **99:** 11381–11386.

16. WESSELLS, H. *et al.* 2003. Ac-Nle-c[Asp-His-DPhe-Arg-Trp-Lys]-NH$_2$ Induces penile erection via brain and spinal melanocortin receptors. Neuroscience **118:** 755–762.

17. BERNABE, J. *et al.* 1999. Intracavernous pressure during erection in rats: an integrative approach based on telemetric recording. Am. J. Physiol. **276:** R441–R449.

18. YAKSH, T.L. & T.A. RUDY. 1976. Analgesia mediated by a direct spinal action of narcotics. Science **192:** 1357–1358.

19. UGWO, S.O. *et al.* 1994. Kinetics of degradation of a cyclic lactam analog of alpha-melanotropin (MT-II) in aqueous solution. Int. J. Pharmaceut. **102:** 193–199.

20. HADLEY, M.E. *et al.* 1996. Melanocortin receptors: identification and characterization by melanotropic peptide agonists and antagonists. Pigm. Cell Res. **9:** 213–234.

21. VEMULAPALLI, R.K., S. KUROWSKI, B. SALISBURY, *et al.* 2001. Activation of central melanocortin receptors by MT-II increases cavernosal pressure in rabbits by the neuronal release of NO. Br. J. Pharmacol. **134:** 1705–1710.

22. ELMQUIST, J.K. 2001. Hypothalamic pathways underlying the endocrine, autonomic, and behavioral effects of leptin. Int. J. Obes. Relat. Metab. Disord. **25** (Suppl. 5): S78–S82.

23. DORR, R.T. *et al.* 1996. Evaluation of melanotan-II, a superpotent cyclic melanotropic peptide in a pilot phase-I clinical study. Life Sci. **58:** 1777–1784.

24. WESSELLS, H. *et al.* 1998. Synthetic melanotropic peptide initiates erections in men with psychogenic erectile dysfunction: double-blind, placebo controlled crossover study. J. Urol. **160:** 389–393.

25. WESSELLS, H. *et al.* 2000. Effect of an alpha-melanocyte stimulating hormone analog on penile erection and sexual desire in men with organic erectile dysfunction. Urology **56:** 641–646.

PT-141: A Melanocortin Agonist for the Treatment of Sexual Dysfunction

P.B. MOLINOFF, A.M. SHADIACK, D. EARLE, L.E. DIAMOND, AND C.Y. QUON

Palatin Technologies, Inc., Cranbury, New Jersey 08512, USA

ABSTRACT: PT-141, a synthetic peptide analogue of α-MSH, is an agonist at melanocortin receptors including the MC3R and MC4R, which are expressed primarily in the central nervous system. Administration of PT-141 to rats and nonhuman primates results in penile erections. Systemic administration of PT-141 to rats activates neurons in the hypothalamus as shown by an increase in c-Fos immunoreactivity. Neurons in the same region of the central nervous system take up pseudorabies virus injected into the corpus cavernosum of the rat penis. Administration of PT-141 to normal men and to patients with erectile dysfunction resulted in a rapid dose-dependent increase in erectile activity. The results suggest that PT-141 holds promise as a new treatment for sexual dysfunction.

KEYWORDS: melanocortin receptors; melanocortin agonist; erectile dysfunction

INTRODUCTION

Melanocortins, including α-MSH, have multiple physiologic effects mediated by both peripheral and central mechanisms.[1] Central administration of melanocortin agonists to experimental animals has been reported to cause stretching, yawning, and penile erection.[2] The subcutaneous administration of a synthetic melanocortin agonist, called MT-II, to patients with erectile dysfunction resulted in clinically significant erections.[3]

PT-141 is a cyclic heptapeptide (FIG. 1). For our studies, PT-141 was chemically synthesized and used in *in vivo* and *in vitro* studies to define its mechanism of action and its effect on erectile function.

RESULTS AND DISCUSSION

PT-141 inhibits the binding of ^{125}I-NDP-α-MSH to human recombinant MC4R and MC3R expressed on transfected cells (FIG. 1). Its affinity for the MC4R is approximately 10 nM. PT-141 also has high affinity for the MC1R. Studies of the effect of PT-141 on cAMP accumulation in cells expressing the human recombinant MC4R

Address for correspondence: A.M. Shadiack, Palatin Technologies, Inc., 4-C Cedar Brook Drive, Cedar Brook Corporate Center, Cranbury, NJ 08512. Voice: 609-495-2207; fax: 609-495-2202.

ashadiack@palatin.com

Ann. N.Y. Acad. Sci. 994: 96–102 (2003). © 2003 New York Academy of Sciences.

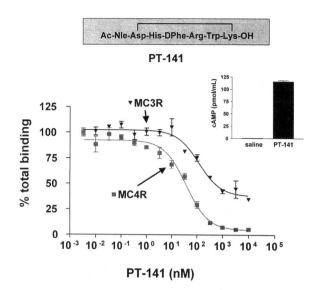

PT-141 (nM)

FIGURE 1. *In vitro* pharmacology of PT-141. PT-141 is a synthetic, cyclic, hepta-peptide with a structure as shown. Affinities for various MCRs were measured using membranes from HEK-293 cells containing cloned human melanocortin receptors. Varying concentrations of PT-141 were incubated with membranes from HEK-293 cells and [^{125}I]-NDP-α-MSH (NEN). The concentration of radioligand was 0.4 nM for studies with MC3R and 0.2 nM for studies with MC4R. Results are expressed as percentage of total binding. PT-141 shows a higher affinity for MC4R than it does for MC3R. Functional evaluation of PT-141 at the MC4R was performed by measuring the accumulation of intracellular cAMP in HEK-293 cells expressing MC4R. cAMP levels were measured by EIA (Amersham, Piscataway, NJ) in cell lysates of transfected HEK-293 cells. PT-141 shows agonist activity at the MC4R.

show that it is an agonist at this receptor (FIG. 1). Administration of PT-141 to rats via a variety of routes of administration results in a significant increase in the number of erections. The mean number of erections per rat is increased (FIG. 2A) as is the percentage of rats showing at least one erection during the 30-min observation period (FIG. 2B). Direct injection of PT-141 into the lateral ventricle was erectogenic at doses 100- to 1000-fold lower than those required when PT-141 was given systemically (data not shown).

The MC3R and MC4R are primarily localized in the hypothalamus and have been shown to be involved in the regulation of food intake and energy homeostasis.[4,5] c-Fos immunoreactivity was used to assess the degree of neuronal activation and the anatomical distribution of activated neurons after the systemic administration of an efficacious dose of PT-141. After administration of PT-141 to adult rats, c-Fos–like immunoreactivity was seen in the paraventricular nucleus and supraoptic nucleus of the hypothalamus (FIG. 3).

Experiments were conducted to attempt to demonstrate an association between nerve endings in the penis, potentially associated with the erectogenic activity of PT-141, and cell bodies in the hypothalamus known to express the MC3R and MC4R and to be activated after the systemic administration of PT-141. For these experi-

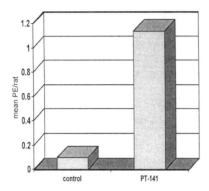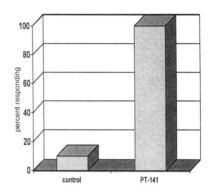

FIGURE 2. PT-141 initiates penile erections in rats. Male Sprague-Dawley rats weighing 200–250 g were dosed with 50 μg/kg of PT-141 intranasally (IN) using a micropipette to deliver 25 μL of solution into one nostril. Immediately after dosing, rats were placed into individual Lucite cages for behavioral observation. Rats were observed for 30 min, and the number of penile erections was recorded. PT-141 at this dose induced an increase in the mean number of penile erections observed for each rat (*left graph*) and caused 100% of the rats to show at least one erectile event during the 30-min observation period (*right graph*).

ments, pseudorabies virus was injected directly into the corpus carvernosum of the penis. After waiting 4 to 5 days for transsynaptic retrograde transport of the virus, the brain and spinal cord were harvested for histology. Pseudorabies virus was detected in the paraventricular nucleus of the hypothalamus using antiserum raised to the virus (FIG. 4). Taken together, these findings suggest that PT-141 exerts its effect by activation of MC3/4R in the hypothalamus.

Based on the erectogenic affects of PT-141 observed in experimental animals, studies were initiated to access the safety, tolerability, pharmacokinetics, and pharmacodynamic effects of PT-141 administered intranasally to normal healthy volunteers. The pharmacodynamic effects were assessed using the RigiScan Plus System and were performed in the absence of visual sexual stimulation. The RigiScan Plus System measures penile rigidity and tumescence. PT-141 was administered intranasally at doses ranging from 4 to 20 mg to 24 healthy male subjects (18–45 years of age) in a randomized double-blind placebo-controlled study. Pharmacokinetic analysis was performed on blood drawn at various time points, and erectile responses were monitored for 6 h after each dose.

All doses were well tolerated and no significant changes in blood pressure, heart rate, or electrocardiographic parameters were identified in response to the administration of PT-141. No serious treatment-related adverse events were noted. No subject experienced priapism and all enrolled subjects completed the study. Pharmacokinetic analysis demonstrated that the T_{max} averaged 30 min (FIG. 5), the C_{max} increased in a dose-dependent manner, and plasma levels declined with a mean terminal elimination half-life of approximately 2 h. The relatively short half-life should result in a reduced risk of drug interactions. The pharmacokinetic properties are consistent with the development of PT-141 as an on-demand treatment for erectile dysfunction.

PVN

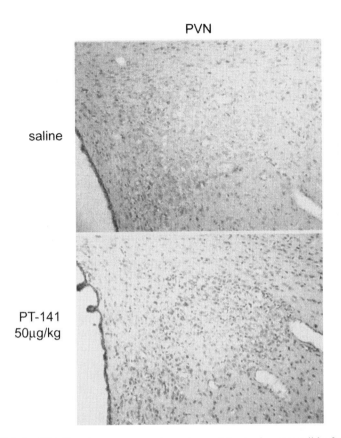

saline

PT-141
50µg/kg

FIGURE 3. PT-141 activates neurons in regions of the brain responsible for sexual function. Male Sprague-Dawley rats were dosed with an efficacious dose (50 µg/kg, IN) of PT-141. Two hours after dosing, they were perfusion fixed and their brains were removed. The tissue was cryosectioned and processed for immunohistochemistry using a goat anti–c-Fos antiserum. The appearance of the product of this immediate-early gene (darkly colored cell nuclei) in the paraventricular nucleus (PVN) of the hypothalamus is a measure of activity in these cells.

The pharmacodynamic effects of PT-141 were assessed using the RigiScan Plus System. Placebo-treated subjects showed very little erectile activity (FIG. 6). The mean cumulative duration of erectile activity, as defined by the duration of erections with greater than 60% base rigidity, was approximately 140 min at the 20 mg dose (FIG. 6) group compared with 22 min in the placebo-treated group. The time to onset of erectile activity ranged from 34 to 63 min after PT-141 administration and showed an inverse correlation with the dose level of PT-141.

After documenting the erectogenic effect of PT-141 in normal subjects, a study was initiated to assess the effect of PT-141 administration in patients with erectile dysfunction. For this study, patients were enrolled who had a clinical diagnosis of erectile dysfunction of at least 6 months' duration and who reported a positive re-

Control   PRV

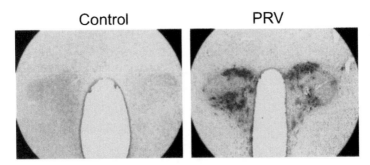

FIGURE 4. Transsynaptic retrograde labeling of neurons in the paraventricular nucleus of the hypothalamus. Pseudorabies virus (PRV) is an effective retrograde transsynaptic marker of autonomic neurons. The PRV was injected into the corpus cavernosum of the penis of the rat, and after a 4- to 6-day survival period the brains were harvested. The brains were processed for immunohistochemistry using an anti-PRV antiserum. The neurons containing the virus in the paraventricular nucleus are cells that, through a chain of synapses, project to the penis. These cells are in the same region of the hypothalamus that is activated by an efficacious dose of PT-141.

20 mg PT-141 Dosed IN

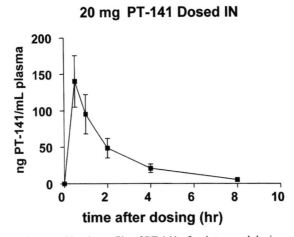

FIGURE 5. Pharmacokinetic profile of PT-141 after intranasal dosing with 20 mg. PT-141 was administered intranasally at a dose of 10 mg into each nostril. Blood samples were taken at various times after dosing, and the concentrations of PT-141 were determined. The C_{max} occurred at approximately 30 min after dosing, and the T1/2 was approximately 120 min.

sponse to the administration of a PDE-5 inhibitor. For this study, 24 patients with mild to moderate erectile dysfunction were enrolled. Each patient was given placebo and two doses of PT-141 by intranasal administration. This study was a double-blind, placebo-controlled, crossover design. For this study, subjects were shown two erotic videos, each of 30-min duration. After administration of PT-141, erectile activity was temporally associated with the viewing of the erotic films (FIG. 7). For this study, like the phase 1 trial described above, efficacy was evaluated using the Rigi-

PT-141: IN Dosing in Normal Subjects

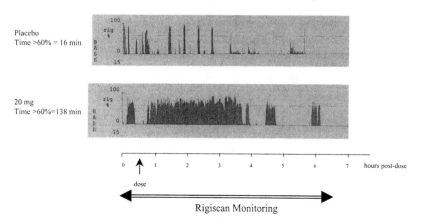

FIGURE 6. Pharmacodynamic effect of PT-141 in normal volunteers. Base rigidity was measured after administration of a placebo or of 20 mg PT-141. For this study, the RigiScan was applied at time T = −30 and continued for 6.5 h, and drug or placebo was administered at time T = 0. This study was performed in the absence of visual sexual stimulation. A representative subject had an erection with greater than 60% base rigidity for 16 min after administration of the placebo. A representative subject treated with 20 mg of PT-141 had an erection with greater than 60% base rigidity for 138 min.

Scan Plus System. PT-141 increased the duration of erectile activity by a factor of three when compared with placebo treatment (FIG. 7). The average time to onset of the first erection was approximately 30 min. The results indicated that PT-141 had a positive effect on both the primary end point of the study, duration of base rigidity greater than 60%, versus placebo ($P < 0.001$) and a more rigorous secondary end point which was duration of base rigidity greater than 80% versus placebo ($P < 0.001$). These results compare favorably with published results of RigiScan studies conducted with a selective PDE-5 inhibitor.[6]

CONCLUSION

PT-141 is erectogenic in both experimental animals and humans. Its intranasal administration to both normal subjects and patients with erectile dysfunction shows the drug to be well tolerated. The administration of PT-141 was not associated with significant cardiovascular liabilities. A significant dose-dependent increase in the duration and extent of penile rigidity, and tumescence was observed in both normal volunteers and in patients with erectile dysfunction. The data suggest a potential role for PT-141 for the treatment of male erectile dysfunction.

PT-141: IN Dosing with VSS in ED Patients

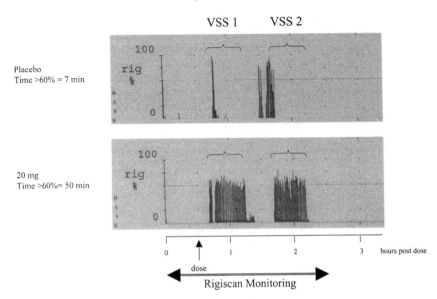

FIGURE 7. Pharmacodynamic effect of PT-141 in patients with mild/moderate erectile dysfunction. Base rigidity was measured after administration of a placebo or of 20 mg of PT-141. For this study, the RigiScan was applied from 30 min before drug or placebo administration until 2.5 h after administration. This study was conducted in the presence of visual sexual stimulation. Visual sexual stimulation took place between 10 and 40 min and 70 and 100 min after drug administration. Each patient received placebo and PT-141 with 3 to 10 days between treatments. In a representative patient, placebo administration resulted in an erection with greater than 60% base rigidity for 7 min; 20 mg of PT-141 resulted in an erection with greater than 60% base rigidity for 50 min, with most erectile activity occurring during periods of visual sexual stimulation.

REFERENCES

1. O'DONAHUE, T.L. & D.M. DORSA. 1982. The opiomelanotropinergic neuronal and endocrine systems. Peptides **3:** 353–395.
2. ARGIOLAS, A. & M.R. MELIS. 1995. Neuromodulation of penile erection. *In* Progress in Neurobiology. G.A. Kerkut & J.W. Phillis, Eds.: 235–255. Pergamon Press. London.
3. WESSELLS, H., K. FUCIARELLI, J. HANSEN, *et al.* 1998. Synthetic melanotropic peptide initiates erections in men with psychogenic erectile dysfunction: double-blind, placebo controlled crossover study. J. Urol. **160:** 389–393.
4. GANTZ, I., H. MIWA, Y. KONDA, *et al.* 1993. Molecular cloning expression and gene localization of a fourth melanocortin receptor. J. Biol. Chem. **268:** 15174–15179.
5. LINDBLOM, J., H.B. SCHIOTH, A. LARSSON, *et al.* 1998. Autoradiographic discrimination of melanocortin receptors indicate that the MC3 subtype dominates in the medial rat brain. Brain Res. **810:** 161–171.
6. KLOTZ, T., A. SACHSE, A. HEIDRICH, *et al.* 2001. Vardenafil increases penile rigidity and tumescence in erectile dysfunction patients: a RigiScan and pharmacokinetic study. World J. Urol. **19:** 32–39.

Body Weight Regulation by Selective MC4 Receptor Agonists and Antagonists

ALAN C. FOSTER, MARGARET JOPPA, STACY MARKISON, KATHY R. GOGAS, BETH A. FLECK, BRIAN J. MURPHY, MEIRA WOLFF, MARY J. CISMOWSKI, NICHOLAS LING, VAL S. GOODFELLOW, CHEN CHEN, JOHN SAUNDERS, AND PAUL J. CONLON

Neurocrine Biosciences Inc., San Diego, California 92121, USA

ABSTRACT: There has been great interest in melanocortin (MC) receptors as targets for the design of novel therapeutics to treat disorders of body weight, such as obesity and cachexia. Both genetic and pharmacological evidence points toward central MC4 receptors as the principal target. Using highly selective peptide tools for the MC4 receptor, which have become available recently, we have provided pharmacological confirmation that central MC4 receptors are the prime mediators of the anorexic and orexigenic effects reported for melanocortin agonists and antagonists, respectively. The current progress with receptor-selective small molecule agonist and antagonist drugs should enable the therapeutic potential of MC4 receptor activation and inhibition to be assessed in the clinic in the near future.

KEYWORDS: melanocortin receptor; MC4; obesity; cachexia; selective agonist; selective antagonist

INTRODUCTION

A key role for the melanocortin (MC) system in the regulation of body weight is apparent from several lines of investigation and is currently the subject of intensive research. The system comprises three endogenous peptide agonists, α-melanocyte stimulating hormone (α-MSH), γ-MSH, and adrenocorticotropic hormone (ACTH), all derived from the precursor proopiomelanocortin (POMC) peptide, an endogenous peptide antagonist, agouti-related peptide (AgRP), and five subtypes of G protein–coupled receptors (GPCRs), MC1, 2, 3, 4, and 5, which are the effectors on the cell surface (for recent reviews, see MacNeil *et al.*[1] and Marks *et al.*[2]). This system has numerous functions in mammalian and nonmammalian species, including the regulation of coat and skin color, secretion of glucocorticoids, and sexual function in addition to the proposed role in body weight regulation. The key components of this system which contribute to body weight regulation are the agonist α-MSH, the endogenous antagonist, AgRP, which act through the MC4 and MC3 receptor sub-

Address for correspondence: Alan C. Foster, Neurocrine Biosciences Inc., 10555 Science Center Drive, San Diego, CA 92121. Voice: 858-658-7760; fax: 858-658-7696.
afoster@neurocrine.com

Ann. N.Y. Acad. Sci. 994: 103–110 (2003). © 2003 New York Academy of Sciences.

types. Most evidence points toward the hypothalamus as the primary site where these players interact to influence body weight and composition.

MELANOCORTIN RECEPTORS AND BODY WEIGHT REGULATION

Genetic Studies

Studies with genetically modified animals have provided a consistent line of evidence that the MC4 receptor, and also to a lesser extent the MC3 receptor, regulates body size and composition. Naturally occurring mouse mutants with a phenotype of a yellow coat color and obesity have overexpression of the agouti protein, which acts as an antagonist at all melanocortin receptor subtypes. The coat color change is caused by the agouti protein's antagonism of MC1 receptors in the skin, and the obesity is caused by centrally expressed agouti protein's antagonism of MC3/4 receptors.[3] Following these studies, AgRP was identified as a naturally occurring centrally expressed homologue for agouti, which acts as an endogenous MC3 and MC4 receptor antagonist.[4,5] Overexpression of AgRP in the mouse produces an obese phenotype.[5] An understanding of the relative roles of MC3 and MC4 receptors has come from studies of knockout mice. The MC4 knockout has a late-onset obesity which is characterized by increased food intake and reduced metabolism.[6] The fact that the homozygotes have greater body weights than the heterozygotes, which, in turn, are heavier than the wild type, indicates that body size is dictated by gene dosage for the MC4 receptor. In contrast, MC3 receptor knockout mice have a mild body weight increase but are characterized by an increase in fat versus lean body mass.[7,8] POMC knockout mice are obese, and their body weight can be reduced by treatment with an α-MSH analogue.[9] Interestingly, the AgRP knockout has normal body weight.[10]

Studies of mutations within the melanocortin system in human subjects have provided a compelling case for the importance of this system in human energy balance. Rare mutations in the POMC gene which eliminate α-MSH secretion produce a severe obesity, ACTH insufficiency, and red hair.[11] Mutations in the MC4 receptor gene which lead to receptor inactivation have been documented in obesity, and it has been estimated that up to 4% of severely obese subjects have MC4 receptor mutations,[12,13] making this the most common genetic aberration in the obese population.

Pharmacological Studies

Pharmacological experiments in animals have provided clear evidence that manipulation of the melanocortin system in the brain leads to acute and sometimes long-term changes in aspects of energy balance. Acute administration of α-MSH or synthetic analogues with improved agonist potency, such as MTII, produce robust inhibition of food intake when injected centrally in rodents and rhesus monkeys.[14,15] Central infusion of MTII via osmotic minipump to rats over a 4-week period led to an initial decrease in food intake which returned to control levels by 2 weeks but a reduction in body weight throughout the infusion period, suggesting that chronic activation of central melanocortin receptors can produce a sustained body weight reduction.[16]

Pharmacological studies also have indicated that antagonism of the central melanocortin system provides a robust increase in food intake and can counteract the anorexia and body weight decrease characteristic of cachexia. The endogenous antagonist, AgRP, is a potent stimulator of food intake when administered intracerebroventricularly to rodents and primates, and a single injection can maintain an orexigenic effect for 24 h or more.[17] The synthetic antagonist, SHU9119, also potentiates acute feeding when given centrally,[14] and infusion of this peptide centrally for 7 days results in a sustained elevation of food intake and an increase in body weight.[18] These peptides also have been shown to be effective in animal models of cachexia. Cachexia occurs in numerous human illnesses, including cancer, AIDS, and renal, pulmonary, liver, and cardiac failure. It is characterized by a pathological decrease in body weight, particularly a loss of skeletal muscle, accompanied by anorexia, and often it is ascribed as a major cause of morbidity and mortality in severe illness.[19] Elevated cytokines have been implicated in many forms of cachexia, particularly IL-1β, IL-6, and TNF-α.[20] Injection of lipopolysaccharide in rats and mice produces anorexia through cytokine-mediated mechanisms, and both AgRP and SHU9119 administered centrally have been shown to improve food intake under these conditions.[21,22] Centrally applied AgRP and SHU9119 also are able to ameliorate the anorexia and body weight decrease which results from tumor growth in mice and rats.[22,23] A specific role for the MC4 receptor subtype has been proposed on the basis that MC4 knockout mice are resistant to the anorexia induced by lipopolysaccharide and tumor growth.[22] The peptide antagonist, HS014, also has been reported to reduce stress-induced anorexia and maintain body weight in the rat when centrally infused for 5 days.[24]

A SPECIFIC ROLE FOR MC4 RECEPTORS IN BODY WEIGHT REGULATION?

Consequently, the view emerging from these human and animal genetic studies and pharmacological experiments is that MC4 is the dominant melanocortin receptor subtype involved in body weight regulation, but that the MC3 receptor also plays a role. This has encouraged efforts in several pharmaceutical companies, including Neurocrine, to develop MC4-selective agonists and antagonists as potentially valuable therapeutic agents for the treatment of disorders of energy balance such as obesity and cachexia. However, knockout, overexpression, and mutation of genes may cause a phenotype with developmental abnormalities caused by the genetic alteration which are unrelated to the specific gene manipulation, and conclusions from such studies are strongest when they can be supported by pharmacological data. As described above, the use of agonists and antagonists in general has provided strong support for the role of the melanocortin system in body weight regulation; however, because of a lack of truly receptor subtype selective ligands, pharmacological studies have shed little light on the relative roles of the MC4 and MC3 receptors. Studies of nonselective ligands in receptor knockout animals can be helpful in this respect. Thus, the anorexic effects of MTII were greatly diminished in MC4 knockout mice.[25] However, pharmacological agents with good selectivity for the MC4 and MC3 receptors would help to clarify their respective roles in body weight regulation.

TABLE 1. Affinity of melanocortin agonists and antagonists for receptor subtypes

Agonists	EC_{50} (nM)				Antagonists	K_i (nM)			
	hMC1	hMC3	hMC4	hMC5		hMC1	hMC3	hMC4	hMC5
α-MSH	9	40	29	97	AgRP (83-132)	>100	0.6	0.3	>40
ACTH	1	8	12	>100	SHU9119	0.2	1	0.1	1.5
γ-MSH	40	6	300	600	HS014	108	54	3	694
MTII	0.2	0.7	0.5	3	HS028	60*	74	1	211*
c(1-6)suc-HfRWK-NH₂	4	1000	1.5	>1000	MBP10	8900	150	0.5	540

NOTE: Values are EC_{50}'s for stimulation of cAMP in cell lines expressing human melanocortin receptors or K_i's for inhibition of ^{125}I-NDP-MSH binding to human melanocortin receptors expressed in cell lines. Values are from Refs. 1 and 28 and unpublished data. *Partial agonist activity observed in functional assays.

Selective Pharmacological Tools

Until recently, compounds with both high affinity and selectivity for the melanocortin receptor subtypes were not available. TABLE 1 summarizes a selection of the compounds reported in the literature. So that pharmacological results can be interpreted with confidence, particularly for *in vivo* experiments, agonists or antagonists should have at least 100-fold (preferably 1000-fold) selectivity for the receptor of interest. Furthermore, careful studies of the compounds activity in a functional assay should be conducted to determine the degree of efficacy at the receptor, because experiments with partial agonists will provide results which are difficult to interpret. Ideally, an antagonist should have an affinity equal to, or greater than, the natural agonist, show no intrinsic efficacy at concentrations 100-fold (preferably 1000-fold) greater than its affinity for the receptor, and exhibit functional antagonism with a Kb value consistent with its receptor binding affinity. An agonist ideally also should have an affinity equal to or greater than the natural agonist, and an equivalent maximal response in a functional assay, by a readout of function which is not too distant from the receptor activation event (e.g., cAMP measurement).

As can be seen from TABLE 1, the most widely used peptide tools for studying melanocortin receptor function, α-MSH, MTII, AgRP, and SHU9119, show inadequate selectivity between receptor subtypes, particularly MC3 and MC4. The peptide analogue, Ro27-3225, has been reported as an agonist with greater than 100-fold selectivity for MC4 over MC3 in functional assays, but equipotent as an agonist at MC1 receptors. However, this compound is apparently less selective for MC4 in radioligand binding assays, and its level of efficacy at MC4 versus α-MSH was not reported.[26] Ro27-3225 suppresses acute food intake in rats which was taken as evidence that the MC4 receptor subtype mediates these effects[26]; however, the poorly characterized nature of its receptor selectivity and efficacy makes this conclusion less certain. Similarly, peptide analogues with MC4 antagonist activity (Ro27-4680; HS014; HS028) have been used in *in vivo* experiments to support the idea that the MC4 receptor mediates the orexigenic effects of melanocortin receptor antagonists.[26–28] Unfortunately, these compounds do not show sufficient selectivity

(TABLE 1) to make this conclusion definitive. Recently, Bednarek *et al.*[29,30] described two cyclic peptides which are highly selective tools for the MC4 receptor subtype. The agonist c(1-6)suc-HfRWK-NH$_2$ has low nanomolar affinity for the MC4 receptor and at least 100-fold selectivity for MC4 over MC3 and MC5, but with agonist activity at MC1.[29] Another cyclic peptide, MBP10, is a low nanomolar affinity antagonist at MC4 with greater than 100-fold selectivity for MC4 over MC3, MC5, and MC1.[1,30] By virtue of their high affinity and selectivity, these peptides are the best tools reported to date for the evaluation of the contribution of the MC4 receptor subtype to body weight regulation.

EXPERIMENTS WITH A POTENT AND SELECTIVE MC4 AGONIST

The profile of c(1-6)suc-HfRWK-NH$_2$ in functional assays for human MC1, 3, 4, and 5 is shown in FIGURE 1. We confirm that this peptide is a full agonist at the MC4 receptor and is approximately 10 times more potent than α-MSH. The compound has equal potency and efficacy to α-MSH as an agonist at the MC1 receptor

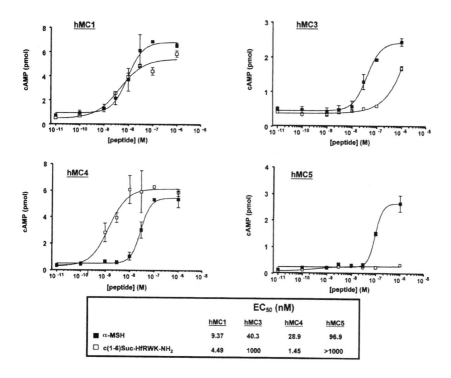

	EC$_{50}$ (nM)			
	hMC1	hMC3	hMC4	hMC5
■ α-MSH	9.37	40.3	28.9	96.9
□ c(1-6)Suc-HfRWK-NH$_2$	4.49	1000	1.45	>1000

FIGURE 1. Comparison of the affinity and efficacy of α-MSH and c(1-6)suc-fRWK-NH$_2$ for human melanocortin receptor subtypes. Measurements of cAMP were made after incubation of HEK293 cells stably expressing human MC1, MC3, MC4, or MC5 with a range of concentrations of α-MSH (■) or c(1-6)suc-fRWK-NH$_2$ (□).

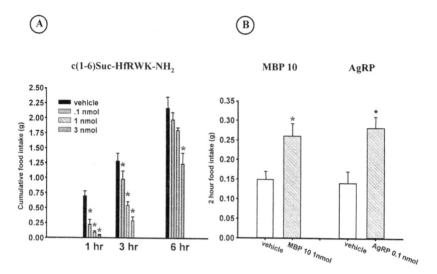

FIGURE 2. Effects of central administration of the selective MC4 receptor agonist, c(1-6)suc-fRWK-NH$_2$, or the selective MC4 receptor antagonist, MBP10, on acute food intake in mice. (**A**) Intracerebroventricular injection of c(1-6)suc-fRWK-NH$_2$ in food-deprived CD-1 mice produced a dose-dependent inhibition of food intake measured up to 6 h. $^*P < 0.05$. (**B**) Intracerebroventricular injection of MBP10 in CD-1 mice which were satiated by habituation to sweetened chow for 2 h produced an increase in food intake that was equivalent to effects observed with AgRP (83-132). $^*P < 0.05$.

and shows 658 and greater than 7000-fold selectivity for MC4 over MC3 and MC5, respectively. This peptide is a potent inhibitor of acute food intake when injected centrally in mice. As shown in FIGURE 2A, a dose-dependent inhibition of food intake was obtained with the lowest dose of 0.1 nmol, giving greater than 50% reduction of food intake at 1 h. By comparison, central administration of α-MSH under the same experimental conditions requires 1 nmol for an equivalent effect. Consequently, the potency of c(1-6)suc-HfRWK-NH$_2$ as an anorexogenic agent provides strong pharmacological evidence that the MC4 receptor subtype plays a dominant role in suppression of food intake (the agonist activity at MC1 is considered to be irrelevant, because MC1 receptors are expressed at very low levels in the brain).

EXPERIMENTS WITH A POTENT AND SELECTIVE MC4 RECEPTOR ANTAGONIST

We have confirmed the data from Bednarek *et al.*[30] that MBP10 is a potent and selective MC4 receptor antagonist in *in vitro* assays of human MC4 receptor subtypes (data not shown). When injected centrally in satiated mice, MBP10 produces an acute increase in food intake which is comparable to that observed with AgRP (83-132) (FIG. 2B), providing pharmacological evidence that selective antagonism of the MC4 receptor is sufficient to reproduce the orexigenic effects of AgRP. We also have examined the relative effects of central administration of MBP10 and

AgRP (83-132) on anorexia induced by central injection of the cytokine, IL-1β, in the mouse as a model relevant for cachexia. Both peptides reduce the inhibition of food intake produced by IL-1b (data not shown), providing further evidence for a key role for the MC4 receptor in cachexia-induced anorexia.

THERAPEUTIC OPPORTUNITIES FOR MC4-SELECTIVE LIGANDS

Given the genetic and pharmacological data pointing toward the MC4 receptor subtype as a key player in body weight regulation, there is considerable interest in developing agonist and antagonist drugs which could be of therapeutic benefit. Peptide therapeutics which do not select between the receptor subtypes are being pursued, although they have the obvious disadvantages of a brief duration of action and poor penetration into the central nervous system in addition to the lack of receptor selectivity. Several companies have published structures of drug-like small molecules which have selectivity for the MC4 receptor (for a review see Goodfellow and Saunders[31]). At Neurocrine, we have successfully identified potent and selective small molecule MC4 receptor agonists and antagonists which are being tested for their potential as treatments for obesity and cachexia, respectively. It is hoped that the approach of manipulating the MC4 receptor through small molecule drugs will one day yield useful therapeutic options for patients suffering from body weight dysfunction.

ACKNOWLEDGMENTS

We are grateful to Drs. Roger Cone, Daniel Marks, and Michael Cowley for their advice, and for the support of SBIR grants 1 R43 DK57969-01 and 1 R43 CA93190-01 from the National Institutes of Health.

REFERENCES

1. MACNEIL, D.J. *et al.* 2002. The role of melanocortins in body weight regulation: opportunities for the treatment of obesity. Eur. J. Pharmacol. **450:** 93–109.
2. MARKS, D.L. *et al.* 2002. Melanocortin pathways: animal models of obesity and disease. Ann. Endocrinol. **63:** 121–124.
3. LU, D. *et al.* 1994. Agouti protein is an antagonist of the melanoortin-stimulating hormone receptor. Nature **371:** 799–802.
4. SHUTTER, J.R. *et al.* 1997. Hypothalamic expression of ART, a novel gene related to agouti, is up-regulated in obese and diabetic mutant mice. Genes Dev. **11:** 593–602.
5. OLLMANN, M.M. *et al.* 1997. Antagonism of central melanocortin receptors in vitro and in vivo by agouti-related protein. Science **278:** 135–138.
6. HUSZAR, D. *et al.* 1997. Targeted disruption of the melanocortin-4 receptor results in obesity in mice. Cell **88:** 131–141.
7. BUTLER, A.A. *et al.* 2000. A unique metabolic syndrome causes obesity in the melanocortin-3 receptor-deficient mouse. Endocrinology **141:** 3518–3521.
8. CHEN, A.S. *et al.* 2000. Inactivation of the mouse melanocortin-3 receptor results in increased fat mass and reduced lean body mass. Nat. Genet. **26:** 97–102.
9. YASWEN, L. *et al.* 1999. Obesity in the mouse model of pro-opiomelanocortin deficiency responds to peripheral melanocortin. Nat. Med. **5:** 1066–1070.

10. QIAN, S. *et al.* 2002. Neither agouti-related protein nor neuropeptide Y is critically required for the regulation of energy homeostasis in mice. Mol. Cell. Biol. **22:** 5027–5035.
11. KRUDE, *et al.* 1998. Severe early-onset obesity, adrenal insufficiency and red hair pigmentation caused by POMC mutations in humans. Nat. Genet. **19:** 155–157.
12. FAROOQI, I.S. *et al.* 2000. Dominant and recessive inheritance of morbid obesity associated with melanocortin 4 receptor deficiency. J. Clin. Invest. **106:** 271–279.
13. VAISSE, C. *et al.* 2000. Melanocortin-4 receptor mutations are a frequent and heterogeneous cause of morbid obesity. J. Clin. Invest. **106:** 253–262.
14. FAN, W. *et al.* 1997. Role of melanocortinergic neurons in feeding and the agouti obesity syndrome. Nature **385:** 165–168.
15. KOEGLER, F.H. *et al.* 2001. Central melanocortin receptors mediate changes in food intake in the rhesus macaque. Endocrinology **142:** 2586–2592.
16. JONSSON, L. *et al.* 2002. Food conversion is transiently affected during 4-week chronic administration of melanocortin agonist and antagonist in rats. J. Endocrinol. **173:** 517–523.
17. LU, X.Y. *et al.* 2001. Time course of short-term and long-term orexigenic effects of agouti-related protein (86-132). Neuroreport **12:** 1281–1284.
18. RAPOSINHO, P.D. *et al.* 2000. Chronic blockade of the melanocortin 4 receptor subtype leads to obesity independently of neuropeptide Y action, with no adverse effects on the gonadotropic and somatotropic axes. Endocrinology **141:** 4419–4427.
19. TISDALE, M.J. 1997. Biology of cachexia. J. Natl. Cancer Inst. **89:** 1763–1773.
20. PLATA-SALAMAN, C.R. 2000. Central nervous system mechanisms contributing to the cancer-anorexia syndrome. Nutrition **16:** 1009–1012.
21. HUANG, Q-H. *et al.* 1999. Role of central melanocortins in endotoxin-induced anorexia. Am. J. Physiol. **276:** R864–R871.
22. MARKS, D.L. *et al.* 2001. Role of central melanocortin system in cachexia. Cancer Res. **61:** 1432–1438.
23. WISSE, B.E. *et al.* 2001. Reversal of cancer anorexia by blockade of central melanocortin receptors in rats. Endocrinology **142:** 3292–3301.
24. VERGONI, A.V. *et al.* 1999. Selective melanocortin MC4 receptor blockade reduces immobilization stress-induced anorexia in rats. Eur. J. Pharmacol. **369:** 11–15.
25. MARSH, D.J. *et al.* 1999. Response of melanocortin-4 receptor-deficient mice to anorectic and orexigenic peptides. Nat. Genet. **21:** 119–122.
26. BENOIT, S.C. *et al.* 2000. A novel selective melanocortin-4 receptor agonist reduces food intake in rats and mice without producing aversive consequences. J. Neurosci. **20:** 3442–3448.
27. VERGONI, A.V. *et al.* 2000. Chronic melanocortin 4 receptor blockage causes obesity without influencing sexual behavior in male rats. J. Endocrinol. **166:** 419–426.
28. SKULADOTTIR, G.V. *et al.* 1999. Long-term orexigenic effect of a novel melanocortin 4 receptor selective antagonist. Br. J. Pharmacol. **126:** 27–34.
29. BEDNAREK, M.A. *et al.* 2001. Potent and selective peptide agonists of α-melanotropin action at human melanocortin receptor 4: their synthesis and biological evaluation in vitro. Biochem. Biophys. Res. Commun. **286:** 641–645.
30. BEDNAREK, M.A. *et al.* 2001. Elective, high affinity peptide antagonists of α-melanotropin action at human melanocortin receptor 4: their synthesis and biological evaluation in vitro. J. Med. Chem. **44:** 3665–3672.
31. GOODFELLOW, V.S. & J. SAUNDERS. 2003. The melanocortin system and its role in obesity and cachexia. Curr. Top. Med. Chem. **3:** 855–883.

Expression, Desensitization, and Internalization of the ACTH Receptor (MC2R)

ADRIAN J.L. CLARK,[a] ASMA H. BAIG,[a] LUKE NOON,[a]
FRANCESCA M. SWORDS,[a] LASZLO HUNYADY,[b] AND PETER J. KING[a]

[a]Department of Endocrinology, Queen Mary University of London,
London EC1A 7BE, United Kingdom

[b]Department of Physiology, Semmelweis University, Faculty of Medicine,
Budapest, Hungary

ABSTRACT: Research into the functions and mechanisms of action of the melanocortin 2 receptor (MC2R) has been severely hampered by difficulties in expressing this gene in heterologous cells. This probably arises because of the need for a cofactor for cell surface expression. Using either the Y1 cell line that expresses endogenous MC2R or the Y6 cell line that expresses this putative expression factor, we have explored the mechanisms of desensitization and internalization after agonist stimulation. Protein kinase A dependence of desensitization has been demonstrated, although internalization is apparently independent of this kinase and dependent on a G protein receptor kinase. Possible underlying reasons for this paradox are discussed.

KEYWORDS: melanocortin 2 receptor; adrenocorticotropin; G protein–coupled receptors; protein kinase A; G protein receptor kinase; desensitization; receptor internalization; molecular chaperones; constitutive activity

INTRODUCTION

The cloning by Mountjoy *et al.* in 1992 of the melanocortin 2 receptor (MC2R) opened the way for a greater understanding of this, the smallest G protein–coupled receptor.[1] The tissue distribution of MC2R gene expression and limited expression data supported the view that this was the receptor for adrenocorticotropin (ACTH).

Our own initial research focused on identification and characterization of mutations of the MC2R gene in familial glucocorticoid deficiency.[2] This rare autosomal recessive disorder is characterized by ACTH insensitivity with consequently subnormal or absent glucocorticoid production from the adrenal and drastically elevated circulating ACTH levels. Genetic segregation of mutant MC2R with disease in many families lent strong support to the view that defects of this gene were responsible for the disease and thus that this was the only gene encoding the ACTH receptor. However, expression of transfected MC2R remained essentially impossible except in two

Address for correspondence: Adrian J.L. Clark, Department of Endocrinology, Barts & the London, Queen Mary University of London, West Smithfield, London EC1A 7BE, UK. Voice: 44-20-7601-7445; fax: 44-20-7601-8468.
a.j.clark@qmul.ac.uk

Ann. N.Y. Acad. Sci. 994: 111–117 (2003). © 2003 New York Academy of Sciences.

circumstances. The transfected gene resulted in ACTH binding or responses if transfected into cells expressing another endogenous MC2R such as the Cloudman M3 cell line,[1,3] or, in our hands, a Cos 7 cell line.[4] However, the function of the additional melanocortin receptors in these cell types obscured proper pharmacological analysis of the transfected gene. The second "special circumstance" was with a stable transformed HeLa cell line transfected with the murine MC2R.[5] Only a single clone exhibited this phenomenon, and it has not been possible to recreate this model. It is assumed that some unique genetic event had occurred in this line that permitted MC2R expression.

A satisfactory model for MC2R characterization was obtained only after the realization that certain mouse Y1 adrenocortical cell line derivatives known as Y6 and OS3 did not express the endogenous MC2R present in Y1 cells, but they presumably still contained whatever additional functions are necessary for cell surface expression of this receptor.[6] Use of these cell lines has enabled our own group and others to characterize the MC2R in greater detail. Studies with this model have confirmed the loss of function resulting from MC2R mutations in familial glucocorticoid deficiency.[7]

RECEPTOR EXPRESSION

The underlying basis for the difficulty in expressing the MC2R remains unclear. We have coupled the green fluorescent protein coding region to the C terminus of the MC2R and expressed it in various cell lines.[8] In the Y6 cell line, fluorescence is visible at the cell surface. In all other cell types tested (CHO-K1, HEK293, SK-N-MC), it tends to accumulate in a perinuclear region that also stains with an anti–KDEL antibody that identifies the endoplasmic reticulum and with $\beta 1,4$-galactosyltransferase coupled to yellow fluorescent protein which identifies the *trans*-medial region of the Golgi apparatus. It therefore seems probable that the receptor becomes arrested at this level in cell types other than the Y6/OS3.

The nature of the impediment to cell surface expression of the MC2R remains uncertain. The block is specific to the MC2R in that there is good cell surface expression of the CCK receptor coupled to GFP when this is cotransfected with the wild-type MC2R.[8] Possible explanations include the requirement for a chaperone-like molecule such as the RAMP[9] or DRiP78 type molecules[10] identified in association with the CGRP and dopamine D1 receptors specifically. The ability of other melanocortin expressing cell types to express the MC2R may imply that this "accessory" factor has specificity for more than one melanocortin receptor subtype and is coexpressed with many of the melanocortin receptors.

DESENSITIZATION

One of our primary aims has been to understand the events that occur in adrenal cells immediately after stimulation by ACTH. It is well established that the MC2R is coupled to $G\alpha_s$ and that activation of adenylate cyclase is a major primary signaling event. There is also significant work over several years describing additional early signaling events, most notably stimulation of calcium entry[11–14] or release of arachidonic acid metabolites.[15]

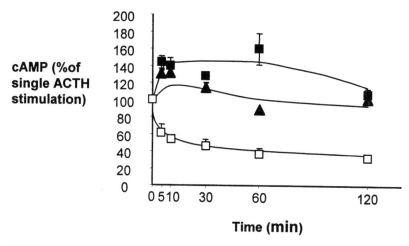

FIGURE 1. Rate of desensitization of the MC2R after different initial stimuli. Y1 cells were stimulated with 10^{-8} M ACTH at the indicated times after a prior stimulation with 10^{-8} M ACTH (□), forskolin, 10^{-5} M (▲), or isoproterenol, 10^{-5}M in cells stably transfected with the β_2-adrenergic receptor (■). cAMP production over a 30-min period in the presence of IBMX was measured by a protein binding assay and expressed as the percentage of the response to a single stimulation. (From Baig *et al.*[16] Reprinted with permission of the American Society for Biochemistry and Molecular Biology.)

As with many other receptor types, the initial ligand receptor interaction and signaling function in Y1 cells expressing an endogenous MC2R is rapidly followed by a desensitization event. We have shown that approximately 60% of this MC2R signaling is desensitized within 30 min of an ACTH stimulus, and that this desensitization persists for several hours.[16] Perhaps surprisingly, other factors that stimulate cAMP production in these cells such as forskolin or isoproterenol (after transfection of the β_2-adrenergic receptor) do not desensitize the MC2R (FIG. 1). Such "homologous" desensitization events usually are mediated by members of the G protein receptor kinase (GRK) family. RT-PCR and immunoblotting studies confirm that Y1 cells express GRK2 and GRK5, but stable transfection of a kinase-deficient dominant negative GRK2 only marginally reduces the desensitization curve normally observed.

The role in desensitization of protein kinase A (PKA), which is clearly activated after ACTH stimulation, was investigated initially using the PKA inhibitor H89. Prior application of this inhibitor in concentrations considered to be specific for PKA resulted in a complete loss of desensitization for the initial 30 to 60 min after a second ACTH stimulation (FIG. 2). Further support for the view that PKA was the primary mediator of desensitization came from site-directed mutagenesis of the single consensus PKA phosphorylation site in the third cytoplasmic loop at serine 208. When this mutant receptor was expressed in Y6 cells and compared with the same cells expressing the wild-type MC2R, desensitization was almost completely lost for the entire 120-min study period.[16]

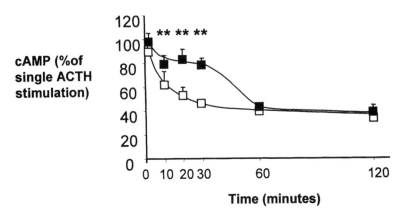

FIGURE 2. Rate of desensitization of the MC2R in the presence of H89. Y1 cells were stimulated with 10^{-8} M ACTH at the indicated times after a prior stimulation with 10^{-8} M ACTH in the presence of either vehicle (□) or H89, 10^{-7}M (■). Results are expressed as in FIGURE 1. $^{**}P < 0.01$. (From Baig et al.[16] Reprinted with permission of the American Society for Biochemistry and Molecular Biology.)

In summary, desensitization of the endogenous MC2R expressed in Y1 cells appears to be incomplete, homologous in character, but mediated by protein kinase A. This is an unusual occurrence and might be explained by tight compartmentalization of the MC2R with protein kinase A at the cell surface.

INTERNALIZATION

The next event that occurs to some G protein–coupled receptors after agonist activation is that they internalize. We investigated the extent and characteristics of internalization using an acid-wash technique which measures the quantity of ^{125}I-ACTH that remains acid soluble on the surface of intact cells in comparison with the quantity that is retained within the cell. These studies show that only approximately 20% of Y1 cell surface MC2R internalizes over a 30-min period. Studies with 0.45 M sucrose or concanavalin A suggest that this internalization is via clathrin-coated pits. Use of filipin or nystatin lent no support to the alternative view that this receptor could internalize via caveolae.[17]

In many G protein–coupled receptor models that have been studied, clathrin-coated pit internalization involves interaction between the receptor, an arrestin molecule and the clathrin triskelion. Using stable Y1 cell lines that overexpressed either the wild-type β–arrestin 1, or a dominant negative β-arrestin 1, we observed no influence on the extent or rate of internalization.[17] However, no positive control was available for these studies which therefore should be treated with caution.

The interaction of an arrestin with receptor normally is greatly enhanced by receptor phosphorylation by a GRK. To our surprise, the dominant negative GRK2 that failed to significantly inhibit desensitization was able to greatly inhibit internalization, implying that phosphorylation by a GRK was needed. However, neither H89 pretreatment or studies with the S208A PKA consensus site MC2R mutant suggested any influence of PKA on internalization.[17]

PHYSIOLOGICAL RELEVANCE

Whereas our data support the view that, at least in the mouse Y1 cell, a fairly active mechanism of receptor desensitization exists, the physiological relevance of this is unclear. In particular, it is not clear whether changes in glucocorticoid production *in vivo* desensitize in the same way that intracellular cAMP levels do. It may be that the stimulation in steroidogenic enzyme gene expression that follows a single ACTH stimulus is so prolonged that short-term desensitization is of no relevance.

An alternative role is suggested by some recent observations on a patient with ACTH-independent Cushing's syndrome with a homozygous germ-line mutation of the MC2R that resulted in the cells exhibiting increased basal cAMP generation. On further investigation, it became apparent that the effect of this mutation which converted Phe 278 to a Cys was to impair desensitization and internalization.[18] The increased basal activity seems to be secondary to this. In fact, the wild-type receptor retains no endogenous basal activity in contrast with many receptors including several members of the melanocortin receptor family. It may be that in the case of the MC2R, any basal activity is undesirable, and thus this receptor has evolved mechanisms for switching off any endogenous activity by an effective receptor desensitization mechanism.

TWO POPULATIONS?

Many of the findings on desensitization appear to be very discordant with those for internalization. Furthermore, it is not clear why all MC2Rs do not desensitize and internalize. A possible explanation is that there exist two distinct populations of MC2R on the Y1 cell surface. Under such a scenario, one larger population of receptor may be tightly compartmentalized with PKA. Such compartmentalization might be achieved by the involvement of an A kinase anchoring protein (AKAP) that tethers the regulatory subunit of PKA to the MC2R. Precedents exist for AKAP interaction with the β_2-adrenergic receptor.[19,20] Alternatively, or in addition, compartmentalization might be achieved within lipid rafts, small regions of cholesterol-rich cell membrane that colocalize certain proteins such as receptors and signaling components.[21] If this were the case, it may be that the compartmentalized MC2R is relatively protected from GRKs and hence would be less likely to desensitize via a GRK-dependent mechanism and less likely to internalize through an arrestin–clathrin complex.

The second smaller population of receptors would be non-compartmentalized in this model and hence GRK would be sensitive and susceptible to internalization. Whether this population could be desensitized is not clear. The aim of some of our studies in the near future will be to investigate this hypothesis. If correct, further implications could be that each population might exhibit different signaling properties, and flux between populations would be a means of regulating receptor behavior under physiological or pathological circumstances.

ACKNOWLEDGMENTS

We are very grateful to Prof. B. Schimmer and Dr. S. S. Ferguson for cell lines, expression vectors containing the wild-type and dominant negative kinases, and ar-restin. This work was supported by a grant and a Ph.D. studentship (to A.H.B.) from the Joint Research Board of St. Bartholomew's Hospital, and a Research Training Fellowship from the Wellcome Trust to F.M.S.

REFERENCES

1. MOUNTJOY, K.G., L.S. ROBBINS, M. MORTRUD, et al. 1992. The cloning of a family of genes that encode melanocortin receptors. Science **257**: 1248–1251.
2. CLARK, A.J.L. & A. WEBER. 1998. Adrenocorticotropin insensitivity syndromes. Endo-crinol. Rev. **19**: 828–843.
3. NAVILLE, D., L. BARJHOUX, C. JAILLARD, et al. 1996. Demonstration by transfection studies that mutations in the adrenocorticotropin receptor gene are one cause of the hereditary syndrome of glucocorticoid deficiency. J. Clin. Endocrinol. Metab. **81**: 1442–1448.
4. WEBER, A., S. KAPAS, J. HINSON, et al. 1993. Functional characterization of the cloned human ACTH receptor: impaired responsiveness of a mutant receptor in familial glu-cocorticoid deficiency. Biochem. Biophys. Res. Commun. **197**: 172–178.
5. KAPAS, S., F.M. CAMMAS, J.P. HINSON, et al. 1996. Agonist and receptor binding prop-erties of adrenocorticotropin peptides using the cloned mouse ACTH receptor expressed in a stably transfected HeLa cell line. Endocrinology **137**: 3291–3294.
6. SCHIMMER, B.P., W.K. KWAN, J. TSAO, et al. 1995. Adrenocorticotropin-resistant mutants of the Y1 adrenal cell line fail to express the adrenocorticotropin receptor. J. Cell. Physiol. **163**: 164–171.
7. ELIAS, L.L.K., A. WEBER, G.D. PULLINGER, et al. 1999. Functional characterization of naturally occurring mutations of the human adrenocorticotropin receptor: poor corre-lation of phenotype and genotype. J. Clin. Endocrinol. Metab. **84**: 2766–2770.
8. NOON, L.A., J.M. FRANKLIN, P.J. KING, et al. 2002. Failed export of the adrenocorticotropin receptor from the endoplasmic reticulum in non-adrenal cells: evidence in support of a requirement for a specific adrenal accessory factor. J. Endocrinol. **174**: 17–25.
9. MCLATCHIE, L.M., N.J. FRASER, M.J. MAIN, et al. 1998. RAMPs regulate the transport and ligand specificity of the calcitonin-receptor-like receptor. Nature **393**: 333–339.
10. BERMAK, J.C., M. LI, C. BULLOCK, et al. 2001. Regulation of transport of the dopamine D1 receptor by a new membrane-associated ER protein. Nat. Cell Biol. **3**: 492–498.
11. KOJIMA, I., K. KOJIMA & H. RASMUSSEN. 1985. Role of calcium and cAMP in the action of adrenocorticotropin on aldosterone secretion. J. Biol. Chem. **260**: 4248–4256.
12. DURROUX,T., N. GALLO-PAYET & M. PAYET. 1991. Effects of adrenocorticotropin on action potential and calcium currents in cultured rat and bovine glomerulosa cells. Endocrinology **129**: 2139–2147.
13. ENYEART, J.J., B. MLINAR & J.A. ENYEART. 1993. T-type Ca^{2+} channels are required for adrenocorticotropin-stimulated cortisol production by bovine adrenal zona fas-ciculata cells. Mol. Endocrinol. **7**: 1031–1040.
14. YAMAZAKI, T., T. KIMOTO, K. HIGUCHI, et al. 1998. Calcium ion as a second messenger for o-nitrophenylsulfenyl-adrenocorticotropin (NPS-ACTH) and ACTH in bovine adrenal steroidogenesis. Endocrinology **139**: 4765–4771.
15. WANG, X-J., L.P. WALSH, A.J. REINHART, et al. 2000. The role of arachidonic acid in steroidogenesis and steroidogenic acute regulatory (StAR) gene and protein expres-sion. J. Biol. Chem. **275**: 20204–20209.
16. BAIG, A.H., F.M. SWORDS, L. NOON, et al. 2001. Desensitization of the Y1 cell adreno-corticotropin receptor: evidence for a restricted heterologous mechanism implying a role for receptor-effector complexes. J. Biol. Chem. **276**: 44792–44797.

17. BAIG, A.H., M. SZASZÁK, P.J. KING, *et al.* 2002. Agonist activated adrenocorticotropin receptor internalizes via a clathrin-mediated G protein receptor kinase dependent mechanism. Endocrinol. Res. **28:** 281–289.

18. SWORDS, F.M., A.H. BAIG, D.M. MALCHOFF, *et al.* 2003. Impaired desensitization of a mutant adrenocorticotropin receptor associated with apparent constitutive activity. Mol. Endocrinol. **16:** 2746–2753.

19. FAN, G.-F., E. SHUMAY, H.-Y. WANG, *et al.* 2001. The scaffold protein gravin (cAMP-dependent protein kinase-anchoring protein 250) binds the b_2 adrenergic receptor via the receptor cytoplasmic Arg 329 to Leu-413 domain and provides a mobile scaffold during desensitization. J. Biol. Chem. 276: 24005–24014.

20. CONG, M., S.J. PERRY, F.-T. LIN, *et al.* 2001. Regulation of membrane targeting of the G protein-coupled receptor kinase 2 by protein kinase A and its anchoring protein AKAP79. J. Biol. Chem. **276:** 15192–15199.

21. ZAJCHOWSKI, L.D. & S.M. ROBBINS. 2001. Lipid rafts and little caves: compartmentalized signaling in membrane microdomains. Eur. J. Biochem. **269:** 737–752.

Identification of a Serine Protease Involved with the Regulation of Adrenal Growth

ANDREW B. BICKNELL

School of Animal and Microbial Sciences, The University of Reading, Whiteknights, P.O. Box 228, Reading, Berkshire, RG6 6AJ, United Kingdom

ABSTRACT: The adrenal cortex is a dynamic organ in which the cells of the outer cortex continually divide. It is well known that this cellular proliferation is dependent on constant stimulation from peptides derived from the ACTH precursor pro-opiomelanocortin (POMC) because disruption of pituitary corticotroph function results in rapid atrophy of the gland. Previous results from our laboratory have suggested that the adrenal mitogen is a fragment derived from the N-terminal of POMC not containing the γ-MSH sequence. Because such a peptide is not generated during processing of POMC in the pituitary, we proposed that the mitogen is generated from circulating pro–γ-MSH by an adrenal protease. Using degenerate oligonucleotides, we identified a secreted serine protease expressed by the adrenal gland that we named adrenal secretory protease (AsP). In the adrenal cortex, expression of AsP is limited to the outer zona glomerulosa/fasciculata, the region where cortical cells are believed to be derived, and is significantly up-regulated during compensatory growth. Y1 adrenocortical cells transfected with a vector expressing an antisense RNA (and thus having reduced levels of endogenous AsP) were found to grow slower than sense controls while also losing their ability to utilize exogenous pro–γ-MSH in the media supporting a role for AsP in adrenal growth. Digestion of an N-POMC peptide substrate encompassing the residues around the dibasic cleavage site at positions 49/50 with affinity-purified AsP showed cleavage not to occur at the dibasic site but two residues downstream leading us to propose the identity of the adrenal mitogen to be N-POMC (1-52).

KEYWORDS: pro-opiomelanocortin; pro–γ-MSH; mitogen; adrenal; protease

INTRODUCTION

The cortex of the mammalian adrenal gland is subdivided into three concentric zones that are functionally and morphologically distinct. The outer zona glomerulosa produces mineralocorticoids, and the inner zona fasciculata and zona reticularis synthesize glucocorticoids. For these reasons, the different adrenocortical zones have been viewed as autonomous entities, each with a distinct stem cell population. It is now becoming clear that all zones almost certainly share a common cellular

Address for correspondence: Andrew B. Bicknell, School of Animal and Microbial Sciences, The University of Reading, Whiteknights, P.O. Box 228, Reading, Berkshire, RG6 6AJ, UK. Voice: 44-0-118-9875123 ext. 7044; fax: 44-0-118-9310180.
a.b.bicknell@rdg.ac.uk

Ann. N.Y. Acad. Sci. 994: 118–122 (2003). © 2003 New York Academy of Sciences.

origin with genesis in the outer region of the gland. This is followed by centripetalT migration, phenotypic switching, first into glomerulosa, and successively into fasciculata and reticularis, and final elimination by apoptosis at the medullary boundary (for review, see Estivariz et al.[1]).

Although several peptides and proteins are able to stimulate adrenal growth (for review, see Estivariz et al.[1]), there is no doubt that the corticotroph cells of the anterior pituitary plays a pivotal role in this respect because any disruption in their function results in rapid atrophy of the adrenal cortex[2–4]). There has been much debate over the possible mitogenic activity of ACTH because it does stimulate an increase in adrenal size, mainly by the process of hypertrophy, but the most convincing evidence against a mitogenic role is the fact that immunoneutralization of ACTH *in vivo* in amounts sufficient to suppress corticosteroid secretion from the adrenal gland does not lead to a reduction in adrenal mass.[5] The fact that long-term steroid therapy, which also results in adrenal atrophy,[4] would also be accompanied in a reduction of the secretion of other peptides derived from the ACTH precursor pro-opiomelanocortin (POMC) encouraged us to explore the adrenal mitogenic effects of peptides derived from the N-terminal region of this precursor. We found, however, that only peptides from the extreme N-terminal (not containing the γ-MSH region) had such activity.[6] Because these shorter peptides are not generated during the normal processing of POMC in the anterior pituitary, it was proposed that the normal corticotroph secretory product pro–γ-MSH (N-POMC 1-74 in the rat) was further processed after secretion by, at that time, an unidentified protease to release the mitogenic activity.

ROLE OF POMC PEPTIDES IN COMPENSATORY ADRENAL GROWTH

Further evidence for the existence of a protease came from experiments investigating the compensatory growth response seen after unilateral adrenalectomy. The adrenal gland is one of the paired glands in which the removal of one gland causes compensatory growth in the remaining gland. In the case of the adrenal gland this results in a drastic (50%) increase in mass within 48 h and is accompanied by similar increases in DNA and RNA content.[7,8] Although there is clear evidence suggesting that the response has a neural element,[8] it has been clearly demonstrated that the peptides derived from the N terminus of POMC play a crucial role in eliciting the compensatory growth response.[7] It was shown using antisera to immunoneutralize specific circulating POMC peptides that the extreme N-terminal peptides and not ACTH were essential in eliciting the growth response adding weight to the argument that the mitogenic POMC peptide resides in the N-terminal fragment. However, a surprising result from this study was the finding that immunoneutralization against N-POMC 50-74 (γ$_3$-MSH) completely inhibited the compensatory growth response. Because this peptide has no mitogenic properties and that its precursor pro–γ-MSH is the major product secreted from the pituitary, a hypothesis was presented to explain the observation. It was postulated that pro–γ-MSH was specifically cleaved after secretion by a protease expressed by the adrenal gland, giving the shorter, mitogenic N-POMC peptides, and it was this cleavage that was being inhibited by the binding of the γ-MSH antibody to pro–γ-MSH.[7]

IDENTIFICATION OF AN ADRENAL SERINE PROTEASE

Indirect evidence for the involvement of a protease came from experiments showing that if the rats were pretreated with the protease inhibitor, aprotinin, before unilateral adrenalectomy, then compensatory growth was reduced. It was also found that aprotinin attenuated the rate of growth of the Y1 adrenocortical cell line when added to the culture medium.[9] Because aprotinin is a peptide of significant size, it appeared unlikely that it exerted its actions within the cell and suggests that the putative protease was likely to be secreted and either released into the microcirculation of the adrenal gland or remain attached to the cell surface.

CLONING OF A NOVEL ADRENAL PROTEASE

To identify candidate proteases, we designed degenerate primers to conserved motifs in the trypsin family of serine proteases, using them to prime PCRs with cDNA from both normal and adrenal glands undergoing compensatory growth as template. Using this approach, we identified a novel serine protease that was upregulated during compensatory growth containing the conserved histidine/aspartic acid/serine catalytic triad motif.[9] The enzyme contains an N-terminal secretory signal sequence, and because of this feature we named it adrenal secretory protease (AsP). Using *in situ* hybridization, we found AsP to be expressed in the capsule and outer cortex which is consistent with a potential role in adrenal proliferation, because this is the region in which cortical cells are believed to be derived.

To investigate whether AsP was necessary for adrenal growth, we transfected Y1 cells with constructs expressing either noncoding sense or antisense AsP mRNA. No difference was observed when these cells were grown in media containing fetal serum because the serum would have been harvested from animals at a time of rapid adrenal growth and would contain high circulating concentrations of N-POMC fragments[10] as well as other growth factors. However, when the cells were grown in media containing a more defined medium (5% Optimem) known to contain intact pituitary POMC peptides, the antisense cells grew significantly slower than the sense cells, providing clear evidence that AsP is involved with the growth of these cells. It was also found that only the sense cells could utilize additional pro–γ-MSH added to the media, suggesting that AsP is required to release its mitogenic activity. This was further supported by the finding that the growth rate of sense cells grown in Optimem medium depleted of γ-MSH peptides was reduced to that of the antisense cells. However, this media had no effect on the growth rate of the antisense cells, again strongly suggesting that AsP specifically activates the mitogenic activity of its specific substrate pro–γ-MSH.[9]

To provide direct evidence that AsP could cleave pro–γ-MSH, we affinity-purified AsP from Y1 cell conditioned media using an AsP antibody linked to solid phase. We then used this enzyme to digest a peptide substrate (named QFS-20 Abz-LTENPRKYVMJS, J is nitrotyrosine and Abz is 2-aminobenzoyl[11]) designed around the dibasic cleavage site at positions 49/50. This peptide was chosen because all the earlier studies[6,7] had strongly indicated that this was the most likely cleavage point. Mass spectrometric analysis of the digested products surprisingly indicated

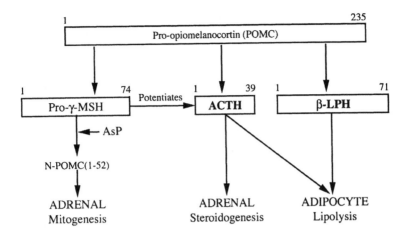

FIGURE 1. The processing of rat POMC in the anterior lobe of the pituitary by prohormone converting enzyme 1 generates pro–γ-melanotropin, corticotropin, and lipotropin. Under normal situations, pro–γ-MSH potentiates the steroidogenic activity of ACTH,[12] and both ACTH and β-LPH have weak lipolytic activity at the adipocyte. During adrenal growth, AsP cleaves pro–γ-MSH in the outer adrenal cortex into N-POMC (1-52), which then acts as the adrenal mitogen.

that the cleavage took place between the Val-Met (and not the expected Arg-Lys) bond, suggesting that the active adrenal mitogen is N-POMC (1-52).

CONCLUSIONS

When taken together, the data present AsP as a strong candidate for the protease that cleaves pro–γ-MSH during adrenal growth, the implications of which are significant because it gives the adrenal gland the ability, by regulation of AsP expression, to control its tonic state independent of the steroidogenic stimuli elicited by co-secreted ACTH (see FIGURE 1). This also poses questions as to the potential role of AsP during development and during pathologic states.

Although the data strongly suggest that the adrenal mitogen is N-POMC (1-52), this needs to be confirmed by digesting the natural peptide with AsP and then analyzing the cleavage products to determine both the cleavage points and their biological actions.

ACKNOWLEDGMENTS

I acknowledge the Wellcome Trust and the British Society for Endocrinology for funding this work. I also thank Professor Philip Lowry for his continuing support and encouragement.

REFERENCES

1. ESTIVARIZ, F.E., P.J. LOWRY & S. JACKSON. 1992. Control of adrenal growth. *In* The Adrenal Gland. V.H.T. James, Ed.: 42–70. Raven Press. New York.
2. VINSON, G.P., J.P. HINSON & P.W. RAVEN. 1985. The relationship between tissue preparation and function: methods for the study of control of aldosterone secretion: a review. Cell Biochem. Funct. **3:** 235–253.
3. YASWEN, L., N. DIEHL, M.B. BRENNAN, *et al.* 1999. Obesity in the mouse model of pro-opiomelanocortin deficiency responds to peripheral melanocortin. Nat. Med. **5:** 1066–1070.
4. WRIGHT, N.A., D.R. APPLETON & A.R. MORLEY. 1974. Effect of dexamethasone on cell population kinetics in the adrenal cortex of the prepubertal male rat. J. Endocrinol. **62:** 527–536.
5. RAO, A.J., J.A. LONG & J. RAMACHANDRAN. 1978. Effects of antiserum to adrenocorticotropin on adrenal growth and function. Endocrinology **102:** 371–378.
6. ESTIVARIZ, F.E., F. ITURRIZA, C. MCLEAN, *et al.* 1982. Stimulation of adrenal mitogenesis by N-terminal proopiocortin peptides. Nature **297:** 419–422.
7. LOWRY, P.J., L. SILAS, C. MCLEAN, *et al.* 1983. Pro-γ-melanocyte stimulating hormone cleavage in adrenal gland undergoing compensatory growth. Nature **306:** 70–72.
8. DALLMAN, M.F., W.C. ENGELAND & J. SHINSAKO. 1976. Compensatory adrenal growth: a neurally mediated reflex. Am. J. Physiol. **231:** 408–414.
9. BICKNELL, A.B., K. LOMTHAISONG, R.J. WOODS, *et al.* 2001. Characterisation of a serine protease that cleaves pro-γ-melanotropin at the adrenal to stimulate growth. Cell **105:** 903–912.
10. SAPHIER, P.W., B.P. GLYNN, R.J. WOODS, *et al.* 1993. Elevated levels of N-terminal proopiomelanocortin peptides in fetal sheep plasma may contribute to fetal adrenal gland development and the preparturient cortisol surge. Endocrinology **133:** 1459–1461.
11. LAZURE, C., D. GAUTHIER, F. JEAN, *et al.* 1998. In vitro cleavage of internally quenched fluorogenic human proparathyroid hormone and proparathyroid-related peptide substrates by furin. Generation of a potent inhibitor. J. Biol. Chem. **273:** 8572–8580.
12. AL DUJAILI, E.A.S., J. HOPE, F.E. ESTIVARIZ, *et al.* 1981. Circulating human pituitary pro-gamma-melanotrophin enhances the adrenal response to ACTH. Nature **291:** 156–159.

γ-MSH Peptides in the Pituitary

Effects, Target Cells, and Receptors

C. DENEF, J. LU, AND E. SWINNEN

Laboratory of Cell Pharmacology, University of Leuven Medical School,
B-3000 Leuven, Belgium

ABSTRACT: The melanocortin (MC) γ3-MSH is believed to signal through the MC3 receptor. We showed that it induces a sustained increase in intracellular free calcium levels ($[Ca^{2+}]_i$) in a subpopulation of pituitary cells. Most of the cells responding to γ3-MSH express more than one pituitary hormone mRNA. The effect of γ3-MSH is blocked by SHU9119, a MC3R and MC4R antagonist, in only 50% of the responsive cells, suggesting that in half of these cells the mediating receptor is not the MC3R. Low picomolar doses of γ3-MSH increase $[Ca^{2+}]_i$ in the growth hormone (GH)- and prolactin (PRL)-secreting GH3 cell line. γ2-MSH and α-MSH display a similar effect. SHU9119 does not affect the γ3-MSH–induced $[Ca^{2+}]_i$ response. MTII, a potent synthetic agonist of the MC3R, MC4R, and MC5R, also shows no or low potency in increasing $[Ca^{2+}]_i$. By means of RT-PCR, the mRNA of the MC2R, MC3R, and MC4R receptors is undetectable. Experiments testing γ2-MSH analogues with single alanine replacements show that, unlike the classic MCRs, the His[5]-Phe[6]-Arg[7]-Trp[8] sequence in γ2-MSH is not a core sequence for activating the γ-MSH receptor in GH3 cells, whereas Met[3] is essential. Low nanomolar doses of γ-MSH increase intracellular cAMP levels. Blockade of protein kinase A abolishes the $[Ca^{2+}]_i$ responses to γ3-MSH. γ2-MSH increases binding of $[S^{35}]GTPγS$ to membrane preparations of GH3 cells. The pharmacological characteristics of γ-MSH peptides and analogues on $[Ca^{2+}]_i$ and the signal-transduction pathways present strong evidence for the expression of a hitherto uncharacterized γ-MSH receptor in GH3 cells, belonging to the G protein–coupled receptor family.

KEYWORDS: pituitary; γ3-MSH; γ2-MSH; SHU9119; MTII; calcium imaging; single-cell RT-PCR; melanocortin receptors

The γ-MSH peptides belong to the family of melanocortin peptides, including α-MSH, β-MSH, γ3-MSH, γ2-MSH, and γ1-MSH. The γ-MSH peptides are generated from the N-terminal domain of proopiomelanocortin [N-POMC; in rat POMC(1–74), in human POMC(1–76)]. N-POMC is also called pro-γ-MSH.[1-6] The C-terminal end of N-POMC is cleaved as γ3-MSH between the two basic amino

Address for correspondence: C. Denef, Laboratory of Cell Pharmacology, University of Leuven Medical School, Campus Gasthuisberg (O&N), B-3000 Leuven, Belgium. Voice: 32-16-345812; fax: 32-16-345699.

Carl.Denef@med.kuleuven.ac.be

Ann. N.Y. Acad. Sci. 994: 123–132 (2003). © 2003 New York Academy of Sciences.

A. γ-MSH peptides

```
                                     glycosyl
                                        |
γ3-MSH:        YVMGHFRWDRFGPRNSSSAGGSAQ
γ2-MSH:        YVMGHFRWDRFG
γ1-MSH:        YVMGHFRWDRF-NH₂
```

B. Allignment of γ-MSH peptides of various species

```
                 1    5    10   15    20   25     30
                 |    |    |    |     |     |      |
MAMMALS
RAT              YVMGHFRWDRFGPRNSSSAGGSAQ
MOUSE            YVMGHFRWDRFGPRNSSSAGSAAQ
CAVIA            YVTGHFRWGRFGRGNSSGASQ
BOVINE           YVMGHFRWDRFGRRNGSSSSGVGGAAQ
PIG              YVMGHFRWDRFGRRNGSSSGGGGGGGGAGQ
SHEEP            YVMGHFRWDRFGRRNGSSSFGAGGAAQ
DOG              YVMGHFRWDRFGRRNSSSSGSAGQ
MONKEY           YVMGHFRWDRFGRRNSSSGSAQ
HUMAN            YVMGHFRWDRFGRRNSSSSGSSGAGQ
BIRDS
OSTRICH          YVMSHFRWNKF??????????????? ?
CHICKEN          YVMSHFRWNKFGRRNSSSGGH
AMPHIBIA
RANA r           YVMSHFRWNKFGRRNS-TSNDNNN--GGY
RANA C           YVMSHFRWNKFGRRNS-TSNDNNNNNGGY
XENOP a          YVMTHFRWNKFGRRNN-TGNDGSS--GGY
XENOP b          YVMTHFRWNKFGRRNS-TGNDGSN--TGY
TOAD             YVMSHFRWNKFGRRNT-TGNEGNS--GS
LUNGFISH
PROTOPTER        YMMTHFRWDKFGRRNNETGN
NEOCERATO        YMMTHFHWDKFGRRNNETGN
CHONDROSTEI
STURGEONa        YVMSHFHWNTFGQRMNGTPGGS
STURGEONb        YIISHFRWNTFGQRVNGTPGGS
POLYODONa        YIMSHFHWNTFGQRMSGTPGGS
POLYODONb        YVMGHFHWNTFGQRMNGTPGGS
SEMIONOTIFORMES
LEPISOST.        YAKSHFRSTALGRRTNGSVGSSKQ
CHONDRICHTYES
STINGRAY         YVMGHFRWNKFGKKRDNSTELSVS...
DOGFISH          YVMGHFRWNKFGKKRGNNTGFSGN...
```

FIGURE 1. (A) Sequence of rat γ-MSH peptides and (B) sequence alignment of γ-MSH peptides of different species. Glycosyl, *N*-glycosylated; polyodon, *Polyodon spathula*; neocerato., *Neoceratodus forsteri*; protopter., *Protopterus annectens*; lepisost., *Lepisosteus osseus*; sturgeon, *Acipenser transmontanus*.

acids Arg^{49} and Lys^{50}, but mainly in the intermediate lobe, and may be further processed to γ2- and γ1-MSH, the latter being a C-amidated peptide. The peptides have been immunologically detected in the pituitary.[3,4,7,8] Unlike γ1- and γ2-MSH, γ3-MSH may be *N*-glycosylated.[9,10] Little is known about the role of the γ-MSH peptides. The striking amino acid sequence homology between species up to ancient fish (FIG. 1), however, predicts an important function, as is the case for the other MSH peptides. Biological effects have been described particularly on the adrenal cortex. The peptides potentiate ACTH-induced steroidogenesis.[11–13] Striking effects of γ-MSH have been reported on the cardiovascular system, a pressor effect concomitant with an increase of heart rhythm and natriuresis.[14] γ-MSH peptides also display behavioral effects including stimulatory effects on sexual behavior.[15,16] N-POMC (pro-γ-MSH) also has an effect on the adrenal cortex. The latter effect is a mitogenic action[17–19] in addition to a potentiating effect on ACTH-induced steroidogenesis.[11,20–26] It recently has been shown that processing of N-POMC to POMC(1–52), catalyzed by a novel serine protease, is a prerequisite for the mitogenic effect on the adrenals.[27] γ-MSH peptides have also an antiinflammatory effect.[28,29]

Melanocortin peptides act through a family of melanocortin receptors (MCRs), five of which are cloned today.[30–35] The presumptive receptor of γ-MSH peptides is the MC3R, one of the five members of MCRs.[32]

Because the POMC gene is highly expressed in the anterior and intermediate lobe of the pituitary and many peptides expressed in the pituitary have a paracrine/autocrine function (reviewed in Schwartz[36] and Houben and Denef[37]) including N-POMC,[38–40] we raised the question as to whether γ-MSH peptides exert some activity on certain pituitary cells, as well as which MCRs are involved in these putative actions. The relevance of the question was supported by the fact that circulating γ-MSH is mainly pro–γ-MSH (N-POMC) (see Bicknell *et al.*[27]), suggesting that processed γ-MSH peptides exist and are functional primarily within the pituitary.

EFFECTS OF γ3-MSH ON INTRACELLULAR FREE CALCIUM LEVELS ($[CA^{2+}]_i$) IN THE PITUITARY

Many peptides are known to increase $[Ca^{2+}]_i$ either as a primary signal-transduction step or as a secondary step in the signal-transduction cascade. We therefore tested whether the peptide was capable of increasing $[Ca^{2+}]_i$ in dispersed pituitary cells (dispersed from 14-day-old female rats and allowed to recover overnight in defined culture medium). Cells were loaded with the fluorescent Ca^{2+} probe Fluo-3, and a field of cells was examined for $[Ca^{2+}]_i$ changes by a confocal scanning laser microscope equipped with software capable to register and quantify the $[Ca^{2+}]_i$ transients in individual cells in the field.

Addition of γ3-MSH to the cells resulted in an increase of $[Ca^{2+}]_i$ within approximately 30 s in part of the pituitary cells.[41] It was found that the peptide induced either a change in the magnitude of the response or an increase of the frequency of $[Ca^{2+}]_i$ spikes or both.[41] Upon testing different doses, it was found that a dose as low as 0.1 nM was effective. Increasing doses, however, did not change the magnitude of the response nor the frequency of the $[Ca^{2+}]_i$ transients.[41] However, increasing doses recruited a higher number of cells responding.[41] At a dose of 10 nM, a maximal number of cells responded, whereas at higher doses the number of responsive cells

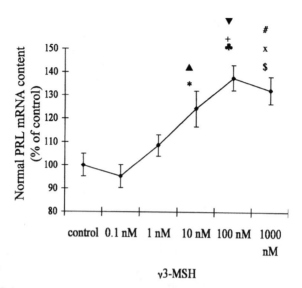

FIGURE 2. Effect of γ3-MSH on PRL mRNA expression in lactotrophs in rat pituitary cell aggregates. PRL mRNA in the aggregates was quantified by means of real-time RT-PCR. (for methods, see Lu *et al.*[51]). Values are means SEM of at least three independent experiments, with each dose of γ3-MSH measured in triplicate culture dishes. *Statistics*: Analysis of variance followed by Fisher's LSD. *$P < 0.01$ vs. control; +$P < 0.001$ vs. control; x = $P < 0.005$ vs. control; (▲) $P < 0.01$ vs. 0.1 nM; (▼) $P < 0.001$ vs. 0.1 nM; #$P < 0.04$ vs. 0.1 nM; (♣) $P < 0.005$ vs. 1 nM; $$P < 0.02$ vs. 1 nM.

again decreased. Importantly, at maximal response only, approximately 15% of the cells responded to γ3-MSH for these [Ca^{2+}]$_i$ changes.[41]

The biological significance of the [Ca^{2+}]$_i$ response to γ3-MSH in pituitary cells remains to be determined. However, in another approach it was found that in re-aggregate cell culture of 14-day-old pituitary γ3-MSH dose-dependently augments mitotic activity in lactotrophs, somatotrophs, and thyrotrophs[40] and that it stimulates PRL mRNA levels as measured by real-time RT-PCR (FIG. 2).

TARGET CELLS

An obvious step to get further insight into the putative physiological role of γ3-MSH in the pituitary is to determine in which cell types the [Ca^{2+}]$_i$ changes occur. This was achieved by two different approaches. One was by determining which among Fluo-3–loaded cells are responsive to γ3-MSH for a [Ca^{2+}]$_i$ increase and subsequently identifying the stored hormone in the responsive cells by immunostaining. In a second approach, the responsive cell was identified by determining the hormone mRNA present in the responsive cells by means of the patch-clamp single-cell RT-PCR method. In the first method, the responsive cells are localizable for identification by immunostaining by examining the cells on a glass slide with numbered grid coordinates. It was found[41] that 53% of the responsive cells showed growth hormone (GH) immunoreactivity (-ir), 12% prolactin (PRL)-ir, 2% TSHβ-ir, and 5% LHβ-ir and 10% showed ACTH-ir, whereas 18% did not express any hormone to a detectable level. Single-cell RT-PCR for the presence of pituitary hormone mRNA showed that 26% of the γ3-MSH–responsive cells contained only GH mRNA, 5% only PRL mRNA, and 4% only TSHβ mRNA. Most of the responsive cells contained two or more hormone mRNAs together in various combinations. Three categories of combinations were observed: (1) cells containing mRNA of GH, PRL, and TSHβ in various dual or triple combinations (22%); (2) cells expressing POMC mRNA together with GH mRNA and/or PRL mRNA and/or TSHβ mRNA (24%); and (3) cells containing LHβ mRNA together with mRNA of GH, PRL, and TSHβ in various combinations (18%). Interestingly, the proportion of cells displaying combinatorial hormone mRNA expression accounted for only approximately 30% of the cells in the total pituitary cell population, suggesting that γ3-MSH preferentially targets cells expressing multiple hormone mRNAs.[41] The specificity of this targeting furthermore was shown by the finding that the combination of mRNAs coexpressed in the cells was highly specific for the targeting peptide. In a pituitary cell population from 14-day-old female rats, GnRH targets cells with mRNA combinations, virtually all comprising LHβ mRNA (FIG. 3).

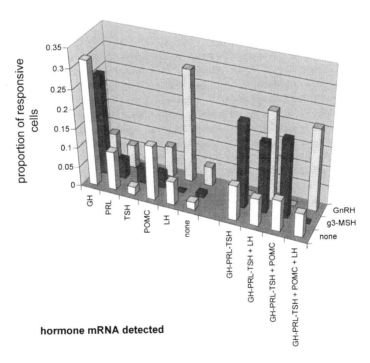

FIGURE 3. Distribution of pituitary hormone mRNAs in single pituitary cells from 14-day-old female rats, showing an increase in $[Ca^{2+}]_i$ in response to γ3-MSH and GnRH (each 10 nM) as determined by patch-clamp single-cell RT-PCR (for methods, see Roudbaraki *et al.*[41]).

These data not only indicate that γ3-MSH may have pleiotropic actions on a sub-population of pituitary cells but also show the existence of cells in the pituitary that express multiple hormone mRNAs. These cells were not described before in normal tissue, although dual colocalization of certain hormones have been observed in basal conditions, and these may expand their number under particular physiological conditions in which a higher demand of hormone output of both hormones concerned is required (see Roudbaraki et al.[41]).

It was also observed that the total number of cells expressing a particular hormone mRNA was significantly higher than the number of cells storing that particular hormone, whereas costorage of multiple hormone proteins in the same cell was not detectable by means of triple immunofluorescence.[41] Thus, cells may exist that co-express more than one hormone mRNA but not the corresponding hormone protein. This seems to suggest that γ3-MSH primarily targets cells that may be reserve cells that contain already various hormone mRNAs that are ready to be translated into protein under conditions of higher physiological demands for particular hormones. The significance of these cells showing combinatorial expression of hormone mRNA is further supported by the finding that the various categories of coexpressions change markedly during fetal and postnatal development[42] and that regulatory factors such as estradiol and gonadotropin-releasing hormone (GnRH) can strongly expand the population of cells expressing a particular hormone mRNA combination, for example, mRNA of PRL and the common glycoprotein hormone α-subunit (αGSU).[43]

RECEPTORS AND SIGNAL TRANSDUCTION

The only known best candidate receptor for mediating the effect of γ3-MSH on $[Ca^{2+}]_i$ is the MC3R.[32] We therefore investigated whether MC3R is expressed in the pituitary. It was found by RT-PCR that MC3R mRNA is present in the anterior and intermediate lobes of the pituitary.[44] The MC5R was also present. SHU9119 is an antagonist of the MC3R.[45] However, it was found that the $[Ca^{2+}]_i$ response to γ3-MSH was blocked only in approximately 50% of the pituitary cells,[41,44] indicating that at least part of the responses were mediated by the MC3R but that in other cells the effect is mediated by another MCR. When the cells on which SHU9119 was effective and not were identified by immunostaining, it was found that SHU9119 did not affect the $[Ca^{2+}]_i$ response in lactotrophs and somatotrophs but did so in corticotrophs and cells not expressing hormones.[41]

Inasmuch as the experiments with SHU9119 indicated that the MC3R does not mediate the $[Ca^{2+}]_i$ response in lactotrophs and somatotrophs, it was hypothesized that in the GH3 cell line, a rat PRL- and GH-secreting cell line, might be responsive to γ3-MSH but that the MC3R would also not be the mediating receptor. It indeed was found that γ3-MSH elicited a $[Ca^{2+}]_i$ response in GH3 cells from low picomolar doses.[46] Increasing doses recruited a higher number of responsive cells. No MC3R mRNA could be detected by RT-PCR.[46] SHU9119 failed to affect the $[Ca^{2+}]_i$ response to γ3-MSH. The peptides γ2-MSH, α-MSH, and NDP-α-MSH stimulated $[Ca^{2+}]_i$ equally well as γ3-MSH but MTII, a potent agonist of MC3R, MC4R, and MC5R, failed to affect $[Ca^{2+}]_i$ even at high doses.[46] The sequence HFRW in melanocortin peptides is the core sequence for activity at the MCRs. Peptides in which one of the amino acids in this sequence was replaced by Ala were equally potent on

$[Ca^{2+}]_t$ as γ2- or γ3-MSH.[47] However, when Meth[3] was replaced by Ala, the peptide lost activity on $[Ca^{2+}]_i$ in GH3 cells.[47] These data clearly show that the receptor-mediating $[Ca^{2+}]_i$ response to γ-MSH peptides in GH3 cells is not the MC3R but a hitherto unidentified MCR or another receptor with high affinity for γ-MSH peptides.

MCRs are G protein–coupled receptors. The main signal-transduction pathway of MCRs is the protein kinase A pathway. In cell lines expressing a MCR, melanocortins stimulate c-AMP accumulation.[30–35] We tested whether γ3-MSH stimulates c-AMP levels in GH3 cells[47] and it was found that it did. Furthermore, the $[Ca^{2+}]_i$ response was inhibited by RP-c-AMPs,[47] an analogue competitively antagonizing activation of protein kinase A by c-AMP. In membrane preparations of GH3 cells, γ3-MSH stimulated GTP-γ-S[35] incorporation,[47] strongly suggesting that the putative novel MCR is a G protein–coupled receptor.

In recent years, different investigators have found pharmacological evidence for the existence in various other biological response systems of other receptors with high affinity for one or more of the γ-MSH peptides and different from the MC3R or other known MCRs.[48–50] However, the pharmacological characteristics of these receptors are different from those of the γ-MSH receptor in GH3 cells.

ACKNOWLEDGMENTS

This investigation was supported by grants from the "Geconcerteerde Onderzoeksacties" (GOA 1997-2001 and 2002-2006) and the "Fonds voor Wetenschappelijk Onderzoek Vlaanderen." The authors express their gratitude to Dr. A. F. Parlow and the National Hormone and Pituitary Program for gifts of anti-PRL, -GH, -TSHβ, -LHβ, and -αGSU antisera.

REFERENCES

1. NAKANISHI, S., A. INOUE, T. KITA, *et al.* 1979. Nucleotide sequence of cloned cDNA for bovine corticotropin-beta-lipotropin precursor. Nature **278:** 423–427.
2. SHIBASAKI, T., N. LING & R. GUILLEMIN. 1980. Pituitary immunoreactive gamma-melanotropins are glycosylated oligopeptides. Nature **285:** 416–417.
3. BLOOM, F.E., E.L. BATTENBERG, T. SHIBASAKI, *et al.* 1980. Localization of gamma-melanocyte stimulating hormone (gamma MSH) immunoreactivity in rat brain and pituitary. Regul. Pept. **1:** 205–222.
4. PELLETIER, G., R. LECLERC, S. BENJANNET, *et al.* 1981. Immunohistochemical localization of gamma-melanocyte-stimulating hormone in the rat pituitary gland. Regul. Pept. **2:** 81–89.
5. SHIBASAKI, T., N. LING & R. GUILLEMIN. 1980. A radioimmunoassay for gamma 1-melanotropin and evidence that the smallest pituitary gamma-melanotropin is amidated at the COOH-terminus. Biochem. Biophys. Res. Commun. **96:** 1393–1399.
6. PEDERSEN, R.C., N. LING & A.C. BROWNIE. 1982. Immunoreactive gamma-melanotropin in rat pituitary and plasma: a partial characterization. Endocrinology **110:** 825–834.
7. BJARTELL, A., R. EKMAN & F. SUNDLER. 1987. gamma 2-MSH-like immunoreactivity in porcine pituitary and adrenal medulla. An immunochemical and immunocytochemical study. Regul. Pept. **19:** 291–306.
8. FENGER, M. 1988. Pro-opiomelanocortin-derived peptides in the pig pituitary: alpha- and gamma 1-melanocyte-stimulating hormones and their glycine-extended forms. Regul. Pept. **20:** 345–357.

9. OKI, S., K. NAKAO, I. TANAKA, et al. 1982. Characterization of gamma-melanotropin-like immunoreactivity and its secretion in an adrenocorticotropin-producing mouse pituitary tumor cell line. Endocrinology 111: 418–424.
10. VAN STRIEN, F.J., B. DEVREESE, J. VAN BEEUMEN, et al. 1995. Biosynthesis and processing of the N-terminal part of proopiomelanocortin in Xenopus laevis: characterization of gamma-MSH peptides. J. Neuroendocrinol. 7: 807–815.
11. SEGER, M.A. & H.P. BENNETT. 1986. Structure and bioactivity of the amino-terminal fragment of pro-opiomelanocortin. J. Steroid Biochem. 25: 703–710.
12. FARESE, R.V., N.C. LING, M.A. SABIR, et al. 1983. Comparison of effects of adrenocorticotropin and Lys-gamma 3-melanocyte-stimulating hormone on steroidogenesis, adenosine 3', 5'-monophosphate production, and phospholipid metabolism in rat adrenal fasciculata-reticularis cells in vitro. Endocrinology 112: 129–132.
13. PEDERSEN, R.C., A.C. BROWNIE & N. LING. 1980. Pro-adrenocorticotropin/endorphin-derived peptides: coordinate action on adrenal steroidogenesis. Science 208: 1044–1046.
14. GRUBER, K.A. & M.F. CALLAHAN. 1989. ACTH(4-10) through gamma-MSH: evidence for a new class of central autonomic nervous system-regulating peptides. Am. J. Physiol. 257: R681–R694.
15. KLUSA, V., S. SVIRSKIS, B. OPMANE, et al. 1999. Behavioural responses of gamma-MSH peptides administered into the rat ventral tegmental area. Acta Physiol. Scand. 167: 99–104.
16. CRAGNOLINI, A., T. SCIMONELLI, M.E. CELIS, et al. 2000. The role of melanocortin receptors in sexual behavior in female rats. Neuropeptides 34: 211–215.
17. LOWRY, P.J., L. SILAS, C. MCLEAN, et al. 1983. Pro-gamma-melanocyte-stimulating hormone cleavage in adrenal gland undergoing compensatory growth. Nature 306: 70–73.
18. ESTIVARIZ, F.E., M. CARINO, P.J. LOWRY, et al. 1988. Further evidence that N-terminal pro-opiomelanocortin peptides are involved in adrenal mitogenesis. J. Endocrinol. 116: 201–206.
19. ESTIVARIZ, F.E., M.I. MORANO, M. CARINO, et al. 1988. Adrenal regeneration in the rat is mediated by mitogenic N-terminal pro-opiomelanocortin peptides generated by changes in precursor processing in the anterior pituitary. J. Endocrinol. 116: 207–216.
20. DURAND, P., A.M. CATHIARD, N.G. SEIDAH, et al. 1984. Effects of proopiomelanocortin-derived peptides, methionine-enkephalin and forskolin on the maturation of ovine fetal adrenal cells in culture. Biol. Reprod. 31: 694–704.
21. LIS, M., P. HAMET, J. GUTKOWSKA, et al. 1981. Effect of N-terminal portion of pro-opiomelanocortin on aldosterone release by human adrenal adenoma in vitro. J. Clin. Endocrinol. Metab. 52: 1053–1056.
22. PEDERSEN, R.C., A.C. BROWNIE & N. LING. 1980. Pro-adrenocorticotropin/endorphin-derived peptides: coordinate action on adrenal steroidogenesis. Science 208: 1044–1046.
23. PEDERSEN, R.C. & A.C. BROWNIE. 1980. Adrenocortical response to corticotropin is potentiated by part of the amino-terminal region of pro-corticotropin/endorphin. Proc. Natl. Acad. Sci. USA 77: 2239–2243.
24. PHAM-HUU-TRUNG, M.T., A. BOGYO, N. DE SMITTER, et al. 1985. Effects of proopiomelanocortin peptides and angiotensin II on steroidogenesis in isolated aldosteronoma cells. J. Clin. Endocrinol. Metab. 61: 467–471.
25. SCHIFFRIN, E.L., M. CHRETIEN, N.G. SEIDAH, et al. 1983. Response of human aldosteronoma cells in culture to the N-terminal glycopeptide of pro-opiomelanocortin and gamma 3-MSH. Horm. Metab. Res. 15: 181–184.
26. SHARP, B. & J.R. SOWERS. 1983. Adrenocortical response to corticotropin is inhibited by gamma 3-MSH antisera in normotensive and spontaneously hypertensive rats. Biochem. Biophys. Res. Commun. 110: 357–363.
27. BICKNELL, A.B., K. LOMTHAISONG, R.J. WOODS, et al. 2001. Characterization of a serine protease that cleaves pro-gamma-melanotropin at the adrenal to stimulate growth. Cell 105: 903–912.
28. XIA, Y., J.E. WIKBERG & T.L. KRUKOFF. 2001. Gamma(2)-melanocyte-stimulating hormone suppression of systemic inflammatory responses to endotoxin is associated

with modulation of central autonomic and neuroendocrine activities. J. Neuroimmunol. **120:** 67–77.

29. GETTING, S.J., G.H. ALLCOCK, R. FLOWER, *et al.* 2001. Natural and synthetic agonists of the melanocortin receptor type 3 possess anti-inflammatory properties. J. Leukoc. Biol. **69:** 98–104.

30. CHHAJLANI, V. & J.E.S. WIKBERG. 1992. Molecular cloning and expression of the human melanocyte stimulating hormone receptor cDNA. FEBS Lett. **309:** 417–420.

31. MOUNTJOY, K.G., L.S. ROBBINS, M.T. MORTRUD, *et al.* 1992. The cloning of a family of genes that encode the melanocortin receptors. Science **257:** 1248–1251.

32. ROSELLI-REHFUSS, L., K.G. MOUNTJOY, L.S. ROBBINS, *et al.* 1993. Identification of a receptor for gamma melanotropin and other proopiomelanocortin peptides in the hypothalamus and limbic system. Proc. Natl. Acad. Sci. USA **90:** 8856–8860.

33. CHHAJLANI, V., R. MUCENIECE & J.E.S. WIKBERG. 1993. Molecular cloning of a novel human melanocortin receptor. Biochem. Biophys. Res. Commun. **195:** 866–873.

34. GANTZ, I., Y. KONDA, T. TASHIRO, *et al.* 1993. Molecular cloning of a novel melanocortin receptor. J. Biol. Chem. **268:** 8246–8250.

35. GANTZ, I., H. MIWA, Y. KONDA, *et al.* 1993. Molecular cloning, expression, and gene localization of a fourth melanocortin receptor. J. Biol. Chem. **268:** 15174–15179.

36. SCHWARTZ, J. 2000. Intercellular communication in the anterior pituitary. Endocrinol. Rev. **21:** 488–513.

37. HOUBEN, H. & C. DENEF. 1994. Bioactive peptides in anterior pituitary cells. Peptides **15:** 547–582.

38. TILEMANS, D., M. ANDRIES, P. PROOST, *et al.* 1994. In vitro evidence that an 11-kilodalton N-terminal fragment of proopiomelanocortin is a growth factor specifically stimulating the development of lactotrophs in rat pituitary during postnatal life. Endocrinology **135:** 168–174.

39. VAN BAEL, A., V. VANDE VIJVER, B. DEVREESE, *et al.* 1996. N-terminal 10 and 12 kDa POMC fragments stimulate differentiation of lactotrophs. Peptides **17:** 1219–1228.

40. TILEMANS, D., D.RAMAEKERS, M. ANDRIES, *et al.* 1997. Effect of POMC1-76, its C-terminal fragment γ3-MSH and anti-POMC1-76 antibodies on DNA replication in lactotrophs in aggregate cell cultures of immature rat pituitary. J. Neuroendocrinol. **9:** 627–637.

41. ROUDBARAKI, M., A. LORSIGNOL, L. LANGOUCHE, *et al.* 1999. Target cells of gamma3-melanocyte-stimulating hormone detected through intracellular Ca^{2+} responses in immature rat pituitary constitute a fraction of all main pituitary cell types, but mostly express multiple hormone phenotypes at the messenger ribonucleic acid level. Refractoriness to melanocortin-3 receptor blockade in the lacto-soma-totroph lineage. Endocrinology **140:** 4874–4885.

42. SEUNTJENS, E., A. HAUSPIE, H. VANKELECOM, *et al.* 2002. Ontogeny of plurihormonal cells in the anterior pituitary of the mouse, as studied by means of hormone mRNA detection in single cells. J. Neuroendocrinol. **14:** 611–619.

43. HAUSPIE, A., E. SEUNTJENS, H. VANKELECOM, *et al.* 2003. Stimulation of combinatorial expression of prolactin and glycoprotein hormone alpha-subunit genes by GnRH and estradiol-17beta in single rat pituitary cells during aggregate cell culture. Endocrinology **144:** 388–399.

44. LORSIGNOL, A., V. VANDE VIJVER, D. RAMAEKERS, *et al.* 1999. Detection of melanocortin-3 receptor mRNA in immature rat pituitary: functional relation to gamma3-MSH-induced changes in intracellular Ca^{2+} concentration? J. Neuroendocrinol. **11:** 171–179.

45. HRUBY, V.J., D. LU, S.D. SHARMA, *et al.* 1995. Cyclic lactam a-melanotropin analogues of Ac-Nle4-cyclo[Asp5,D-Phe7,Lys10] a-melanocyte-stimulating hormone-(4–10)-NH2 with bulky aromatic amino acids at position 7 show high antagonist potency and selectivity at specific melanocortin receptors. J. Med. Chem. **38:** 3454–3461.

46. LANGOUCHE, L., M. ROUDBARAKI, K. PALS, *et al.* 2001. Stimulation of intracellular free calcium in GH3 cells by gamma3-melanocyte-stimulating hormone. Involvement of a novel melanocortin receptor? Endocrinology **142:** 257–266.

47. LANGOUCHE, L., K. PALS & C. DENEF. 2002. Structure-activity relationship and signal transduction of gamma-MSH peptides in GH3 cells: further evidence for a new melanocortin receptor. Peptides **23:** 1077–1086.

48. NIJSEN, M.J., G.J. DE RUITER, C.M. KASBERGEN, *et al.* 2000. Relevance of the C-terminal Arg-Phe sequence in gamma(2)-melanocyte-stimulating hormone (gamma(2)-MSH) for inducing cardiovascular effects in conscious rats. Br. J. Pharmacol. **131:** 1468–1474.
49. LI, S.J., K. VARGA, P. ARCHER, *et al.* 1996. Melanocortin antagonists define two distinct pathways of cardiovascular control by alpha- and gamma-melanocyte-stimulating hormones. J. Neurosci. **16:** 5182–5188.
50. KLUSA, V., S. GERMANE, S. SVIRSKIS, *et al.* 2001. The gamma(2)-MSH peptide mediates a central analgesic effect via a GABA-ergic mechanism that is independent from activation of melanocortin receptors. Neuropeptides **35:** 50–57.
51. LU, J., E. SWINNEN, P. PROOST, *et al.* 2002. Isolation and structure-bioactivity characterization of glycosylated N-POMC isoforms. J. Neuroendocrinol. **14:** 869–879.

New Insights into the Functions of α-MSH and Related Peptides in the Immune System

THOMAS A. LUGER, THOMAS E. SCHOLZEN, THOMAS BRZOSKA, AND MARKUS BÖHM

Department of Dermatology and Ludwig Boltzmann Institute for Cell Biology and Immunobiology of the Skin, University of Münster, 48149 Münster, Germany

ABSTRACT: There is a substantial body of evidence that the tridecapeptide α-melanocyte-stimulating hormone (α-MSH) functions as a mediator of immunity and inflammation. The immunomodulating capacity of α-MSH is primarily because of its effects on melanocortin receptor (MC-1R)–expressing monocytes, macrophages, and dendritic cells (DCs). α-MSH down-regulates the production of proinflammatory and immunomodulating cytokines (IL-1, IL-6, TNF-α, IL-2, IFN-γ, IL-4, IL-13) as well as the expression of costimulatory molecules (CD86, CD40, ICAM-1) on antigen-presenting DCs. In contrast, the production of the cytokine synthesis inhibitor IL-10 is up-regulated by α-MSH. At the molecular level, these effects of α-MSH are mediated via the inhibition of the activation of transcription factors such as NFκB. Not only α-MSH but also its C-terminal tripeptide (α-MSH 11-13, KPV) was able to bind to MC-1R and to modulate the function of APCs. *In vivo*, using a mouse model of contact hypersensitivity (CHS) systemic and topical application of α-MSH or KPV inhibited the sensitization and the elicitation phase of CHS and was able to induce hapten-specific tolerance. To investigate the underlying mechanisms of tolerance induction, we have performed *in vivo* transfer experiments. Treatment of naive mice with bone marrow–derived immature haptenized and α-MSH–pulsed DCs resulted in a significant inhibition of CHS. Furthermore, tolerance induction was found to be mediated by the generation of CTLA4$^+$ and IL-10–producing T lymphocytes. The potent capacity of α-MSH to modulate the function of antigen-presenting cells (APCs) has been further supported in another experimental approach. *In vitro*, by activating APCs, α-MSH has been shown to modulate IgE production by IL-4 and anti-CD40 stimulated B lymphocytes. Moreover, in a murine model of allergic airway inflammation, systemic treatment with α-MSH resulted in a significant reduction of allergen-specific IgE production, eosinophil influx, and IL-4 production. These effects were mediated via IL-10 production, because IL-10 knockout mice were resistant to α-MSH treatment. Therefore, therapeutic application of α-MSH or related peptides (KPVs) as well as α-MSH/KPV–pulsed DCs may be a useful approach for the treatment of inflammatory, autoimmune, and allergic diseases in the future.

KEYWORDS: skin; immune reaction; inflammation; neuropeptide

Address for correspondence: Thomas A. Luger, Department of Dermatology and Ludwig Boltzmann Institute for Cell Biology and Immunobiology of the Skin, University of Münster, 48149 Münster, Germany. Voice: 49-251-835-6504; fax: +49-251-835-6522.
luger@uni-muenster.de

Ann. N.Y. Acad. Sci. 994: 133–140 (2003). © 2003 New York Academy of Sciences.

INTRODUCTION

The cross-talk between immunocompetent cells as well as their ability to mount different types of immune responses is largely dependent on the presence of costimulatory signals, soluble mediators, and the expression of their respective receptors. Accordingly, a complex network of communicating signals comprising cytokines, chemokines, and growth factors are crucial components in the regulation of immunity and inflammation.[1] These mediators are extremely potent at picomolar concentrations and act by binding to specific cell surface receptors. In addition to these mediators of an immune response, it became evident that many neuropeptides exert cytokine-like activities and may influence the outcome of an immune response in a critical manner.[2,3] Among several neuropeptides, the proopiomelanocortin (POMC)–derived melanocortin α-melanocyte-stimulating hormone (α-MSH) recently has been recognized to exhibit antiinflammatory as well as immunomodulating activities.[3–5] In addition to their pituitary origin, POMC peptides also have been detected in several other organs and cells including epithelial cells, endothelial cells, and immunocompetent cells.[3,4] POMC peptides exert their various biological effects via binding to specific receptors belonging to the group of G-protein–coupled receptors with seven transmembrane domains.[6,7] This article briefly summarizes the role of α-MSH as well as α-MSH–derived peptides in the regulation of immune and inflammatory responses (TABLE 1).

EXPRESSION OF MELANOCORTIN RECEPTORS ON ANTIGEN-PRESENTING CELLS

Several studies have addressed the question whether any of the known melanocortin (MC) receptors is expressed on antigen-presenting cells (APCs) required for the initiation of an immune response.[8] Human as well as murine peripheral blood–derived monocytes and macrophages as well as monocytic cell lines (THP-1, RAW 264.7) have been shown to express the MC-1R but none of the other known MC-Rs. Usually, MC-1R expression on monocytic cells is low and significantly enhanced upon encountering inflammatory stimuli such as bacterial endotoxins and mitogens.[9–11] However, there is recent evidence for MC-3R expression on murine macrophages and its possible role in the down-regulation of an inflammatory reaction.[12,13] Moreover,

TABLE 1. Immunomodulating effects of α-MSH

- Down-regulation of proinflammatory cytokines: IL-1, IL-6, TNF-α
- Down-regulation of immunomodulating cytokines: IL-2, IL-4, IL-13, IFN-γ
- Up-regulation of IL-10
- Down-regulation of costimulatory molecules: CD40, CD86
- Down-regulation of MHC class I expression
- Down-regulation of adhesion molecule expression: ICAM-1, VCAM-1, E-selectin
- Down-regulation of IgE production
- Down-regulation of NO production

human peripheral blood–derived dendritic cells (DCs), when cultured in the presence of GM-CSF, IL-4, and monocyte conditioned medium, were found to express MC-1R. In comparison with immature DCs after 5 days of culture, matured DCs at day 8 expressed the highest levels of MC-1R.[14] In addition, MC-1R also was detected on other cells being involved in the modulation of immune and inflammatory responses such as neutrophils, mast cells, keratinocytes, and endothelial cells.[15–19] The observation that immunocompetent cells are capable of expressing MC-1R with α-MSH as the major ligand indicates that this melanocortin may affect their functions and thereby play a role in regulation of an immune response. In contrast, melanocortins such as β-MSH and γ-MSH which require the expression of other MC-Rs exert a less pronounced immunomodulating activity. The binding of melanocortins to MC-1R is known to be blocked by the agouti signaling protein (ASIP) which is encoded by the agouti locus. Accordingly, ASIP inhibits the effects of α-MSH on melanocytes,[19] but it is not yet fully elucidated whether it also neutralizes α-MSH–mediated functions of immunocompetent cells.

EFFECT OF α-MSH ON ANTIGEN-PRESENTING CELLS

There is evidence from several studies that α-MSH is capable of modulating the function of MC-1R–expressing APCs (TABLE 1). α-MSH in a dose-dependent manner was found to up-regulate the mRNA expression and release of the cytokine synthesis inhibitor IL-10 in monocytes.[20] In contrast, α-MSH down-regulates the synthesis and release of proinflammatory cytokines such as IL-1, IL-6, and TNF-α[3] as well as the production of proinflammatory nitric oxide and neopterin by macrophages.[11] α-MSH also turned out to be a potent inhibitor of IL-1–mediated effects such as thymocyte proliferation and fever induction.[21] By modulating the secretion of chemokines such as IL-8 and Gro-α, α-MSH might be able to regulate the migration of monocytes and other inflammatory cells.[22]

In addition to its effects on cytokine production and activities, there is also evidence that α-MSH affects the expression of MHC antigens and costimulatory molecules which are required to mount an effective immune response. Accordingly, α-MSH in a dose-dependent manner down-regulates the expression of MHC class I molecules on monocytes and significantly suppresses the expression of CD86 and CD40 on monocytes and DCs, whereas CD80 expression was not substantially altered.[9,14] To further elucidate the mechanism of α-MSH–mediated immunomodulation, we performed expression monitoring by hybridrization of cRNA samples from murine bone marrow–derived immature DCs to high-density oligonucleotide arrays. A 2-h stimulation with the soluble antigen DNBS resulted in induction and/or up-regulation of many genes encoding several proinflammatory cytokines (e.g., IL-1α), growth factors (e.g., hepatoma-derived growth factor), signal transduction intermediates (e.g., MAP kinase kinase 3), and transcription factors (e.g., heat shock transcription factor 1). α-MSH treatment of DCs significantly reduced the DNBS-driven induction or up-regulation of the above gene products and modulated the amount of mRNA for several novel regulatory molecules whose exact role in α-MSH–mediated immunomodulation awaits further confirmation. On the basis of the previously reported findings, suppression of costimulatory signals such as CD86 and CD40 and induction of suppressor factors such as IL-10, however, may represent some of the α-MSH–mediated signals required for the down-regulation of an immune response and possibly the induction of tolerance (TABLE 1).

The molecular mechanisms underlying the immunomodulating functions of α-MSH are not yet completely understood. Among several intracellular signaling pathways, the activation of the transcription factor NFκB is a crucial event in any immune and inflammatory response. NFκB is activated by proinflammatory cytokines such as IL-1, TNF-α, and endotoxin and other stimuli.[23] It controls the expression of many genes involved in inflammation including cytokines, chemokines MHC class I, adhesion molecules or nitric oxide synthase. Upon treatment with α-MSH, the TNF-α, IL-1, liposaccharide, and ceramide-mediated activation of NFκB were significantly down-regulated in a dose- and time-dependent manner. Moreover, the TNF-α–mediated degradation of the inhibitory subunit IκBα and the nuclear translocation of the p65 subunit of NFκB was inhibited.[24,25] The α-MSH–mediated inhibition of NFκB activation appears to be dependent on cAMP.[26] Suppression of NFκB activation was not cell specific and could be observed in monocytes, monocytic cell lines, keratinocytes, melanocytes, and endothelial cells.[24,25,27] Therefore, α-MSH appears to exert its antiinflammatory and immunomodulating effects by generally inhibiting NFκB activation. This has been supported by a recent observation indicating that in HIV-infected macrophages α-MSH and its C-terminal tripeptide reduce viral expression as well as replication, most likely via inhibition of NF-κB activation that is known to enhance viral expression.[28]

The *in vitro* immunomodulating capacity of α-MSH was supported by a murine model of cutaneous hypersensitivity (CHS). Systemic or topical application of α-MSH before sensitization with potent contact allergens inhibits both the sensitization as well as the elicitation of CHS and induces hapten-specific tolerance. In addi-

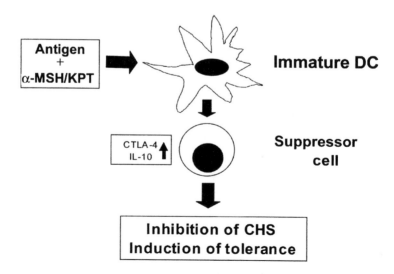

FIGURE 1. Possible role of α-MSH in the treatment of allergic and autoimmune diseases.

tion, the C-terminal tripeptide of α-MSH (KPV) turned out to be responsible for its immunosuppressive activity.[29,30] The α-MSH–mediated tolerance induction appears to be mediated via the up-regulation of suppressor factors such as IL-10 because the CHS inhibition by α-MSH partly could be blocked by the subcutaneous administration of IL-10 antibodies. Furthermore, a single intravenous injection of α-MSH into mice resulted in a significant elevation of IL-10 plasma levels which lasted for more than 1 week (unpublished observation). Therefore, α-MSH may impair, either directly or indirectly, via IL-10 induction, the function of APCs and direct the outcome of an immune response toward tolerance.

To further evaluate the effect of α-MSH on APCs *in vivo*, we generated bone marrow–derived immature murine DCs and treated them *in vitro* with the soluble antigen DNBS alone or in the presence of α-MSH or KPVs. Subsequently, cells were injected intravenously into naive syngenic mice. Upon epicutaneous challenge with the same hapten (DNFB), the ear swelling response was significantly reduced in mice treated with haptenized and α-MSH or KPV-pulsed DCs. These animals could not be resensitized with the same hapten after 2 weeks, suggesting the induction of immune tolerance. Furthermore, CHS inhibition and tolerance induction was demonstrated to be caused by the generation of a subpopulation of IL-10–producing CTLA4+, CD4+, or CD8+ suppressor lymphocytes[31] (FIG. 1).

EFFECT OF α-MSH IN IMMEDIATE-TYPE IMMUNE REACTIONS

There is evidence from several investigations that α-MSH also seems to function as a modulator of immediate-type immune reactions. Accordingly, α-MSH directly affects mast cell functions and down-regulates histamine release as well as mRNA expression of IL-1, TNF-α, and lymphotactin.[32,33] α-MSH also has been found to play a role in the regulation of the antibody synthesis by B lymphocytes. This melanocortin appears to exert its activities mainly in concert with other B-cell stimulatory cytokines. Depending on which stimuli or cytokines are present, α-MSH either inhibits or enhances antibody production. Accordingly, α-MSH was found to modulate the IL-4 and anti-CD40–mediated IgE release by human peripheral blood mononuclear cells. At low physiological concentrations, α-MSH increases IgE synthesis, whereas higher concentrations significantly inhibit IgE production by B lymphocytes.[34,35] As is the case with T cells, α-MSH seems to have no direct effect on B cells. Accordingly, the α-MSH–mediated IgE production was strictly dependent on the presence of APCs in the culture, which are known to regulate IgE synthesis via the release of soluble mediators such as IL-4 and IL-10.

The *in vivo* functional role of α-MSH in immediate-type hypersensitivity was investigated utilizing a well-characterized model of murine bronchial asthma that is defined by the development of airway hyperresponsiveness and inflammation with Th-2–cell influx, eosinophilia as well as IgE/IgG1 antibody production.[36] In the bronchoalveolar lavage (BAL) fluid of mice with allergic airway inflammation, significantly decreased levels of α-MSH were detected. To investigate the effect of α-MSH on the allergic response, we treated mice intravenously with α-MSH during sensitization with the allergen. Markedly decreased levels of allergen-specific IgE, but not total IgE, were detected in α-MSH/allergen–treated mice as compared with controls treated with allergen alone. Moreover, α-MSH was able to prevent the influx of eosinophils and significantly decreased IL-4 and IL-13 levels in the BAL. Interestingly, IL-10 levels in BAL were up-regulated upon exposure to α-MSH. To analyze whether the inhibitory effect of α-MSH on allergic airway inflammation is mediated via the enhanced IL-10 production, we examined IL-10 knockout animals. In contrast with wild-type mice, α-MSH had no effect on allergen-specific IgE production, eosinophil influx, or BAL cytokine levels in IL-10 knockout animals.[36] These findings indicate that similar to CHS the immunomodulating effects of α-MSH in immediate-type immune reactions such as allergic airway inflammation appear to depend on the production of IL-10.

CONCLUSION

The POMC-derived tridecapeptide α-MSH appears to exert multiple immunomodulating effects which play a role in the regulation of T- and B-cell functions. Using animal models, we found that α-MSH prevents sensitization with potent contact allergens, induces immune tolerance, and prevents immediate-type airway hypersensitivity. These immunomodulating effects of α-MSH appear to be predominantly mediated by its effect on MC-1R–expressing antigen-presenting cells. Another important underlying mechanism seems to be the capacity of α-MSH to function as a strong inducer of IL-10. First clinical trials using IL-10 for the treatment of autoimmune-mediated diseases such as rheumatoid arthritis or psoriasis are promising. Therefore, α-MSH or α-MSH–derived peptides such as KPVs with the advantage of their smaller size and lower antigenicity may be developed in the future as immunomodulating drugs which will open new therapeutic strategies for the treatment of autoimmune and allergic diseases.

ACKNOWLEDGMENTS

This work was supported by grants of the Deutsche Forschungsgemeinschaft (Scho 629/1-1, 629/1-3), Interdisziplinäres Zentrum für klinische Forschung (IZKF/D 15 and 16), the Centre de Recherche et Investigations Epidermiques et Sensorielles C.E.R.I.E.S. Paris, and the Integrated Functional Genomics service unit of the IZKF Münster. We are grateful for technical assistance by Kerstin Klimmek.

REFERENCES

1. BEISSERT, S. & T. SCHWARZ. 1999. Mechanisms invoved in ultraviolet light-induced immunosuppression. J. Invest. Dermatol. Symp. Proc. **4:** 61–64.
2. BLALOCK, J.E. 1994. The syntax of immune-neuroendocrine communication. Immunol. Today **15:** 504–511.
3. LIPTON, J.M. & A. CATANIA. 1997. Antiinflammatory actions of the neuroimmunomodulator α-MSH. Immunol.Today **18:** 140–145.
4. LUGER, T.A., T. SCHOLZEN & S. GRABBE. 1997. The role of α-melanocyte stimulating hormone in cutaneous biology. J. Invest. Dermatol. Symp. Proc. **2:** 87–93.
5. EBERLE, A.N. 1988. The Melanotropins. Karger. Basel.
6. PASTERNAK, G.W. 1988. Multiple morphine and enkephalin receptors and the relief of pain. JAMA **259:** 1362–1367.

7. WIKBERG, J.E., R. MUCENIECE, I. MANDRIKA, *et al.* 2000. New aspects on the melanocortins and their receptors. Pharmacol. Res. **42:** 393–420.

8. JONULEIT, H., E. SCHMITT, K. STEINBRINK, *et al.* 2001. Dendritic cells as a tool to induce anergic and regulatory T cells. Trends Immunol. **22:** 394–400.

9. BHARDWAJ, R.S., E. BECHER, K. MAHNKE, *et al.* 1997. Evidence for the differential expression of the functional alpha melanocyte stimulating hormone receptor MC-1 on human monocytes. J. Immunol. **158:** 3378–3384.

10. STAR, R.A., N. RAJORA, J. HUANG, *et al.* 1995. Evidence of autocrine modulation of macrophage nitric oxide synthase by alpha-MSH. Proc. Natl. Acad. Sci. USA **92:** 8016–8020.

11. RAJORA, N., G. CERIANI, A. CATANIA, *et al.* 1996. alpha-MSH production, receptors, and influence on neopterin in a human monocyte/macrophage cell line. J. Leukoc. Biol. **59:** 248–253.

12. GETTING, S.J., L. GIBBS, A.J. CLARK, *et al.* 1999. POMC gene-derived peptides activate melanocortin type 3 receptor on murine macrophages, suppress cytokine release, and inhibit neutrophil migration in acute experimental inflammation. J. Immunol. **162:** 7446–7453.

13. GETTING, S.J., G.H. ALLCOCK, R. FLOWER, *et al.* 2001. Natural and synthetic agonists of the melanocortin receptor type 3 possess anti-inflammatory properties. J. Leukoc. Biol. **69:** 98–104.

14. BECHER, E., K. MAHNKE, T. BRZOSKA, *et al.* 1999. Human peripheral blood-derived dendritic cells express functional melanocortin receptor MC-1R. Ann. N. Y. Acad. Sci. **885:** 188–195.

15. CATANIA, A., N. RAJORA, F. CAPSONI, *et al.* 1996. The neuropeptide alpha-MSH has specific receptors on neutrophils and reduces chemotaxis in vitro. Peptides **17:** 675–679.

16. HARTMEYER, M., T. SCHOLZEN, E. BECHER, *et al.* 1997. Human microvascular endothelial cells (HMEC-1) express the melanocortin receptor type 1 and produce increased levels of IL-8 upon stimulation with α-MSH. J. Immunol. **159:** 1930–1937.

17. BRZOSKA, T., T. SCHOLZEN, E. BECHER, *et al.* 1997. Effect of UV light on the production of proopiomelanocortin-derived peptides and melanocortin receptors in the skin. *In* Skin Cancer and UV-Radiation. P. Altmeyer, K. Hoffmann & M. Stücker, Eds.: 227–237. Springer-Verlag. Berlin.

18. ARTUC, M., A. GRUTZKAU, T. LUGER, *et al.* 1999. Expression of MC1- and MC5-receptors on the human mast cell line HMC-1. Ann. N. Y. Acad. Sci. **885:** 364–367.

19. SUZUKI, I., A. TADA, M.M. OLLMANN, *et al.* 1997. Agouti signaling protein inhibits melanogenesis and the response of human melanocytes to alpha-melanotropin. J. Invest. Dermatol. **108:** 838–842.

20. BHARDWAJ, R.S., A. SCHWARZ, E. BECHER, *et al.* 1996. Pro-opiomelanocortin-derived peptides induce IL-10 production in human monocytes. J. Immunol. **156:** 2517–2521.

21. HUANG, Q.H., V.J. HRUBY & J.B. TATRO. 1998. Systemic alpha-MSH suppresses LPS fever via central melanocortin receptors independently of its suppression of corticosterone and IL-6 release. Am. J. Physiol. **275:** R524–R530.

22. SCHOLZEN, T.E., T. BRZOSKA, D.H. KALDEN, *et al.* 1999. Expression of functional melanocortin receptors and proopiomelanocortin peptides by human dermal microvascular endothelial cells. Ann. N. Y. Acad. Sci. **885:** 239–253.

23. BAEUERLE, P.A. & D. BALTIMORE. 1996. NF-kappa B: ten years after. Cell **87:** 13–20.

24. BRZOSKA, T., D.H. KALDEN, T. SCHOLZEN, *et al.* 1999. Molecular basis of α-MSH/IL-1 antagonism. Ann. N. Y. Acad. Sci. **885:** 230–238.

25. MANNA, S.K. & B.B. AGGARWAL. 1998. Alpha-melanocyte-stimulating hormone inhibits the nuclear transcription factor NF-kappa B activation induced by various inflammatory agents. J. Immunol. **161:** 2873–2880.

26. MANDRIKA, I., R. MUCENIECE & J.E. WIKBERG. 2001. Effects of melanocortin peptides on lipopolysaccharide and interferon gamma-induced NFκB DNA binding and nitric oxide production in macrophage-like RAW 2647 cells: evidence for dual mechanisms of action. Biochem. Pharmacol. **61:** 613–621.

27. KALDEN, D.H., T. SCHOLZEN, T. BRZOSKA, *et al.* 1999. Mechanisms of the antiinflammatory effects of α-MSH: role of transcription factor NF- κB and adhesion molecule expression. Ann. N. Y. Acad. Sci. **885:** 254–261.

28. BARCELLINI, W., G. COLOMBO, L. LA MAESTRA, *et al.* 2000. Alpha-melanocyte-stimulating hormone peptides inhibit HIV-1 expression in chronically infected promonocytic U1 cells and in acutely infected monocytes. J. Leukoc. Biol. **68:** 693–699.
29. GRABBE, S., R.S. BHARDWAJ, M. STEINERT, *et al.* 1996. Alpha-melanocyte stimulating hormone induces hapten-specific tolerance in mice. J. Immunol. **156:** 473–478.
30. SLOMINSKI, A., J. WORTSMAN, T. LUGER, *et al.* 2000. Corticotropin releasing hormone and proopiomelanocortin involvement in the cutaneous response to stress. Physiol. Rev. **80:** 979–1020.
31. BRZOSKA, T., A. SCHWARZ, M. MOELLER, *et al.* 2001. Treatment of murine BMDC with α-MSH results in the generation of T-suppressor cells. J. Invest. Dermatol. **117:** 453.
32. ADACHI, S., T. NAKANO, H. VLIAGOFTIS, *et al.* 1999. Receptor-mediated modulation of murine mast cell function by alpha-melanocyte stimulating hormone. J. Immunol. **163:** 3363–3368.
33. GRUTZKAU, A., B.M. HENZ, L. KIRCHHOF, *et al.* 2000. alpha-Melanocyte stimulating hormone acts as a selective inducer of secretory functions in human mast cells. Biochem. Biophys. Res. Commun. **278:** 14–19.
34. AEBISCHER, I., M.R. STÄMPFLI, A. ZÜRCHER, *et al.* 1994. Neuropeptides are potent modulators of human in vitro immunoglobulin E synthesis. Eur. J. Immunol. **24:** 1908–1913.
35. AEBISCHER, I., M. STÄMPFLI, S. MIESCHER, *et al.* 1996. Neuropeptides accentuate interleukin-4 induced human immunoglobulin E synthesis in vitro. Exp. Dermatol. **5:** 38–44.
36. RAAP, U., T. BRZOSKA, G. PÄTH, *et al.* 2001. α-melanocyte stimulating hormone (α-MSH) is a modulator of immune responses in allergen-induced airway inflammation in vivo. Allergy **56:** 21.

Regulation of PPARγ and Obesity by Agouti/Melanocortin Signaling in Adipocytes

RANDALL L. MYNATT[a,b] AND JACQUELINE M. STEPHENS[a,b]

[a]Pennington Biomedical Research Center, Baton Rouge, Louisiana 70808, USA

[b]Louisiana State University, Department of Biological Sciences, Baton Rouge, Louisiana 70803, USA

ABSTRACT: To study the potential biological role of agouti/melanocortin signaling in human adipose tissue, we engineered transgenic mice to overexpress agouti in adipose tissue. The aP2-agouti transgenic mice become significantly heavier than littermates. The increased body weight is maintained at approximately 15% above nontransgenic mice through 20 weeks and is caused by increased fat mass. The obesity is increased by a high-fat diet. There is no change in food intake in the aP2-agouti mice suggesting changes in energy utilization. A possible mechanism is that the agouti/melanocortin signaling regulates levels of PPARγ. PPARγ functions as a major regulator of adipocyte differentiation and as a receptor for the antidiabetic thiazolidinediones. Agouti increases PPARγ protein levels in differentiated 3T3-L1 adipocytes, and PPARγ expression is elevated in the fat pads of the aP2-agouti transgenic mice. The modest weight gain observed in the transgenic mice suggests that hypothalamic pathways regulating food intake are intact and the observed adiposity is within ranges that can be achieved by a paracrine mechanism at the adipocyte level.

KEYWORDS: agouti; adipocytes; PPARγ; obesity

The agouti peptide (131 amino acids) is synthesized in the hair follicles and acts in a paracrine fashion on melanocytes where it antagonizes the binding of α-melanocyte-stimulating hormone (α-MSH) to melanocortin-1 receptor (MC1-R).[1] The transcription of the wild-type mouse *agouti* is temporally regulated, being expressed solely in the skin during part of the hair-growth cycle. However, several dominant alleles at the mouse *agouti* locus cause a widespread, ectopic *agouti* pattern of expression and are associated phenotypically with yellow fur, late-onset obesity, hyperphagia, increased linear growth, non–insulin-dependent diabetes, and increased tumorogenesis.[2–6] The human homologue of the *agouti* gene is 85% identical to the mouse gene and encodes a protein of 132 amino acids.[7] The major difference between mouse and human *agouti* is the expression pattern. Although mouse *agouti* is transiently expressed only in the hair follicle,[3] human *agouti* is expressed

Address for correspondence: Randall L. Mynatt, Pennington Biomedical Research Center, 6400 Perkins Rd., Baton Rouge, LA 70808. Voice: 225-763-3100; fax: 225-763-2525.
mynattrl@pbrc.edu

Ann. N.Y. Acad. Sci. 1: 141–146 (2003). © 2003 New York Academy of Sciences.

in diverse tissues: adipose tissue, testis, heart, liver, kidney, ovary, and foreskin.[7,8] Aside from analysis of the structure of the human *agouti* gene, there are no publications on its regulation or why the tissue expression pattern is so different between man and mice. One can speculate that humans have lost the need for camouflage and agouti has a different role in humans.

ACTH and α-MSH are potent lipolytic hormones with high-affinity binding sites on adipocytes.[9] By using RT-PCR and Northern blot hybridization, Boston and Cone[10] showed that both MC2 and MC5 receptors are expressed in differentiated 3T3-L1 adipocytes and mouse adipose tissues, but not in preadipocytes. Mountjoy *et al.*[11] have shown expression of MC1, MC2, and MC5 receptor mRNA in differentiated 3T3F442A adipocytes using RT-PCR, and in both white adipose tissue (MC1-R and MC2-R) and brown adipose tissue (MC2-R and MC5-R) in mice.[11] The mRNA for all five melanocortin receptors has been detected by RT-PCR in human subcutaneous adipose tissue.[12] Expression of melanocortin receptors and agouti in human fat raises questions as to their normal function in adipocyte biology and whether perturbation of the agouti/melanocortin system in fat might contribute to obesity and diabetes.

To understand the role of agouti in human fat, we made transgenic mice (aP2-agouti) to express *agouti* in adipose tissue.[13] FIGURE 1A compares body weights between transgenic mice and wild-type littermates from 4 to 20 weeks of age. The BAP-agouti transgenic mice express agouti ubiquitously and are a model for systemic melanocortin receptor antagonism and morbid obesity. The aP2-agouti transgenic mice were heavier than littermates at all times, and the difference became statistically significant by 8 to 10 weeks. FIGURE 1B compares individual fat depots between aP2-agouti mice and their littermates. The increased body weight correlated with increased mass of all fat depots. Combined fat depot weight increased from 30% to 50% at 10 weeks in individual transgenic mice compared with littermates. Total RNA was extracted from the combined fat depots from 10-week-old mice and leptin mRNA levels were compared by Northern blot analysis (data not shown). In accordance with increased fat mass, leptin mRNA levels were significantly higher in the 10-week-old transgenic mice. Circulating levels of leptin are 2.3-fold higher ($P = 0.02$) in 10-week-old transgenic mice. These data indicate that the correlation of fat mass and plasma leptin is normal in aP2-agouti mice and the increased weight gain cannot be attributed to decreased leptin secretion by adipocytes.

Mutations that interfere with the neuronal control of body weight homeostasis generally cause morbid obesity and may not reflect the modest increase in body mass index seen in the population as a whole. Alternatively, autocrine/paracrine effects in adipose tissue may cause changes in body weight that are clinically significant for the comorbidities that are associated with obesity. The degree of obesity observed in the aP2-agouti transgenic mice suggests that the normal hypothalamic pathways regulating body weight are intact. A hallmark feature in hypothalamic-induced obesity in mice is hyperphagia. Food intake was measured daily from weaning until 10 weeks of age in aP2-agouti transgenic ($n = 8$) and wild-type mice ($n = 9$) fed a 45% kcal fat diet (Research Diets no. D12451). After a 1-week acclimation period to metabolic cages, food intake and body weight were measured every other day for 5 weeks. There were no differences in food intake between transgenic and nontransgenic mice (FIG. 1D). However, the transgenic mice gained significantly more body weight (FIG. 1C). These data demonstrate that the increased fat mass and obesity in

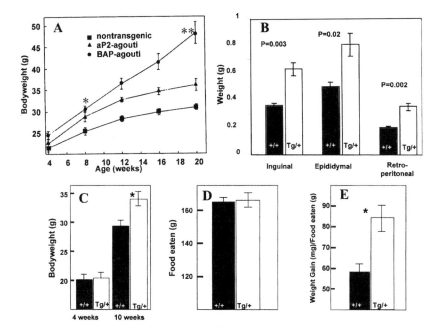

FIGURE 1. (A) Comparison of body weights between transgenic and nontransgenic mice. All mice were maintained on the FVB/N background, fed a diet containing 11% fat by weight, and weaned at 21 to 25 days of age. Data are from mice that are hemizygous for the transgene or their nontransgenic littermates. There were between 12 and 28 mice per data point. Data are presented as mean ± SEM. *Single asterisk* indicates that the value is significantly different ($P \geq 0.01$) from that of nontransgenic mice. *Double asterisks* indicate that the value is significantly different ($P \geq 0.01$) from that of nontransgenic and aP2-agouti mice. (B) Comparison of fat pad weight between aP2-agouti mice and nontransgenic littermates at 10 weeks of age. Data are presented as mean ± SEM; $n = 10$. (C–E) Food intake and weight gain on high-fat diet. Weanling aP2-agouti transgenic ($n = 8$) and wild-type mice ($n = 9$) were fed a 45% kcal fat diet (Research Diets no. D12451) for 6 weeks. Food intake and body weight were measured daily. Data are from mice that are hemizygous for the transgene or their nontransgenic littermates. Data are presented as mean ± SEM. *Asterisk* indicates that the value is significantly different ($P \geq 0.01$) from that of nontransgenic mice.

aP2-agouti transgenic mice are not caused by changes in food intake. Therefore, changes in energy expenditure are the likely causes for the increased adiposity in the aP2-agouti mice. One interpretation is that agouti is affecting nutrient partitioning into adipose tissue possibly by inhibiting melanocortin-stimulated lipolysis. In addition, agouti may be decreasing energy expenditure by inhibiting thermogenesis in brown adipose tissue (BAT). It will be interesting to determine overall energy expenditure and BAT thermogenesis in aP2-agouti mice.

Several studies have shown that PPARγ is an essential transcription factor for differentiation and maturation of adipocytes.[14–20] In addition, ectopic expression of PPARγ in nonadipogenic fibroblast promotes lipid accumulation and characteristics of mature adipocytes.[21] Therefore, protein levels of PPARγ in retroperitoneal fat

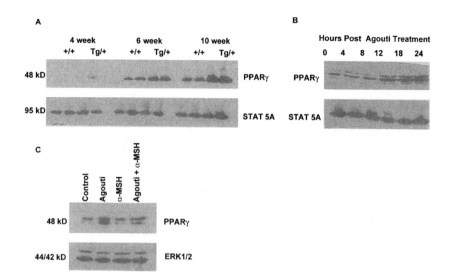

FIGURE 2. Effects of agouti on PPARγ. (**A**) Retroperitoneal fat pad extracts were prepared from aP2-agouti and nontransgenic mice and subjected to Western blot analysis. (**B**) Whole-cell extracts were prepared from fully differentiated 3T3-L1 adipocytes after a treatment with agouti for 0, 4, 8, 12, 18, and 24 h. (**C**) Whole-cell extracts were prepared from fully differentiated 3T3-L1 adipocytes after a treatment with agouti, NDP–α-MSH, or both for 12 h.

pads from transgenic mice and wild-type littermates between 4 and 10 weeks of age were examined. We previously had shown that PPARγ levels were elevated in aP2-agouti transgenic mice at 10 weeks of age.[22] In this study, the levels of this transcription factor in retroperitoneal fat pads from transgenic mice and wild-type littermates between 4 and 10 weeks of age were examined. FIGURE 2A is a representative Western blot showing PPARγ levels from 4 to 10 weeks of age. We observed relatively low levels of PPARγ at 4 weeks of age with no detectable difference in PPARγ levels between wild-type and transgenic mice. However, by 6 weeks of age PPARγ levels had increased significantly in both genotypes of mice, and there was a detectable, but not statistically significant, increase in PPARγ expression in the transgenic mice compared with wild-type controls. However, by 10 weeks of age, the expression of PPARγ was significantly elevated in transgenic animals compared with nontransgenic littermates. The PPARγ levels are not significantly different in nontransgenic mice between 6 and 10 weeks of age, suggesting the PPARγ levels reach maximum levels by 6 weeks in nontransgenic mice. In contrast, PPARγ levels continually increase in aP2-agouti mice with age through 10 weeks.

To examine the direct effects of agouti on adipocytes, we treated mature 3T3-L1 adipocytes with recombinant murine agouti. As with the adipose tissue from transgenic mice, PPARγ was elevated in the agouti-treated 3T3-L1 adipocytes (FIG. 2B). STAT 5A levels were unchanged. Dose-response curves from 0.5 to 200 nM agouti demonstrated a near maximum effect of agouti at 50 nM (data not shown), which is in agreement with inhibition of NDP–α-MSH binding to cells expressing melanocortin receptors.[23-25]

In contrast with the immense interest in ligands and antagonists for melanocortin receptors, information on melanocortin signaling has been slow to develop. To date, most of agouti's effects on obesity and coat color have been attributed to inhibition of α-MSH binding. Experiments from Michael Zemel's laboratory[26] demonstrate that agouti stimulates intracellular calcium levels in cells cultured in serum-free media. Also, data from Naima Moustaid's laboratory[27] has shown that agouti increases fatty acid synthetase and increases lipid accumulation in differentiated 3T3-L1 adipocytes in serum-free media. Genetic experiments from Greg Barsh's laboratory suggest that α-MSH and agouti protein function as independent ligands that inhibit each other's binding and transduce opposite signals through a single receptor to regulate coat color.[28] Treatment of the 3T3-L1 adipocytes with NDP–α-MSH did not affect PPARγ but blocked the effects of agouti on PPARγ (FIG. 2C), suggesting that these effects are mediated through melanocortin receptors and that agouti may be the agonist and NDP–α-MSH the antagonist for PPARγ regulation.

It is well recognized that the agouti/melanocortin system is a critical component of body weight homeostasis. Many genetic and pharmacological studies have shown that agouti and agouti-related protein compete with proopiomelanocortin-derived peptides for binding sites on melanocortin receptors to regulate several biological functions, including body weight. The majority of obesity-related agouti/melanocortin research has focused on central nervous system control of food intake and energy expenditure. Data collected from transgenic mice that overexpress agouti in adipose tissue and adipocyte cell culture experiments lead us to propose that agouti/melanocortins are part of a communication circuit between the brain and adipose tissue which conveys centrally mediated signals (ACTH and α-MSH) and modulates adipocyte responsiveness to these signals (agouti).

ACKNOWLEDGMENTS

This work was supported by American Heart Association Scientist Development Grant 9630120N (R.L.M.) and R01DK52968-02 from the National Institutes of Health (J.M.S.).

REFERENCES

1. LU, D. et al. 1994. Agouti protein is an antagonist of the melanocyte-stimulating-hormone receptor. Nature 371: 799–802.
2. DUHL, D.M. et al. 1994. Pleiotropic effects of the mouse lethal yellow (Ay) mutation explained by deletion of a maternally expressed gene and the simultaneous production of agouti fusion RNAs. Development 120: 1695–1708.
3. BULTMAN, S.J., E.J. MICHAUD & R.P. WOYCHIK. 1992. Molecular characterization of the mouse agouti locus. Cell 71: 1195–1204.
4. MICHAUD, E.J. et al. 1993. The embryonic lethality of homozygous lethal yellow mice (Ay/Ay) is associated with the disruption of a novel RNA-binding protein. Genes Dev. 7: 1203–1213.
5. MICHAUD, E.J. et al. 1994. Differential expression of a new dominant agouti allele (Aiapy) is correlated with methylation state and is influenced by parental lineage. Genes Dev. 8: 1463–1472.

6. MILLER, M.W. *et al.* 1993. Cloning of the mouse agouti gene predicts a secreted protein ubiquitously expressed in mice carrying the lethal yellow mutation. Genes Dev. **7:** 454–467.

7. KWON, H.Y. *et al.* 1994. Molecular structure and chromosomal mapping of the human homolog of the agouti gene. Proc. Natl. Acad. Sci. USA **91:** 9760–9764.

8. WILSON, B.D. *et al.* 1995. Structure and function of ASP, the human homolog of the mouse agouti gene. Hum. Mol. Genet. **4:** 223–230.

9. BOSTON, B.A. 1999. The role of melanocortins in adipocyte function. Ann. N. Y. Acad. Sci. **885:** 75–84.

10. BOSTON, B.A. & R.D. CONE. 1996. Characterization of melanocortin receptor subtype expression in murine adipose tissues and in the 3T3-L1 cell line. Endocrinology **137:** 2043–2050.

11. MOUNTJOY, K.G. & J. WONG. 1997. Obesity, diabetes and functions for proopiomelanocortin-derived peptides. Mol. Cell. Endocrinol. **128:** 171–177.

12. CHAGNON, Y.C. *et al.* 1997. Linkage and association studies between the melanocortin receptors 4 and 5 genes and obesity-related phenotypes in the Quebec Family Study. Mol. Med. **3:** 663–673.

13. MYNATT, R.L. *et al.* 1997. Combined effects of insulin treatment and adipose tissue-specific agouti expression on the development of obesity. Proc. Natl. Acad. Sci. USA **94:** 919–922.

14. HAMM, J.K. *et al.* 1999. Role of PPAR gamma in regulating adipocyte differentiation and insulin-responsive glucose uptake. Ann. N. Y. Acad. Sci. **892:** 134–145.

15. MORRISON, R.F. & S.R. FARMER. 1999. Insights into the transcriptional control of adipocyte differentiation. J. Cell. Biochem. Suppl. **32–33:** 59–67.

16. MORRISON, R.F. & S.R. FARMER. 1999. Role of PPARgamma in regulating a cascade expression of cyclin-dependent kinase inhibitors, p18(INK4c) and p21(Waf1/Cip1), during adipogenesis. J. Biol. Chem. **274:** 17088–17097.

17. SPIEGELMAN, B.M. 1997. Peroxisome proliferator-activated receptor gamma: a key regulator of adipogenesis and systemic insulin sensitivity. Eur. J. Med. Res. **2:** 457–464.

18. STEPHENS, J.M. *et al.* 1999. PPARgamma ligand-dependent induction of STAT1, STAT5A, and STAT5B during adipogenesis. Biochem. Biophys. Res. Commun. **262:** 216–222.

19. TONTONOZ, P., E. HU & B.M. SPIEGELMAN. 1995. Regulation of adipocyte gene expression and differentiation by peroxisome proliferator activated receptor gamma. Curr. Opin. Genet. Dev. **5:** 571–576.

20. WU, Z. *et al.* 1999. Cross-regulation of C/EBP alpha and PPAR gamma controls the transcriptional pathway of adipogenesis and insulin sensitivity. Mol. Cell **3:** 151–158.

21. TONTONOZ, P., E. HU & B.M. SPIEGELMAN. 1994. Stimulation of adipogenesis in fibroblasts by PPAR gamma 2, a lipid-activated transcription factor [published erratum appears in Cell 1995; 80: 957]. Cell **79:** 1147–1156.

22. MYNATT, R.L. & J.M. STEPHENS. 2001. Agouti regulates adipocyte transcription factors. Am. J. Physiol. **280:** C954–C961.

23. YANG, Y.K. *et al.* 1997. Effects of recombinant agouti-signaling protein on melanocortin action. Mol. Endocrinol. **11:** 274–280.

24. KIEFER, L.L. *et al.* 1998. Melanocortin receptor binding determinants in the agouti protein. Biochemistry **37:** 991–997.

25. KIEFER, L.L. *et al.* 1997. Mutations in the carboxyl terminus of the agouti protein decrease agouti inhibition of ligand binding to the melanocortin receptors. Biochemistry **36:** 2084–2090.

26. ZEMEL, M.B. *et al.* 1995. Agouti regulation of intracellular calcium: role in the insulin resistance of viable yellow mice. Proc. Natl. Acad. Sci. USA **92:** 4733–4737.

27. JONES, B.H. *et al.* 1996. Upregulation of adipocyte metabolism by agouti protein: possible paracrine actions in yellow mouse obesity. Am. J. Physiol. **270:** E192–E196.

28. BARSH, G.S. *et al.* 1999. Molecular pharmacology of agouti protein in vitro and in vivo. Ann. N. Y. Acad. Sci. **885:** 143–152.

A Polymorphic Form of Steroidogenic Factor 1 Associated with ACTH Receptor Deficiency in Mouse Adrenal Cell Mutants

BERNARD P. SCHIMMER,[a,b] MARTHA CORDOVA,[a] JENNIVINE TSAO,[a] AND CLAUDIA FRIGERI[a,b]

[a]Banting and Best Department of Medical Research, University of Toronto, Toronto, Ontario, Canada

[b]Department of Pharmacology, University of Toronto, Toronto, Ontario, Canada

ABSTRACT: We have described a family of adrenocortical tumor cell mutants (including clones OS3, Y6, and 10r9) that are resistant to ACTH because they fail to express the gene encoding the ACTH receptor (MC2R). The MC2R deficiency results from a mutation that impairs the activity of the nuclear receptor steroidogenic factor 1 (SF1) at the MC2R promoter. In this report, we show that ACTH resistance in the mutant clones is associated with a *Sf1* gene that has Ser at codon 172 instead of Ala. In two of the three mutant clones, this *Sf1* allele is amplified together with flanking DNA from chromosome 2 that includes the genes encoding germ cell nuclear factor and the beta-type proteosome subunit Psmb7. $SF1^{A172}$ and $SF1^{S172}$ exhibit little or no difference in transcriptional activity in SF1-dependent reporter gene assays, suggesting that $SF1^{S172}$ per se is not directly responsible for the loss of MC2R expression. Instead, the *Sf1^{S172}* allele appears to be a marker of ACTH resistance in this family of adrenocortical tumor cell mutants, possibly reflecting the activity of a neighboring gene.

KEYWORDS: ACTH receptor; chromosome 2; extrachromosomal DNA; germ cell nuclear factor; gene amplification; MC2 receptor; polymorphism; proteosome subunit Psmb7; steroidogenic factor 1; Y1 adrenocortical tumor cells

INTRODUCTION

Several ACTH-resistant mutants isolated from the ACTH-responsive Y1 mouse adrenocortical tumor cell line have proved useful in understanding the signaling mechanisms involved in ACTH action in the adrenal cortex. Among these are a family that harbors dominant inhibitory mutations in the regulatory subunit of the type 1 cAMP-dependent protein kinase, rendering the protein kinase resistant to activation by cAMP[1,2]; a family that prevents ACTH-induced uncoupling of receptor–G protein complexes, leading to resistance to ACTH-induced desensitization of adenylyl cyclase[3]; and a family that fails to express the gene encoding the MC2R, leading to

Address for correspondence: Bernard P. Schimmer, Banting and Best Department of Medical Research, University of Toronto, Toronto, Ontatio, Canada. Voice: 416-978-6088; fax: 416-978-8528.

bernard.schimmer@utoronto.ca

Ann. N.Y. Acad. Sci. 944: 147–153 (2003). © 2003 New York Academy of Sciences.

MC2R deficiency.[4,5] These mutants have provided insights into the roles of cAMP and cAMP-dependent protein kinase in ACTH action,[1,6,7] into the contributions of G proteins to homologous desensitization,[3] and into the factors that govern the expression of MC2R and other genes required for adrenal steroidogenesis.[8,9]

In this report, we explore the molecular basis for ACTH resistance among three MC2R-deficient adrenal cell mutants (clones10r9, OS3, and Y6). Previously, we demonstrated that the MC2R deficiency in these mutants was caused by a defect that compromised the activity of SF1 at the *MC2R* promoter, such that the *MC2R* gene was not expressed. The defect also affected the expression of several other SF1-dependent genes required for steroid hormone synthesis and thus caused global defects in hormone-regulated steroidogenesis.[9] The defect did not affect the DNA binding activity of SF1; rather, it affected the ability of SF1 to interact with transcriptional coactivators and to activate transcription.[9,10] We now demonstrate that the MC2R-deficient mutants contain a form of SF1 that differs from the SF1 found in ACTH-responsive Y1 cells. Furthermore, the gene encoding the variant form of SF1 is amplified in two of the three mutants along with other genes closely linked to *Sf1* on mouse chromosome 2. The two forms of SF1 do not differ in activity when assayed in transfection experiments using SF1-dependent reporter genes, leading us to suggest that the variant form of SF1 serves only as a marker of ACTH resistance in this family of adrenocortical tumor cell mutants. Instead, the ACTH resistance may result from the activity of a gene linked to *Sf1* on chromosome 2. The observations presented here are extensions of results reported elsewhere.[11,12]

METHODS

RT-PCR, PCR, and Southern blot hybridization conditions were as described previously.[11] PCR primers used to estimate the relative levels of the following genes were: AGTGAGTGAGGCTATTGCAGC and AGCACCTCAATCTCCAGAGG from exons 7 and 8 of the proteosome *Psmb7* gene (accession number SEG_D85563S); TGCGTGCTGATCGAATGC and CCAGTCGACAATG-GAGATAAAGG from exon 3 of the *Sf1* gene (accession number S65919); TGGAT-TCATGGCATTCAGG and TGCTTGCTGTACACTGTGAGG from exons 7 and 8 of the *Gcnf* gene (accession number AF254820); ATAGTTGGTTTTTGTTGCTAT-TGC and TGCATTTGCAACATACCCC for D2Mit58 (UniSTS:129453); TAGC-CTACAGAGTGGACAGCC and CTAGGCTTTATGTAGCCTCTTTGC for D2Mit80 (UniSTS:129453).

RESULTS AND DISCUSSION

A Restriction Fragment Length Polymorphism Distinguishes the Genes Encoding SF1 from Y1 and Mutant Cells

Sf1 genes from ACTH-responsive Y1 and ACTH-resistant mouse adrenocortical tumor cells were compared to explore the basis for the observed differences in SF1 function. As summarized in FIGURE 1A, SF1 transcripts from Y1 and mutant cells differed in sequence at the third base in codon 3 which did not cause an amino acid

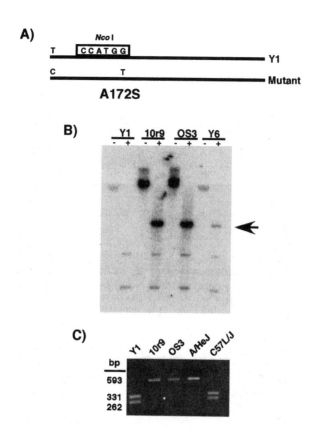

FIGURE 1. Identification of SF1 polymorphisms in Y1 and mutant cells and in mouse strains. (**A**) Nucleotide sequence differences at position 3 of codon 3 and position 1 of codon 172 in the SF1 transcripts isolated from Y1 (GenBank accession number AF511594) and mutant cells. The *boxed sequence* represents the unique *Nco*I restriction site in the transcript from Y1 cells. (**B**) Genomic DNA (10 μg) from Y1 and mutant cells was incubated in the absence (−) or presence (+) of *Nco*I and analyzed for *Sf1* by Southern blot hybridization using an exon 3 probe. The *arrow* denotes the restriction fragment that distinguishes the *Sf1^{S172}* allele associated with mutant cells. (**C**) *Nco*I digestion of Exon 3 of the *Sf1* gene amplified from the genomic DNA of Y1 and mutant cells and A/HeJ and C57L/J mouse strains.

change, and in the first base of codon 172 that substituted a Ser for Ala. The base substitution at codon 172 also disrupted a unique *Nco*I restriction site, creating a restriction fragment length polymorphism (RFLP). Transcripts from the 10r9 and OS3 mutants encoded primarily SF1^{S172}, whereas transcripts from mutant clone Y6 encoded both SF1^{A172} and SF1^{S172} in approximately equal amounts.

The basis for this SF1 RFLP was explored at the genomic level by Southern blot hybridization. Strikingly, *Sf1* signals in genomic DNA from mutant clones 10r9 and OS3 were considerably stronger on Southern blots than *Sf1* signals in the genomic DNA from Y1 or Y6 cells (FIG. 1B). Furthermore, digestion of Y1 DNA to completion with *Nco*I produced two *Sf1* fragments, whereas digestion of mutant DNA to completion with *Nco*I produced three *Sf1* fragments—two fragments that matched those in Y1 both in size and relative intensity and a larger fragment (FIG. 1B, arrow) that was much more intense in the mutants 10r9 and OS3. As determined from fluorescent *in situ* hybridization experiments,[11] Y1 and ACTH-resistant mutant cells each contained *Sf1* genes situated near the centromeric ends of a chromosome 2 pair; the mutants 10r9 and OS3 additionally contained multiple copies of *Sf1* on acentric extrachromosomal fragments. Together, these observations indicated that Y1 cells were homozygous diploid for *Sf1*^{A172}, whereas Y6 cells were *Sf1*^{A172}/*Sf1*^{S172} heterozygous diploid. The mutants 10r9 and OS3 were diploid for chromosome 2 but contained multiple copies of *Sf1*^{S172} on small fragments of DNA that were not otherwise visible at the light microscope level.

Origins of the SF1 RFLP

The Y1 cell line and the ACTH-resistant mutant clones trace their lineage to an adrenal tumor originating in a C57L/J x A/HeJ mouse.[13] Inasmuch as the ACTH-resistant mutants all contained the same *Sf1*^{S172} allele, including the silent mutation at codon 3 (FIG. 1A), we considered the possibility that this allele represented a naturally occurring polymorphism arising from one of the mouse strains rather than a *de novo* mutation. Accordingly, we examined the genomic DNA from the two mouse strains for a RFLP representing the *Sf1*^{S172} allele. Exon 3 of the *Sf1* gene, the exon that includes codon 172 and the diagnostic *Nco*I restriction site, was amplified from Y1 and mutant genomic DNA and from the genomic DNA of the two mouse strains by PCR with gene-specific primers. As shown in FIGURE 1C, the SF1 product from the C57L/J strain yielded two fragments (331 and 262 bp) after *Nco*I digestion. The same restriction pattern was observed after amplification and *Nco*I digestion of exon 3 of the *Sf1* gene from Y1 cells. These results demonstrate that the *Sf1*^{A172} allele was derived from the C57/L mouse. The *Sf1*^{S172} allele, in contrast, was clearly derived from the A/HeJ mouse, because the corresponding 593-bp SF1 products from this strain and from mutant 10r9 and OS3 cells were resistant to *Nco*I digestion. Direct sequencing of the regions surrounding codons 3 and 172 confirmed the identities of the *Sf1*^{A172} and *Sf1*^{S172} alleles in these mouse strains.[11]

Transcriptional Activities of SF1^{A172} and SF1^{S172}

We next evaluated whether the impaired SF1 activity and consequent ACTH resistance seen in the mutant clones were caused by differences in the transcriptional activities of the two forms of SF1. For these experiments, COS-1 African green monkey cells were transfected with a luciferase reporter plasmid driven by five copies of a SF1 site from the *Cyp*21 gene together with expression vectors encoding SF1^{A172} or SF1^{S172} and assayed for SF1-dependent luciferase activity as described.[11] Under these conditions, SF1^{S172} had 117% ± 20% of the activity of SF1^{A172} ($n = 3$), indicating that the activities of the two alleles are similar. To the extent that the transfection experiments accurately reflect the activities of the two forms of SF1, these observations indicate that SF1^{S172} is not directly responsible for the phenotype of the mutant clones; rather, SF1^{S172} appears to be a marker for another gene that compromises SF1 function and confers ACTH resistance in the mutants.

Other Genes from Chromosome 2 Are Amplified Together with Sf1 in Mutant 10r9 and OS3 Cells

Extrachromosomal fragments like those observed in the mutants 10r9 and OS3 are commonly found in tumor cells and are thought to arise from chromosome breakage, followed by asymmetric segregation of the chromosome and its fragments during mitosis. These fragments often contain several genes from the originating chromosome[14] and thus can be examined to identify genes that are in close proximity. To identify candidate genes responsible for the ACTH-resistant phenotype, we sought to identify genes originating from chromosome 2 that were close to *Sf1* based on their coamplification with *Sf1* in the 10r9 and OS3 mutants. Two such genes, proteosome subunit *Psmb7* and the transcription factor *Gcnf* (germ cell nuclear factor), were more abundant in mutant 10r9 and OS3 cells than in Y1 or mutant Y6 cells, indicating that these genes were amplified together with *Sf1* in the two mutants (FIG. 2). Based on an alignment of contiguous DNA sequences in the ENSEMBL database,[15] *Gcnf* and *Psmb7* flank *Sf1* on the telomeric and centromeric sides, respectively, and span approximately 300 kb of DNA (FIG. 2). The gene encoding Gsα, which resides at the telomeric end of chromosome 2, was not amplified. LIM homeobox 2 (*Lh-X2*), which is situated upstream of *Psmb7*, is amplified in some preparations[12] but not in others (data not shown), suggesting a lack of stability of *Lh-X2* in the amplicon, possibly because it is near one end of the extrachromosomal fragment.

We also identified two informative PCR polymorphisms on chromosome 2, D2Mit80 and D2Mit58. These markers produced PCR products of different sizes when amplified from the genomic DNA of C57L/J and A/HeJ mice (FIG. 2) and thus are useful in distinguishing the chromosomes bearing the two *Sf1* alleles. Amplification of these two markers from Y1 genomic DNA produced products that were the same size as that produced from C57L/J genomic DNA. Amplification of these markers from the three ACTH-resistant mutants in each case produced two products, one that matched the product from C57L/J genomic DNA and one that matched the product from A/HeJ genomic DNA; neither marker was present in multiple copies in clones 10r9 and OS3. These results demonstrate that the three mutants are heterozygous for both the D2Mit80 and D2Mit58 markers. Furthermore, the chromosome 2 fragments amplified in clone 10r9 and OS3 do not extend to the regions demarcated by the two D2Mit markers. It thus would appear that the mutants contain one copy of chromosome 2 from the C57L/J mouse bearing the *Sf1^{A172}* allele and one copy of the chromosome from the A/HeJ mouse bearing the *Sf1^{S172}* allele, consistent with their C57L/J × A/HeJ origin. In contrast, the homozygosity of the ACTH-responsive Y1 cell for the C57L/J-derived chromosome 2 supports the conclusion that the chromosome 2 pair in Y1 cells was derived from the C57L/J mouse strain and probably reflects the asymmetric segregation of chromosome 2 pairs in this clonal isolate after the fragmentation of the A/HeJ-derived chromosome.

The association of the *Sf1^{S172}* allele with altered SF1 function, MC2R deficiency, and global defects in steroidogenesis in our mutant clones raises the possibility of a linkage between *Sf1* genotype and phenotype. This linkage is strengthened by the correlation of *Sf1* genotype with steroidogenic capacity among mouse strains; strains with high steroidogenic capacity carry the *Sf1^{A172}* allele, whereas strains with low steroidogenic capacity carry the *Sf1^{S172}* allele.[11] The *Sf1* alleles per se seem not to be directly responsible for the phenotype because SF1^{A172} and SF1^{S172} show little

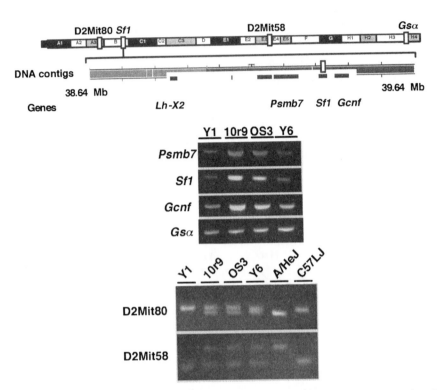

FIGURE 2. Genes on chromosome 2 amplified together with *Sf1*. Chromosome 2 and the expanded B region from 38.64 to 39.64 Mb are depicted with the centromere to the left (*top*). Genomic DNA (0.03 μg) from the cell lines and mouse strains indicated were subjected to PCR in the presence of gene-specific primers to amplify fragments of *Psmb7*, *Sf1*, and *Gcnf* (*middle*) and the chromosome 2 markers D2Mit80 and D2Mit58 (*bottom*).

or no difference activity in SF1-dependent reporter gene assays. Instead, we propose that *Sf1*[S172] serves as a marker for another linked gene that affects SF1 function. The orphan nuclear receptor GCNF is an interesting candidate in this regard because this transcription factor has been shown to bind to SF1 recognition sites and antagonize SF1 action. Although GCNF expression reportedly is restricted to gonads,[16] preliminary RT-PCR experiments in our laboratory indicate that GCNF is indeed expressed in our adrenal cell lines (unpublished observations). It would be interesting to determine whether the GCNF gene from our mutant clones can confer a mutant phenotype on ACTH-responsive Y1 cells in transfection experiments.

ACKNOWLEDGMENTS

This work was supported by research grants from the Canadian Institutes of Health Research. We acknowledge Keith Parker for helpful discussions and Rachel Bloch for assistance with some of the work.

REFERENCES

1. OLSON, M.F. *et al.* 1993. Molecular basis for the 3′,5′-cyclic adenosine monophosphate resistance of kin mutant Y1 adrenocortical tumor cells. Mol. Endocrinol. **7:** 477–487.
2. WONG, M., A.J. KROLCZYK & B.P. SCHIMMER. 1992. The causal relationship between mutations in cAMP-dependent protein kinase and the loss of adrenocorticotropin-regulated adrenocortical functions. Mol. Endocrinol. **6:** 1614–1624.
3. COLANTONIO, C.M. *et al.* 1998. Altered G protein activity in a desensitization-resistant mutant of the Y1 adrenocortical tumor cell line. Endocrinology **139:** 626–633.
4. SCHIMMER, B.P. *et al.* 1995. ACTH-receptor deficient mutants of the Y1 mouse adrenocortical tumor cell line. Endocr. Res. **21:** 139–156.
5. QIU, R. *et al.* 1996. Mutations to forskolin resistance result in loss of adrenocorticotropin receptors and consequent reductions in levels of G protein α-subunits. Mol. Endocrinol. **10:** 1708–1718.
6. RAE, P.A., J. TSAO & B.P. SCHIMMER. 1979. Evaluation of receptor function in ACTH-responsive and ACTH-insensitive adrenal tumor cells. Can. J. Biochem. **57:** 509–516.
7. LE, T. & B.P. SCHIMMER. 2001. The regulation of mitogen-activated protein kinases in Y1 mouse adrenocortical tumor cells. Endocrinology **142:** 4282–4287.
8. WONG, M. *et al.* 1989. The roles of cAMP and cAMP-dependent protein kinase in the expression of cholesterol side chain cleavage and steroid 11β-hydroxylase genes in mouse adrenocortical tumor cells. J. Biol. Chem. **264:** 12867–12871.
9. FRIGERI, C. *et al.* 2000. Impaired steroidogenic factor 1 (NR5A1) activity in mutant Y1 mouse adrenocortical tumor cells. Mol. Endocrinol. **14:** 535–544.
10. FRIGERI, C. & B.P. SCHIMMER. 2000. The activation function of steroidogenic factor-1 is impaired in ACTH-resistant Y1 mutants. Endocrinol. Res. **26:** 1005–1009.
11. FRIGERI, C. *et al.* 2002. A polymorphic form of steroidogenic factor-1 is associated with ACTH resistance in Y1 mouse adrenocortical tumor cell mutants. Endocrinology **143:** 4031–4037.
12. SCHIMMER, B.P. *et al.* 2002. SF1 polymorphisms in the mouse and steroidogenic potential. Endocrinol. Res. **28:** 523–529.
13. SCHIMMER, B.P. 1979. Adrenocortical Y1 cells. Methods Enzymol. **52:** 570–574.
14. HAHN, P.J. 1993. Molecular biology of double-minute chromosomes. Bioessays **15:** 477–484.
15. HUBBARD, T. *et al.* 2002. The Ensembl genome database project. Nucleic Acids Res. **30:** 38–41.
16. CHEN, F. *et al.* 1994. Cloning of a novel orphan receptor (GCNF) expressed during germ cell development. Mol. Endocrinol. **8:** 1434–1444.

Proopiomelanocortin Peptides and Sebogenesis

LI ZHANG, MIKE ANTHONAVAGE, QIULING HUANG, WEN-HWA LI, AND
MAGDALENA EISINGER

*Skin Research Center, Johnson & Johnson Consumer and Personal Care Group,
Skillman, New Jersey 08558, USA*

ABSTRACT: Previous animal studies have demonstrated that α-melanocyte-stimulating hormone (α-MSH) is a sebotropic hormone in rats and that targeted disruption of melanocortin 5 receptor (MC5-R) can down-regulate sebum output in mice. To study the role of proopiomelanocortin (POMC) peptides in the regulation of human sebaceous lipid production and sebocyte differentiation, we established a primary human sebocyte culture system. Sebocytes were derived from normal human facial skin. Differentiation of sebocytes, induced by POMC-derived peptides such as MSH, adrenocorticotropic hormone (ACTH), or bovine pituitary extract (BPE), resulted in the appearance of prominent cytoplasmic lipid droplets. Partial induction of sebocyte differentiation also was observed in serum-depleted cultures, but there was very limited spontaneous differentiation in serum-containing medium. Analysis by high-performance thin-layer chromatography (HPTLC) of ^{14}C-acetate–labeled lipids showed a dose-dependent increase in synthesis of sebaceous-specific lipid (i.e., squalene) induced by NDP α-MSH. Molecular studies using RT-PCR showed a low level of human MC5-R expression under serum-free condition but a substantial increase after treatment with NDP α-MSH or BPE. In contrast, MC1-R expression remained the same, independent of treatment. Our data indicate that expression of MC5-R correlates with sebocyte differentiation and suggest a regulatory role for MC5-R in human sebaceous lipid production.

KEYWORDS: melanocortin; melanocortin receptor; sebocyte; differentiation; lipid; skin

INTRODUCTION

Sebogenesis is the process of producing sebum-specific lipids within the sebaceous gland. The gland and the hair follicle form the pilosebaceous unit of mammalian skin. After its production, sebum is released from cells by a holocrine secretion into the infundibulum and hair canal. It is delivered to the hair and skin surface for protective coating and moisturization.[1] Human sebum consists of lipids, such as squalene, wax esters, triglycerides, and fatty acids, which are not secreted by most other mammalian cells.[2–4] Sebaceous glands have long been recognized as the target of androgens and the melanocortins such as α-MSH.[5]

Address for correspondence: Dr. Magdalena Eisinger, Johnson & Johnson Skin Research Center, Consumer and Personal Care Group, RG-4, 199 Grandview Road, Skillman, NJ 08558. Voice: 908-874-1204; fax: 908-874-1254.

meising@cpcus.jnj.com

Ann. N.Y. Acad. Sci. 994: 154–161 (2003). © 2003 New York Academy of Sciences.

Melanocortins comprise a family of peptides coded by a single proopiomelano-cortin gene (POMC). POMC gene product is proteolytically cleaved to yield partic-ular peptide species including adrenocorticotropic hormone (ACTH), α-, β-, and γ-melanocyte-stimulating hormone (MSH). Melanocortins elicit their diverse biologi-cal effects by binding and activating their receptors on plasma membrane. Melano-cortin receptors belong to the superfamily of G-protein–coupled receptors with seven transmembrane domains. They stimulate cAMP production upon activation. The structures, biological functions, and pharmacological properties of melanocort-in peptides and their receptors have been extensively reviewed.[6,7] Among the five distinct melanocortin receptors that have been identified, both MC1-R and MC5-R are expressed in human skin. These receptors have been shown to play important roles in skin homeostasis including the regulation of pigmentation, immune and in-flammatory responses, hair growth, extracellular matrix interactions, and exocrine gland functions.[8]

The first elucidation of functional importance of MC5-R came from studies by Chen *et al.*[9] It was found that deletion of MC5-R in mice led to severe defects in water repulsion and thermoregulation due to decreased production of particular sebaceous lipids. The above studies were complemented by the findings of Thody and his col-leagues. In their studies, removal of the neurointermediate lobe of the pituitary led to decreased sebum production in rats.[10] They also showed that application of α-MSH (acting on lipogenesis) and androgen (promoting sebaceous cell proliferation and turnover) fully restored sebum production.[5] Moreover, POMC-deficient mice also had a similar phenotype as seen in mutant mice lacking MC5-R.[10] The results from these animal studies provide support that α-MSH may stimulate sebaceous lip-id production through MC5-R; however, direct evidence is still to be furnished.

In human, it was shown that MC5-R[12] and MC1-R[13] were expressed in human sebaceous glands and cultured human sebocytes at both message and protein levels. To study regulation of human sebaceous glands, many attempts have been made to-ward the development of a human sebocyte culture model[14–18] aimed at the difficult task of controlling differentiation. Recently, an immortalized human sebaceous gland cell line was established,[19] but only MC1-R and not MC5-R was detected in these cells.[13]

To understand the role of POMC peptides in direct regulation of human seba-ceous lipid production and sebocyte differentiation, we established a three-step primary human sebocyte culture system in which sebocyte differentiation can be in-duced by MSH and ACTH. Here, we also present evidence that melanocortins stim-ulate sebaceous lipid production and that expression of MC5-R, but not MC1-R, is associated with the process of sebocyte differentiation.

METHODS

Sebocyte Isolation and Culture

Human facial skin was removed during cosmetic facelift surgeries by criteria de-scribed previously,[20] and sebaceous cells were isolated by the method of Karasek.[14] *In vitro* human sebocyte culture was performed as described (M. Eisinger *et al.*, manuscript in preparation).

Sebaceous Lipid Analysis Using HPTLC

At the end of the culture, human sebocytes were labeled with ^{14}C-acetic acid for 24 h. Cells were harvested and processed for lipid extraction. Neutral lipids were resolved using HPTLC analysis.[20]

RT-PCR

Total RNA was isolated from cultured sebocytes at the end of the culture, and cDNA was prepared by reverse transcription using 10 mg DNA-free total RNA. Primers for MC5-R were MC5-1Fa (5′ CTC AAC CTG AAT GCC ACA GAG 3′) and MC5-1R (5′ GTC TGC TAT CAC TAG GTG CTT G 3′). Primers for MC1R were MC1-1F (5′ CTG CCA ACC AGA CAG GAG C 3′) and MC1-1R (5′ CGT GCT GAA GAC GAC ACT GG 3′). Glyceraldehyde 3-phosphate dehydrogenase (GAPDH) was used as a loading control. PCR conditions were 94°C, 12 min; 94°C, 1.5 min; 58°C, 1 min; 72°C, 1.5 min; 72°C, 10 min. A total of 35 and 20 cycles were performed for MC-Rs and GAPDH, respectively. The PCR products were separated by 2% agarose gel electrophoresis.

RESULTS

Induction of Human Sebocyte Differentiation by POMC Peptides

To investigate the effect of melanocortins on human sebocyte differentiation, we established a primary human sebocyte culture system. As indicated in FIGURE 1, differentiation of human sebocytes, resulting in the appearance of prominent cytoplasmic lipid droplets, could be induced in vitro by POMC-derived peptides, such as NDP α-MSH, ACTH, or bovine pituitary extract. Partial induction of sebocyte differentiation was also observed in serum-free cultures, and there was limited spontaneous differentiation in serum-containing medium (data not shown).

Stimulation of Squalene Synthesis by POMC Peptides

Lipid analysis by HPTLC in FIGURE 2 demonstrated that both bovine pituitary extract and NDP α-MSH induced squalene synthesis. NDP α-MSH from 0.001 to 100 nM stimulated squalene production in a dose-dependent manner with the peak of squalene synthesis at 10 nM. Although human sebum was found to contain 8% to 13% squalene and 20% to 30% wax and cholesterol esters,[21] the levels of these marker lipids found in our sebocyte cultures were lower. Squalene represented only 0.1% of the total lipids isolated from sebaceous cells in vitro. Detection of wax esters was only occasional, but production of cholesterol esters was very prominent.

Association of MC5-R with Sebocyte Differentiation

It was found that both MC5-R[12] and MC1-R[13] were expressed in human sebocytes, but not the other three melanocortin receptor subtypes. Our interest was in determining which receptor(s) was associated with sebocyte differentiation. As shown in FIGURE 3, expression of MC5-R was up-regulated when sebocytes were treated with either bovine pituitary extract or NDP α-MSH, compared with the very low level of

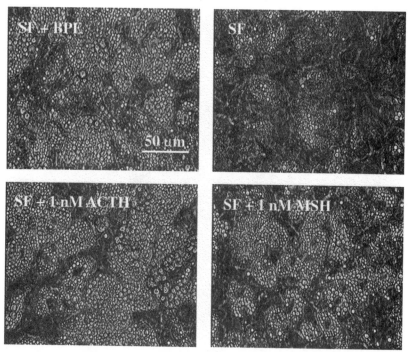

FIGURE 1. Induction of human sebocyte differentiation by melanocortins. Human sebocytes were plated onto a feeder layer of mitomycin C–treated 3T3 fibroblasts in serum-containing medium for 3 days. Before induction, the cells were grown in serum-reduced medium for another 3 days. Sebocytes then were induced to differentiation in serum-free medium with or without POMC peptides for 9 days. Differentiation of sebocytes, resulting in lipid droplet appearance in the cytoplasm, was induced by bovine pituitary extract (BPE, *top left panel*), ACTH (*bottom left panel*), or NDP α-MSH (*bottom right panel*). Serum-free (SF, *top right panel*) medium alone can only partially induce sebocyte differentiation. Bar = 50 mm.

MC5-R expression under serum-free condition. In contrast, MC1-R was detected constitutively, independent of treatment of melanocortins. These results indicate that expression of MC5-R, but not MC1-R, correlates with sebocyte differentiation and suggest a regulatory role for MC5-R in human sebaceous lipid production.

DISCUSSION

Our study is the first to show direct evidence that POMC peptides stimulate sebocyte differentiation *in vitro* and that human sebaceous lipid production may be regulated by MC5-R. We have also found the presence of MC5-R in human sebaceous glands by Northern blot, RT-PCR, and immunohistochemical staining (our unpublished data). The suggestion that MC5-R is involved in sebogenesis comes from the finding that its induction is concomitant with the initiation of the sebocyte differentiation, whose hallmark is the appearance of sebaceous lipids. A recent study

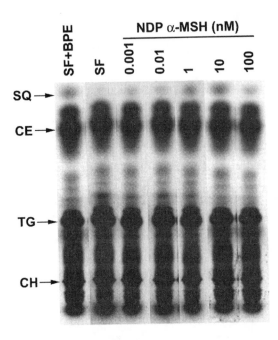

FIGURE 2. Stimulation of human sebaceous lipid synthesis by NDP α-MSH. Analysis of ^{14}C-acetate labeled lipids by HTPLC was performed to investigate sebaceous lipids induced by bovine pituitary extract in serum-free medium (SF + BPE), NDP α-MSH at from 0.001 to 100 nM in serum-free medium versus serum-free medium only (SF). Both BPE and NDP α-MSH increased synthesis of squalene (SQ), comparing with serum-free medium only. Note the dose-dependent effect of NDP α-MSH on squalene synthesis. CE, cholesterol esters; TG, triglycerides; CH, cholesterol.

indicates that MC5-R could be detected in human skin melanocytes using a microarray assay. The receptor may mediate melanocytes response to α-MSH by increase in intracellular calcium ion concentrations.[22] This finding is of interest in view of a recent report by the same group showing presence of amelanotic melanocytes within the sebaceous gland.[23] Melanocytes are well known to express MC1-R; thus, their presence in sebaceous gland may contribute to the finding of MC1-R expression in sebaceous glands and cultures.

In this study, we also observed constitutive expression of MC1-R by RT-PCR in our cultured sebocytes, but its expression was independent of sebocyte differentiation. In contrast with the above, MC5-R expression was increased only at the time of sebocyte differentiation. However, in the immortalized human sebaceous gland cell line,[19] only MC1-R has been detected.[13] In human skin, MC1-R is localized in a variety of cell types including sebaceous glands, hair follicle epithelia, secretory and ductal epithelia of sweat glands, melanocytes, keratinocytes, fibroblasts, endothelial cells, monocytes, and some periadnexal mesenchymal cells.[24] So far, we have

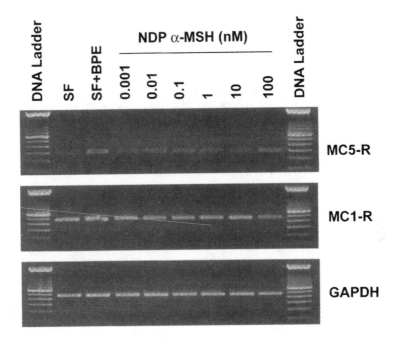

FIGURE 3. Expression of MC5-R is associated with sebocyte differentiation. RT-PCR amplification of MC5-R, MC1-R, and glyceraldehyde 3-phosphate dehydrogenase (GAPDH) was used as a loading control. The results indicate that expression of MC5-R, but not MC1-R, correlates with sebocyte differentiation. SF, serum-free medium only; SF + BPE, serum-free medium containing bovine pituitary extract.

no direct evidence to support that MC1-R is involved in sebaceous lipid production and sebocyte differentiation. It is important to realize that the sebocyte culture system utilized may contain cells other than sebocytes that may express MC1-R. However, thus far we could not detect MC1-R protein in our culture using immunohistochemical stainings.

We showed in this study that NDP α-MSH stimulated the synthesis of squalene, one of the human sebaceous lipids, in a dose-dependent manner. Because both squalene production and MC5-R induction were associated with sebocyte differentiation, this suggested an active role for MC5-R in sebaceous lipid production. Understanding the regulation of sebum production is of importance because an increase in sebum production leads to a variety of dermatological disorders, that is, acne, seborrheic dermatitis, and androgenetic alopecia. Based on clinical evidence that sebum production is associated with puberty, or dehydroepiandrosterone (DHEA) production in prepubertal children,[25] androgens were thought to be the main stimulators of sebaceous gland activity. Our findings, combined with previous evidence by others,[5,9] implicate melanocortins as inducers of sebaceous lipid production. Melanocortins, in particular ACTH and MSH, are important mediators of the stress response in the classical hypothalamic-pituitary-adrenal (HPA) axis. Both of these hormones have been shown to be produced by skin cells, that is, keratinocytes and melanocytes.[26] Moreover, other components of HPA axis can also be produced by skin cells, including corticotropin-releasing hormone, the most proximal inducer of HPA, which is also produced in sebaceous cells.[27] We have shown that MSH and ACTH are produced by cultured sebocytes (our unpublished data), thus establishing a possible link between stress-induced increase in sebum production and melanocortins.

In this chapter, we discussed the use of a human sebocyte culture system for studying the regulation of human sebaceous lipid production and sebocyte differentiation by POMC peptides via melanocortin receptors. This *in vitro* culture model offers unique opportunities for future investigations of the sebaceous gland physiology, sebum-related skin disorders, and evaluation of pharmaceutical agents.

REFERENCES

1. STRAUSS, J.S. *et al.* 1991. Sebaceous glands. *In* Physiology, Biochemistry and Molecular Biology of the Skin. G. La, Ed.: 712–740. Oxford University Press. New York.
2. NICOLAIDES, N. 1974. Skin lipids: their biochemical uniqueness. Science **186:** 19–26.
3. NIKKARI, T. 1974. Comparative chemistry of sebum. J. Invest. Dermatol. **62:** 257–267.
4. THODY, A.J. & S. SHUSTER. 1989. Control and function of sebaceous glands. Physiol. Rev. **69:** 383–416.
5. THODY, A.J. *et al.* 1976. Effect of alpha-melanocyte-stimulating hormone and testosterone on cutaneous and modified sebaceous glands in the rat. J. Endocrinol. **71:** 279–288.
6. CONE, R.D. *et al.* 1996. The melanocortin receptors: agonists, antagonists, and the hormonal control of pigmentation. Recent Prog. Horm. Res. **51:** 287–317.
7. EBERLE, A.N. 2000. Proopiomelanocortin and the melanocortin peptides. *In* The Melanocortin Receptors. R.D. Cone, Ed.: 3–67. Humana Press. Totowa, NJ.
8. BÖHM, M. & T.A. LUGER. 2000. The role of melanocortins in skin homeostasis. Horm. Res. **54:** 287–293.
9. CHEN, W. *et al.* 1997. Exocrine gland dysfunction in MC5-R-deficient mice: evidence for coordinated regulation of exocrine gland function by melanocortin peptides. Cell **91:** 789–798.
10. THODY, A.J. & S. SHUSTER. 1973. Possible role of MSH in the mammal. Nature **245:** 207–209.
11. YASWEN, L. *et al.* 1999. Obesity in the mouse model of pro-opiomelanocortin deficiency responds to peripheral melanocortin. Nat. Med. **5:** 1066–1070.
12. THIBOUTOT, D. *et al.* 2000. The melanocortin 5 receptor is expressed in human sebaceous glands and rat preputial cells. J. Invest. Dermatol. **115:** 614–619.
13. BÖHM, M. *et al.* 2002. Evidence for expression of melanocortin-1 receptor in human sebocytes in vitro and in situ. J. Invest. Dermatol. **118:** 533–539.
14. KARASEK, M. 1986. Isolation and characterization of cells from the human sebaceous gland. In Vitro **22:** 22.
15. XIA, L.Q. *et al.* 1989. Isolation of human sebaceous glands and cultivation of sebaceous gland-derived cells as an in vitro model. J. Invest. Dermatol. **93:** 315–321.
16. DORAN, T.I. *et al.* 1991. Characterization of human sebaceous cells in vitro. J. Invest. Dermatol. **96:** 341–348.
17. ZOUBOULIS, C.C. *et al.* 1991. Culture of human sebocytes and markers of sebocytic differentiation in vitro. Skin Pharmacol. **4:** 74–83.
18. FUJIE, T. *et al.* 1996. Culture of cells derived from the human sebaceous gland under serum-free conditions without a biological feeder layer or specific matrices. Arch. Dermatol. Res. **288:** 703–708.
19. ZOUBOULIS, C.C. *et al.* 1999. Establishment and characterization of an immortalized human sebaceous gland cell line (SZ95). J. Invest. Dermatol. **113:** 1011–1020.

20. Pappas, A., M. Anthonavage & J.S. Gordon. 2002. Metabolic fate and selective utilization of major fatty acids in human sebaceous gland. J. Invest. Dermatol. **118:** 164–171.
21. Greene, R.S. *et al.* 1970. Anatomical variation in the amount and composition of human skin surface lipid. J. Invest. Dermatol. **54:** 240–247.
22. Hoogduijn, M.J. *et al.* 2002. Does α-MSH act via the MC5R to increase $[Ca^{2+}]_i$ in human melanocytes? Abstracts of the 5th International Melanocortin Meeting. Sunriver, OR. August. Abstr. P095.
23. Ancans, J., I. Suzuki & A.J. Thody. 2002. Identification of cells that express POMC, PC-2 and melanocyte specific markers in human sebaceous glands [abstr.]. Pigm. Cell Res. **15:** 82.
24. Böhm, M. *et al.* 1999. Detection of melanocortin-1 receptor antigenicity on human skin cells in culture and in situ. Exp. Dermatol. **8:** 453–461.
25. Thiboutot, D. *et al.* 1999. Androgen metabolism in sebaceous glands from subjects with and without acne. Arch. Dermatol. **135:** 1041–1045.
26. Slominski, A. & J. Wortsman. 2000. Neuroendocrinology of the skin. Endocr. Rev. **21:** 457–487.
27. Zouboulis, C.C. *et al.* 2002. Corticotropin-releasing hormone: an autocrine hormone that promotes lipogenesis in human sebocytes. Proc. Natl. Acad. Sci. USA **99:** 7148–7153.

The Gut Hormone Peptide YY
Regulates Appetite

RACHEL L. BATTERHAM AND STEPHEN R. BLOOM

Imperial College Faculty of Medicine, Hammersmith Campus,
London, W12 0NN, United Kingdom

ABSTRACT: The gut hormone peptide YY (PYY) belongs to the pancreatic polypeptide (PP) family along with PP and neuropeptide Y (NPY). These peptides mediate their effects through the NPY receptors of which there are several subtypes (Y1, Y2, Y4, and Y5). The L cells of the gastrointestinal tract are the major source of PYY, which exists in two endogenous forms: PYY_{1-36} and PYY_{3-36}. The latter is produced by the action of the enzyme dipeptidyl peptidase-IV (DPP-IV). PYY_{1-36} binds to and activates at least three Y receptor subtypes (Y1, Y2, and Y5), whereas PYY_{3-36} is more selective for Y2 receptor (Y2R). The hypothalamic arcuate nucleus, a key brain area regulating appetite, has access to nutrients and hormones within the peripheral circulation. NPY neurons within the arcuate nucleus express the Y2R. In response to food ingestion plasma PYY_{3-36} concentrations rise within 15 min and plateau by approximately 90 min. The peak PYY_{3-36} level achieved is proportional to the calories ingested, suggesting that PYY_{3-36} may signal food ingestion from the gut to appetite-regulating circuits within the brain. We found that peripheral administration of PYY_{3-36} inhibited food intake in rodents and increased C-Fos immunoreactivity in the arcuate nucleus. Moreover, direct intra-arcuate administration of PYY_{3-36} inhibited food intake. We have shown that Y2R null mice are resistant to the anorectic effects of peripherally administered PYY_{3-36}, suggesting that PYY_{3-36} inhibits food intake through the Y2R.

In humans, peripheral infusion of PYY_{3-36}, at a dose which produced normal postprandial concentrations, significantly decreased appetite and reduced food intake by 33% over 24 h. These findings suggest that PYY_{3-36} released in response to a meal acts via the Y2R in the arcuate nucleus to physiologically regulate food intake.

KEYWORDS: peptide YY (PYY); Y2 receptor; neuropeptide Y (NPY); proopiomelanocortin (POMC); arcuate nucleus; appetite

INTRODUCTION

In response to a meal, hunger is reduced for several hours. However, the mechanisms responsible for this prolonged suppression of appetite are unknown. Intravenous infusion of nutrients does not have this long-lasting effect, suggesting that gut-derived factors are important. The hypothalamic region of the brain receives and integrates neural, metabolic, and endocrine signals from the periphery and orches-

Address for correspondence: Rachel L. Batterham, Imperial College Faculty of Medicine, Hammersmith Campus, Du Cane Road, London, W12 0NN, United Kingdom. Voice: 44 (0) 20 8383 3242; fax: 44 (0) 20 8282 3142.
r.batterham@ic.ac.uk

Ann. N.Y. Acad. Sci. 994: 162–168 (2003). © 2003 New York Academy of Sciences.

trates appropriate changes in appetite and energy expenditure.[1] Within the hypothalamus, the arcuate nucleus plays a key role in the regulation of ingestive behavior. The arcuate nucleus is a circumventricular organ, with access to the third ventricle and the portal vasculature and is responsive to a wide range of peripheral hormones and nutrients.[2] Two distinct subsets of neurons controlling food intake are found within the arcuate nucleus: the neuropeptide Y (NPY)/agouti-related peptide (Agrp) neurons and the pro-opiomelanocortin (POMC) neurons. Activation of the NPY/Agrp neurons causes increased food intake and decreased energy expenditure, whereas activation of the POMC neurons decreases food intake and increases energy expenditure.[3] These two neuronal subsets act as sensors, responding to circulating hormones that reflect body energy stores, such as leptin.[4] However, the identity of peripheral factors signaling food ingestion to these feeding circuits remains largely unclear.

PEPTIDE YY

Peptide YY (PYY) is a 36-amino-acid gastrointestinal hormone first isolated from porcine small intestine by Tatemoto in 1980[5] and named PYY because of the presence of an amino acid terminal (Y) tyrosine and a carboxyl terminal tyrosine amide (Y). PYY belongs to the pancreatic polypeptide (PP) family of peptides together with NPY and PP. The L cells of the gastrointestinal tract are the major source of PYY of which there are two main endogenous forms: PYY_{1-36} and PYY_{3-36}.[6] The latter is produced by the action of the enzyme dipeptidyl peptidase-IV (DPP-IV), which hydrolyzes PYY at the Pro^2-Ile^3 bond.[7] PYY_{1-36} binds to and activates at least three Y receptor subtypes in rats and humans (Y1, Y2, and Y5). Removing the first two amino acids at the N-terminal changes the receptor selectivity such that PYY_{3-36} is more selective for Y2 receptor.[8] The percentage of these two forms in human blood has been reported to differ according to the feeding status. In the fasted state, the concentration of PYY_{1-36} predominates over that of PYY_{3-36}. In contrast, after a meal, PYY_{3-36} is the major circulating form.[9] Following ingestion of food, plasma levels increase within 15 min, reach a peak at approximately 90 min, and then remain elevated for up to 6 h.[10] Interestingly, PYY levels reflect meal size and the nature of the food, fat being the most potent nutrient in releasing PYY.

There have been conflicting reports regarding the distribution of PYY from immunocytochemical studies, in part due to cross-reactivity of antibodies with NPY and PP, whereas *in situ* hybridization studies have identified PYY messenger RNA (mRNA) only in the gastrointestinal tract, the pancreas, and the brainstem.[11] PYY has been shown to have several biological actions, including vasoconstriction, inhibition of gastric acid secretion, reduction of pancreatic and intestinal secretion, and inhibition of gastrointestinal motility. When injected into the cerebral ventricles, the paraventricular nucleus (PVN) of the hypothalamus, or hippocampus, PYY increases food intake. Indeed, among all the orexigenic peptides and neurotransmitters described to date, PYY is the most potent stimulator of food intake.

Y2 RECEPTOR

The NPY Y2 receptor (Y2R) is a 381-amino-acid, 7-transmembrane-spanning, G-protein-coupled receptor, which inhibits the activation of adenylyl cyclase via

G_i.[12] While it has low homology with other known NPY receptors, there is a high degree of conservation between rat and human Y2 receptors with 98% amino acid identity.[13] The Y2R is widely distributed within the central nervous system in both rodents and man. In the hypothalamus, Y2 mRNA is localized in the arcuate nucleus, preoptic nucleus, and dorsomedial nucleus.[14] Other forebrain areas that contain substantial Y2R mRNA include the posterior hypothalamic nuclei, medial nucleus of the amygdala, parabrachial area, substantia nigra, and the paraventricular thalamic nucleus.[15] Brainstem regions containing Y2R mRNA include the nucleus of the solitary tract and the lateral reticular nucleus, providing both ascending innervation to the hypothalamus and descending projections to the spinal cord. In the human brain the Y2R is the predominant Y receptor subtype in the brain. In addition to the areas described above, Y2R mRNA is also found in the dentate gyrus and the cerebral cortex.[16] Within the arcuate nucleus, over 80% of the NPY neurons co-express Y2R mRNA.[17] Application of Y2-selective agonists have been shown to reduce the release of NPY from hypothalamic slices *in vitro*, whereas the Y2 non-peptide antagonist BIIE0246 increases NPY release.[18] These findings support the role of the Y2R as a presynaptic autoreceptor that regulates the NPY release and hence may be involved in the regulation of feeding.

ROLE OF ARCUATE NUCLEUS Y2R IN FEEDING

To examine the role of the NPY Y2 receptor in the regulation of feeding, we utilized a Y2-selective agonist (Y2A), a C-terminal analogue of NPY (*N*-acetyl [Leu[28], Leu[31]] neuropeptide Y-24–36).[19] Receptor binding studies performed upon distinct cells lines expressing NPY Y1, Y2, and Y5 receptors confirmed that this Y2 agonist

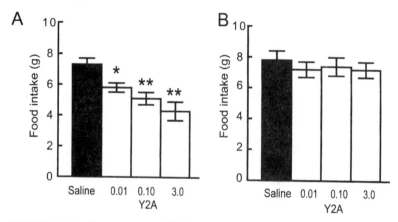

FIGURE 1. Feeding response to Y2A in rats. (**A**) Fasted rats were injected with saline or Y2A into the arcuate nucleus at the doses indicated. Post injection, 2-h food intake was measured. Each result is expressed as a mean ± SEM ($N = 12$ per group), ** = $P < 0.01$ compared to saline. (**B**) Fasted rats were injected with saline or Y2A into the paraventricular nucleus at the doses indicated (nmols). Post injection, 2-h food intake was measured. Each result is expressed as a mean ± SEM.

selectively bound to the Y5 receptor but not the Y1 and Y5 receptors (Y2: EC_{50} of 0.3 nM vs. Y1: $EC_{50} > 5000$ nM and Y2: $EC_{50} > 5000$ nM). This clear selectivity for the Y2 receptor provided a useful tool to examine the role of Y2R *in vivo*. Injection of the Y2 agonist into the arcuate nucleus of rats at the onset of the dark-phase caused a reduction in food intake. This effect was seen over a dose range from 100 fmol to 3 nmol and persisted for as long as 8 h post administration. Similarly, intra-arcuate administration the Y2A caused a comparable decrease in re-feeding following a fast (FIG. 1A). In contrast, injection of the Y2A into the PVN had no effect on feeding (FIG. 1B). Using *in vitro* studies on hypothalamic explants, we examined the effects of Y2A on release of the orexigenic peptide NPY and the anorectic POMC gene product α-melanocyte-stimulating hormone (α-MSH), both potent regulators of feeding *in vivo*. Y2A reduced the release of NPY while increasing the release of the α-MSH from POMC neurons, consistent with the proposed role as an inhibitory autoreceptor on NPY neurons.

THE ROLE OF PYY$_{3-36}$ IN THE REGULATION OF FOOD INTAKE

PYY$_{3-36}$ is a Y2R agonist and has been shown to be released by food intake. Therefore, we examined the effects of PYY$_{3-36}$ upon feeding when administered peripherally to rodents. These studies demonstrated that PYY$_{3-36}$, when given at doses that achieved plasma levels within the normal postprandial range, reduced both dark-phase food intake and re-feeding following a fast. Twice daily administration of PYY$_{3-36}$ for 8 days revealed no attenuation of the inhibitory effect on food intake and resulted in decreased cumulative food intake and reduced body weight gain compared with saline-treated animals (FIG. 2).[20] Taken together these findings suggest that the PYY$_{3-36}$ released postprandially regulates food intake in rodents.

To establish whether the anorectic effect of PYY$_{3-36}$ required the Y2R, we examined the response of mice with targeted gene deletion of the Y2R to peripheral ad-

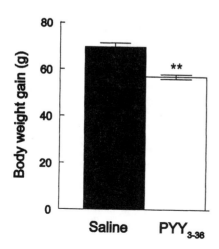

FIGURE 2. The effects of chronic PYY$_{3-36}$ treatment on body weight gain. Rats were injected intraperitoneally with PYY$_{3-36}$ (5 μg/100 g) or saline twice daily for 8 days. Each day, body weight was measured, and the total weight gain for each group was calculated (saline: *filled bar*; PYY$_{3-36}$: *open bar*). Each result is expressed as a mean ± SEM ($N = 12$ per group), $** = P < 0.01$ compared to saline.

ministration of PYY_{3-36}.[21] Peripheral administration of PYY_{3-36} resulted in a dose-dependent inhibition of food intake in wild-type littermate control mice. In contrast, PYY_{3-36} had no effect on food intake in Y2R knockout mice, suggesting that the anorectic effects of PYY_{3-36} require the presence of the Y2R.

Peripheral administration of PYY_{3-36} resulted in an increase C-Fos expression in the arcuate nucleus and in particular increased the number of POMC neurons that co-localized with C-Fos, suggesting that POMC neurons were being activated. Furthermore, direct intra-arcuate injection of PYY_{3-36} resulted in decreased food intake in fasted rodents, again supporting the hypothesis of the arcuate nucleus as a site of action. Using transgenic mice with targeted expression of green fluorescent protein in POMC neurons,[4] we were able to demonstrate that both the Y2A and PYY_{3-36} depolarized and thus activated POMC neurons.[20]

HUMAN STUDIES

To extend these observations made in rodent models and to study the physiological role of PYY_{3-36} in man, we investigated the effects of a 90-min infusion of PYY_{3-36} (0.8 pmol/kg/min) or saline, on appetite and food intake in 12 healthy subjects (6 males and 6 females; mean age: 26.7 ± 0.7 years: body mass index = 24.6 ± 0.94 kg/m^2) in a randomized crossover study (FIG. 3). During the PYY_{3-36} infusion, plasma levels reached a plateau by 30 min, and peak PYY_{3-36} concentrations were within the normal postprandial range, returning to baseline by 120 min. Food intake assessed from a free choice buffet lunch, 2 h after the termination of the infusion, was decreased $36 \pm 7.4\%$ compared with saline (FIG. 4). Volunteers continued to record their food intake for 24 h post infusion. Food intake as assessed by food diaries remained significantly reduced in the post-infusion period in the PYY_{3-36} treated group, resulting in a 33% reduction in total 24-h food intake.

CONCLUSIONS

Our findings suggest that PYY_{3-36}, released in proportion to calories ingested, regulates subsequent food intake by modulating the activity of the NPY and POMC neurons in the arcuate nucleus of the hypothalamus. The kinetics of PYY_{3-36} secre-

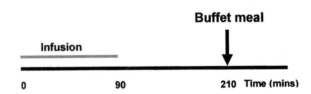

FIGURE 3. Infusion protocol. Fasted subjects received a 90-min infusion of either saline or PYY_{3-36} in a randomized, double-blind, crossover study. Two hours after the termination of the infusion, subjects were offered a buffet meal.

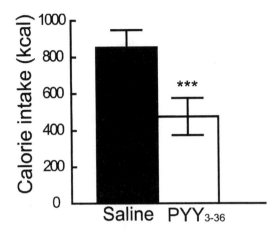

FIGURE 4. Effect of PYY_{3-36} infusion on food intake in humans. Calorie intake from "free-choice" buffet meal 2 h after infusion with saline or PYY_{3-36}. Each result is expressed as a mean ± SEM ($N = 12$ per group), *** = $P < 0.001$ compared to saline.

tion and duration of action differentiate it from other classical meal terminating signals such as cholecystokinin (CCK).[22] Previously, factors controlling the regulation of food intake have arbitrarily divided into "short-term regulators" (such as CCK) that act rapidly to influence the termination of individual meals and "long-term regulators" (such as leptin and insulin) that reflect body energy stores. Unlike CCK, the levels of which peak within 30 min of food ingestion, the levels of PYY_{3-36} peak later and remain elevated for several hours following a meal. Together with the long-lasting inhibition of food intake that we have observed, these findings led us to suggest that PYY_{3-36} is involved in the "intermediate" term regulation of food intake.

Our findings suggest that the PYY system may be a therapeutic target for the treatment of obesity. Furthermore, abnormalities of the PYY system may be involved in the pathogenesis of this condition.

ACKNOWLEDGMENTS

We would like to thank Dr. Herbert Herzog, Dr. Michael Cowley, Dr. Roger Cone, and Dr. Malcolm Low.

REFERENCES

1. SCHWARTZ, M.W., S.C. WOODS, D. PORTE, JR., *et al.* 2000. Central nervous system control of food intake. Nature **404:** 661–671.
2. CONE, R.D., M.A. COWLEY, A.A. BUTLER, *et al.* 2001. The arcuate nucleus as a conduit for diverse signals relevant to energy homeostasis. Int. J. Obes. Relat. Metab. Disord. **25**(Suppl 5): S63–S67.

3. BARSH, G.S., I.S. FAROOQI & S. O'RAHILLY. 2000. Genetics of body-weight regulation. Nature **404:** 644–651.
4. COWLEY, M.A., J.L. SMART, M. RUBINSTEIN, *et al.* 2001. Leptin activates anorexigenic POMC neurons through a neural network in the arcuate nucleus. Nature **411:** 480–484.
5. TATEMOTO, K. & V. MUTT. 1980. Isolation of two novel candidate hormones using a chemical method for finding naturally occurring polypeptides. Nature **285:** 417-418.
6. GRANDT, D., M. SCHIMICZEK, K. STRUK, *et al.* 1994. Characterization of two forms of peptide YY, PYY(1-36) and PYY(3-36), in the rabbit. Peptides **15:** 815–820.
7. MEDEIROS, M.D. & A.J. TURNER. 1994. Processing and metabolism of peptide-YY: Pivotal roles of dipeptidylpeptidase-IV, aminopeptidase-P, and endopeptidase-24.11. Endocrinology **134:** 2088–2094.
8. DUMONT, Y., A. FOURNIER, S. ST-PIERRE, *et al.* 1995. Characterization of neuropeptide Y binding sites in rat brain membrane preparations using [^{125}I][Leu31,Pro34]peptide YY and [^{125}I]peptide YY3-36 as selective Y1 and Y2 radioligands. J. Pharmacol. Exp. Ther. **272:** 673-680.
9. GRANDT, D., M. SCHIMICZEK, C. BEGLINGER, *et al.* 1994. Two molecular forms of peptide YY (PYY) are abundant in human blood: Characterization of a radioimmunoassay recognizing PYY 1-36 and PYY 3-36. Regul. Pept. **51:**151–159.
10. ADRIAN, T.E., G.L. FERRI, A.J. BACARESE-HAMILTON, *et al.* 1985. Human distribution and release of a putative new gut hormone, peptide YY. Gastroenterology **89:** 1070–1077.
11. BROOME, M., T. HOKFELT & L. TERENIUS. 1985. Peptide YY (PYY)-immunoreactive neurons in the lower brain stem and spinal cord of rat. Acta Physiol. Scand. **125:** 349–352.
12. INGENHOVEN, N., C.P. ECKARD, D.R. GEHLERT, *et al.* 1999. Molecular characterization of the human neuropeptide Y Y2-receptor. Biochemistry **38:** 6897–6902.
13. GERALD, C., M.W. WALKER, P.J. VAYSSE, *et al.* 1995. Expression cloning and pharmacological characterization of a human hippocampal neuropeptide Y/peptide YY Y2 receptor subtype. J. Biol. Chem. **270:** 26758–26761.
14. GUSTAFSON, E.L., K.E. SMITH, M.M. DURKIN, *et al.* 1997. Distribution of the neuropeptide Y Y2 receptor mRNA in rat central nervous system. Brain Res. Mol. Brain Res. **46:** 223–235.
15. DUMONT, Y., D. JACQUES, P. BOUCHARD, *et al.* 1998. Species differences in the expression and distribution of the neuropeptide Y Y1, Y2, Y4, and Y5 receptors in rodents, guinea pig, and primates brains. J. Comp. Neurol. **402:** 372–384.
16. CABERLOTTO, L., K. FUXE, J.M. RIMLAND, *et al.* 1998. Regional distribution of neuropeptide Y Y2 receptor messenger RNA in the human post mortem brain. Neuroscience **86:** 167–178.
17. BROBERGER, C., M. LANDRY, H. WONG, *et al.* 1997. Subtypes Y1 and Y2 of the neuropeptide Y receptor are respectively expressed in pro-opiomelanocortin- and neuropeptide-Y-containing neurons of the rat hypothalamic arcuate nucleus. Neuroendocrinology **66:** 393–408.
18. KING, P.J., G. WILLIAMS, H. DOODS, *et al.* 2000. Effect of a selective neuropeptide Y Y(2) receptor antagonist, BIIE0246 on neuropeptide Y release. Eur. J. Pharmacol. **396:** R1–R3.
19. POTTER, E.K. & M.J. MCCLOSKEY. 1992. [Leu31, Pro34] NPY, a selective functional postjunctional agonist at neuropeptide-Y receptors in anaesthetised rats. Neurosci. Lett. **134:** 183–186.
20. BATTERHAM, R.L., M.A. COWLEY, C.J. SMALL, *et al.* 2002. Gut hormone PYY(3-36) physiologically inhibits food intake. Nature **418:** 650–654.
21. SAINSBURY, A., C. SCHWARZER, M. COUZENS, *et al.* 2002. Important role of hypothalamic Y2 receptors in body weight regulation revealed in conditional knockout mice. Proc. Natl. Acad. Sci. USA **99:** 8938–8943.
22. MORAN, T.H. 2000. Cholecystokinin and satiety: Current perspectives. Nutrition **16:** 858–865.

Central Serotonin and Melanocortin Pathways Regulating Energy Homeostasis

LORA K. HEISLER,[a] MICHAEL A. COWLEY,[b,c] TOSHIRO KISHI,[a]
LAURENCE H. TECOTT,[d] WEI FAN,[c] MALCOLM J. LOW,[c] JAMES L. SMART,[c]
MARCELO RUBINSTEIN,[e] JEFFREY B. TATRO,[f] JEFFREY M. ZIGMAN,[a]
ROGER D. CONE,[c] AND JOEL K. ELMQUIST[a]

[a]Division of Endocrinology, Diabetes, and Metabolism, Departments of Medicine and
Neurology, Beth Israel Deaconess Medical Center and Program in Neuroscience,
Harvard Medical School, Boston, Massachusetts 02215, USA

[b]Division of Neuroscience, Oregon Regional Primate Research Center,
Oregon Health and Science University, Beaverton, Oregon 97006, USA

[c]The Vollum Institute, Oregon Health and Science University,
Portland, Oregon 97201, USA

[d]Department of Psychiatry and Center for Neurobiology and Psychiatry,
University of California at San Francisco, San Francisco, California 94117, USA

[e]Instituto de Investigaciones en Ingeniería Genética y Biología Molecular (CONICET)
and Department of Biological Sciences, FCEyN, Universidad de Buenos Aires, Argentina

[f]Division of Endocrinology, Diabetes, Metabolism, and Molecular Medicine,
Tufts–New England Medical Center and Department of Neuroscience and Pharmacology
and Experimental Therapeutics, Tufts University School of Medicine,
Boston, Massachusetts 02111, USA

ABSTRACT: It is now established that the hypothalamus is essential in coordinating endocrine, autonomic, and behavioral responses to changes in energy availability. However, the interaction of key peptides, neuropeptides, and neurotransmitters systems within the hypothalamus has yet to be delineated. Recently, we investigated the mechanisms through which central serotonergic (5-hydroxytryptamine, 5-HT) systems recruit leptin-responsive hypothalamic pathways, such as the melanocortin systems, to affect energy balance. Through a combination of functional neuroanatomy, feeding, and electrophysiology studies in rodents, we found that 5-HT drugs require functional melanocortin pathways to exert their effects on food intake. Specifically, we observed that anorectic 5-HT drugs activate pro-opiomelanocortin (POMC) neurons in the arcuate nucleus of the hypothalamus (Arc). We provide evidence that the serotonin 2C receptor (5-HT$_{2C}$R) is expressed on POMC neurons and contributes to this effect. Finally, we found that 5-HT drug-induced hypophagia is attenuated by pharmacological or genetic blockade of downstream melanocortin

Address for correspondence: Joel K. Elmquist, Division of Endocrinology, Diabetes, and Metabolism, Departments of Medicine and Neurology, Beth Israel Deaconess Medical Center and Program in Neuroscience, Harvard Medical School, Boston, MA 02215. Voice: 617-667-0845; fax: 617-667-2927.

jelmquis@bidmc.harvard.edu

Ann. N.Y. Acad. Sci. 994: 169–174 (2003). © 2003 New York Academy of Sciences.

3 and 4 receptors. We review candidate brain regions expressing melanocortin 3 and 4 receptors that play a role in energy balance. A model is presented in which activation of the melanocortin system is downstream of 5-HT and is necessary to produce the complete anorectic effect of 5-HT drugs. The data reviewed in this paper incorporate the central 5-HT system to the growing list of metabolic signals that converge on melanocortin neurons in the hypothalamus.

KEYWORDS: serotonin; melanocortin; pro-opiomelanocortin (POMC); serotonin 2C receptor (5-HT$_{2C}$R); melanocortin 4 receptor (MC4-R); food intake; body weight; hypothalamus; and arcuate nucleus

Obesity and type II diabetes are rising at alarming rates in the United States.[1] Elucidating the basic neurobiology of energy homeostasis is paramount in the prevention and treatment of these conditions.[2] Over the past decade, remarkable progress has been made in the understanding of how the central nervous system (CNS), especially the hypothalamus, controls food intake and body weight homeostasis. In the short review that follows, we outline some recent advances in the hypothalamic mechanisms involved in regulating energy balance.

The central serotonin (5-hydroxytryptamine, 5-HT) system has been long been associated with food intake and body weight regulation. Pharmacological agents that increase 5-HT activity in the CNS inhibit food intake and promote weight loss.[3–5] However, the 5-HT neural circuits underlying energy homeostasis have been difficult to discern because of the widespread nature of 5-HT neuronal projections, the identification of at least 14 distinct 5-HT receptors, and the relative paucity of receptor-selective drugs.[6] The inability to pharmacologically target specific 5-HT pathways and receptors has contributed to the incidence of unwanted side-effects with anorectic 5-HT indirect agonists. A notable example is d-fenfluramine (d-Fen), a drug that blocks the reuptake of 5-HT and stimulates its release.[7,8] In the mid-1990s, d-Fen was prescribed to millions of people in the United States for weight loss, frequently in combination with phentermine (fen/phen), but was withdrawn from clinical use in 1997 by the Food and Drug Administration due to reports of adverse cardiopulmonary events.[9]

Recently, we have undertaken a series of experiments in an effort to determine specific CNS pathways through which d-Fen selectively reduces food intake and body weight. d-Fen dose-dependently induces Fos-like immunoreactivity (FOS-IR), a commonly used marker of neuronal activation, in many brain regions associated with energy regulation.[10] We observed significant FOS-IR induction in the lateral, but not medial, arcuate nucleus of the hypothalamus (Arc) of the rat.[10] This pattern of FOS-IR induction is consistent with the distribution of neurons containing the anorectic neuropeptides pro-opiomelanocortin (POMC) and cocaine and amphetamine regulated transcript (CART).[11] Additional evidence that 5-HT acts on POMC neurons is found when examining 5-HT neuronal projections and receptor expression patterns. Specifically, 5-HT-immunoreactive nerve terminals contact POMC neurons in the Arc.[12] 5-HT drugs also cause the release of the protein product of POMC, α-melanocyte stimulating hormone (α-MSH) from superfused hypothalamic slices.[13] 5-HT receptor distribution studies indicate that at least four 5-HT receptors are expressed in the Arc.[14–16] These data suggest that the central 5-HT system is positioned to act on POMC neurons.

Activation of POMC neurons causes the release of the endogenous melanocortin receptor agonist α-MSH. Centrally administered α-MSH reduces food intake, while treatment with the endogenous melanocortin receptor antagonist agouti-related protein (AgRP) increases feeding.[17,18] α-MSH and AgRP are expressed in distinct Arc neuronal populations, both of which project to regions implicated in energy homeostasis and autonomic outflow that contain melanocortin 4 receptor (MC4-R) expressing neurons.[11,19–22] Pharmacological blockade or genetic inactivation of MC4-Rs in laboratory animals and human subjects produces hyperphagia, reduced energy expenditure, obesity, and insulin resistance.[23–29] The recent surge of data on the CNS melanocortin pathways indicates that this system is a critical regulator of energy homeostasis.

We propose that 5-HT drugs, such as d-Fen, reduce food intake and body weight by engaging these melanocortin pathways. Support for this hypothesis was found through both *in vivo* and electrophysiological techniques. Specifically, rats treated with anorectic doses of d-Fen exhibited significant FOS-IR induction in POMC neurons at all rostrocaudal levels of the Arc.[10] We also performed electrophysiological recordings in transgenic mice expressing green fluorescent protein under the control of POMC promoter (GFP-POMC).[10,30] These experiments indicate that d-Fen and 5-HT induce highly consistent depolarization of Arc POMC neurons.[10] Taken together, these data provide strong support for a mechanism of d-Fen hypophagia via 5-HT-induced activation of Arc POMC neurons.

Of the 5-HT receptors identified in the Arc, the $5\text{-HT}_{2C}R$ is the most intriguing candidate contributing to the anorectic actions of d-Fen. $5\text{-HT}_{2C}Rs$ have been implicated in the modulation of food intake, body weight, and autonomic function through both transgenic and pharmacological studies.[3,5] $5\text{-HT}_{2C}R$ deficient (–/–) mice are hyperphagic, obese, and display blunted responses to anorectic properties of d-Fen.[31,32] We determined that $5\text{-HT}_{2C}R$ mRNA is co-expressed at a rate up to 80% with α-MSH-containing neurons in the Arc.[10] If d-Fen and 5-HT activate POMC neurons via action at $5\text{-HT}_{2C}Rs$, then treatment with a $5\text{-HT}_{2C}R$ agonist should mimic the effects of d-Fen and 5-HT in this population of neurons. We found that the $5\text{-HT}_{2C/1B}R$ agonist 1-(3-chlorophenyl)piperazine (mCPP) induces dose-dependent FOS-IR in Arc POMC neurons in a similar manner produced by d-Fen.[10] mCPP and the $5\text{-HT}_{2C/2A}R$ agonist MK212 also dose-dependently depolarize POMC neurons in GFP-POMC transgenic mice.[10] These results strongly support the hypothesis that $5\text{-HT}_{2C}Rs$ expressed on POMC neurons are involved in hypophagia induced by 5-HT drugs like d-Fen.

If the melanocortin pathway is a critical mediator of 5-HT drug action on energy homeostasis, then it could be hypothesized that blockade of a downstream melanocortin receptor will attenuate these effects. Supporting this hypothesis, experiments with mice ectopically overexpressing the endogenous melanocortin receptor antagonist agouti (Ay mice) show blunted responses to threshold anorectic doses of d-Fen.[10] Similarly, rats pretreated with the MC3-R/MC4-R antagonist SHU9119 also display attenuated responses to d-Fen-induced hypophagia.[10] These results are remarkable in light of the efficacy of d-Fen in other rodent models of obesity, such as the ob/ob mouse.[33] These data demonstrate that the anorectic properties of d-Fen require functional melanocortin pathways. On the other hand, $5\text{-HT}_{2C}R$-deficient (–/–) mice treated with the MC3-R/MC4-R agonist MTII exhibit reductions in food intake comparable to their wild-type littermates. This latter observation suggests that $5\text{-HT}_{2C}R$

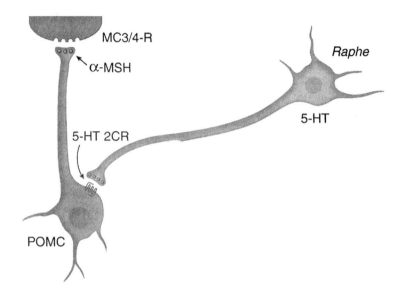

FIGURE 1. Schematic of proposed hypophagic action of 5-HT drugs involving the recruitment of central melanocortin pathways.

action is upstream of MC3-R/MC4-Rs.[10] Taken together, these data lead us to offer a model of the mechanisms of d-Fen-induced hypophagia (FIG. 1). This model predicts that d-Fen increases the availability of 5-HT in the Arc, which acts on POMC neurons through 5-HT$_{2C}$Rs. This activation of POMC neurons causes the release of α-MSH, which then acts on neurons expressing MC3-Rs and/or MC4-Rs in downstream target CNS regions regulating energy homeostasis.

Candidate CNS regions implicated in feeding behavior that receive dense inputs from Arc POMC and AgRP neurons and contain MC4-Rs and/or MC3-Rs are the paraventricular nucleus of the hypothalamus (PVH) and lateral hypothalamic area (LHA).[34–36] Localized PVH injections of the MC3-R/MC4-R agonist MTII inhibit feeding, while injections of the MC3-R/MC4-R antagonist SHU9119 or MC4-R antagonist HS014 increases food intake.[37–39] Relatively few neurons express MC4-R mRNA in the LHA.[20,21] Most likely, either the MC3-R is a predominant melanocortin receptor regulating feeding behavior in the LHA,[40] or MC4-Rs are expressed on axon terminals (from certain CNS sites projecting to the LHA) and act presynaptically. The dorsal motor nucleus of the vagus (DMV) is another region that contains a high degree of MC4-R mRNA expression that may contribute to the melanocortin regulation of food intake and body weight.[20,21,41]

In summary, the pathophysiological basis for aberrant feeding behavior, body weight regulation, and endocrine function has yet to be determined. Pharmacological agents targeting both the serotonergic and melanocortin pathways are very effective in altering energy balance. In this paper, we propose that the central 5-HT system converges on melanocortin pathways as a final common output mechanism to promote energy homeostasis. We propose that a serotonergic receptor critically involved in this pathway is the $5\text{-HT}_{2C}R$, and that the MC4-R is a crucial downstream melanocortin receptor affecting energy balance. Future work on this model of energy homeostasis may delineate necessary pathways of serotonergic and melanocortin drug action, and thereby may identify potential selective targets for the prevention and treatment of obesity and type II diabetes.

REFERENCES

1. FREEDMAN, D.S., L.K. KHAN, M.K. SERDULA, *et al.* 2002. Trends and correlates of class 3 obesity in the United States from 1990 through 2000. J. Am. Med. Assoc. **288:** 1758–1761.
2. HILL, J.O. & J.C. PETERS. 1998. Environmental contributions to the obesity epidemic. Science **280:** 1371–1374.
3. HEISLER, L.K., H.-M. CHU & L.H. TECOTT. 1998. Epilepsy and obesity in serotonin 5-HT_{2C} receptor mutant mice. Ann. N.Y. Acad. Sci. **861:** 74–87.
4. HEISLER, L.K., R.B. KANAREK & B.A. HOMOLESKI. 1999. Acute and chronic fluoxetine attenuates fat and protein but not carbohydrate consumption in female rats. Pharm. Biochem. Behav. **63:** 377–385.
5. SIMANSKY, K.J. 1996. Serotonergic control of the organization of feeding and satiety. Behav. Brain Res. **73:** 37–42.
6. PEROUTKA, S.I. 1994. Molecular biology of serotonin (5-HT) receptors. Synapse **18:** 241–260.
7. ROWLAND, N.E. & J. CARLTON. 1986. Neurobiology of an anorectic drug: Fenfluramine. Prog. Neurobiol. **27:** 13–62.
8. GUY-GRAND, B. 1992. Clinical studies with d-fenfluramine. Am. J. Clin. Nutr. **55:** 173S–176S.
9. CONNOLLY, H.M., J.L. CRARY, M.D. MCGOON, *et al.* 1997. Valvular heart disease associated with fenfluramine-phentermine. N. Engl. J. Med. **337:** 581–588.
10. HEISLER, L.K., M.A. COWLEY, L.H. TECOTT, *et al.* 2002. Activation of central melanocortin pathways by fenfluramine. Science **297:** 609–611.
11. ELIAS, C.F., C.E. LEE, J. KELLY, *et al.* 1998. Leptin activates hypothalamic CART neurons projecting to the spinal cord. Neuron **21:** 1375–1385.
12. KISS, I., C.S LERANTH & B. HALASZ. 1984. Serotonergic endings on VIP-neurons in the suprachiasmatic nucleus and on ACTH-neurons in the arcuate nucleus of the hypothalamus: a combination of high resolution autoradiography and electron microscopic immunocytochemistry. Neurosci. Lett. **44:** 119–124.
13. TILIGADA, E. & I.F. WILSON. 1989. Regulation of alpha-melanocyte-stimulating hormone release from superfused slices of rat hypothalamus by serotonin and the interaction of serotonin with the dopaminergic system inhibiting peptide release. Brain Res. **503:** 225–228.
14. WRIGHT, D.E., K.B. SEROOGY, K.H. LUNDGREN, *et al.* 1995. Comparative localization of serotonin 1A, 1C, and 2 receptor subtypes in the rat brain. J. Comp. Neurol. **351:** 357–373.
15. MOLINEAUX, S.M., T.M. JESSELL, R. AXEL, *et al.* 1989. 5-HT1c receptor is a prominent serotonin receptor subtype in the central nervous system. Proc. Natl. Acad. Sci. USA **86:** 6793–6797.
16. BRUINVELS, A.T., B. LANDWHRMEYER, E.L. GUSTAFSON, *et al.* 1994. Localization of 5-HT1B, 5-HT1D, 5-HT1E, and 5-HT1F receptor mRNA in rodent and primate brain. Neuropharmacology **33:** 367–386.
17. MCMINN, J.E., C.W. WILKINSON, P.J. HAVEL, *et al.* 2000. Effect of intracerebroventricular alpha-MSH on food intake, adiposity, c-Fos induction, and neuropeptide expression. Am. J. Physiol. Regul. Integr. Comp. Physiol. **279:** R695–R703.
18. HAHN, T.M., J.F. BREININGER, D.G. BASKIN, *et al.* 1998. Coexpression of Agrp and NPY in fasting-activated hypothalamic neurons. Nat. Neurosci. **1:** 271–272.
19. ELIAS, C.F., C.B. SAPER, E. MARATOS-FLIER, *et al.* 1998. Chemically defined projections linking the mediobasal hypothalamus and the lateral hypothalamic area. J. Comp. Neurol. **402:** 442–459.

20. MOUNTJOY, K.A., M.T. MORTRUD, M.J. LOW, et al. 1994. Localization of the melano-cortin-4 receptor (MC4-R) in neuroendocrine and autonomic control circuits in the brain. Mol. Endocrinol. **8:** 1298–1308.
21. KISHI, T., C.J. ASCHKENASI, C.E. LEE, et al. 2003. Expression of melanocortin 4 receptor mRNA in the central nervous system of the rat. J. Comp. Neurol. **457:** 213–235.
22. ELMQUIST, J.K., C.F. ELIAS & C.B. SAPER. 1999. From lesions to leptin: Hypothalamic control of food intake and body weight. Neuron **22:** 221–232.
23. HUSZAR, D., C.A. LYNCH, V. FAIRCHILD-HUNTRESS, et al. 1997. Targeted disruption of the melanocortin-4 receptor results in obesity in mice. Cell **88:** 131–141.
24. YEN, T.T., A.M. GILL, L.A. FRIGERI, et al. 1994. Obesity, diabetes and neoplasia in yellow Ay mice: ectopic expression of the agouti gene. FASEB J. **8:** 479–488.
25. YEO, G.S., I.S. FAROOQI, S. AMINIAN, et al. 1998. A frameshift mutation in MC4R associated with dominantly inherited human obesity. Nat. Genet. **20:** 111–112.
26. HINNEY, A., A. SCHMIDT, K. NOTTEBOM, et al. 1999. Several mutations in the melano-cortin-4 receptor gene including a nonsense and frameshift mutation associates with dominantly inherited obesity in humans. J. Clin. Endocrinol. Metab. **84:** 1483–1486.
27. TOOQI, I.S., G.S. YEO, J.M. KEOGH, et al. 2000. Dominant and recessive inheritance of morbid obesity associated with melanocortin 4 receptor deficiency. J. Clin. Invest. **106:** 271–279.
28. FAN, W., D.M. DINULESCU, A.A. BUTLER, et al. 2000. The central melanocortin system can directly regulate serum insulin levels. Endocrinology **141:** 3072–3079.
29. FAN, W., B.A. BOSTON, R.A. KESTERSON, et al. 1997. Role of melanocortinergic neurons in feeding and the agouti obesity syndrome. Nature **385:** 165–168.
30. COWLEY, M.A., I.L. SMART, M. RUBINSTEIN, et al. 2001. Leptin activates anorexigenic POMC neurons through a neural network in the arcuate nucleus. Nature **411:** 480–484.
31. TECOTT, L.H., L.M. SUN, S.F. AKANA, et al. 1995. Eating disorder and epilepsy in mice lacking 5-HT2C serotonin receptors. Nature **374:** 542–546.
32. VICKERS, S.P., P.G. CLIFTON, C.T. DOURISH, et al. 1999. Reduced satiating effect of d-fenfluramine in serotonin 5-HT(2C) receptor mutant mice. Psychopharmacology **143:** 309–314.
33. ROWLAND, N.E. 1994. Long-term administration of dexfenfluramine to genetically obese (ob/ob) and lean mice: Body weight and brain serotonin changes. Pharmacol. Biochem. Behav. **49:** 287–294.
34. BAGNOL, D., X. LU, C.B. KAELIN, et al. 1999. Anatomy of an endogenous antagonist: relationship between agouti-related protein and proopiomelanocortin in brain. J. Neurosci. **19:** RC26.
35. HASKELL-LUEVANO, C., P. CHEN, C. LI, et al. 1999. Characterization of the neuroanatomical distribution of agouti-related protein immunoreactivity in the rhesus monkey and the rat. Endocrinology **140:** 1408–1415.
36. JACOBOWITZ, D.M. & T.L. O'DONOHUE. 1978. α-Melanocyte stimulating hormone: immunohistochemical identification and mapping in neurons of rat brain. Proc. Natl. Acad. Sci. USA **75:** 6300–6304.
37. GIRAUDO, S.Q., C.J. BILLINGTON & A.S. LEVINE. 1998. Feeding effects of hypothalamic injection of melanocortin 4 receptor ligands. Brain Res. **809:** 302–306.
38. WIRTH, M.M., P.K. OLSZEWSKI, C. YU, et al. 2001. Paraventricular hypothalamic α-melanocyte-stimulating hormone and MTII reduce feeding without causing aversive effects. Peptides **22:** 129–134.
39. KASK, A. & H.B. SCHIÖTH. 2000. Tonic inhibition of food intake during inactive phase is reversed by the injection of the melanocortin receptor antagonist into the paraventricular nucleus of the hypothalamus and central amygdala of the rat. Brain Res. **887:** 460–464.
40. ROSELLI-REHFUSS, L., K.G. MOUNTJOY, L.S. ROBBINS, et al. 1993. Identification of a receptor for α melanotropin and other proopiomelanocortin peptides in the hypothalamus and limbic system. Proc. Natl. Acad. Sci. USA **90:** 8856–8860.
41. GRILL, H.J., A.B. GINSBERG, R.J. SEELEY, et al. 1998. Brainstem application of melano-cortin receptor ligands produces long-lasting effects on feeding and body weight. J. Neurosci. **18:** 10128–10135.

Electrophysiological Actions of Peripheral Hormones on Melanocortin Neurons

MICHAEL A. COWLEY,[a] ROGER D. CONE,[b] PABLO ENRIORI,[a] INGRID LOUISELLE,[a] SARAH M. WILLIAMS,[a] AND ANNE E. EVANS[a]

[a]Division of Neuroscience, Oregon National Primate Research Center, Oregon Health and Sciences University, Beaverton, Oregon 97006, USA

[b]Vollum Institute, Oregon Health and Sciences University, Portland, Oregon 97239, USA

ABSTRACT: Neurons of the arcuate nucleus of the hypothalamus (ARH) appear to be sites of convergence of central and peripheral signals of energy stores, and profoundly modulate the activity of the melanocortin circuits, providing a strong rationale for pursuing these circuits as therapeutic targets for disorders of energy homeostasis. Recently, tremendous advances have been made in identifying genes and pathways important to regulating energy homeostasis, particularly the hormone leptin and its receptor. This hormone/receptor pair is expressed at high levels in the so-called satiety centers in the hypothalamus, and at lower levels elsewhere in the body. Recent studies in our lab and those of our collaborators have shown that leptin modulates different populations of hypothalamic cells in different ways, rapidly activating POMC neurons and inhibiting NPY/AgRP neurons.

In this report, we outline an integrated model of leptin's action in the arcuate nucleus of the hypothalamus, derived from our electrophysiological studies of brain slice preparations taken from transgenic mice that have been bred to express a variety of fluorescent proteins in specific cell types. We also discuss the recently withdrawn obesity drug fenfluramine, which appears to act on POMC neurons via the serotonin 2C receptor. Nutrient-sensing serotonin neurons may project from the raphe nuclei in the brainstem to the hypothalamus; within the arcuate nucleus, serotonin signals are integrated with others such as leptin, ghrelin, and peptide YY_{3-36} from the gut, to produce a coordinated response to nutrient state. Finally, we review the current inquiries into the ability of the hormone ghrelin to stimulate appetite by its action of NPY neurons and inhibition of POMC neurons.

KEYWORDS: neuropeptides; energy homeostasis; food intake; electrophysiology; gut; hypothalamus; pro-opiomelanocortin (POMC); neuropeptide Y (NPY); agouti-related peptide (AgRP); transgene

INTRODUCTION

Obesity is one of the major health challenges facing the developed world, accounting for 280,000 deaths annually in the United States.[1] Being obese or overweight decreases life expectancy 3–13 years,[1,2] and obesity significantly increases

Address for correspondence: Michael A. Cowley, Division of Neuroscience, Oregon National Primate Research Center, Oregon Health and Sciences University, Beaverton, Oregon 97006. Voice: 503-533-2421; fax: 503-690-5384.

cowleym@ohsu.edu

Ann. N.Y. Acad. Sci. 994: 175–186 (2003). © 2003 New York Academy of Sciences.

the risk of coronary disease, diabetes, and breast cancer. Also, significant economic costs are associated with healthcare for such an obese population. An estimated 9.4% of U.S. healthcare expenditure is directly related to "obesity and inactivity," while recent costs due to diabetes were estimated at $98 billion *per annum*.[3]

Early studies identified the hypothalamus as an important center of energy homeostatic control,[4] and recently the role of the arcuate nucleus has been highlighted.[5] There have also been tremendous advances identifying genes and pathways important for regulating energy homeostasis,[6] particularly the hormone-receptor pair leptin (Lep^{ob})[7] and leptin receptor (*Lepr*, LR).[8]

White adipose cells produce leptin[7] and the absence of leptin or functional LRs causes morbid obesity.[6] The "long" LRb has a 301 amino acid intracellular tail[8–10] and is expressed at high levels in so-called "satiety" centers in the hypothalamus as well as in some non-neuronal tissues, such as pancreatic β cells.[8,11–14] A point mutation in the LRb splice junction results in the premature truncation of LR in *db* mice.[8,9,15,16] Thus, while the role of the short LR forms remains unclear, LRb function is critical for maintenance of endocrine function and energy balance.

NEURONS IN THE ARH ARE A MAJOR TARGET OF PERIPHERAL LEPTIN

LRb is expressed highly in the hypothalamus, in the arcuate nucleus of the hypothalamus (ARH), in the ventromedial hypothalamic nucleus (VMH), and in the dorsomedial hypothalamic nucleus, with lower expression in the paraventricular hypothalamic nucleus (PVH) and the lateral hypothalamus.[12,13,17,18] These regions have all been implicated in the control of energy homeostasis.[14] One of the major targets for circulating leptin is the ARH. Leptin receptors were first identified on cells in the ARH and VMH[19] and radiolabeled leptin binds in the ARH after peripheral injection.[20] Furthermore, peripheral injection of leptin stimulates robust expression of the immediate-early gene c-fos (a marker of neuronal activity) and expression of the leptin marker SOCS-3[21] in neurons within this nucleus.[22] The paucity of the blood-brain barrier here means that circulating leptin has better access to leptin receptors in this region.[20] Leptin is specifically transported across the blood-brain barrier[20] to reach sites distal to the circumventricular organs, and saturation of these transporters is hypothesized to be part of the cause of diet-induced obesity. Leptin exerts some of its effects on energy homeostasis by complementary actions on both neuropeptide Y/agouti-related peptide (NPY/AgRP) and pro-opiomelanocortin (POMC) neurons, and both these neuron types express leptin receptors.[23,24]

GENOMIC ACTIONS OF LEPTIN ON NPY/AgRP AND POMC NEURONS

NPY has well-described effects on energy homeostasis (for a review, see Ref. 25). NPY is produced in hypothalamic nuclei known to regulate appetite and metabolism, and NPY is one of the most potent orexigens. Intracerebroventricular injection of NPY will profoundly stimulate feeding[26,27] and will reduce energy expenditure.[26] It has been demonstrated that leptin modulates NPY levels: leptin deficiency results in a significant increase in NPY mRNA expression, while leptin administration causes a decrease in expression.[28] Furthermore, genetic deletion of NPY in leptin-deficient animals resulted in a significant decrease in their degree of obesity.[29] Treatment of *Lep^{ob}/Lep^{ob}* mice with leptin reduced the hyperphagia and normalized NPY expression, while NPY synthesis and secretion are increased in most models of energy deficiency or increased metabolic demand.[30] Interestingly, NPY neurons co-express AgRP,[31] a potent antagonist of the melanocortin-3 and -4 receptors (MC3-R and MC4-R). AgRP expression is robustly stimulated by fasting and inhibited by leptin,[31-33] and leptin-deficient mice have elevated AgRP mRNA levels.[32,33]

Leptin modulates POMC mRNA,[34] although to a much lower extent than leptin modulates AgRP mRNA, and leptin-deficient mice have reduced levels of POMC transcript.[35] The cloning of the agouti gene[36] and the discovery of the mechanism of agouti's action[37,38] led to the proposal that the melanocortin system, specifically the POMC neurons acting via α-melanocyte-stimulating hormone (α-MSH) and the MC4-R, acts centrally to limit adipose stores and to stimulate energy expenditure. This hypothesis was supported by the demonstration of hyperphagia and obesity in MC4-R–deficient mice,[39] by the characterization of AgRP as an endogenous hypothalamic melanocortin receptor antagonist,[40,41] and by the obesity evident in transgenic mice that ectopically overexpress AgRP.[41]

Thus, it appears that leptin inhibits the expression of orexigenic neuropeptides and increases the expression of anorexigenic neuropeptides. However, it is important to remember that the brain is not a "neuropeptide soup" and that neuropeptides only modulate neurotransmission. To better understand how leptin acts on the brain, it is necessary to understand the effects of leptin on neurotransmission and neuronal activity.

LEPTIN RAPIDLY ACTIVATES SOME HYPOTHALAMIC NEURONS

Soon after its discovery, leptin was shown to cause an increase in the expression of immunoreactive c-fos in the mediobasal hypothalamus, especially the PVH, but also the ARH and the VMH of *Lep^{ob}/Lep^{ob}* mice.[42] This suggested an increase in the activity of neurons in these regions, although the neuropeptide identity of the activated cells was not determined. This work was further refined when it was shown that peripheral injection of leptin rapidly induced c-fos expression in POMC neurons, while it did not induce c-fos in NPY neurons.[22] Both POMC and NPY neurons showed increased expression of mRNA for SOCS-3. The investigators proposed that leptin was acting differentially on the two neuronal populations, to activate POMC neurons and to inhibit the NPY/AgRP neurons. Leptin rapidly stimulates the secretion of α-MSH from a hypothalamic slice preparation.[43] Leptin can also depolarize parvocellular PVH neurons,[44] through activation of a depolarizing, nonspecific cation current, and leptin has been shown to have effects on sympathetic outflow within 15 min,[45] demonstrating rapid (nongenomic) effects of leptin.

LEPTIN RAPIDLY INHIBITS SOME HYPOTHALAMIC NEURONS

Miller and colleagues determined that leptin inhibited the electrophysiological activity of some neurons in the ARH[46] by performing whole cell recordings. These responses were absent in Zucker fatty rats ($Lepr^{fa}/Lepr^{fa}$), which lack functional leptin receptors. Using a similar hypothalamic slice preparation, leptin was shown to hyperpolarize and decrease the action potential firing rate of ARH and VMH neurons.[47] The effect on neurons was due to activation of an ATP-sensitive potassium channel (K_{ATP}). The consequent efflux of K^+ ions from the cell reduced the membrane potential, and this hyperpolarization led to less frequent action potentials. There was no effect of leptin in obese Zucker rats.

Interestingly, the neurons that were hyperpolarized by leptin were also hyperpolarized by glucose, also acting through the K_{ATP} channel. The convergence of glucose and leptin signals onto hypothalamic neurons has been confirmed.[48] In particular, it appears that modulation of the K_{ATP} channel is a common mechanism used by glucose, insulin, and leptin[47,49–56] to change cellular activity.

INTEGRATED MODEL OF LEPTIN ACTION IN THE ARH

Leptin may thus exert diverse effects on neuronal activity depending upon the neuropeptide phenotype of the neuron that is responding. To understand leptin actions on the brain, it is necessary to know the neuropeptide phenotype of the leptin-sensitive neurons. In a recent publication,[57] we have offered a model by which leptin directly and differentially acts on neurons in the ARH. We formulated the model of arcuate nucleus circuitry based on electrophysiological data we generated by recording from identified POMC neurons in brains slices taken from transgenic mice. Our collaborators, Malcolm Low and Jim Smart, developed a transgenic mouse termed the POMC-EGFP mouse; these mice express a transgene incorporating 11 kb of the POMC gene promoter and a gene for EGFP (enhanced green fluorescent protein) (FIG. 1a) to cause eutopic expression of EGFP in POMC neurons (FIG. 1b).

To determine the effects of specific hormones on the activity of POMC neurons, we cut coronal sections of the brains of POMC-EGFP transgenic mice and performed standard electrophysiological recordings from neurons that showed EGFP fluorescence (FIG. 1c). Application of leptin to the tissue bath depolarized POMC neurons (FIG. 1d) from a resting potential of –40 to –45 mV in a concentration-responsive manner (FIG. 1e).

Two components contributed to the depolarization of the POMC neurons. Leptin increased the conductance of a nonspecific cation channel on POMC neurons, an effect that other investigators had seen with leptin before,[44] and leptin decreased the frequency of inhibitory currents onto POMC neurons. In studies with Tamas Horvath, we showed that the majority of the GABAergic inputs to POMC neurons also expressed NPY, suggesting that there was a local circuit within the ARH and that NPY neurons innervated and provided a dominant inhibitory tone onto POMC neurons. We also showed that leptin inhibited the release of GABA and NPY from NPY terminals in the ARH. Thus, leptin was having different actions on different classes of neurons, and the net effect was to increase the tone of anorectic POMC neurons and decrease the tone of the orexigenic NPY/AgRP neurons. We then went on to test if this model circuit (FIG. 1f) was sensitive to other signals that regulate energy homeostasis.

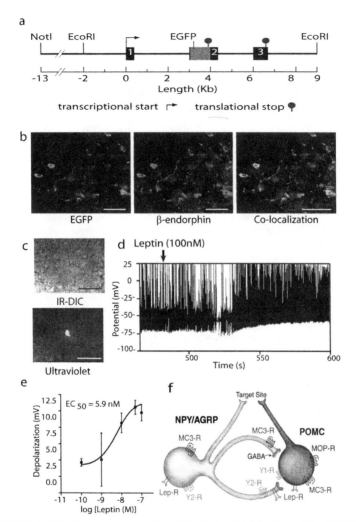

FIGURE 1. POMC neurons are activated by leptin. (**a**) Structure of the POMC-EGFP transgene. (**b**) Co-localization of EGFP and β-endorphin immunoreactivity in ARH POMC neurons in brain slices of POMC-EGFP transgenic mice. Scale bars in **b** and **c**: 50 μm. (**c**) Identification of a single POMC neuron (*arrowhead* on recording electrode tip) by EGFP fluorescence (*upper*) and IR-DIC microscopy (*lower*) in a living ARH slice prior to electrophysiological recordings. (**d**) Leptin depolarizes POMC neurons and increases the frequency of action potentials within 1 to 10 min of addition. The figure is a representative example of recordings made from 77 POMC neurons. (**e**) Leptin causes a concentration-dependent depolarization of POMC cells. The depolarization caused by leptin was determined at 0.1, 1, 10, 50, and 100 nM (EC_{50} = 5.9 nM) in 8, 7, 9, 3, and 45 cells, respectively. (**f**) Model of leptin regulation of NPY/GABA and POMC neurons in the ARH. Leptin directly depolarizes the POMC neurons, simultaneously hyperpolarizes the somata of NPY/GABA neurons, and diminishes release from NPY/GABA terminals. This diminished GABA release disinhibits the POMC neurons. Together, the direct and indirect effects of leptin result in an activation of POMC neurons and an increased frequency of action potentials.

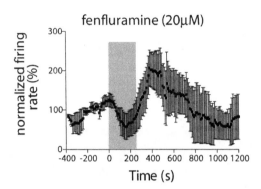

FIGURE 2. Fenfluramine increases the activity of POMC neurons. Fenfluramine (20 μM) caused a twofold increase in the spontaneous action potential frequency of POMC neurons from POMC-EGFP mice when recorded in the loose-cellê–attached mode. (Adapted from Heisler et al.[58])

FENFLURAMINE ACTION SHOWS HOW SEROTONIN AFFECTS ENERGY HOMEOSTASIS

Before its withdrawal, because of complications in a subset of patients, fenfluramine was one of the most successful therapies for obesity. Fenfluramine increased the concentration of serotonin (5-HT) at synapses, but the mechanism by which it caused weight loss was unknown. We investigated the effect of fenfluramine on POMC neurons and other compounds that activate the serotonergic system as part of an effort to determine how fenfluramine caused weight loss.

We showed that fenfluramine increased the firing rate of POMC neurons (FIG. 2) and that this was due to depolarization of the neurons via activation of the serotonin 2C receptor (5-HT$_{2C}$R). This was in accordance with functional anatomical and physiological data generated by our collaborators in Joel Elmquist's laboratory, showing that fenfluramine increased the expression of c-fos in POMC neurons that expressed the 5-HT$_{2C}$R and that fenfluramine required a functional melanocortin system to inhibit feeding. These combined findings led us to deduce that fenfluramine acted through 5-HT$_{2C}$Rs and the melanocortin system (see the article by Heisler et al.[58]) to inhibit feeding and ultimately to reduce body weight. We hope that an understanding of the role of the 5-HT$_{2C}$R in the control of feeding and energy expenditure will lead to better therapies for obesity.

These findings also suggest that nutrient-sensing serotonin neurons may project from the raphe nuclei in the brainstem to the hypothalamus. Within the ARH, serotonin signals are integrated with other signals, such as leptin and (as will be shown later) ghrelin and PYY$_{3-36}$ from the gut, to produce a coordinated response to a nutrient state. This signal is likely propagated through the brain by the coordinated effects of NPY, AgRP, and α-MSH.[59]

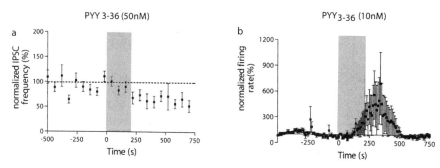

FIGURE 3. The gut hormone PYY$_{3-36}$ inhibits NPY neurons and thus activates POMC neurons. (a) PYY$_{3-36}$ (50 nM) decreased the frequency of spontaneous mini-IPSCs onto POMC neurons when recorded in voltage-clamp configuration. (b) PYY$_{3-36}$ (10 nM) caused a fourfold increase in the frequency of spontaneous action potentials in POMC neurons from POMC-EGFP mice when recorded in the loose-cell-attached configuration. (Adapted from Batterham *et al.*[60])

NPY-SENSITIVE GABA RELEASE HIGHLIGHTS A NEW ROLE FOR PYY$_{3-36}$

To establish the identity of the GABAergic inputs to POMC neurons, we analyzed the NPY sensitivity of the GABA release. We reasoned that a subtype of the NPY receptor, the Y2 receptor (NPY$_{Y2}$R), was expressed on NPY neurons in the ARH and may act as an autoreceptor to limit the activity of NPY neurons when the local concentrations of NPY rise too high. We found that the GABA release onto POMC neurons was extraordinarily sensitive to NPY and later to NPY$_{Y2}$R-specific agonists. This led us to propose that NPY$_{Y2}$R agonists would decrease the activity of NPY/ AgRP neurons and thus increase the activity of POMC neurons, causing similar effects on feeding and energy homeostasis to leptin. Because the ARH blood-brain barrier is more permeable than other parts of the brain, we reasoned that circulating hormones might have access to these Y2 receptor sites.

One candidate ligand to act on the NPY$_{Y2}$R is peptide YY$_{3-36}$ (PYY$_{3-36}$). It had long been known that PYY$_{3-36}$ was released after a meal, but its function was not clear. We showed that PYY$_{3-36}$ decreased the frequency of inhibitory postsynaptic currents (IPSCs, due to GABA release) onto POMC neurons (FIG. 3a), an effect that we have previously shown to be due to activity of NPY neurons, indicating that PYY$_{3-36}$ inhibited NPY neurons. We also showed that PYY$_{3-36}$ activated POMC neurons (FIG. 3b), increasing the neurons' spontaneous activity. In collaboration with us, Rachel Batterham in the Bloom laboratory showed that the effect of this dual inhibition of NPY neurons and activation of POMC neurons was that Y2 receptor agonists and PYY$_{3-36}$ inhibit feeding in rats and mice. Furthermore, PYY$_{3-36}$ had no effect on NPY$_{Y2}$R-deficient mice. It was also shown that infusion of postprandial concentrations of PYY$_{3-36}$ caused reduced appetite and food consumption in human volunteers (see the article by Batterham *et al.*[60]). These findings have led to renewed interest in PYY$_{3-36}$ and related compounds as possible therapies for obesity. Strategies that activate these neurons in a similar manner to (but independently of) leptin may bypass the resistance to leptin that is seen in most human obesity. These findings are also an example of the predictive power of the melanocortin circuitry model we have devised. Observations about the nature of the circuitry led us to hypothesize and demonstrate a new gut-brain axis that limits food intake.

GHRELIN ACTS OPPOSITE PYY$_{3-36}$ ON MELANOCORTIN CIRCUITS

The findings reported so far all show how hormones can inhibit food intake and increase energy expenditure by activating POMC neurons and stimulating the melanocortin system. It is reasonable to wonder whether this system can be modified in the other direction, that is, can hormones stimulate feeding by inhibiting the melanocortin system? There is more than an academic interest in this, because cachexia and wasting disorders are significant medical problems (see the article by Marks *et al.*[61]). Ghrelin is a hormone that is secreted by A-like cells of the stomach,[62] stimulates growth hormone secretion, and independently stimulates appetite.[63,64] There is also evidence that ghrelin may be an important factor in the long-term success of gastric bypass procedures for weight control.[65] Previous data suggest that ghrelin acts in the ARH; however, the mechanism of that effect is unclear. We sought to determine the mechanism by which ghrelin stimulates feeding. In a recent publication, we demonstrate that ghrelin activates NPY neurons and causes secondary inhibition of POMC neurons.[66]

To determine the mechanism of ghrelin action on hypothalamic neuronal activity, we used two models: the POMC-EGFP mouse described above and another trans-

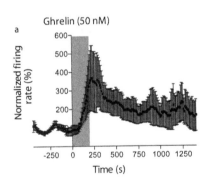 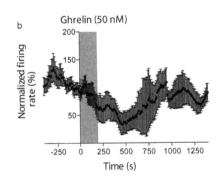

FIGURE 4. The stomach hormone ghrelin activates NPY neurons and thus inhibits POMC neurons. (**a**) Ghrelin (50 nM) caused a fourfold increase in the spontaneous action potential frequency of NPY neurons, measured using loose-cell–attached patch recordings. (**b**) Ghrelin (50 nM) caused a 50% decrease in the spontaneous activity of POMC neurons, using loose-cell–attached patch recordings. (From Cowley *et al.*[59] Reprinted with permission from *Neuron.*)

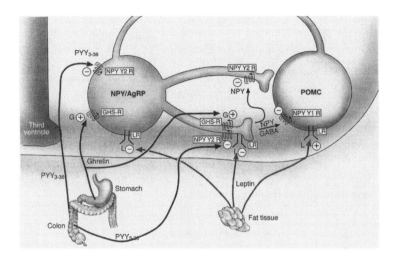

FIGURE 5. A model of convergence and action of peripheral hormones on melano-cortin circuits in the ARH.

genic mouse, which expresses sapphire fluorescent protein (SFP) under the control of the NPY promoter. The NPY-SFP mouse was developed by our collaborators in the Friedman laboratory and directs eutopic expression of SFP to NPY neurons, including ARH NPY neurons.

Bath application of ghrelin increased the activity of ARH NPY neurons (FIG. 4a), stimulating the release of GABA, NPY, and AgRP. The coordinated actions of these signals hyperpolarized and reduced the spontaneous activity of the POMC neurons (FIG. 4b). The effect of ghrelin on POMC neurons was secondary to the effect on NPY neurons; in this manner, ghrelin is acting like PYY_{3-36}, but causing the opposite effects.

Our collaborators in the Horvath and Smith laboratories went on to identify a population of neurons in the hypothalamus that express ghrelin and send fibers to contact NPY neurons in many regions of the hypothalamus, and Heiman and co-workers showed that ghrelin binds to NPY neurons. Ghrelin fibers also make contact with other kinds of hypothalamic neurons. In the same publication, the Colmers laboratory showed that ghrelin caused a decreased release of GABA onto cortico-trophin-releasing hormone (CRH) neurons and that this effect required NPY actions. This suggests that ghrelin was stimulating the release of NPY adjacent to CRH neurons, and we have previously shown that NPY reduces GABAergic IPSCs[59] in the PVH. Furthermore, this finding provides a potential mechanism for the previously demonstrated stimulation of adrenocorticotrophic hormone by ghrelin.[67]

CONCLUSION

These data show that ARH neurons are a site of convergence of central and peripheral signals of energy stores. Signals from the body and the brain converge on the melanocortin circuits here and profoundly modulate the activity of the melanocortin circuits. The new findings reviewed here also point to the importance of the inhibitory tone of the NPY/GABA synapses onto POMC neurons. Both the gut hormones characterized here act on the NPY neuron and cause a reciprocal change in the activity of the POMC neurons. The data described here also provide strong rationale for pursuing melanocortin circuits as therapeutic targets for disorders of energy homeostasis.

ACKNOWLEDGMENTS

We thank Joel Ito for assistance with illustrations. This work was supported by the National Institutes of Health (Grant DK62202).

REFERENCES

1. PEETERS, A. et al. 2003. Obesity in adulthood and its consequences for life expectancy: a life-table analysis. Ann. Intern. Med. **138:** 24-32.
2. FONTAINE, K.R. et al. 2003. Years of life lost due to obesity. J. Am. Med. Assoc. **289:** 187–193.
3. MOKDAD, A.H. et al. 2001. The continuing epidemics of obesity and diabetes in the United States. J. Am. Med. Assoc. **286:** 1195–1200.
4. HETHERINGTON, A.W. & S.W. RANSON. 1940. Hypothalamic lesions and obesity on the rat. Anat. Rec. **78:** 149–172.
5. CONE, R.D. et al. 2001. The arcuate nucleus as a conduit for diverse signals relevant to energy homeostasis. Int. J. Obes. Relat. Metab. Disord. **25**(Suppl 5): S63–S67.
6. SPIEGELMAN, B.M. & J.S. FLIER. 2001. Obesity and the regulation of energy balance. Cell **104:** 531–543.
7. ZHANG, Y. et al. 1994. Positional cloning of the mouse obese gene and its human homologue. Nature **372:** 425–432.
8. TARTAGLIA, L.A. et al. 1995. Identification and expression cloning of a leptin receptor, OB-R. Cell **83:** 1263–1271.
9. CHUA, S.C., JR. et al. 1996. Phenotypes of mouse diabetes and rat fatty due to mutations in the OB (leptin) receptor [see comments]. Science **271:** 994–996.
10. CHUA, S.C., JR. et al. 1997. Fine structure of the murine leptin receptor gene: Splice site suppression is required to form two alternatively spliced transcripts. Genomics **45:** 264–270.
11. EMILSSON, V. et al. 1997. Expression of the functional leptin receptor mRNA in pancreatic islets and direct inhibitory action of leptin on insulin secretion. Diabetes **46:** 313–316.
12. ELMQUIST, J.K. et al. 1998. Distributions of leptin receptor mRNA isoforms in the rat brain. J. Comp. Neurol. **395:** 535–547.
13. FEI, H. et al. 1997. Anatomic localization of alternatively spliced leptin receptors (Ob-R) in mouse brain and other tissues. Proc. Natl. Acad. Sci. USA **94:** 7001–7005.
14. ELMQUIST, J.K. et al. 1998. Unraveling the central nervous system pathways underlying responses to leptin. Nat. Neurosci. **1:** 445–450.
15. LEE, G.H. et al. 1996. Abnormal splicing of the leptin receptor in diabetic mice. Nature **379:** 632–635.
16. CHEN, H. et al. 1996. Evidence that the diabetes gene encodes the leptin receptor: Identification of a mutation in the leptin receptor gene in db/db mice. Cell **84:** 491–495.
17. MERCER, J.G. et al. 1996. Localization of leptin receptor mRNA and the long form splice variant (Ob-Rb) in mouse hypothalamus and adjacent brain regions by in situ hybridization. FEBS Lett. **387:** 113–116.

18. SCHWARTZ, M.W. *et al.* 1996. Identification of targets of leptin action in rat hypothalamus. J. Clin. Invest. **98:** 1101–1106.
19. MERCER, J.G. *et al.* 1996. Coexpression of leptin receptor and preproneuropeptide Y mRNA in arcuate nucleus of mouse hypothalamus. J. Neuroendocrinol. **8:** 733–735.
20. BANKS, W.A. *et al.* 1996. Leptin enters the brain by a saturable system independent of insulin. Peptides **17:** 305–311.
21. BJORBAEK, C. *et al.* 1998. Identification of SOCS-3 as a potential mediator of central leptin resistance. Mol. Cell **1:** 619–625.
22. ELIAS, C.F. *et al.* 1999. Leptin differentially regulates NPY and POMC neurons projecting to the lateral hypothalamic area. Neuron **23:** 775–786.
23. CHEUNG, C.C., D.K. CLIFTON & R.A. STEINER. 1997. Proopiomelanocortin neurons are direct targets for leptin in the hypothalamus. Endocrinology **138:** 4489–4492.
24. HAKANSSON, M.L., A.L. HULTING & B. MEISTER. 1996. Expression of leptin receptor mRNA in the hypothalamic arcuate nucleus—relationship with NPY neurones. Neuroreport **7:** 3087–3092.
25. KALRA, S.P. *et al.* 1999. Interacting appetite-regulating pathways in the hypothalamic regulation of body weight. Endocr. Rev. **20:** 68–100.
26. BILLINGTON, C.J. *et al.* 1991. Effects of intracerebroventricular injection of neuropeptide Y on energy metabolism. Am. J. Physiol. **260:** R321–R327.
27. STANLEY, B.G., A.S. CHIN & S.F. LEIBOWITZ. 1985. Feeding and drinking elicited by central injection of neuropeptide Y: evidence for a hypothalamic site(s) of action. Brain Res. Bull. **14:** 521–524.
28. STEPHENS, T.W. *et al.* 1995. The role of neuropeptide Y in the antiobesity action of the obese gene product. Nature **377:** 530–532.
29. ERICKSON, J.C., G. HOLLOPETER & R.D. PALMITER. 1996. Attenuation of the obesity syndrome of ob/ob mice by the loss of neuropeptide Y. Science **274:** 1704–1707.
30. INUI, A. 1999. Neuropeptide Y: A key molecule in anorexia and cachexia in wasting disorders? Mol. Med. Today **5:** 79–85.
31. BROBERGER, C. *et al.* 1998. The neuropeptide Y/agouti gene-related protein (AGRP) brain circuitry in normal, anorectic, and monosodium glutamate-treated mice. Proc. Natl. Acad. Sci. USA **95:** 15043–15048.
32. MIZUNO, T.M. & C.V. MOBBS. 1999. Hypothalamic agouti-related protein messenger ribonucleic acid is inhibited by leptin and stimulated by fasting. Endocrinology **140:** 814–817.
33. SHUTTER, J.R. *et al.* 1997. Hypothalamic expression of ART, a novel gene related to agouti, is up-regulated in obese and diabetic mutant mice. Genes Dev. **11:** 593–602.
34. THORNTON, J.E. *et al.* 1997. Regulation of hypothalamic proopiomelanocortin mRNA by leptin in ob/ob mice. Endocrinology **138:** 5063–5066.
35. WILDING, J.P. *et al.* 1993. Increased neuropeptide-Y messenger ribonucleic acid (mRNA) and decreased neurotensin mRNA in the hypothalamus of the obese (ob/ob) mouse. Endocrinology **132:** 1939–1944.
36. BULTMAN, S.J., E.J. MICHAUD & R.P. WOYCHIK. 1992. Molecular characterization of the mouse agouti locus. Cell **71:** 1195–1204.
37. FAN, W. *et al.* 1997. Role of melanocortinergic neurons in feeding and the agouti obesity syndrome. Nature **385:** 165–168.
38. LU, D. *et al.* 1994. Agouti protein is an antagonist of the melanocyte-stimulating-hormone receptor. Nature **371:** 799–802.
39. HUSZAR, D. *et al.* 1997. Targeted disruption of the melanocortin-4 receptor results in obesity in mice. Cell **88:** 131–141.
40. FONG, T.M. *et al.* 1997. ART (protein product of agouti-related transcript) as an antagonist of MC-3 and MC-4 receptors. Biochem. Biophys. Res. Commun. **237:** 629–631.
41. OLLMANN, M.M. *et al.* 1997. Antagonism of central melanocortin receptors in vitro and in vivo by agouti-related protein. Science **278:** 135–138.
42. WOODS, A.J. & M.J. STOCK. 1996. Leptin activation in hypothalamus. Nature **381:** 745.
43. KIM, M.S. *et al.* 2000. The central melanocortin system affects the hypothalamo-pituitary thyroid axis and may mediate the effect of leptin. J. Clin. Invest. **105:** 1005–1011.

44. Powis, J.E., J.S. Bains & A.V. Ferguson. 1998. Leptin depolarizes rat hypothalamic paraventricular nucleus neurons. Am. J. Physiol. **274:** R1468–R1472.
45. Haynes, W.G. *et al.* 1997. Receptor-mediated regional sympathetic nerve activation by leptin. J. Clin. Invest. **100:** 270–278.
46. Glaum, S.R. *et al.* 1996. Leptin, the obese gene product, rapidly modulates synaptic transmission in the hypothalamus. Mol. Pharmacol. **50:** 230–235.
47. Spanswick, D. *et al.* 1997. Leptin inhibits hypothalamic neurons by activation of ATP-sensitive potassium channels. Nature **390:** 521–525.
48. Shiraishi, T. *et al.* 1999. Leptin effects on feeding-related hypothalamic and peripheral neuronal activities in normal and obese rats. Nutrition **15:** 576–579.
49. Spanswick, D. *et al.* 2000. Insulin activates ATP-sensitive K^+ channels in hypothalamic neurons of lean, but not obese rats. Nat. Neurosci. **3:** 757–758.
50. Levin, B.E., A.A. Dunn-Meynell & V. H. Routh. 1999. Brain glucose sensing and body energy homeostasis: Role in obesity and diabetes. Am. J. Physiol. **276:** R1223–R1231.
51. Levin, B.E., A.A. Dunn-Meynell & V.H. Routh. 2001. Brain glucosensing and the K(ATP) channel. Nat. Neurosci. **4:** 459–460.
52. Harvey, J. & M.L. Ashford. 1998. Insulin occludes leptin activation of ATP-sensitive K^+ channels in rat CRI-G1 insulin secreting cells. J. Physiol. **511:** 695–706.
53. Harvey, J. & M.L. Ashford. 1998. Role of tyrosine phosphorylation in leptin activation of ATP-sensitive K^+ channels in the rat insulinoma cell line CRI-G1. J. Physiol. **510:** 47–61.
54. Harvey, J. & M.L. Ashford. 1998. Insulin occludes leptin activation of ATP-sensitive K^+ channels in rat CRI-G1 insulin secreting cells. J. Physiol. (Lond.) **511:** 695–706.
55. Harvey, J. *et al.* 2000. Essential role of phosphoinositide 3-kinase in leptin-induced K(ATP) channel activation in the rat CRI-G1 insulinoma cell line. J. Biol. Chem. **275:** 4660–4669.
56. Harvey, J. *et al.* 1997. Leptin activates ATP-sensitive potassium channels in the rat insulin-secreting cell line, CRI-G1. J. Physiol. (Lond.) **504:** 527–535.
57. Cowley, M.A. *et al.* 2001. Leptin activates anorexigenic POMC neurons through a neural network in the arcuate nucleus. Nature **411:** 480–484.
58. Heisler, L.K. *et al.* 2002. Activation of central melanocortin pathways by fenfluramine. Science **297:** 609–611.
59. Cowley, M.A. *et al.* 1999. Integration of NPY, AGRP, and melanocortin signals in the hypothalamic paraventricular nucleus: Evidence of a cellular basis for the adipostat. Neuron **24:** 155–163.
60. Batterham, R.L. *et al.* 2002. Gut hormone PYY(3-36) physiologically inhibits food intake. Nature **418:** 650–654.
61. Marks, D.L., N. Ling & R.D. Cone. 2001. Role of the central melanocortin system in cachexia. Cancer Res. **61:** 1432–1438.
62. Dornonville de la Cour, C. *et al.* 2001. A-like cells in the rat stomach contain ghrelin and do not operate under gastrin control. Regul. Pept. **99:** 141–150.
63. Kojima, M. *et al.* 1999. Ghrelin is a growth-hormone-releasing acylated peptide from stomach. Nature **402:** 656–660.
64. Cummings, D.E. *et al.* 2001. A preprandial rise in plasma ghrelin levels suggests a role in meal initiation in humans. Diabetes **50:** 1714–1719.
65. Cummings, D.E. *et al.* 2002. Plasma ghrelin levels after diet-induced weight loss or gastric bypass surgery. N. Engl. J. Med. **346:** 1623–1630.
66. Cowley, M.A. *et al.* 2003. The distribution and mechanism of action of ghrelin in the CNS demonstrates a novel hypothalamic circuit regulating energy homeostasis. Neuron **39:** 647–661.
67. Arvat, E. *et al.* 2001. Endocrine activities of ghrelin, a natural growth hormone secretagogue (GHS), in humans: Comparison and interactions with hexarelin, a nonnatural peptidyl GHS, and GH-releasing hormone. J. Clin. Endocrinol. Metab. **86:** 1169–1174.

Agouti-Related Protein: Appetite or Reward?

PAWEL K. OLSZEWSKI,[a,b,c] KATHIE WICKWIRE,[d] MICHELLE M. WIRTH,[e]
ALLEN S. LEVINE,[a,b,f] AND SILVIA Q. GIRAUDO[d]

[a]Research Service, Veterans Affairs Medical Center, Minneapolis, Minnesota 55417, USA

[b]Department of Medicine, University of Minnesota, Minneapolis, Minnesota 55455, USA

[c]University of Minnesota College of Veterinary Medicine, St. Paul, Minnesota 55108, USA

[d]Department of Foods and Nutrition, University of Georgia, Athens, Georgia 30602, USA

[e]Department of Psychology, University of Michigan, Ann Harbor, Michigan 48109, USA

[f]Department of Food Science and Nutrition, University of Minnesota,
St. Paul, Minnesota 55108, USA

ABSTRACT: Agouti-related protein (AgRP) is an orexigenic peptide that acts as
an antagonist of the melanocortin-3 and -4 receptors in the hypothalamus.
Studies suggest that the melanocortin and opioid systems interact in the control
of ingestive behavior. Also, AgRP has been shown to especially increase intake
of a palatable diet. Given these observations, we wished to examine whether the
effects of AgRP on ingestive behavior resemble those of opioids. AgRP was injected into the hypothalamic paraventricular nucleus in animals given a choice
between a palatable sucrose solution and calorically dense chow. As a result of
AgRP injection, animals increased intake of chow but not sucrose relative to
controls, in contrast to what has been seen with opioid agonists. These results
together with prior findings suggest that the primary effect of AgRP is to cause
an increase in food intake to satisfy energy needs, though AgRP also has opioid-
like effects, possibly due to melanocortin-opioid interactions.

KEYWORDS: agouti-related protein (AgRP); melanocortins; opioids; paraventricular nucleus (PVN); food intake

INTRODUCTION

Agouti-related protein (AgRP) is an endogenous, high-affinity antagonist of the
melanocortin-3 and -4 receptors (MC3-R and MC4-R). Together, AgRP and α-
melanocyte-stimulating hormone (α-MSH), an agonist of these receptors, have been
identified as important in the regulation of food intake and energy homeostasis.
Administration of AgRP to the cerebral ventricle (ICV) or in specific sites expressing MC3/4-R such as the paraventricular (PVN) and dorsomedial hypothalamic nuclei (DMH) result in increases in feeding.[1–5]
Comparisons have been made between the effects on feeding of AgRP and opioid
receptor agonists. Unlike some orexigenic peptides such as neuropeptide Y (NPY),

Address for correspondence: Silvia Q. Giraudo, Department of Foods and Nutrition, University
of Georgia, Athens, GA 30602. Voice: 706-542-697; fax: 706-542-5059.
sgiraudo@uga.edu

Ann. N.Y. Acad. Sci. 994: 187–191 (2003). © 2003 New York Academy of Sciences.

but similar to opioids, AgRP appears to be involved in the maintenance rather than the initiation of feeding.[4] Also, AgRP was found to have a more powerful influence on the intake of preferred versus non-preferred diets.[5] This result parallels the feeding effect observed following injection of opioids, known for their role in rewarding aspects of food ingestion.[6,7]

In addition, a growing body of evidence suggests that the melanocortin and opioid systems interact with one another. Chronic opiate administration has been found to downregulate gene expression of MC4-R,[8] and melanocortins antagonize feeding induced by opioid agonists.[9] Furthermore, we recently observed that peripherally administered naltrexone causes activation of α-MSH–containing neurons in the arcuated nucleus (Arc); naltrexone was also able to suppress the increase in feeding caused by centrally administered AgRP.[10] These findings suggest that opioid peptides exert their effect on feeding partly by suppressing α-MSH–dependent satiety mechanisms and that, in turn, the orexigenic action of AgRP may be mediated by opioid-dependent circuitry.

In addition to increasing intake of preferred foods, opioids display other characteristic effects on ingestive behavior. It has been shown in our laboratory that opioid receptor agonists can block the acquisition of lithium chloride (LiCl)-induced conditioned taste aversion, an effect that appears to be mediated in part by inhibition of activation in oxytocin-containing neurons.[11] Recently, we have shown that AgRP delivered to the lateral cerebral ventricle in a dose known to increase feeding, partially blocks LiCl-induced conditioned taste aversion.[12] AgRP also decreased the percentage of c-Fos–positive oxytocin neurons induced by LiCl in the PVN and supraoptic nucleus (SON).[12] These inhibitory effects of AgRP on acquisition of conditioned taste aversion and aversion-associated activation of oxytocin parallel what has previously been shown with opioid receptor agonists.

Given the evidence for a relationship between melanocortins and opioids in the control of food intake, along with our prior finding that AgRP has an especially strong influence on the intake of a palatable diet, questions arise as to whether AgRP exerts similar effects to opioids on consummatory behavior. The present study investigates whether AgRP's orexigenic effects are driven more by reward-seeking than by satisfaction of energy needs, elucidated by offering animals a choice between calorically dense chow and a palatable sucrose solution. Using this paradigm, we showed previously that NPY and the μ-opioid receptor agonist DAMGO have different effects on food intake. Rats injected with NPY into the PVN ingested an equal amount of energy from the 10% sucrose solution and laboratory chow, whereas those injected with DAMGO into the PVN ingested more kilocalories from the 10% sucrose solution than laboratory chow.[13] These data support the notion that opioids are involved in the rewarding aspects of feeding, whereas NPY is involved in the energy needs of the animal.

MATERIAL AND METHODS

Animals

Male Sprague-Dawley rats (Charles River Laboratories, Wilmington, MA) were individually housed in conventional hanging cages in a temperature- and humidity-controlled room, LD 12:12 (lights on at 07:00). Animals had *ad libitum* access to water and chow (Certified Rodent Chow, Teklad, Indianapolis, IN) unless otherwise specified.

Surgery and Injections

Each animal was anesthetized with Nembutal (40 mg/kg body weight) and fitted with a 26-gauge (PVN) stainless steel guide cannula (Plastics One, Austin, TX). Stereotaxic coordinates were determined from the rat brain atlas by Paxinos and Watson (Academic Press, 1986) and were as follows: 0.5 mm lateral, 1.9 mm posterior to bregma, and 7.3 mm below the skull surface. Animals were allowed at least 7 days recovery post surgery before beginning experiments. Following recovery and after the completion of experiments, animals were sacrificed and brains were dissected out to verify cannula placement by histologic examination. Data from animals with incorrect cannula placements were discarded.

Injections were performed with Hamilton syringes (Hamilton, Reno, NV) over 30 seconds in a volume of 0.5 µL. Drugs were dissolved in 0.9% saline. AgRP (83–132)-NH_2 was purchased from Phoenix Pharmaceuticals (Belmont, CA).

Twelve PVN-cannulated animals were given access to a 10% sucrose solution in their home cages in the week prior to the experimental trials to avoid neophobia. A 10% sucrose solution was selected based on previous food/fluid intake studies from our laboratory.[13,14] On the day of the experimental trial, animals received via cannula 100 pmol AgRP or vehicle (injections at 11:00). AgRP administered to the PVN at this dose has been shown to significantly stimulate food intake.[4] Following injections, animals were allowed free access to chow, 10% sucrose solution, and water. Intake of all three were recorded at 2, 4, and 24 h post injection; calculation of chow intake included a correction for spillage. Following 3 days without access to sucrose, animals underwent the same paradigm with the treatment groups reversed. Intakes of chow and sucrose are expressed in terms of kilocalories (chow: 3.93 kcal/g; sucrose solution: 0.4 kcal/g).

Data Analysis

Each data point is expressed as a mean ± standard error of the mean. Data were analyzed using one-factor analysis of variance (ANOVA) with repeated measures, and means were compared using the least-significance difference test.

RESULTS

AgRP exerted its main effect on chow intake at the 0–4 h time period, during which control animals ate 1.93 ± 1.05 kcal of chow and AgRP-treated rats ate 7.97 ± 2.56 kcal of chow ($F_{(1,11)} = 5.09$, $P = 0.0454$) (FIG. 1). Sucrose consumption did not differ between groups ($F_{(1,11)} = 0.006$, $P = 0.9402$). Therefore, the difference in total kilocalories consumed between the two groups was due to the increased chow consumption in the AgRP-treated rats ($F_{(1,11)} = 5.141$, $P = 0.0445$). During the 4–24 h period (overnight), AgRP-injected animals also consumed significantly more kilocalories of chow than did controls (90.16 ± 5.07 versus 70.95 ± 4.90 kcal; $F_{(1,11)}$

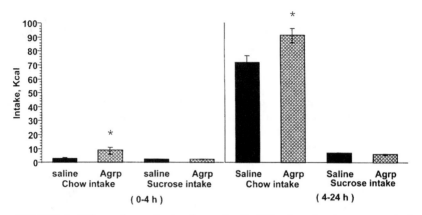

FIGURE 1. Effect of agouti-related protein (AgRP) given in the paraventricular nucleus (PVN) on cumulative chow and sucrose intake. AgRP was dissolved in 0.5 μL 0.9% saline; injections were performed at 11:00. Each bar represents a mean ± SEM. *Asterisks* indicate values that are significantly different from saline within a time course ($P < 0.05$).

= 21.841, $P = 0.0007$) (FIG. 1). Neither sucrose solution intake, water intake, nor total fluid intake differed between treatments at any time point.

DISCUSSION

In the current study, we once again noted that injection of AgRP into the PVN increases feeding in rats.[4,15] When AgRP-injected rats were given a choice between laboratory chow and a relatively energy-dilute (but palatable) 10% sucrose solution, they ingested many more kilocalories from the chow than the sucrose solution. This is the opposite of what we previously noted with DAMGO-injected rats; that is, they ingested most of their energy from the 10% sucrose solution rather than the chow. This suggests that AgRP is involved in eating related to energy needs rather than reward.

This result may seem to contradict the earlier finding that AgRP induces a greater increase in consumption of a palatable (sweet) diet than of a less-preferred diet.[5] However, in that case, both diets were equally calorically dense, whereas the differing caloric densities in the present study forced animals to choose between reward and energy needs. The effect on the intake of palatable foods appears to be important, but secondary to the effect of this peptide on satisfying energy needs. Hagan *et al.*[16] demonstrated a similar effect as seen here with third ventricle administration of AgRP; importantly, the investigators also tested the effect of AgRP on intake of a sucrose solution when chow was not available and found no effect. We have observed the same with PVN-administered AgRP (unpublished data).

Overall, it appears that the primary effect of AgRP on an animal is to signal a negative energy state and, thus, to stimulate food intake. Interestingly, NPY, a peptide that extensively colocalizes with AgRP, also powerfully stimulates caloric intake, and NPY-treated animals choose calorically dense chow over a sucrose solution.

However, it should be noted that both AgRP and NPY appear to affect reward in some fashion, perhaps through their interactions with opioids.

REFERENCES

1. KIM, M.S. *et al.* 2000. Hypothalamic localization of the feeding effect of Agouti-related peptide and α-melanocyte-stimulating hormone. Diabetes **49:** 177–182.
2. ROSSI, M. *et al.* 1998. A C-terminal fragment of Agouti-related protein increases feeding and antagonizes the effect of alpha-melanocyte stimulating hormone in vivo. Endocrinology **139:** 4428–4431.
3. HAGAN, M.M. *et al.* 2000. Long-term orexigenic effects of AgRP-(83–132) involve mechanisms other than melanocortin receptor blockade. Am. J. Physiol. Regul. Integr. Comp. Physiol. **279:** R47–R52.
4. WIRTH, M.M. & S.Q. GIRAUDO. 2000. Agouti-related protein in the hypothalamic paraventricular nucleus: effect on feeding. Peptides **21:** 1371–1377.
5. WIRTH, M.M. & S.Q. GIRAUDO. 2001. Effect of Agouti-related protein delivered to the dorsomedial nucleus of the hypothalamus on intake of a preferred versus a non-preferred diet. Brain Res. **897:** 169–174.
6. LEVINE, A.S. & C.J. BILLINGTON. 1989. Opioids: are they regulators of feeding? Ann. N.Y. Acad. Sci. **575:** 209–219.
7. WELDON, D.T. *et al.* 1996. Effect of naloxone on intake of cornstarch, sucrose, and Polycose diets in nonrestricted rats. Am. J. Physiol. **270:** R1183-R1188.
8. ALVARO, J.D., J.B. TATRO & R.S. DUMAN. 1997. Melanocortins and opiate addiction. Life Sci. **61:** 1–9.
9. CONTRERAS, P.C. & A.E. TAKEMORI. 1984. Antagonism of morphine-induced analgesia, tolerance and dependence by α-melanocyte-stimulating hormone. J. Pharmacol. Exp. Ther. **229:** 21–26.
10. OLSZEWSKI, P.K. *et al.* 2001. Evidence of interactions between melanocortin and opioid systems in regulation of feeding. Neuroreport **12:** 1727–1730.
11. OLSZEWSKI, P.K. *et al.* 2000. Opioids affect acquisition of LiCl-induced conditioned taste aversion: Involvement of OT and VP systems. Am. J. Physiol. Regul. Integr. Comp. Physiol. **279:** R1504–R1511.
12. WIRTH, M.M. *et al.* 2002. Effect of agouti-related protein on the development of conditioned taste aversion and oxytocin neuronal activation. Neuroreport **13:** 1355–1358.
13. GIRAUDO, S.Q. *et al.* 1999. Differential effects of neuropeptide Y and the mu-agonist DAMGO on 'palatability' vs. 'energy'. Brain Res. **834:** 160–163.
14. LYNCH, W.C. *et al.* 1993. Effects of neuropeptide Y on ingestion of flavored solutions in nondeprived rats. Physiol. Behav. **54:** 877–880.
15. GIRAUDO, S.Q., C.J. BILLINGTON & A.S. LEVINE. 1998. Feeding effects of hypothalamic injection of melanocortin 4 receptor ligands. Brain Res. **809:** 302–306.
16. HAGAN, M.M. *et al.* 2001. Opioid receptor involvement in the effects of AgRP-(83–132) on food intake and food selection. Am. J. Physiol. Regul. Integr. Comp. Physiol. **280:** R814–R821.

State-Dependent Modulation of Feeding Behavior by Proopiomelanocortin-Derived β-Endorphin

MALCOLM J. LOW,[a,b] MICHAEL D. HAYWARD,[a] SUZANNE M. APPLEYARD,[a] AND MARCELO RUBINSTEIN[c]

[a]Vollum Institute, Oregon Health and Science University, Portland, Oregon 97239-3098, USA

[b]Department of Behavioral Neuroscience, Oregon Health and Science University, Portland, Oregon 97239-3098, USA

[c]Instituto de Investigaciones en Ingeneria Genética y Biología Molecular, CONICET, Department of Physiology and Molecular and Cellular Biology, School of Sciences, University of Buenos Aires, Buenos Aires, Argentina 1428

ABSTRACT: Feeding behavior can be divided into appetitive and consummatory phases, differing in neural substrates and effects of deprivation. Opioids play an important role in the appetitive aspects of feeding, but they also have acute stimulatory effects on food consumption. Because the opioid peptide β-endorphin is co-synthesized and released with melanocortins from proopiomelanocortin (POMC) neuronal terminals, we examined the physiological role of β-endorphin in feeding and energy homeostasis using a strain of mutant mice with a selective deficiency of β-endorphin. Male β-endorphin–deficient mice unexpectedly became obese with ad libitum access to rodent chow. Total body weight increased by 15% with a 50–100% increase in the mass of white fat. The mice were hyperphagic with a normal metabolic rate. Despite the absence of endogenous β-endorphin, the mutant mice did not differ from wild-type mice in their acute feeding responses to β-endorphin or neuropeptide Y administered intracerebroventricularly or naloxone administered intraperitoneally. Additional mice were studied using an operant behavioral paradigm to examine their acquisition of food reinforcers under increasing work demands. Food-deprived, β-endorphin–deficient male mice emitted the same number of lever presses under a progressive ratio schedule compared to wild-type mice. However, the mutant mice worked significantly less than did the wild-type mice for food reinforcers under nondeprived conditions. Controls for nonspecific effects on acquisition of conditioned learning, activity, satiety, and resistance to extinction revealed no genotype differences, supporting our interpretation that β-endorphin selectively affects a motivational component of reward behavior under nondeprived conditions. Therefore, we propose that β-endorphin may function in at least two primary modes to modulate feeding. In the appetitive phase, β-endorphin release increases the incentive value of food as a primary reinforcer. In contrast, it appears that endogenous β-endorphin may inhibit

Address for correspondence: Malcolm J. Low, Vollum Institute, Oregon Health and Science University, Portland, Oregon 97239-3098. Voice: 503-494-4672; fax: 503-494-4976.
low@ohsu.edu

Ann. N.Y. Acad. Sci. 994: 192–201 (2003). © 2003 New York Academy of Sciences.

food consumption in parallel with melanocortins and that the orexigenic properties previously ascribed to it may actually be due to other classes of endogenous opioid peptides.

KEYWORDS: β-endorphin; opioid peptides; operant conditioning; deprivation state; motivation; reinforcer; hyperphagia; knockout mice; metabolic rate; sexual dimorphism

INTRODUCTION

The endogenous opioid system influences incentive-motivation in a number of different tests. Naloxone, a non-subtype–selective opiate receptor antagonist, reduces consumption of a variety of palatable foods and decreases operant responding for food reinforcers.[1,2] The endogenous opioid system also modulates self-administration of alcohol, benzodiazepines, psychostimulants, narcotics, and intracranial electrical self-stimulation.[3,4] In fact, a role for enkephalin in reward behavior was suggested soon after the first identification of an endogenous opioid.[5] Thus, a general role for the endogenous opioid system may be to enhance the incentive value of rewarding stimuli.

Many previous experiments have studied the role of the endogenous opioid system in reward-related behaviors by using subtype-selective opioid receptor antagonists. However, the endogenous opioid peptides interact relatively nonspecifically with the different opioid receptors, making it difficult to draw conclusions as to which endogenous opioids are involved in behaviors such as positive reinforcement.[6] β-Endorphin has nearly equal affinity for the mu and delta opioid receptor, and enkephalin preferentially binds to the delta receptor, although it also has physiologically relevant affinity for the mu receptor.[6] Agonists for all three opioid receptors can stimulate feeding to varying degrees, but agonists for the mu and delta receptor are thought to be intrinsically rewarding, whereas agonists for the kappa receptor have been shown to actually be aversive.[7] Thus, β-endorphin and enkephalin are the most likely opioid peptides to be involved in positively reinforced operant behavior. To test the function of one of these specific endogenous opioid peptides on energy homeostasis and feeding behavior in mice, we mutated the proopiomelanocortin (POMC) gene so that it does not express β-endorphin.[8]

UNCONDITIONED FEEDING BEHAVIOR IN β-ENDORPHIN-DEFICIENT MICE

To examine the effect of β-endorphin deficiency on baseline weight homeostasis, we analyzed sibling wild-type, heterozygous, and homozygous mice reared by heterozygous breeding pairs. Growth curves demonstrated that male β-$END^{-/-}$ mice weighed significantly more than male β-$END^{+/+}$ mice, starting at 4 weeks of age and continuing into adulthood (FIG. 1). By contrast, the weight of female β-$END^{-/-}$ mice differed only transiently from that of female β-$END^{+/+}$ mice between 4 and 8 weeks of age. β-$END^{+/-}$ mice did not differ from wild-type mice. Body length and the weight of various organs such as liver, spleen, kidney, heart, and gonads were not changed in either sex of β-$END^{-/-}$ mice. However, both the inguinal/gonadal and the

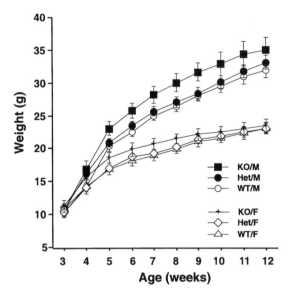

FIGURE 1. Growth curves of β-$END^{+/+}$ (WT), β-$END^{+/-}$ (Het), and β-$END^{-/-}$ (KO) male (M) and female (F) mutant mice reared by β-$END^{+/-}$ parents and fed a standard fat composition rodent chow (4% by weight) after weaning ($N = 11$–29 per group with the genotypes and sexes randomly distributed among a total of 19 separate litters with an average litter size of 6 pups).

retroperitoneal/perirenal white fat stores of male β-$END^{-/-}$ mice were twofold heavier than those of male β-$END^{+/+}$ mice, whereas intrascapular brown fat was not altered. Furthermore, male β-$END^{-/-}$ mice had 50% greater total body fat as measured by dual energy x-ray absorptiometry (DEXA)-scan, and histological examination suggested hypertrophy of adipocyte tissue.[9] The fat stores of female β-$END^{-/-}$ mice were not increased. An identical sexually dimorphic phenotype with equivalent or greater male-pattern obesity was observed in β-$END^{-/-}$ mice crossed onto either a 129S6/SV or an outbred Swiss albino background. These data indicate that development of obesity in β-endorphin–deficient mice is independent of the known genetic predisposition of C57BL/6 mice to gain excessive weight and adipose mass.

To determine the underlying metabolic reason for the increased weight and adiposity of the β-$END^{-/-}$ mice, we examined their food intake and basal metabolic rate. The average daily food intake of male β-$END^{-/-}$ mice was significantly increased compared to that of β-$END^{+/+}$ males. In contrast to the augmented food intake, no significant differences were noted between genotypes in their basal metabolic rate, as measured by oxygen consumption, respiratory quotient, basal core temperature, activity levels, or serum thyroxine levels.[9] Male β-$END^{-/-}$ mice gained more weight than did controls when fed a high-fat diet (9% vs. 5% fat content), but genotype differences paralleled changes on the regular chow (FIG. 2).

We examined whether the response of the β-$END^{-/-}$ mice to food intake stimulated by opioids or neuropeptide Y (NPY) was altered. β-Endorphin injected intra-

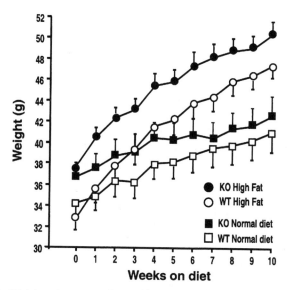

FIGURE 2. Weight gain of adult β-*END*$^{+/+}$ (WT) and β-*END*$^{-/-}$ (KO) male mice fed a normal-fat (4% by weight) and high-fat (9% by weight) chow (N = 6–8 per group).

cerebroventricularly stimulated an equivalent increase in food intake in both wild-type and β-*END*$^{-/-}$ mice.[9] Interestingly, the orexigenic effects of NPY were slightly increased in male β-*END*$^{-/-}$ mice, but unaltered in female β-*END*$^{-/-}$ mice, consistent with their normal weight and feeding behavior. However, NPY stimulation of food intake was inhibited equivalently by the nonselective opioid antagonist naloxone in both genotypes. Furthermore, naloxone also inhibited feeding to the same extent in previously food-restricted β-*END*$^{-/-}$ and wild-type mice. The retained actions of naloxone in β-endorphin–deficient mice do not appear to be explained by a compensatory increase in the expression of enkephalin or dynorphin in brain nuclei associated with feeding or reward behavior (FIG. 3).

OPERANT CONDITIONED FEEDING BEHAVIOR IN β-ENDORPHIN–DEFICIENT MICE

We tested for changes in the incentive value of rewarding stimuli by quantifying the reinforcing efficacy of food pellets using operant responding under a progressive ratio (PR) schedule, which requires additional bar presses of a defined number for each subsequent reinforcer.[10] PR schedules have been widely used to quantify the value an animal places on a commodity by measuring the effort it will expend to receive that reinforcer. In fact, this procedure has been successfully used to measure naloxone effects on motivation to obtain food reinforcers in rats and mice.[11–14] The incentive value of food varies with motivational states so that the neurobiological substrate underlying the instrumental behavior may be different in freely fed states from food-deprived states.[14,15] For example, the hedonic value of food may be the

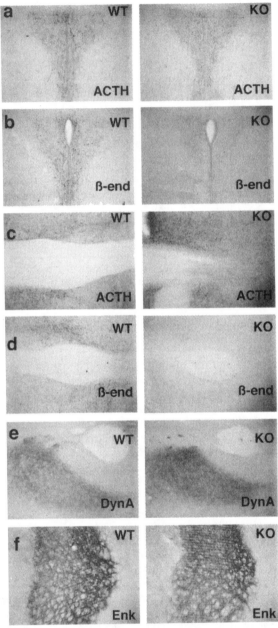

FIGURE 3. Expression of ACTH and opioid peptides in the brains of adult β-$END^{+/+}$ (WT) and β-$END^{-/-}$ (KO) male mice. ACTH, β-endorphin, dynorphin A, and Leu-enkephalin immunoreactive axons and terminals were detected with specific antisera and the ABC/diaminobenzidine technique on 50-μm coronal vibratome sections. (**a,b**) Paraventricular nucleus of the hypothalamus. (**c,d**) Bed nuclei of the stria terminalis dorsal and ventral to the anterior commissure. (**e**) Nucleus accumbens shell/ventral pallidum. (**f**) Globus pallidus. All digital photomicrographs were obtained at the identical magnification with a 10× objective and constant illumination.

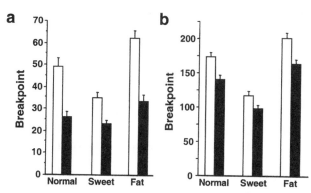

FIGURE 4. Operant responding for food reinforcers under a PR3 schedule and different states of food deprivation. (**a**) A significant difference was noted between genotypes in their breakpoints when the mice have *ad libitum* access to food in their home cages between operant sessions. (**b**) Food restriction to maintain mice at 75–80% of their normal body weight abolishes the breakpoint difference between the two genotypes, but it does not change the relative differences among the three types of food reinforcers. Note the large difference in scales on the *y*-axis between panels. β-$END^{+/+}$ (WT, *open bars*, N = 9) and β-$END^{-/-}$ (KO, *solid bars*, N = 10) male mice were used.

primary motivator in food-reinforced operant behavior under free-feeding conditions. Caloric imbalance would be the significant contributor to the incentive value of food reinforcers under food-deprived conditions.

β-$END^{+/+}$ and β-$END^{-/-}$ male mice were shaped initially to lever press for food reinforcers in an operant conditioning chamber under restricted feeding conditions. This training period consisted of a number of fixed ratio (FR) sessions first under an FR1, then under an FR5 before introducing the PR schedule. During the FR portion of the training period under restricted feeding conditions, the total number of reinforcers earned did not differ significantly between genotypes. Similarly, response rates on both active and inactive levers during the final FR5 session under *ad libitum* feeding conditions did not vary between genotypes. No main effect of genotype was detected by one-factor analysis of variance (ANOVA) on the active lever or inactive lever. Thus, all of the subjects performed at the same level before being introduced to the PR3 schedule.[16]

We compared the reinforcement efficacy between the genotypes by conducting PR3 sessions on *ad libitum* feeding subjects and measured the breakpoints (last ratio completed for a reinforcer) for each of three formulas of reinforcers: normal, sweet, and fat chow. Breakpoints were significantly higher for wild-type mice responding for all three formulas of reinforcers compared to the mutant genotype (FIG. 4a). Breakpoints of the β-$END^{-/-}$ male mice for the normal and fat chow reinforcers were only approximately half those of the wild-type mice. A main effect of chow formula and a chow formula by genotype interaction was also detected. However, separate repeated measures ANOVAs conducted on wild-type or β-endorphin–deficient mice detected main effects by chow formula for both genotypes. Thus, the rank order of preference was the same in all mice (fat chow > normal chow > sweet chow), but the level of behavior supported by these reinforcers differed between genotypes.

For mice in a nondeprived state, the endogenous opioids clearly modulated the efficacy of food reinforcers. However, in a food-deprived state induced by restricted access, the relative reinforcer efficacy was indistinguishable between the genotypes (FIG. 4b). Breakpoints did not differ significantly, and no main effect of genotype was detected under a PR3 schedule when mice were given restricted access to food. However, a main effect of chow formula was still detected. The rank order of preference was the same as in the *ad libitum* fed condition, but breakpoints were substantially higher under the food-restricted condition, and there was no genotype difference. This finding supports our argument that the difference shown in FIGURE 4a was not due to a motor deficit and suggests a state-dependent difference in motivation.

Differences in satiation between the genotypes under free-feeding conditions and a PR3 schedule could confound our interpretation of that data. Consequently, we examined operant behavior under an FR5, a relatively unchallenging and consistent schedule. The average duration of a PR3 session under *ad libitum* feeding conditions was approximately 1 h, so we used an FR5 schedule for 1 h to compare to the data gathered under a PR3 schedule during *ad libitum* feeding conditions. No main effect of genotype was detected for the number of pellets received under an FR5 schedule when mice were fed *ad libitum*, and there was no genotypic interaction by chow formula. However, a main effect of chow formula was detected, and the rank order of preference for the different reinforcers appeared to be the same as that found in the PR3 sessions, but there was no response difference between the genotypes under an FR5 schedule. In addition to response rates, we measured the number of pellets eaten during the 1-h FR5 sessions and compared these data to those of the PR3 sessions performed under *ad libitum* feeding conditions. During an FR5 schedule, mice from both genotypes ate significantly more reinforcers of all three formulas than they did during their PR3 testing. A significant main effect of schedule was detected by repeated measures ANOVA conducted on these data. This is unlikely to be an artifact of an increased workload under the PR3 schedule, because the average number of lever presses under the PR3 schedule was only roughly twice that under the FR5 (active lever presses by β-$END^{+/+}$ male mice for normal chow: PR3 = 497.6 ± 34.5 vs. FR5 = 253.6 ± 8.1). These data demonstrated that mice of both genotypes could eat significantly more pellets in 1 h than they ate during the PR3 sessions under *ad libitum* feeding conditions, suggesting that β-$END^{-/-}$ mice did not satiate earlier than did wild-type mice.[16]

Because a PR schedule uses an extinction criterion as an endpoint, we also examined the resistance to extinction by the two genotypes. Extinction sessions were conducted when the mice were being fed *ad libitum* using the PR3 schedule and a 15-minute limit to reach the next ratio, but with no reinforcers (PR3-EXT). The endpoint for extinction trials was determined by a *post-hoc* analysis and a criterion of 3 consecutive days that were not significantly different from each other (days 6–8). Extinction curves were generated using the data from days 1–6. These data confirmed that resistance to extinction was equivalent between genotypes and was not likely a contributing factor in the reduced breakpoints of the opioid mutant mice.[16]

While breakpoints under the PR3 schedule varied with the formula of reinforcer, we also noted an interaction between genotype and chow formula under the PR3 schedule in *ad libitum* fed mice. This interaction suggested that although the mutant genotypes varied their instrumental behavior for different formulas of reinforcers with the same rank order, they did not vary their response to the same degree as did wild-type mice. Therefore, we determined if preference was altered in these subjects by testing consummatory behavior independently of instrumental behavior. Two-bottle free-choice experiments were conducted in the mice home cages using four concentrations of sucrose versus water and four concentrations of saccharin versus water.[16] Sucrose and saccharin were chosen because they allow a comparison of sweet compounds with and without caloric value and because an abundant number of studies have shown that opioid antagonists will decrease preference for these two compounds. Both genotypes increased their preference for the sucrose-containing bottle in an identical concentration-dependent pattern. Similarly, both genotypes had identical concentration-dependent changes in preference ratios for the saccharin-containing bottle. Basal water consumption in the home cage did not differ between the genotypes.

Mice normally balance their caloric intake with energy expenditure. When a highly palatable substance such as sucrose-flavored water is introduced into their diet, they will generally decrease their food intake to compensate for the extra caloric intake from sucrose. As shown above, the male β-*END*[−/−] mice are slightly hyperphagic at 1–4 months of age, so we determined if the caloric balance of intake was altered in the older mice used for the two-bottle, free-choice experiment. Food consumption in the home cages was measured while the two-bottle, free-choice experiments were conducted. Both genotypes decreased their food consumption with increasing concentrations of sucrose, and we detected no main effect of genotype on the amount of food eaten across four concentrations of sucrose, but a main effect of sucrose concentration was detected. Using saccharin, a noncaloric tastant, we also did not detect a main effect by genotype on the amount of food eaten across four concentrations. However, we detected a main effect of saccharin concentration on the amount of food eaten but no genotype by saccharin concentration interaction. The change in feeding during saccharin presentation was likely due to a change in the volume of liquid consumed, because the amount of food eaten did not decrease with higher concentrations of saccharin. Overall, it appeared that regulation of energy homeostasis was largely intact in the opioid mutant mice at older ages.[16]

CONCLUSIONS

Opioids have generally been shown to increase caloric intake, particularly the intake of highly palatable foods.[17–19] Similarities among the functions of opioids in the modulation of behaviors related to food intake, sexual activity, and drug abuse suggest a common action in the brain's reward circuits.[3] However, the role of each distinct endogenous opioid gene in these circuits is unknown. β-Endorphin is particularly intriguing because it is synthesized from a common prohormonal precursor together with the anorexigenic melanocortin peptides.[20,21] Our studies were designed to test the possible role of β-endorphin using a genetic approach that leaves intact the expression of all other opioid and opioid receptor genes.

Our data suggest that the effects of POMC-derived β-endorphin on feeding behavior are complex and dependent on sex steroid hormones, age of the animal, and deprivational state.[9,16] The development of obesity in young adult male β-endorphin–deficient mice indicates that endogenous β-endorphin may normally suppress, rather than stimulate feeding. Although this result at first appears paradoxical,[22] pharmacologic studies in the mutant mice showed a normal stimulatory effect of exogenous β-endorphin and an inhibitory effect of a nonspecific opioid agonist on food intake. It is possible that an unknown compensatory change has occurred during brain development in the β-endorphin–deficient mice, but studies to date indicate normal expression of enkephalin and dynorphin and the mu, delta, and kappa opioid receptors.[23,24] Furthermore, enkephalin-deficient mice do not develop obesity,[16] whereas mu-receptor–deficient mice exhibit a similar sexually dimorphic increase in body mass of males compared to the β-endorphin–deficient strain (Dr. B. Keiffer, personal communication). We therefore favor the interpretation that peptides other than β-endorphin are responsible for the physiological endogenous opioid tone in wild-type rodents that can be blocked with opioid receptor antagonists to decrease food intake.

By contrast, our data from the operant studies of lever pressing for food reinforcers strongly support the hypothesis that β-endorphin is an essential neurochemical component of the brain's reward circuitry. Intriguingly, the effect is only apparent under nondeprived conditions for food availability. Chronically food-restricted mice clearly have a strong incentive to work to obtain food reinforcers, but under these extreme conditions the absence of β-endorphin did not alter the motivated behavior. We also found that the absence of enkephalin and, in fact, the double mutation causing the absence of both β-endorphin and enkephalin produced essentially the same results on the operant tests for self-administration of food reinforcers.[16] This latter result is consistent with the idea that the two different opioid peptides converge on a common node in the neural circuitry underlying reward.

What are the implications of our data concerning the presumptive co-release of β-endorphin and melanocortin peptides from common POMC nerve terminals? One interpretation from an ethological perspective bears on the fact that typically a wild animal is more likely to be in caloric deficit than surfeit, and under this condition of deprivation the activity of POMC neurons and the expression of POMC are suppressed. Under the fortunate circumstances of locating an abundant source of calorie-dense food, POMC neuronal activity would increase. Although the acute actions of melanocortins, possibly in combination with those of β-endorphin, are to increase metabolic rate and decrease feeding, the longer-term actions of β-endorphin in the lateral hypothalamus, ventral tegmental area, and nucleus accumbens might be to strengthen the neural associations among the sensory, environmental, and nutritional qualities of the food cache to support the survival of the animal. Further experimental work is necessary to determine the relevant sites of action of β-endorphin and its interaction with the other opioid peptides and melanocortin peptides in this reward circuit.

ACKNOWLEDGMENTS

We thank Juan Young for his efforts in the initial metabolic characterization of the $\beta\text{-}END^{-/-}$ mice. This work was supported by grants from the National Institutes of Health (DK55819, DA14203, DA05841, DK10082, and TW01233).

REFERENCES

1. MORLEY, J.E. 1987. Neuropeptide regulation of appetite and weight. Endocr. Rev. **8:** 256–287.
2. GLASS, M.J., C.J. BILLINGTON & A.S. LEVINE. 1999. Opioids and food intake: distributed functional neural pathways? Neuropeptides **33:** 360–368.
3. VAN REE, J.M., M.A. GERRITS & L.J. VANDERSCHUREN. 1999. Opioids, reward, and addiction: an encounter of biology, psychology, and medicine. Pharmacol. Rev. **51:** 341–396.
4. TRUJILLO, K.A., J.D. BELLUZZI & L. STEIN. 1989. Opiate antagonists and self-stimulation: Extinction-like response patterns suggest selective reward deficit. Brain Res. **492:** 15–28.
5. BELLUZZI, J.D. & L. STEIN. 1977. Enkephaline may mediate euphoria and drive-reduction reward. Nature **266:** 556–558.
6. REISINE, T. & G.W. PASTERNAK. 1996. Opioid analgesics and antagonists. *In* Goodman and Gilman's Pharmacologic Basis of Therapeutics. J.G. Hardman, A.G. Gilman & L.E. Limbird, Eds. : 521–555. McGraw-Hill. New York.
7. MUCHA, R.F. & A. HERZ. 1985. Motivational properties of kappa and mu opioid receptor agonists studied with place and taste preference conditioning. Psychopharmacology (Berlin) **86:** 274–280.
8. RUBINSTEIN, M. *et al.* 1996. Absence of opioid stress-induced analgesia in mice lacking beta-endorphin by site-directed mutagenesis. Proc. Natl. Acad. Sci. USA **93:** 3995–4000.
9. APPLEYARD, S.M., *et al.* 2003. A role for the endogenous opioid β-endorphin in energy homeostasis. Endocrinology **144:** 1753–1760.
10. HODOS, W. 1961. Progressive ratio as a measure of reward strength. Science **134:** 943–944.
11. HAYWARD, M.D. & M.J. LOW. 2001. The effect of naloxone on operant behavior for food reinforcers in DBA/2 mice. Brain Res. Bull. **56:** 537–543.
12. GLASS, M.J. *et al.* 1999. The effect of naloxone on food-motivated behavior in the obese Zucker rat. Psychopharmacology (Berlin) **141:** 378–384.
13. CLEARY, J. *et al.* 1996. Naloxone effects on sucrose-motivated behavior. Psychopharmacology (Berlin) **126:** 110–114.
14. RUDSKI, J.M., C.J. BILLINGTON & A.S. LEVINE. 1994. Naloxone's effects on operant responding depend upon level of deprivation. Pharmacol. Biochem. Behav. **49:** 377–383.
15. NADER, K., A. BECHARA & D. VAN DER KOOY. 1997. Neurobiological constraints on behavioral models of motivation. Annu. Rev. Psychol. **48:** 85–114.
16. HAYWARD, M.D., J.E. PINTAR & M.J. LOW. 2002. Selective reward deficit in mice lacking beta-endorphin and enkephalin. J. Neurosci. **22:** 8251–8258.
17. KALRA, S.P. *et al.* 1999. Interacting appetite-regulating pathways in the hypothalamic regulation of body weight. Endocr. Rev. **20:** 68–100.
18. GLASS, M.J. *et al.* 1996. Potency of naloxone's anorectic effect in rats is dependent on diet preference. Am. J. Physiol. **271:** R217–R221.
19. WELDON, D.T. *et al.* 1996. Effect of naloxone on intake of cornstarch, sucrose, and polycose diets in restricted and nonrestricted rats. Am. J. Physiol. **270:** R1183–R1188.
20. PRITCHARD, L.E., A.V. TURNBULL & A. WHITE. 2002. Pro-opiomelanocortin processing in the hypothalamus: impact on melanocortin signalling and obesity. J. Endocrinol. **172:** 411–421.
21. CASTRO, M.G. & E. MORRISON. 1997. Post-translational processing of proopiomelanocortin in the pituitary and in the brain. Crit. Rev. Neurobiol. **11:** 35–57.
22. SILVA, R.M. *et al.* 2001. Beta-endorphin-induced feeding: pharmacological characterization using selective opioid antagonists and antisense probes in rats. J. Pharmacol. Exp. Ther. **297:** 590–596.
23. SLUGG, R.M. *et al.* 2000. Effect of the mu-opioid agonist DAMGO on medial basal hypothalamic neurons in beta-endorphin knockout mice. Neuroendocrinology **72:** 208–217.
24. MOGIL, J.S. *et al.* 2000. Disparate spinal and supraspinal opioid antinociceptive responses in beta endorphin-deficient mutant mice. Neuroscience **101:** 709–717.

Lack of Proopiomelanocortin Peptides Results in Obesity and Defective Adrenal Function but Normal Melanocyte Pigmentation in the Murine C57BL/6 Genetic Background

JAMES L. SMART[a] AND MALCOLM J. LOW[a,b]

[a]Vollum Institute and [b]Department of Behavioral Neuroscience, Oregon Health and Science University, Portland, Oregon 97239-3098, USA

ABSTRACT: Mice deficient in proopiomelanocortin peptides ($Pomc^{-/-}$) generated on a 129 (A^w/A^w) genetic background were back-crossed onto the C57BL/6 (a/a) genetic background. These mice exhibited most of the phenotypic characteristics previously reported on the 129 genetic background (Yaswen et al. 1999. Nat. Med. 5: 1066–1070). Adult mice became obese, their adrenals were atrophied, and they had undetectable plasma corticosterone in basal and stressed states. The partial perinatal lethality previously reported was also present on the C57BL/6 background. In addition, we found that both male and female homozygote (–/–) adults were fertile, but when homozygous males were intercrossed with homozygous females, all the pups died in the perinatal period. Attempts to rescue the perinatal lethality of pups from homozygous breeder pairs by supplementing the mother's drinking water with glucocorticoids were unsuccessful. Furthermore, failure to stimulate adrenal development and corticosterone production/release with daily exogenous adrenocorticotropin-stimulating hormone (ACTH) injections indicates an adrenal dependence on POMC peptides for normal development and function. While the original $Pomc^{-/-}$ mice, bred on a mixed white-bellied agouti (A^w/A^w) 129 genetic background, had patchy alternations in their coat color, they clearly were not a uniform yellow like the lethal yellow (A^y/a) mice. Our $Pomc^{-/-}$ mice bred onto the C57BL/6 (a/a) genetic background had a black coat color indistinguishable from that of the wild-type C57BL/6 mice, further suggesting that the POMC peptide melanocyte-stimulating hormone (α-MSH) is not essential for the production of eumelanin (black/brown) pigmentation.

KEYWORDS: proopiomelanocortin (POMC); obesity; adrenal; hypothalamic-pituitary-adrenal axis (HPA); corticosterone; pigmentation

INTRODUCTION

Over the last ten years, the role of proopiomelanocortin (POMC) peptides as central regulators of energy homeostasis has become very apparent.[1] Early evidence centered on the detection of alpha-melanocyte-stimulating hormone (α-MSH) bind-

Address for correspondence: James L. Smart, Vollum Institute, Oregon Health and Science University, Portland, Oregon 97239-3098. Voice: 503-494-4675; fax: 503-494-4976.
smartj@ohsu.edu

Ann. N.Y. Acad. Sci. 994: 202–210 (2003). © 2003 New York Academy of Sciences.

ing to hypothalamic nuclei involved in regulating energy homeostasis.[2] Since then, a variety of mutant mouse models have shown central α-MSH to be an anorexigenic peptide, whereas the POMC opioid β-endorphin may play a more complex role in a reward pathway that reinforces feeding behavior.[3–7] In addition to the energy homeostatic role of central POMC peptides, a large body of evidence implicates central POMC involvement in stress responses, reproduction, and ontogeny of the hypothalamus, pituitary, and adrenal glands.[8–14] However, further experiments are needed to conclusively establish these hypothesized roles of central POMC peptides from other peptidergic systems. Additionally, *in vivo* experiments using POMC transgenes are currently underway to delineate the roles of central POMC peptides from those of peripheral POMC peptides released from the pituitary gland corticotrophs and melanotrophs. This manuscript identifies specific physiological alterations in the *Pomc*$^{-/-}$ mutant mice on the C57BL/6 genetic background.

ORIGINAL POMC NULL MUTATION

A mouse strain deficient in all POMC peptides was generated by the Hochgeschwender and Brennan laboratories using homologous recombination of a replacement-type *Pomc* allele containing a deletion of exon 3 and insertion of *neo*. This mutant strain was generated on a mixed white-bellied agouti 129 genetic background ($A^w/A^w\ pTyr^C/pTyr^{C-Ch}$) and develops hyperphagia and obesity similar to those seen in the *lethal yellow* (A^y/a) mice and melanocortin receptor-4 (MC4-R)–deficient mice.[15] However, in contrast to these and other previously described obesity models, *Pomc*$^{-/-}$ mice have no detectable levels of circulating corticosterone or epinephrine. Analysis of adult suprarenal fat pads revealed what appeared to be an adrenal remnant but no clearly identifiable adrenal glands. A possible result of the adrenal absence was the partial perinatal lethality seen in offspring of heterozygous breeder pairs. The loss of circulating α-MSH, the melanocortin receptor-1 (MC1-R) agonist, was predicted to eliminate the production of eumelanin (black/brown) pigmentation in melanocytes of the *Pomc*$^{-/-}$ mice; however, they had only modest increases in phaeomelanin (yellow) pigmentation, most noticeable on their bellies.

INCREASED ADIPOSITY AND LINEAR GROWTH

To better understand the contribution of POMC peptides to hair follicle pigmentation, we back-crossed the *Pomc*$^{-/-}$ mutant allele from the agouti (A^w/A^w) 129 genetic background on to the C57BL/6 (*a/a*) genetic background for two successive generations. Additionally, the C57BL/6 genetic background allows us to study proposed POMC physiological functions on a more commonly used inbred mouse line and one that is predisposed to diet-induced obesity, thus eliminating confounds due to genetic heterogeneity when comparing data to the melanocortin receptor knockout and other spontaneous mouse obesity models.

Analysis of growth curves of *Pomc*$^{-/-}$ mice (FIG. 1) revealed an increase in body mass very similar to that seen in the *lethal yellow* (A^y/a) and MC4-R–deficient mice.[3,16] Further analysis showed that the increased size of the *Pomc*$^{-/-}$ mice was due to both increased linear growth and increased mass of white fat depots, again

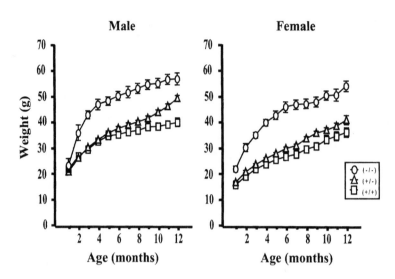

FIGURE 1. Male and female mice were group housed, fed a standard diet of rodent chow (5% fat, 19% protein, and 5% fiber by weight; 3.4 kcal/g), and weighed monthly. $Pomc^{-/-}$ mice became significantly heavier than wild-type littermates at 6–8 weeks old for males and 4–6 weeks old for females. Heterozygous males developed later onset obesity at about 8–10 months, whereas heterozygous females had a smaller increase in body mass at 8–10 months. (○, homozygous; △, heterozygous; □, wild-type; $N = 25$ to 105.)

similar to those of the *lethal yellow* and MC4-R–deficient mice. There were no obvious differences in the growth curves of $Pomc^{-/-}$ mice on the 129 versus C57BL/6 genetic background. However, all murine mutant models besides the $Pomc^{-/-}$ mice have circulating glucocorticoids. In addition, when the *lethal yellow* or the *leptin-deficient ob/ob* mice are adrenalectomized, these mice lose a significant amount of their white-fat mass, suggesting that glucocorticoids play a significant role in the development of obesity in these mutant mouse models.[16] $Pomc^{-/-}$ mice develop an obesity phenotype in the absence of circulating glucocorticoids, suggesting a role for POMC peptides regulating energy homeostasis independent of corticosterone.

ADRENAL ATROPHY AND HORMONE DEFICIENCIES

The adrenal gland, composed of an outer cortical shell and an inner medullary core, secretes the steroid hormones corticosterone, aldosterone, and androgens together with the catecholamines norepinephrine and epinephrine. Although the $Pomc^{-/-}$ mutant mice on the 129 genetic background fail to develop any distinguishable adrenal glands, the C57BL/6 mutant mice have easily identifiable, but severely hypoplastic adrenals. $Pomc^{-/-}$ mice adrenal glands were approximately 15% (in females) to 30% (in males) the weight of their respective wild-type littermates (FIG. 2). The greater reduction in the size of female adrenal glands compared with male adrenal glands eliminates the sexual dimorphism normally found in wild-type

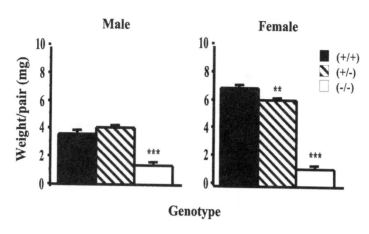

FIGURE 2. Adrenals were removed, trimmed of fat, and weighed as a pair from each mouse. Male *Pomc*$^{-/-}$ mice adrenals were 70% smaller than those from their wild-type and heterozygous littermates. Female *Pomc*$^{-/-}$ mice adrenals were 85% smaller than those from their wild-type and heterozygous littermates. In addition, female *Pomc*$^{+/-}$ heterozygous mice adrenals were slightly, but significantly, smaller than those from wild-type littermates. (***P <0.0001 and **P <0.01 compared to wild-type; N = 5 to 15.)

mice. Despite the presence of these hypoplastic adrenals on the C57BL/6 genetic background, plasma collected from unstressed, restraint-stressed, or lipopolysaccharide-challenged *Pomc*$^{-/-}$ mice still had no detectable levels of corticosterone (data not shown). Furthermore, treatment of *Pomc*$^{-/-}$ mice with 1 µg adrenocorticotropin-stimulating hormone (ACTH) (1–24) twice daily for 2 weeks failed to elicit any corticosterone secretory response, whereas wild-type mice responded promptly to the same dose of exogenous ACTH 3 h after dexamethasone suppression of their hypothalamic-pituitary-adrenal (HPA) axis (FIG. 3). The adrenal deficiencies in these mice strengthen the hypothesis that normal POMC expression is needed for complete adrenal maturation; however, the role of NH$_2$-terminal peptides in addition to ACTH remains an open question.[10]

We hypothesize that the lack of adrenal hormones, and not POMC peptides *per se*, is responsible for the non-Mendelian production of viable *Pomc*$^{-/-}$ offspring (FIG. 4). Both heterozygous and homozygous male and female mice were shown to be fertile. Further evaluation of heterozygous breeder pairs revealed that transmission of the *Pomc*$^{-/-}$ allele was normal but that 50–70% of the homozygous *Pomc*$^{-/-}$ mice die in the perinatal period, typically within the first hour after birth. Homozygous breeder pairs produced full-term litters with no pup surviving the perinatal period. Mice lacking corticotrophin-releasing hormone (CRH) are also unable to make corticosterone and exhibit a similar perinatal lethality.[17] The CRH (−/−) lethality can be rescued by administration of glucocorticoids to the mother's drinking water. However, an identical attempt to rescue the lethality in the homozygous *Pomc*$^{-/-}$ breeder pairs proved unsuccessful, suggesting the *Pomc*$^{-/-}$ perinatal lethality may be a result of the combined lack of glucocorticoids and epinephrine in the

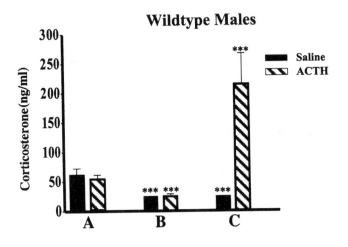

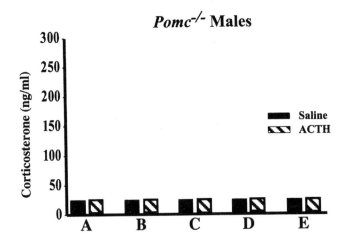

FIGURE 3. EDTA-plasma corticosterone levels were measured using a ^{125}I-labeled corticosterone radioimmunoassay kit from ICN. Dexamethasone (10 µg/mouse) was given ip immediately following blood draw at baseline time point A. At all other time points, mice received either saline or 1 µg ACTH (1–24) ip 30 min prior to blood collection from a tail vein. All null male mouse EDTA-plasma samples at all time points were below the assay sensitivity (25 ng/mL). (**A**) Dexamethasone injection (time 0). (**B**) 2.5 h after dexamethasone (saline or ACTH injection). (**C**) 3.0 h after dexamethasone and 30 min after saline or ACTH. (**D**) 7 days after twice daily ACTH injections (1 µg/injection). (**E**) 14 days after twice daily ACTH injections (1 µg/injection). (***P <0.001 compared to baseline values at point A; $N = 2$ to 6.)

Breeder Pair: (+/-) X (+/-)
Offspring: (+/+)   (+/-)   (-/-)
Number: 83   158   31
Percentage: 30.5%   58.1%   11.4%

Breeder Pair: (-/-) X (+/-)
Offspring: (+/+)   (+/-)   (-/-)
Number: 0   93   27
Percentage: 0%   77.5%   22.5%

FIGURE 4. Offspring numbers and percentages of total offspring for each genotype are shown for two types of breeder pairs: heterozygous $Pomc^{+/-}$ male X heterozygous $Pomc^{+/-}$ female, and homozygous $Pomc^{-/-}$ male X heterozygous $Pomc^{+/-}$ female.

$Pomc^{-/-}$ pups. $Pomc^{-/-}$ mice expressing a transgene that rescues pituitary POMC peptides and therefore adrenal corticosterone production, but that are still void of central POMC peptides are currently being evaluated.

PIGMENTATION IS UNALTERED IN THE C57BL/6 GENETIC BACKGROUND

Information gained from the *recessive yellow* ($Mc1r^e/Mc1r^e$) and the autosomal-dominant *lethal yellow* (A^y/a) mice would predict an inability to make eumelanin pigmentation in the $Pomc^{-/-}$ mice.[18] This prediction for the $Pomc^{-/-}$ mice is based on the absence of stimulation of the MC1-R by its natural POMC peptide agonist, α-MSH. In the *lethal yellow* mice, the constitutively activated expression of agout-ipeptide antagonizes the MC1-R, preventing stimulation by α-MSH and inhibiting eumelanin synthesis. The original $Pomc^{-/-}$ mice on an agouti (A^w/A^w) 129 genetic background resulted in only subtle alteration of yellow hair pigmentation on the belly, much different from the coat color seen in the *lethal yellow* (A^y/a) mice. The presence of agouti in the 129 genetic background produces a banded hair shaft with both eumelanin and phaeomelanin pigmentation, complicating the analysis of coat color on this genetic background. To more closely evaluate the potential alterations in coat color in the absence of peripheral α-MSH, the $Pomc^{-/-}$ allele was evaluated on an agouti-less genetic background, C57BL/6 (a/a), to eliminate all phaeomelanin caused by antagonism of α-MSH binding. Our studies show no distinguishable difference in $Pomc^{-/-}$ black coat color from wild-type (+/+) or heterozygous (+/–)

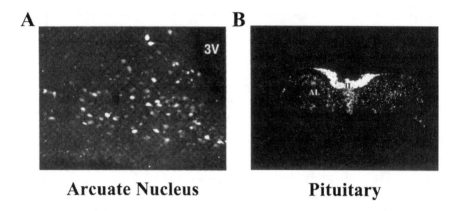

A   **B**

Arcuate Nucleus   **Pituitary**

FIGURE 5. (A) Coronal vibratome sections (50-μm thick) from a compound $Pomc^{-/-}$, $Pomc$-EGFP$^{Tg/+}$ mouse were viewed under UV light at 10× magnification on a Zeiss microscope with an FITC filter cube. Neurons expressing EGFP are brightly fluorescent and distributed in the arcuate nucleus of the basal hypothalamus in the same location as POMC neurons. **(B)** A pituitary section (12 μm) exhibiting robust fluorescence in the intermediate lobe (IL) and the anterior lobe (AL) under UV light at 3× magnification. Fluorescent cells seen in both intermediate and anterior lobes closely resemble those seen in melanotroph and corticotroph distributions.

littermates on the C57BL/6 background, including the presence of a few yellow hairs behind the ears, in the perianal area and around the female nipples seen on wild-type C57BL/6 mice. Possible explanations for the black coat color in the absence of α-MSH include an endogenous constitutively active MC1-R, compensatory pathways that stimulate cAMP production and tyrosinase activity, thereby bypassing the MC1-R altogether, and alternative endogenous agonist ligands for the receptor.

POMC NEURONS, CORTICOTROPHS, AND MELANOTROPHS ARE STILL PRESENT IN $Pomc^{-/-}$ MICE

The inability to immunohistochemically detect POMC cells in $Pomc^{-/-}$ mice makes it difficult to know if the cells, void of the POMC peptides, are still present. To identify and study POMC neurons *in vivo*, we previously generated a transgenic C57BL/6 mouse line expressing enhanced green fluorescent protein (EGFP, Clontech) under the transcriptional control of POMC genomic regulatory elements.[19] The transgene was designed to only express EGFP under the *Pomc* promoter, but not to express any of the POMC peptides. We back-crossed this POMC-EGFP transgene onto the $Pomc^{-/-}$ mice to determine if the POMC neurons and POMC pituitary cells developed in their normal numbers and locations. Our analysis revealed that POMC neurons along with pituitary corticotrophs and melanotrophs remain intact in the $Pomc^{-/-}$ mice (FIG. 5). POMC neurons have been found to coexpress another peptide called cocaine-amphetamine–regulated transcript (CART); however, no other peptide has been shown to be coexpressed in corticotrophs and melanotrophs with

POMC peptides.[20] Whether CART neuron physiology has been altered and/or whether pituitary corticotrophs and melanotrophs void of POMC peptides have any physiological function in the $Pomc^{-/-}$ mouse has yet to be determined.

CONCLUSION

The heterogeneity found among the many inbred mouse lines can result in an epistatic modification of phenotypes when the same genetic mutations are analyzed.[21,22] Therefore, the back-crossing and analysis of the $Pomc^{-/-}$ allele are essential for understanding the function of POMC peptides on the C57BL/6 genetic background, an inbred strain that has been used for functional analysis of melanocortin-receptor–deficient mice and other spontaneous obesity models. In addition, C57BL/6 mice are sensitive to diet-induced obesity and have been the subject of extensive behavioral and energy homeostatic studies. Although much of the data originally obtained from the mutant POMC allele on the 129 genetic background has been recapitulated in the C57BL/6 genetic background, we have found certain discrepancies in the phenotypes. Most notably, we find that the coat color of $Pomc^{-/-}$ on the C57BL/6 background is indistinguishable from that of wild-type littermates. In addition, we find that the adrenals are present in adult $Pomc^{-/-}$ mice on the C57BL/6 background, although they are very much reduced in size and still incapable of secreting corticosterone or responding to exogenous ACTH stimulation.

ACKNOWLEDGMENTS

We thank Dr. Ute Hochgeschwender for her generous gift of a heterozygous $Pomc^{+/-}$ mouse from which we initiated our breeding colony. This work was supported by the National Institutes of Health (Grants F31 HG00201 and P01DK55819).

REFERENCES

1. ELMQUIST, J.K., C.F. ELIAS & C.B. SAPER. 1999. From lesions to leptin: hypothalamic control of food intake and body weight. Neuron **22:** 221–232.
2. TATRO, J.B. 1990. Melanotropin receptors in the brain are differentially distributed and recognize both corticotropin and alpha-melanocyte stimulating hormone. Brain Res. **536:** 124–132.
3. HUSZAR, D. *et al.* 1997. Targeted disruption of the melanocortin-4 receptor results in obesity in mice. Cell **88:** 131–141.
4. BUTLER, A.A. *et al.* 2000. A unique metabolic syndrome causes obesity in the melanocortin-3 receptor-deficient mouse. Endocrinology **141:** 3518–3521.
5. CHEN, A.S. *et al.* 2000. Inactivation of the mouse melanocortin-3 receptor results in increased fat mass and reduced lean body mass. Nat. Genet. **26:** 97–102.
6. APPLEYARD, S.M. *et al.* 2003. A role for the endogenous opioid β-endorphin in energy homeostasis. Endocrinology **144:** 1753–1760.
7. HAYWARD, M.D., J.E. PINTAR & M.J. LOW. 2002. Selective reward deficit in mice lacking beta-endorphin and enkephalin. J. Neurosci. **22:** 8251–8258.
8. ALLEN, R.G. *et al.* 1984. Biosynthesis and processing of pro-opiomelanocortin-derived peptides during fetal pituitary development. Dev. Biol. **102:** 43–50.

9. BERRY, S. & L.W. HAYNES. 1989. The opiomelanocortin peptide family: neuronal expression and modulation of neural cellular development and regeneration in the central nervous system. Comp. Biochem. Physiol. A **93:** 267–272.
10. BICKNELL, A.B. *et al.* 2001. Characterization of a serine protease that cleaves pro-gamma-melanotropin at the adrenal to stimulate growth. Cell **105:** 903–912.
11. ESTIVARIZ, F.E. *et al.* 1988. Adrenal regeneration in the rat is mediated by mitogenic N-terminal pro-opiomelanocortin peptides generated by changes in precursor processing in the anterior pituitary. J. Endocrinol. **116:** 207–216.
12. ESTIVARIZ, F.E. *et al.* 1988. Further evidence that N-terminal pro-opiomelanocortin peptides are involved in adrenal mitogenesis. J. Endocrinol. **116:** 201–206.
13. MARSH, D.J. *et al.* 1999. Response of melanocortin-4 receptor-deficient mice to anorectic and orexigenic peptides. Nat. Genet. **21:** 119–122.
14. ZAGON, I.S. & P.J. MCLAUGHLIN. 1987. Endogenous opioid systems regulate cell proliferation in the developing rat brain. Brain Res. **412:** 68–72.
15. YASWEN, L. *et al.* 1999. Obesity in the mouse model of pro-opiomelanocortin deficiency responds to peripheral melanocortin. Nat. Med. **5:** 1066–1070.
16. BOSTON, B.A. *et al.* 1997. Independent and additive effects of central POMC and leptin pathways on murine obesity. Science **278:** 1641–1644.
17. MUGLIA, L. *et al.* 1995. Corticotropin-releasing hormone deficiency reveals major fetal but not adult glucocorticoid need. Nature **373:** 427–432.
18. OLLMANN, M.M. *et al.* 1998. Interaction of Agouti protein with the melanocortin 1 receptor in vitro and in vivo. Genes Dev. **12:** 316–330.
19. COWLEY, M.A. *et al.* 2001. Leptin activates anorexigenic POMC neurons through a neural network in the arcuate nucleus. Nature **411:** 480–484.
20. ELIAS, C.F. *et al.* 1998. Leptin activates hypothalamic CART neurons projecting to the spinal cord. Neuron **21:** 1375–1385.
21. BUTLER, A.A. & R.D. CONE. 2002. The melanocortin receptors: lessons from knockout models. Neuropeptides **36:** 77–84.
22. KOZAK, L.P. & M. ROSSMEISL. 2002. Adiposity and the development of diabetes in mouse genetic models. Ann. N.Y. Acad. Sci. **967:** 80–87.

Putative Targets of CNS Melanocortin Receptor Activity

CLIFFORD R. LAMAR, WENDI GARDNER, AMY BRAZDA, AND
ROBERT A. KESTERSON

*Department of Molecular Physiology and Biophysics, Vanderbilt University,
Nashville, Tennessee 37232-0615, USA*

ABSTRACT: Chronic antagonism of hypothalamic melanocortin receptors,
primarily melanocortin-4 receptor (MC4R), is the molecular basis for "agouti
obesity syndrome," whereas suppression of MC4R gene activity due to genetic
mutations induces obesity in both rodents and humans. However, little is
known about the neurocircuitry of MC4R-mediated control of energy balance,
the regulation of MC4R gene expression, or how suppression of MC4R activity
leads to differential expression of potential downstream central nervous system
(CNS) targets or effectors of melanocortin signaling. This paper focuses on
strategies for mapping CNS melanocortin circuits using transgenic mouse
models for conditional expression of MC4R and MC3R as well as progress in
characterizing the murine MC4R promoter. Additionally, preliminary studies
that focus on putative targets of melanocortinergic signaling will include a
discussion of CD81, a gene identified using the polymerase chain reaction-
based method of suppression subtractive hybridization. CD81, first described
as TAPA-1 (target of antiproliferative antibody), is a member of the tetraspanin family of cell surface proteins believed to function in cell-cell adhesion,
signal transduction, and possibly neuronal plasticity. Elevated expression of
CD81 mRNA in hypothalamic regions of obese yellow mice suggests that loss of
MC4R activity may lead to altered neuronal function via modulation of the cell
surface protein CD81.

KEYWORDS: melanocortin; proopiomelanocortin (POMC); agouti-related peptide (AgRP); melanocortin-4 receptor (MC4R); MC3R; agouti; obesity; transgene; hypothalamus; promoter; CD81; tetraspanin

INTRODUCTION

Recently, central nervous system (CNS) melanocortin pathways have emerged as
key coordinators of metabolic status.[1–4] The activities of "neural" melanocortin
receptors (MC3R and MC4R)[5] are tightly regulated by ligands secreted by distinct
populations of anorexigenic and orexigenic arcuate nucleus (ARC) neurons that
project throughout the brain.[6,7] The anorexigenic ARC neurons produce the agonists
α-melanocyte-stimulating hormone (α-MSH) and γ-melanocyte-stimulating hor-

Address for correspondence: Robert A. Kesterson, Department of Molecular Physiology and
Biophysics, Vanderbilt University, Nashville, TN 37232-0615. Voice: 615-936-3723; fax: 615-
343-0490.
   bob.kesterson@mcmail.vanderbilt.edu

Ann. N.Y. Acad. Sci. 994: 211–217 (2003). © 2003 New York Academy of Sciences.

mone (γ-MSH), which are derived from the proopiomelanocortin (POMC) gene; the orexigenic ARC neurons produce an antagonist, that is, agouti-related protein (AgRP). The activities of both POMC- and AgRP-specific neurons are regulated by the adipocyte-specific hormone leptin,[7–9] thereby providing a mechanism for CNS melanocortin sensing of peripheral metabolic stores. While a great deal of progress has been made in understanding how leptin relays orexigenic signals to AgRPergic neurons while simultaneously stimulating anorexigenic POMCergic neurons to suppress feeding, little is known about how, or even where, melanocortin receptor activity functions to propagate these feeding signals.

MAPPING OF MC4R-DEPENDENT METABOLIC PATHWAYS

Pharmacological approaches for determining the site of action for MC4R-dependent control of feeding have left many questions unanswered. For instance, acute injections of melanocortin receptor agonists and antagonists (MTII and SHU9119, respectively) into the paraventricular nucleus of the hypothalamus (PVN) will modulate feeding behavior[10,11]; however, injections of the same compounds into the fourth ventricle[12] or the medullary dorsal-vagal complex[13] produce similar effects, thus implicating involvement of brainstem melanocortin receptors. The outcome of a detailed "sensitivity" feeding experiment by Kim *et al.*[10] indicates that the PVN, dorsal medial hypothalamus (DMH), and medial preoptic area are the most sensitive to orexigenic actions of AgRP 8 h after administration, whereas the central nucleus of the amygdala, anterior hypothalamic area, and ARC are relatively sensitive and the ventral medial hypothalamus is totally insensitive to AgRP. Although it is tempting to conclude from these data that the PVN is an important melanocortin site of action (which may be the case), these experiments also point out major limitations to pharmacologic approaches in that little is learned about the location of the cell bodies or the number of neurons that facilitate melanocortin action.

Mapping of POMCergic and AgRPergic neuronal projections has led to several proposed models of downstream melanocortin signaling pathways.[14] In one model, melanocortins regulate basal metabolic rate via modulation of the thyroid hormone axis at the level of the hypothalamus. Dense innervation of PVN thyroid-releasing hormone (TRH) neurons by POMC and AgRP projections,[15,16] coexpression of MC4R on PVN-TRH neurons,[17] and α-MSH stimulation of the pro-TRH gene promoter via a cAMP response element provide strong support for this model.[17] In another model, neurons located in the lateral hypothalamic area that express orexigenic neuropeptides may be regulated by melanocortin receptor activity since MC4R mRNA is expressed in this brain region.[18] Distinct lateral hypothalamic area neuronal populations express either orexins[19] (also known as hypocretins[20]) or melanin-concentrating hormone.[21] Other potential effectors of melanocortin action include the anorexigenic hormones corticotropin-releasing hormone (CRH) and urocortin[22,23] as well as GABAergic interneurons that may integrate multiple orexigenic and anorexigenic signals.[24] Although these models point out many potential interactions between melanocortin pathways and putative feeding circuits, the necessary tools to address the mechanism(s) by which MC4R regulates energy balance are limited.

To specifically address many of these questions and identify neurons within the CNS that depend on normal MC4R activity to maintain metabolic homeostasis, our

TABLE 1. Transgenic mouse lines that express *Cre* recombinase in neural tissues

Transgenic name	Promoter	Expression pattern
Synapsin-*Cre*	Synapsin I	Neurons (brain and spinal cord)
HoxB6-*Cre*	Limb/LPM enhancer of HoxB6	Extraembryonic mesoderm, lateral plate, and limb mesoderm, midbrain/hindbrain junction
Six3-*Cre*	Six3	Retina and ventral forebrain
Nestin-*Cre*	Nestin	Neural precursor cells
TRH-*Cre*	Thyrotropin-releasing hormone	Paraventricular hypothalamic neurons
MCH-*Cre*	Melanin-concentrating hormone	Lateral hypothalamic neurons

laboratory has taken a genetic approach to create cell-specific deletions of the MC4R gene using the *cre/loxP* system. The *cre* recombinase protein will excise a DNA segment located between specific 34-bp recognition sequences known as loxP sites in transgenic mice. Therefore, the selective targeting of *cre* expression to distinct brain regions can be used to either (1) excise engineered transcriptional STOP sequences from inactive transgenes, or (2) remove essential gene regions from active genes, to induce or repress gene function, respectively. Using the latter approach, a mouse model for the conditional deletion of the single coding exon for the MC4R gene has been created. Ongoing experiments to determine the relative contribution of MC4R expression in various regions of the brain involve breeding MC4RcondKO mice to transgenic animals that express *cre* in specific brain regions (TABLE 1). For instance, MC4RcondKO mice will be crossed with HoxB6-*cre* transgenic mice to evaluate the role of MC4R in potential hindbrain regions of the CNS, whereas genetic crosses to TRH-*cre* animals will allow for a more specific hypothesis to be tested. In particular, by deleting MC4R only in hypothalamic TRH neurons, a determination of whether the thyroid hormone axis mediates any, or part, of melanocortinergic control of energy balance through PVN-TRH expression of MC4R can be made. Additionally, more refined mapping of MC4R-dependent sites of expression will be established using cell-specific recombination and deletion of the floxed MC4R locus using an adenoviral vector that expresses *cre* (that is, Adeno/*cre*).[25] Using a stereotaxic apparatus for delivery of Adeno/*cre*, we demonstrated the feasibility of transducing lateral hypothalamic neurons of ROSA26a mice, which harbor an enhanced recombination-sensitive β-galactosidase reporter gene. A similar approach is being taken to create a model for conditional deletion of the MC3R gene to map neural pathways that may specifically regulate differential partitioning of energy stores.

MC4R PROMOTER ANALYSIS

While the CNS expression pattern of MC4R mRNA has been delineated, the regulatory elements of the MC4R gene that control expression have yet to be fully characterized. Using 5′-RACE (rapid amplification of cDNA ends) of mouse brain RNA,

Dumont et al.[26] recently demonstrated that a major transcriptional start site lies 429-bp upstream of the start of translation. Analysis of a series of mouse MC4R 5'-flanking region fragments fused to a luciferase reporter gene established that fragments up to 3.3 kb would function as basal promoters when transfected into HEK293, UMR106, and GT1-7 cell lines, but not Neuro 2A neuronal cells.[26]

Using a similar approach, we further characterized the mouse MC4R promoter region; however, while our preliminary studies are in general agreement with those of Dumont et al.,[26] there are several subtle differences. Using hypothalamic GT1-1 and GT1-7 cells (originally derived from tumors from mice harboring a GnRH promoter/SV40 T antigen transgene[27]), we find that MC4R promoter constructs consisting of 1600 bp of 5'-flanking sequences confer maximal basal promoter activity, whereas a 3.3-kb construct is nearly as active. Shorter constructs with as little as 180 bp of putative 5'-untranslated region of the mRNA transcript will also function as a promoter in GT1-1 cells. However, a more extensive construct containing 7.9 kb of 5'-untranslated region displays markedly reduced promoter activity, indicating that one or more potential negative regulatory enhancer elements likely reside distally. A comparison between mouse and human distal promoter regions indicates that several "islands" of conserved sequences indeed reside 3.3 to 7.9 kb upstream of the putative transcriptional start site. In contrast to previous results, a similar profile of expression is also seen with each of the promoter fragments transfected into Neuro 2A cells, albeit at reduced overall levels of activity. This latter result, however, does support a previous report of MC4R expression in Neuro 2A cells.[28] Overall, these results are surprising in that substantial promoter activity was not expected in any of the neuronal cell lines, because detection of endogenous MC4R mRNA and protein, both in tissue culture models and also in vivo, has been relatively difficult to date.

IDENTIFICATION OF CD81 AS A PUTATIVE TARGET OF MELANOCORTINERGIC SIGNALING

As just discussed, neuroanatomical data have suggested various molecular targets and thereby models by which melanocortinergic signals may be propagated. To further uncover downsteam mediators of melanocortin action, we utilized a powerful cloning technique, known as suppression subtractive hybridization (SSH), that relies on enhancing differentially expressed cDNA fragments, while simultaneously suppressing those of equal abundance.[29] Using SSH to enrich for genes expressed in the CNS of obese A^y mice compared to control non-obese C57B16 littermates, we identified the CD81 gene as a candidate target of melanocortinergic signaling. Differential expression of CD81 mRNA in hypothalamic regions was confirmed via Northern blot analysis, which indicated an approximate threefold induction in obese A^y mice.

To establish the expression pattern of CD81 in the mouse CNS and determine if stimulated expression was generalized or localized to specific brain regions, in situ hybridization analyses were conducted using antisense riboprobes corresponding to the 5' end of the CD81 gene. As previously seen in the rat,[30] CD81 mRNA is most highly expressed in primarily glial structures including circumventricular organs such as the subfornical organ and choroid plexus. However, non-glial expression was confirmed in many regions of the brain, such as the hippocampus, and in several thalamic and hypothalamic nuclei. CD81 mRNA expression within the hypothalamus is diffuse in many areas; however, numerous structures were extensively labeled, including the PVH, anterior hypothalamic area, posterior part, ventromedial nucleus, DMH, and ARC. Relatively high signal intensity is found in the PVH as well as in a unique staining pattern in the caudal region of the hypothalamus consisting of the median eminence, posterior ARC, and DMH. Elevated expression associated with the A^y genotype was seen throughout the thalamus and hypothalamus wherever CD81 is normally expressed.

CD81 is a member of the tetraspanin family of cell surface proteins characterized by four transmembrane domains with intracellular N- and C-terminal regions along with one large extracellular loop[31] and it may be a co-receptor for hepatitis C virus.[32] A current model of tetraspanin function predicts that homodimers and heterodimers facilitate interactions with other membrane proteins to create a purported microdomain or "tetraspanin web" that may in turn sustain signaling processes.[33] The crystal structure of human CD81 large extracellular domain suggests that while one subdomain is necessary for dimerization, another may facilitate the known interactions with integrins or other adaptor molecules such as syndecans, which also can interact with integrins.[34] Therefore, CD81 may be part of a larger signaling complex that contains, among other things, syndecans and mahogany[35,36] (or possibly mahogany-like[37]) proteins that are necessary for normal melanocortin receptor signaling. Interestingly, CD81 expression was recently demonstrated to be upregulated in the nucleus accumbens of the rat after cocaine administration,[30] thus indicating a potential role for CD81 in dopaminergic reward pathways. Supporting evidence from CD81-deficient mice indicates altered behavioral and locomotor responses in animals treated with cocaine when compared to wild-type controls.[38] Current efforts in our laboratory include genetic crosses of CD81-deficient mice to numerous mouse models of altered melanocortin signaling to determine a functional role for CD81 in melanocortin signaling. For example, CD81 may be required for dominant *agouti* alleles to exert metabolic or pigmentation phenotypes, as is the case for *mahogany* and *mahoganoid* genes.

In summary, unanswered questions concerning CNS melanocortin receptor (MC4R and MC3R) function are currently being addressed using a variety of molecular techniques. Transgenic animal models for the conditional expression of MC4R and MC3R are being bred to transgenic mice that harbor neural-specific *cre* transgenes to specifically delete melanocortin receptors from distinct subsets of neurons. Tissue culture models to characterize the murine MC4R promoter indicate that proximal enhancer elements (located within 1600 bp of the translational start site) control basal expression of the promoter, whereas distal repressor elements are located between 3300 and 7900 bp upstream of the promoter. Lastly, we have shown that the expression of CD81 is elevated in obese yellow mice with chronic blockade of melanocortinergic signaling due to ectopic expression of *agouti* signaling protein. Thus, tetraspanins are a potential target of melanocortin signaling and may play a role in creating multiprotein signaling complexes located at the cell surface.

ACKNOWLEDGMENTS

These studies were supported by the National Institutes of Health (Grant DK20593), the Whitehall Foundation, and the American Diabetes Association (a Career Development Award provided to R.A.K.).

REFERENCES

1. BUTLER, A.A. & R.D. CONE. 2002. The melanocortin receptors: lessons from knockout models. Neuropeptides **36:** 77–84.
2. FAN, W. *et al.* 1997. Role of melanocortinergic neurons in feeding and the agouti obesity syndrome. Nature **385:** 165–168.
3. HUSZAR, D. *et al.* 1997. Targeted disruption of the melanocortin-4 receptor results in obesity in mice. Cell **88:** 131–141.
4. MARKS, D.L. *et al.* 2002. Melanocortin pathway: animal models of obesity and disease. Ann. Endocrinol. (Paris) **63:** 121–124.
5. LOW, M.J. *et al.* 1994. Receptors for the melanocortin peptides in the central nervous system. Curr. Opin. Endocrinol. Diabetes : 79–88.
6. BAGNOL, D. *et al.* 1999. Anatomy of an endogenous antagonist: relationship between agouti-related protein and proopiomelanocortin in brain. J. Neurosci. **26:** 1–7.
7. WILSON, B.D. *et al.* 1999. Physiological and anatomical circuitry between agouti-related protein and leptin signaling. Endocrinology **140:** 2387–2397.
8. CHEUNG, C.C. *et al.* 1997. Proopiomelanocortin neurons are direct targets for leptin in the hypothalamus. Endocrinology **138:** 4489–4492.
9. ELIAS, C.F. *et al.* 1999. Leptin differentially regulates NPY and POMC neurons projecting to the lateral hypothalamic area. Neuron **23:** 775–786.
10. KIM, M.S. *et al.* 2000. Hypothalamic localization of the feeding effect of agouti-related peptide and α-melanocyte-stimulating hormone. Diabetes **49:** 177–182.
11. GIRAUDO, S.Q. *et al.* 1998. Feeding effects of hypothalamic injection of melanocortin 4 receptor ligands. Brain Res. **809:** 302–306.
12. GRILL, H.J. *et al.* 1998. Brainstem application of melanocortin receptor ligands produces long-lasting effects on feeding and body weight. J. Neurosci. **18:** 10128–10135.
13. WILLIAMS, D.L. *et al.* 2000. The role of the dorsal vagal complex and the vagus nerve in feeding effects of melanocortin-3/4 receptor stimulation. Endocrinology **141:** 1332–1337.
14. SPIEGELMAN, B.M. & J.S. FLIER. 2001. Obesity and the regulation of energy balance. Cell **104:** 531–543.
15. FEKETE, C. *et al.* 2000. α-Melanocyte-stimulating hormone is contained in nerve terminals innervating thyrotropin-releasing hormone-synthesizing neurons in the hypothalamic paraventricular nucleus and prevents fasting-induced suppression of prothyrotropin-releasing hormone gene expression. J. Neurosci. **20:** 1550–1558.
16. LEGRADI, G. & R.M. LECHAN. 1999. Agouti-related protein containing nerve terminals innervate thyrotropin-releasing hormone neurons in the hypothalamic paraventricular nucleus. Endocrinology **40:** 3643–3652.
17. HARRIS, M. *et al.* 2001. Transcriptional regulation of the thyrotropin-releasing hormone gene by leptin and melanocortin signaling. J. Clin. Invest. **107:** 111–120.
18. MOUNTJOY, K.G. *et al.* 1994. Localization of the melanocortin-4 receptor (MC4-R) in neuroendocrine and autonomic control circuits in the brain. Mol. Endocrinol. **8:** 1298–1308.
19. SAKURAI, T. *et al.* 1998. Orexins and orexin receptors—A family of hypothalamic neuropeptides and G protein-coupled receptors that regulate feeding behavior. Cell **92:** 573–585.
20. DELECEA, L. *et al.* 1998. The hypocretins—hypothalamus-specific peptides with neuroexcitatory activity. Proc. Natl. Acad. Sci. USA **95:** 322–327.
21. QU, D. *et al.* 1996. A role for melanin-concentrating hormone in the central regulation of feeding behaviour. Nature **380:** 243–247.
22. MORIN, S.M. *et al.* 1999. Differential distribution of urocortin- and corticotropin-releasing factor-like immunoreactivities in the rat brain. Neuroscience **92:** 281–291.
23. FEKETE, C. *et al.* 2000. α-Melanocyte stimulating hormone prevents fasting-induced suppression of corticotropin-releasing hormone gene expression in the rat hypothalamic paraventricular nucleus. Neurosci. Lett. **289:** 152–156.
24. COWLEY, M.A. *et al.* 1999. Integration of NPY, AGRP, and melanocortin signals in the hypothalamic paraventricular nucleus: evidence of a cellular basis for the adipostat. Neuron **24:** 155–163.

25. WANG, Y. *et al.* 1996. Targeted DNA recombination in vivo using an adenovirus carrying the cre recombinase gene. Proc. Natl. Acad. Sci. USA **93:** 3932–3936.
26. DUMONT, L.M. *et al.* 2001. Mouse melanocortin-4 receptor gene 5'-flanking region imparts cell-specific expression *in vitro*. Mol. Cell. Endocrinol. **184:** 173–185.
27. MELLON P.L. *et al.* 1990. Immortalization of hypothalamic GnRH neurons by genetically targeted tumorigenesis. Neuron **1:** 1–10.
28. ADAN, R.A. *et al.* 1996. Melanocortin receptors mediate α-MSH-induced stimulation of neurite outgrowth in Neuro 2a cells. Brain Res. Mol. Brain Res. **36:** 37–44.
29. DIATCHENKO, L. *et al.* 1996. Suppression subtractive hybridization: a method for generating differentially regulated or tissue-specific CDNA probes and libraries. Proc. Natl. Acad. Sci. USA **93:** 6025–6030.
30. BRENZ VERCA, M.S. *et al.* 2001. Cocaine-induced expression of the tetraspanin CD81 and its relation to hypothalamic function. Mol. Cell. Neurosci. **17:** 303–316.
31. LEVY, S. *et al.* 1998 CD81 (TAPA-1): a molecule involved in signal transduction and cell adhesion in the immune system. Annu. Rev. Immunol. **16:** 89–109.
32. PILERI, P. *et al.* 1998. Binding of hepatitis C virus to CD81. Science **282:** 938–941.
33. VOGT, A.B. *et al.* 2002. Clustering of MHC-peptide complexes prior to their engagement in the immunological synapse: lipid raft and tetraspan microdomains. Immunol. Rev. **189:** 136–151.
34. WOODS, A. & J.R. COUCHMAN. 2001. Syndecan-4 and focal adhesion function. Curr. Opin. Cell. Biol. **13:** 578–583.
35. GUNN, T.M. *et al.* 1999. The mouse mahogany locus encodes a transmembrane form of human attractin. Nature **398:** 152–156.
36. NAGLE, D.L. *et al.* 1999. The mahogany protein is a receptor involved in suppression of obesity. Nature **398:** 148–152.
37. HAQQ, A.M. *et al.* 2002. Characterization of a novel binding partner of the melanocortin 4 receptor (MC4R): attractin-like protein (ALP). Paper presented at the Fifth International Melanocortin Meeting.
38. MICHNA, L. *et al.* 2001. Altered sensitivity of CD81-deficient mice to neurobehavioral effects of cocaine. Brain Res. Mol. Brain Res. **90:** 68–74.

Oxytocin Released from Magnocellular Dendrites

A Potential Modulator of α-Melanocyte–Stimulating Hormone Behavioral Actions?

NANCY SABATIER, CÉLINE CAQUINEAU, ALISON J. DOUGLAS, AND GARETH LENG

University Medical School, Edinburgh, United Kingdom

ABSTRACT: α-Melanocyte–stimulating hormone (α-MSH) is implicated in a variety of behavioral processes that are remarkably similar to those behaviors in which centrally acting oxytocin has been implicated. Central oxytocin derives in part from centrally projecting parvocellular neurons of the paraventricular nucleus, but large amounts of oxytocin are also released from dendrites of magnocellular oxytocin neurons in the supraoptic and paraventricular nuclei of the hypothalamus. Oxytocin release from dendrites is semi-independent of electrical activity and can be modulated by peptidergic signals independently of release from nerve terminals. Oxytocin is released from dendrites by stimuli that mobilize intracellular calcium stores, and such stimuli also prime dendritic stores of oxytocin, making them available for subsequent activity-dependent secretion. Evidence exists for efferent projections to the supraoptic nucleus from the arcuate nucleus where α-MSH neurons are located, and the supraoptic and paraventricular nuclei show high levels of expression of mRNA for the melanocortin receptor MC4R. These projections may be involved specifically in the regulation of dendritic oxytocin release.

KEYWORDS: α-melanocyte–stimulating hormone (α-MSH); intracellular calcium; Fos; dendritic release; supraoptic nucleus

In the rat, α-melanocyte–stimulating hormone (α-MSH) is produced by neurons in the arcuate nucleus and in the dorsolateral hypothalamus, and fibers containing α-MSH have been found in many brain areas. Centrally, α-MSH has been implicated in various behavioral processes, including the stretching-yawning syndrome, grooming, pain perception, feeding behavior, and male and female sexual behavior. The central actions of α-MSH are thought to be mediated primarily by melanocortin (MC3 and MC4) receptors, and mRNAs for these are expressed in various regions of the brain.[1] In particular, the supraoptic and paraventricular nuclei of the hypothal-

Address for correspondence: Gareth Leng, University Medical School, Edinburgh, UK. Voice: 0131 650 6811; fax: 0131 651 1287.

g.leng@ed.ac.uk

Ann. N.Y. Acad. Sci. 994: 218–224 (2003). © 2003 New York Academy of Sciences.

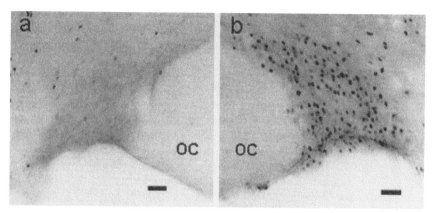

FIGURE 1. Fos immunoreactivity in the rat supraoptic nucleus after i.c.v. injection of (**a**) 5 μL saline or (**b**) 1 μg/5 μL α-MSH (unpublished experiments). OC, optic chiasm. Bar: 100 μm.

amus show high levels of expression of mRNA for the MC4 receptor. Moreover, intracerebroventricular (i.c.v.) injection of α-MSH induces Fos expression in supraoptic nucleus neurons (FIG. 1).

The behaviors with which α-MSH has been associated are remarkably similar to those behaviors in which oxytocin has been implicated. After central but not peripheral administration of α-MSH, the "stretching-yawning reflex" was observed in many species such as rats, mice, cats, dogs, and rabbits. This yawning behavior is likely to be mediated by the MC4 receptors.[2] Central oxytocin also is remarkably efficient at inducing yawning in rats,[3] even at doses as low as 3 pmol when injected into the paraventricular nucleus (PVN). Lesions of the PVN prevent oxytocin-induced yawning,[4] and oxytocin- and α-MSH–induced yawning are similarly modulated by muscarinic M1 receptors and β-adrenoreceptors.[5]

Both oxytocin and α-MSH have facilitatory effects on sexual behavior. For example, i.c.v. injections or injections in the PVN, MPOA, or VMN induce penile erection in many species.[2,3] Oxytocin and α-MSH also enhance female sexual behavior. Injected into the lateral ventricle, MPOA, or VMN, they both increase sexual motivation and receptivity with a long-lasting effect.[6–8]

Finally, both oxytocin and α-MSH induce anorectic responses after central injection.[9,10] Oxytocin reduces food consumption and the time spent eating, and it increases the latency to the first meal in fasted rats. α-MSH inhibits food intake and is described as a satiety mediator by Olszewski *et al.*[11]

Oxytocin is produced in parvocellular neurons of the PVN that project centrally; fibers are present in many brain areas, including the hypothalamus, amygdala, septum, brainstem, and spinal cord. Specific oxytocin receptors are also widely expressed in the central nervous system (CNS), including in these particular areas. The receptors appear to be functional, and their activation appears to be associated with specific behaviors. It is therefore natural to think that central oxytocin pathways are the mediators of behaviors with which oxytocin has been implicated.

Oxytocin is also produced by magnocellular neurosecretory neurons. These neurons project to the neurohypophysis, where oxytocin is released into the systemic circulation, and oxytocin released into the blood does not re-enter the brain in significant amounts. Magnocellular neurons have very few axon collaterals that terminate within the CNS, and activation of this system is not consistently associated with oxytocin-dependent behaviors. Therefore, it has been natural to assume also that the magnocellular system is *not* functionally involved in oxytocin-dependent behaviors.

These arguments, however, deserve testing in the light of accumulating evidence and some striking recent findings.

(1) Expression of a peptide in a given neuron is weak evidence that it has a functional role. In addition to oxytocin and vasopressin, the magnocellular neurons themselves express, often in relative abundance, dynorphin, met-enkephalin, cholecystokinin, neuropeptide Y, neuropeptide FF, endothelin I, cocaine- and amphetamine-regulated transcript (CART), neurotensin, angiotensin II, galanin, vasoactive intestinal peptide, thyroid-releasing hormone (TRH), corticotropin-releasing hormone (CRH), and many other peptides. Some of these may have a functional role, though only for dynorphin is the evidence strong. However, it is difficult to escape the conclusion that most of these peptides have no important functional role in the magnocellular neurons.

(2) Peptides in centrally projecting neurons generally coexist with a conventional neurotransmitter. Descriptions of "oxytocinergic" synapses appear typical of peptidergic terminals in the brain: large clusters of clear, small, immunonegative vesicles abut the presynaptic density, while a very few immunopositive dense-cored vesicle are also present.[12] To call any centrally projecting neuron an "oxytocin neuron" may be acceptable shorthand, but these neurons have other secretory products, including conventional neurotransmitters, that may be more important as signaling molecules.

(3) There is no consistent relationship between abundance of receptors and density of innervation, and there is a very high density of receptor expression at some sites where there is a virtual absence of innervation. This may reflect redundant expression of receptors at some sites, just as there may be redundant expression of oxytocin in some neurons.

The foregoing arguments are reasons to be cautious about assuming functional importance on the basis of some selected anatomical appearances. At some sites, notably the NTS, there is a relatively high density of oxytocin innervation, complemented by electrophysiological evidence implicating synaptically released oxytocin as a functional neuromodulator implicated in a clear physiological reflex. It would be reckless to argue that oxytocin expression in centrally projecting neurons has no functional significance.

Indeed, oxytocin has been measured in the extracellular fluid in many brain areas, and the measured concentrations change with physiological state. However, it is not transparently clear that the oxytocin measured in brain regions derives from terminals in that region or diffuses from release elsewhere. Studies on release from microdissected brain regions show that basal release of oxytocin per 10 min is typically much less than 1% of the local tissue content, and the release of oxytocin stimulated by depolarization with high K^+ represents ~3% of content.[13,14] However, concentrations measured in many brain regions seem much higher than would be expected from local release. In push-pull perfusion studies *in vivo*, Landgraf *et al.*[15] found a basal level of oxytocin in the mediolateral septum of 0.56 pg/10 min, which represents ~3.6%/10 min of the oxytocin content of the lateral/dorsal septum. In the dorsal hippocampus, the basal release of oxytocin is reported to be 0.84 pg/10 min, 7.6%/10 min of content.

The major source of oxytocin in the brain is the magnocellular system. Buijs *et al.*[16] measured the total tissue content of oxytocin in various regions of the rat brain and reported 15.6 pg in the lateral/dorsal septum, 35 pg in the amygdaloid nucleus, 11 pg in the dorsal hippocampus, and 188 pg in the NTS. By contrast, the supraoptic nucleus contained 1,800 pg of oxytocin. The supraoptic nuclei contain only about 30% of the magnocellular oxytocin neurons, the rest being present in the PVN, in accessory magnocellular groups, particularly the anterior commissural nuclei, and in individual cells in the anterior hypothalamus.[17] This oxytocin is not all obviously destined for release from the neural lobe; up to 90% of the content in the supraoptic nucleus is located within neurosecretory granules in the dendrites, rather than in the soma or axons.

Oxytocin can be released from dendrites by exocytosis.[18] The oxytocin concentration is particularly high in the extracellular fluid of the supraoptic nucleus (1,000 to 10,000 pg/mL, that is, 100- to 1,000-fold higher than the plasma concentration[15]), but dendritic release is not subject to the same rules that govern release from nerve terminals. Differential regulation of central and peripheral release clearly occurs in the case of osmotic challenge, as shown by microdialysis studies.[19] Differential release also occurs when rats are exposed to various stressors.[20] Oxytocin is released within the supraoptic nucleus in response to social defeat, but it is not released into the blood. During social defeat, oxytocin release is 10 times higher in the supraoptic nucleus than in the mediolateral septum, which is yet one of the major areas where oxytocin influences behaviors. Thus, oxytocin in the extracellular fluid in areas in or adjacent to the anterior hypothalamus may derive from the diffusion of oxytocin released from dendrites of magnocellular neurons. Furthermore, peptides are also very likely to be released into the cerebrospinal fluid (CSF) by neurons located in the PVN as oxytocin and vasopressin-containing fibers run close or even penetrate the ependymal lining of the third ventricle.[21] The basal levels of oxytocin in the CSF range from 0.5 to 0.7 pg/mL, that is, 5 to 7 pg in the total ~100 μL of CSF.

Exocytosis from nerve terminals is directly linked to electrical activity, resulting from depolarization following action potential activity. By contrast, oxytocin release from dendrites is semi-independent of electrical activity. Some peptide messengers, notably oxytocin itself[19] but also α-MSH (unpublished data), can trigger dendritic release without any change in the electrical activity of oxytocin neurons and therefore without any oxytocin release into plasma. Electrophysiological and morphological studies have reported evidence for efferent projections from the arcuate nucleus to the supraoptic nucleus.[22,23] In addition, expression of MC4 receptor mRNA has been detected in the supraoptic nucleus.[24]

Lambert *et al.*[25] showed that in supraoptic neurons, oxytocin mobilizes intracellular Ca^{2+} from thapsigargin-sensitive stores (FIG. 2a). Electrical stimulation of axons in the neural stalk, which triggers a large secretion of oxytocin in plasma, does not affect intranuclear release of oxytocin (FIG. 2b) despite antidromic propagation

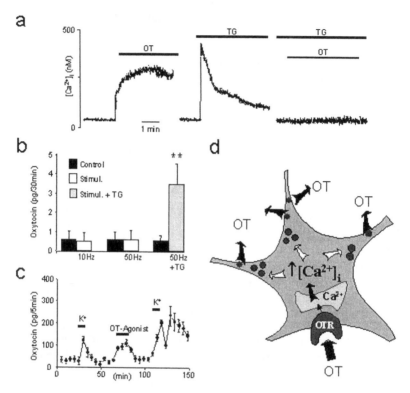

FIGURE 2. Intracellular Ca^{2+} stores regulate dendritic release of oxytocin. (**a**) In isolated supraoptic oxytocin neurons, the Ca^{2+} response to oxytocin is abolished after emptying intracellular Ca^{2+} stores by thapsigargin (TG). (**b**) *In vivo*, electrical stimulation of neural stalk induced oxytocin dendritic release in TG-pretreated rats but not in control rats. (**c**) In isolated supraoptic nuclei *in vitro*, oxytocin agonist induced oxytocin release and potentiated subsequent K^+ depolarization-induced release. The *in vivo* concentrations of oxytocin shown in **b** were measured in microdialysates, and for comparison with concentrations shown in **c**, they should be corrected for the recovery rate (1%) of the probes. (**d**) Schematic representation of putative mechanisms of oxytocin dendritic release: the binding of oxytocin on its receptor triggers Ca^{2+} release from intracellular stores, which then "prime" the secretory vesicles to the membrane and make them ready to be released in response to subsequent stimulation. (Adapted from Lambert *et al.*[25] and Ludwig *et al.*[26])

of action potentials in cell bodies.[26] However, when thapsigargin, an agent that mobilizes Ca^{2+} from intracellular stores, is previously applied to the supraoptic nucleus, the same electrical stimulation then evokes dendritic release of oxytocin within the supraoptic nucleus (FIG. 2b). Like thapsigargin, an oxytocin agonist potentiated the dendritic release of oxytocin stimulated by K^+-depolarization (FIG. 2c). Thus, agents that induce large increases in $[Ca^{2+}]$, from intracellular stores, not only trigger the dendritic release of oxytocin that is dissociated from electrical activity in cell bodies and peripheral release, but also seem to prime dendritic release in preparation for subsequent increase of electrical activity (FIG. 2d).

CONCLUSIONS

It is difficult to understand how a peptide injected into the brain can produce coherent behavioral responses; if peptides were released at synapses in an activity-dependent manner, then we would expect that to mimic the effect of a peptidergic pathway, we would have to reproduce the temporal and spatial specificity of the endogenous signal. However, peptides injected into the brain may work not by mimicking a complex spatiotemporal pattern of endogenous peptide release that carries specific information across synapses, but by mimicking a neurohormonal peptide that has long-lasting effects on the functional architecture of neuronal networks. Strong circumstantial evidence indicates that the behavioral response to central injections of α-MSH may be mediated by oxytocin; the activation of Fos expression in magnocellular neurons by α-MSH raises the possibility that the behavioral responses may be mediated by dendritic release of oxytocin from these neurons rather than by activation of parvocellular neurons.

REFERENCES

1. ADAN, R.A.H. & W.H. GISPEN. 1997. Brain melanocortin receptors: from cloning to function. Peptides **18:** 1279–1287.
2. ARGIOLAS, A., M.R. MELIS, S. MURGIA & H.B. SCHIÖTH. 2000. ACTH- and α-MSH-induced grooming, stretching, yawning and penile erection in male rats: site of action in the brain and role of melanocortin receptors. Brain Res. Bull. **51:** 425–431.
3. MELIS, M.R., A. ARGIOLAS & G.L. GESSA. 1986. Oxytocin-induced penile erection and yawning: site of action in the brain. Brain Res. **398:** 259–265.
4. ARGIOLAS, A., M.R. MELIS, A. MAURI & G.L. GESSA. 1987. Paraventricular nucleus lesion prevents yawning and penile erection induced by apomorphine and oxytocin but not by ACTH in rats. Brain Res. **421:** 349–352.
5. FUJIKAWA, M., K. YAMADA, M. NAGASHIMA & T. FURKAWA. 1995. Involvement of β-adrenoceptors in regulation of the yawning induced by neuropeptides, oxytocin and α-melanocyte-stimulating hormone, in rats. Pharmacol. Biochem. Behav. **50:** 339–343.
6. GONZALEZ, M., M.E. MELIS, D.R. HOLE & C.A. WILSON. 1993. Interaction of oestradiol, melanotropin and NA within the VMN in control of female sexual behaviour. Neuroendocrinology **58:** 218–226.
7. CALDWELL, J.D. 1992. Central oxytocin and female sexual behavior. Ann. N.Y. Acad. Sci. **652:** 167–179.
8. CRAGNOLINI, A., T. SCIMONELLI, M. ESTER CELIS & H.B. SCHIÖTH. 2000. The role of melanocortin receptors in sexual behavior in female rats. Neuropeptides **34:** 211–215.
9. MURPHY, B., C.N. NUNES, J.J. RONAN, *et al.* 1998. Melanocortin mediated inhibition of feeding behavior in rats. Neuropeptides **32:** 491–497.
10. RICHARD, P., F. MOOS & M.-J. FREUND-MERCIER. 1991. Central effects of oxytocin. Physiol. Rev. **71:** 331–370.
11. OLSZEWSKI, P.K., M.M. WIRTH, T.J. SHAW, *et al.* 2001. Role of alpha-MSH in the regulation of consummatory behavior: immunohistochemical evidence. Am. J. Physiol. Regul. Integrat. Comp. Physiol. **28:** P673–P680.
12. VOORN, P. & R.M BUIJS. 1983. An immuno-electronmicroscopical study comparing vasopressin, oxytocin, substance P and enkephalin containing nerve terminals in the nucleus of the solitary tract of the rat. Brain Res. **270:** 169–173.
13. MOOS, F., M.-J. FREUND-MERCIER, Y. GUERNE, *et al.* 1984. Release of oxytocin and vasopressin by magnocellular nuclei *in vitro*: specific facilitatory effect of oxytocin on its own release. J. Endocrinol. **102:** 63–72.

14. Buijs, R.M. & J.J. Van Heerikhuize. 1982. Vasopressin and oxytocin release in the brain: a synaptic event. Brain Res. **252:** 71–76.
15. Landgraf, R., I. Neumann, J.A. Russell & Q.J. Pittman. 1992. Push-pull perfusion and microdialysis studies on central oxytocin and vasopressin release in freely moving rats during pregnancy, parturition, and lactation. Ann. N.Y. Acad. Sci. **652:** 326–339.
16. Buijs, R.M., G.J. De Vries & F.W. Van Leeeuwen. 1985. The distribution and synaptic release of oxytocin in the central nervous system. *In* Oxytocin: Clinical and Laboratory Studies. J.A. Amico & A.G. Robinson, Eds.: 77–86. Elsevier Science Publishers. UK.
17. Rhodes, C.H., J.I. Morrell & D.W. Pfaff. 1981. Immunohistochemical analysis of magnocellular elements in rat hypothalamus: distribution and numbers of cells containing neurophysin, oxytocin, and vasopressin. J. Comp. Neurol. **198:** 45–64.
18. Pow, D.V. & J.F. Morris. 1989. Dendrites of hypothalamic magnocellular neurons release neurohypophysial peptides by exocytosis. Neuroscience **32:** 435–439.
19. Ludwig, M. 1998. Dendritic release of vasopressin and oxytocin. J. Neuroendocrinol. **10:** 881–895.
20. Engelmann, M., C.T. Wotjak, K. Ebner & R. Landgraf. 2000. Behavioural impact of intraseptally released vasopressin and oxytocin in rats. Exp. Physiol. **85S:** 125S–130S.
21. Robinson, I.C.A.F. 1983. Neurohypophysial peptides in cerebrospinal fluid. *In* The Neurohypophysis: Structure, Function and Control, Progress in Brain Research. B.A. Cross & G. Leng, Eds.: 129–145. Elsevier Science Publishers. UK.
22. Leng, G., H. Yamashita, R.E. Dyball & R. Bunting. 1988. Electrophysiological evidence for a projection from the arcuate nucleus to the supraoptic nucleus. Neurosci. Lett. **29:** 146–151.
23. Douglas, A.J., R.J Bicknell, G. Leng, *et al.* 2002. β-Endorphin cells in the arcuate nucleus: projections to the supraoptic nucleus and changes in expression during pregnancy and parturition. J. Neuroendocrinol. **14:** 768–777.
24. Mountjoy, K.G., M.T. Mortrud, M.J. Low, *et al.* 1994. Localization of the melanocortin-4 receptor (MC4-R) in neuroendocrine and autonomic control circuits in the brain. Mol. Endocrinol. **8:** 1298–1308.
25. Lambert, R.C., G. Dayanithi, F.C. Moos & P. Richard. 1994. A rise in intracellular Ca^{2+} concentration of isolated rat supraoptic cells in response to oxytocin. J. Physiol. **478:** 275–288.
26. Ludwig, M., N. Sabatier, P.M. Bull, *et al.* 2002. Intracellular calcium stores regulate activity-dependent neuropeptide release from dendrites. Nature **418:** 85–89.

Engineering the Melanocortin-4 Receptor to Control G$_S$ Signaling *in Vivo*

SUPRIYA SRINIVASAN,[a] CHRISTIAN VAISSE,[b] AND BRUCE R. CONKLIN[a]

[a]*Gladstone Institute of Cardiovascular Disease and*
[b]*The Diabetes Center, University of California, San Francisco,*
San Francisco, California 94141, USA

ABSTRACT: G-protein–coupled receptors (GPCRs) are the largest known family of cell surface receptors, and they control many important physiological events, including sensory perception, chemotaxis, neurotransmission, and energy homeostasis. However, GPCR signaling can be difficult to study *in vivo* because of the multitude of GPCRs, the lack of specific synthetic agonists, and the fact that some GPCRs activate multiple signaling pathways. One method to circumvent these problems is to develop an engineered receptor that is unresponsive to its endogenous agonist, yet can be fully activated by synthetic, small-molecule drugs. Such a receptor, called a receptor activated solely by a synthetic ligand (RASSL), can be rapidly and reversibly activated by a small-molecule drug and would be a powerful tool to control G-protein signaling *in vivo*. Here we present the development of a G$_s$-coupled RASSL based on the melanocortin-4 receptor (MC4R). MC4R couples exclusively to G$_s$ at physiologically relevant concentrations of its endogenous ligand, α-melanocyte–stimulating hormone (α-MSH). Data from human patients and structure-activity studies have shown that several mutations in MC4R cause a decreased affinity for α-MSH and can be exploited for RASSL development. Synthetic, small-molecule agonists of MC4R are now available and can be used to activate mutated receptors *in vivo*. We are engineering a series of mutations in MC4R to remove the peptide-binding site while retaining small-molecule binding and activation. The MC4R G$_s$ RASSL could be used to control many physiological responses associated with G$_s$ signaling such as heart rate, energy homeostasis, and cell proliferation.

KEYWORDS: G-protein–coupled receptor (GPCR); signal transduction; engineered receptor; embryonic stem cells

INTRODUCTION

G-protein–coupled receptors (GPCRs) are the largest known family of cell-surface receptors, consisting of ~370 hormone receptors and ~350 sensory receptors. GPCRs can be stimulated by a variety of natural ligands, including peptide hormones, odorants, light, biogenic amines, and lipids. Through the actions of G pro-

Address for correspondence: Supriya Srinivasan, Gladstone Institute of Cardiovascular Disease, University of California, San Francisco, CA 94141. Voice: 415-695-3760; fax: 415-285-5632.
ssrinivasan@gladstone.ucsf.edu

Ann. N.Y. Acad. Sci. 994: 225–232 (2003). © 2003 New York Academy of Sciences.

teins, GPCRs activate signal transduction cascades that control diverse cellular responses. For example, GPCR-mediated activation of G_s leads to an increase in intracellular cAMP by activating adenylyl cyclase. The rise of cAMP levels in the cell triggers a cascade of cellular and physiological responses, including an increase in heart rate, thyroid hormone synthesis, cortisol secretion, bone resorption, triglyceride breakdown, long-term effects on learning and memory, and changes in energy homeostasis.

Given the biological importance of GPCRs, the ability to stimulate a specific GPCR in a tissue of choice *in vivo* would be useful towards understanding the resultant changes in downstream signaling. For example, one can inject a specific GPCR ligand into a tissue of interest and study the changes in physiology or behavior of the animal. However, there is no way to restrict receptor activation to a specific sub-population of cells. Other factors may also complicate the interpretation of such experiments, because many GPCRs belong to large families of closely related subtypes with different coupling specificities (G_s, G_i, G_q, or $G_{12/13}$) and may be activated by the same ligand. In addition, the presence of the endogenous ligand would activate the receptor of interest in an uncontrolled manner.

Two Ligand-Binding Domains of a G Protein-Coupled Receptor

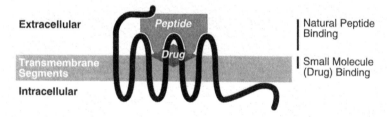

Receptor Activated Solely by a Synthetic Ligand (RASSL)

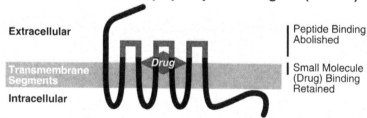

FIGURE 1. Principle of RASSL design. Peptide hormone receptors are activated by the binding of endogenous hormones on the extracellular surface of the receptor, whereas activation by small molecule synthetic ligands typically occurs by binding in the transmembrane segments of the receptor. Site-directed mutagenesis to eliminate the peptide-binding regions of the receptor would result in an engineered receptor that is activated solely by a synthetic ligand.

To circumvent these problems, we are developing engineered receptors that are activated solely by synthetic ligands (RASSLs). These genetically engineered receptors are designed to be insensitive to their natural, endogenous ligand(s), but can still be fully activated by synthetic, small-molecule drugs (see Scearce-Levie *et al.*[1] for a review). RASSL design is based on the premise that peptide hormone receptors are activated by the binding of endogenous hormones on the extracellular surface of the receptor, whereas activation by small molecules typically occurs by binding in the transmembrane segments of the receptor (FIG. 1). Such engineered receptors can be used to activate a G-protein pathway of interest rapidly and reversibly, mimicking the speed, localization, regulation, and amplification of endogenous GPCR signals. The resulting cellular, biochemical, and physiological changes would then be clearly attributable to the specific G-protein pathway activated and could be studied in detail.

A G_i-COUPLED RASSL FOR CONTROL OF G_i SIGNALING

The κ–opioid receptor was used to construct the first G_i-coupled RASSL.[2] The importance of the opioid receptor family in pain modulation has resulted in the generation of several small, high-affinity molecules that activate these receptors. One such molecule, spiradoline, can be used to selectively activate the κ–opioid receptor and is thought to bind in the transmembrane regions of the receptor. The second extracellular loop of the receptor is known to be important for binding of the endogenous ligand dynorphin. Thus, substitution of the second extracellular loop of the κ opioid receptor with that of the δ–opioid receptor (which is not activated by dynorphin) resulted in a receptor that is activated solely by spiradoline, called Ro1

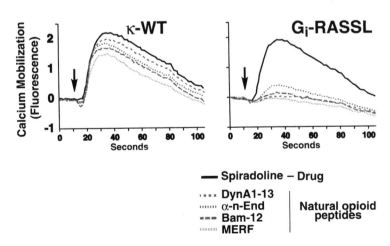

FIGURE 2A. Drug-induced signaling is preserved while natural opioid peptide signaling is decreased in the G_i-coupled RASSL. Within 20 s after spiradoline injection, a sharp increase in calcium mobilization is seen, corresponding to activation of a chimeric G_q protein that couples to G_i-coupled receptors.

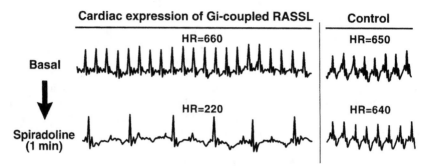

FIGURE 2B. Activation of the G_i-coupled RASSL slows heart rate *in vivo*. ECG measurements were performed on transgenic mice expressing the Ro1 transgene in the heart. One minute after spiradoline injection to activate the transgene, a decrease in heart rate from 660 bpm to 220 bpm was observed, demonstrating that the G_i-coupled RASSL can be used to control heart rate *in vivo*. (From Redfern *et al.*[3] Reprinted, with permission, from *Nature Biotechnology*.)

(RASSL opioid 1).[2] In transfected cells, Ro1 signals specifically in response to spiradoline and is unresponsive to dynorphin and other related peptides (FIG. 2A).

Although Ro1 clearly behaved as a RASSL in cultured cells, the key demonstration of its ability to control G_i-modulated physiology *in vivo* came from experiments in which Ro1 was expressed in the adult mouse heart.[3] The G_i pathway is known to control the sympathetic decrease in heart rate. Spatial and temporal control of Ro1 expression was obtained by using the tetracycline transactivator system. Activation of Ro1 by spiradoline decreased the heart rate within 1 min (FIG. 2B). In addition, long-term expression of Ro1 in the heart caused ventricular conduction delay and a lethal cardiomyopathy.[4] These experiments provided the critical proof-of-concept that a RASSL can be used to control G-protein signaling *in vivo*.

DESIGN OF THE G_s-COUPLED RASSL

Although the G_i-coupled RASSL has many uses for the study of physiological changes caused by G_i, we are now specifically interested in developing a G_s-coupled RASSL to study the *in vivo* responses to G_s signaling. Our strategy to develop a G_s-coupled RASSL is based on the melanocortin-4 receptor (MC4R). MC4R is primarily expressed in the mammalian brain and plays an important role in body weight regulation.[5] Activation of the central leptin receptor by the hormone leptin (released from adipose tissue) results in upregulation of the proopiomelanocortin peptide (POMC). α-Melanocyte–stimulating hormone (α-MSH), the natural ligand for MC4R, is generated from the POMC peptide and activates MC4R to increase the local concentration of cAMP through the effects of G_s. The resulting changes in physiology are the inhibition of food intake and the promotion of weight loss. Conversely, agouti-related peptide (AgRP), the natural MC4R inverse agonist, increases food intake and promotes weight gain.[6,7] The identification of inactivating point mutations in MC4R in obese human subjects[8] has lent support to the hypothesis that

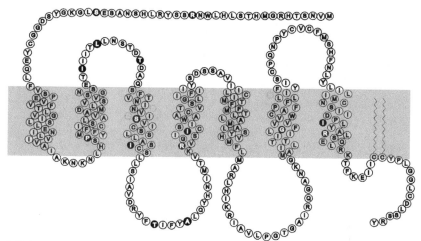

● Point-to-point contacts with α–MSH
● Mutations in human subjects; normal receptor expression

FIGURE 3. Model of the melanocortin-4 receptor. Amino acids (*highlighted in gray*) identify regions of point-to-point contacts with the endogenous hormone α-MSH. Amino acids (*highlighted in black*) represent a subset of data from human patients that carry mutations in the corresponding amino acids. These mutations still allow cell-surface expression of the receptor.

MC4R plays a key role in weight regulation. In addition, mice lacking both copies of the MC4R gene are severely obese.[9]

The discovery that MC4R plays an important role in the regulation of body weight has generated the interest of pharmaceutical companies that have created potent, small-molecule agonists of MC4R to treat obesity,[10,11] a growing health concern in the western world. Knowledge of point mutations that uncouple MC4R activation from the natural agonist α-MSH as well as the availability of nonpeptide, small-molecule agonists has allowed us to design a G_s-coupled RASSL based on MC4R (FIG. 3). As with the G_i-coupled RASSL, the design of the MC4 RASSL is based on the premise that peptide hormone receptors have different binding sites for endogenous hormones and small-molecule drugs (FIG. 1). Thus, we predict that the MC4 RASSL can be generated by mutagenesis and removal of the peptide-binding site on the first extracellular loop and the N-terminus, which have been well-established as the primary regions of α-MSH binding.[12,13] The transmembrane segments where the small molecule can bind will be preserved. The response of multiple mutated receptors to the endogenous hormone α-MSH and the small-molecule synthetic agonist can then be measured by generating cAMP dose-response curves to screen for receptors that behave as RASSLs. The mutations in MC4R that may be used to develop RASSLs are shown in FIGURE 3.

We have tagged the MC4R with the FLAG epitope and the N-terminal green fluorescent protein (GFP) to allow easy detection and visualization in cells. The GFP-tagged receptor is expressed normally at the cell surface in HEK293 cells

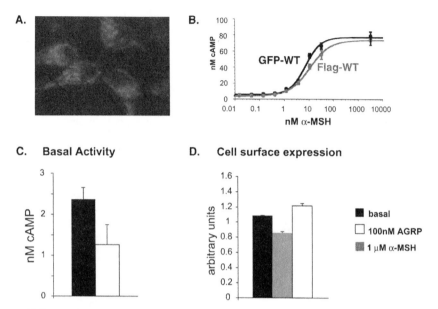

FIGURE 4. (A) The GFP-tagged MC4R is expressed on the cell surface in HEK 293 cells. (B) The GFP- and FLAG-tagged receptors signal normally in response to α-MSH. (C) Basal activity of the receptor can be decreased by treatment with the inverse agonist, AgRP. (D) Cell-surface expression and internalization of MC4R are shown in response to α-MSH and AgRP.

(FIG. 4A) and signals normally compared to the FLAG-tagged receptor (FIG. 4B). The basal activities of wild-type and mutant receptors can be assessed by measuring the responses to the inverse agonist, AgRP (FIG. 4C). Ideally, the basal activity of the RASSL should be the same as or less than that of the wild-type receptor. We have measured cell-surface expression of the receptor by enzyme-linked immunosorbent assays (ELISAs), using the FLAG epitope tag. These studies showed that loss of the cell-surface receptor in response to agonist stimulation can be used as a measure of receptor internalization or desensitization (FIG. 4D).

APPLICATIONS OF A G_s-COUPLED RASSL

One of the most immediate applications of RASSL technology lies in its use as a G_s switch for the study of food intake and body weight regulation in mice. Experiments based on peripherally administered α-MSH have delineated a role for MC4R in these processes. However, a detailed, mechanistic understanding of the cellular and biochemical changes in response to MC4R signaling has been difficult to elucidate because of the uncontrolled presence of the endogenous hormone, α-MSH. To circumvent this problem, a targeted gene knock-in approach can be used to replace the genomic copies of the endogenous receptor with the MC4-RASSL. The RASSL would then be expressed in place of the endogenous receptor, but would remain functionally silent until it is activated by administration of the small-molecule drug. Cellular, physiological, and behavioral changes that result from activation of the RASSL would then be directly attributable to the G_s pathway in regions of RASSL expression.

The G_s signaling pathway controls many other physiological responses, including changes in heart rate, chemotaxis, cell proliferation, and endocrine hormone secretion (see Neves *et al.*[14] for a review). Studies in adult mice have shown that the G_s pathway is crucial for the maintenance of heart rate (see Post *et al.*[15] for a review). However, very little is known about the precise role of this pathway in the early specification and differentiation of the cardiac lineage in early cardiomyogenesis. This is primarily because it is difficult to selectively activate the G_s pathway in developing cardiomyocytes. These problems can be overcome by the use of RASSL technology in the study of the G_s pathway in embryonic stem (ES) cells that can be differentiated into cardiomyocytes. In addition, an understanding of the G_s pathway in ES cells would be valuable, because they are being increasingly used in transplantation therapies in heart failure.[16] A critical limiting factor in these studies is the inability to control cells once they have been transplanted *in vivo*. Engineered receptors such as the G_s-coupled RASSL described here offer an important advantage in this regard. Activation of the G_s pathway in implanted cells by administration of the RASSL agonist can be used to control both the electrical activity of transplanted cells and the balance between proliferation and differentiation.

SUMMARY AND PERSPECTIVES

An engineered receptor that controls G_s signaling has enormous potential. The RASSL can be expressed in ES cells or *in vivo* in any tissue of choice. The tetracycline transactivator system can be used for the temporal control of RASSL expression. In conjunction with other designer signaling systems, engineered receptors offer a tremendous opportunity for the study of signaling pathways that underlie the cellular, physiological, and behavioral changes in mammals.

REFERENCES

1. SCEARCE-LEVIE, K., P. COWARD, C.H. REDFERN, *et al.* 2001. Engineering receptors activated solely by synthetic ligands (RASSLs). Trends Pharmacol. Sci. **22:** 414–420.
2. COWARD, P., H.G. WADA, M.S. FALK, *et al.* 1998. Controlling signaling with a specifically designed G_i-coupled receptor. Proc. Natl. Acad. Sci. USA **95:** 352–357.
3. REDFERN, C.H., P. COWARD, M.Y. DEGTYAREV, *et al.* 1999. Conditional expression and signaling of a specifically designed G_i-coupled receptor in transgenic mice. Nat. Biotechnol. **17:** 165–169.
4. REDFERN, C.H., M.Y. DEGTYAREV, A.T. KWA, *et al.* 2000. Conditional expression of a Gi-coupled receptor causes ventricular conduction delay and a lethal cardiomyopathy. Proc. Natl. Acad. Sci. USA **97:** 4826–4831.
5. CONE, R.D. 2000. The melanocortin-4 receptor. *In* The Melanocortin Receptors. R.D. Cone, Ed.: 405–447. Humana. Totowa, NJ.
6. LU, D., D.I. VAGE & R.D. CONE. 1998. A ligand-mimetic model for constitutive activation of the melanocortin-1 receptor. Mol. Endocrinol. **12:** 592–604.
7. YANG, Y.K., M.M. OLLMANN, B.D. WILSON, *et al.* 1997. Effects of recombinant agouti-signaling protein on melanocortin action. Mol. Endocrinol. **11:** 274–280.

8. VAISSE, C., K. CLEMENT, E. DURAND, *et al.* 2000. Melanocortin-4 receptor mutations are a frequent and heterogeneous cause of morbid obesity. J. Clin. Invest. **106:** 253–262.
9. HUSZAR, D., C.A. LYNCH, V. FAIRCHILD-HUNTRESS, *et al.* 1997. Targeted disruption of the melanocortin-4 receptor results in obesity in mice. Cell **88:** 131–141.
10. SEBHAT, I.K., W.J. MARTIN, Z. YE, *et al.* 2002. Design and pharmacology of N-[(3R)-1,2,3,4-tetrahydroisoquinolinium-3-ylcarbonyl]-(1R)-1-(4-chlorobenzyl)-2-[4-cyclo-hexyl-4-(1H-1,2,4-triazol-1-ylmethyl)piperidin-1-yl]-2-oxoethylamine (1), a potent, selective, melanocortin subtype-4 receptor agonist. J. Med. Chem. **45:** 4589–4593.
11. MACNEIL, D.J., A.D. HOWARD, X. GUAN, *et al.* 2002. The role of melanocortins in body weight regulation: opportunities for the treatment of obesity. Eur. J. Pharmacol. **450:** 141–157.
12. YANG, Y.K., T.M. FONG, C.J. DICKINSON, *et al.* 2000. Molecular determinants of ligand binding to the human melanocortin-4 receptor. Biochemistry **39:** 14900–14911.
13. YANG, Y.K., C.J. DICKINSON, Q. ZENG, *et al.* 1999. Contribution of melanocortin receptor exoloops to Agouti-related protein binding. J. Biol. Chem. **274:** 14100–14106.
14. NEVES, S.R., P.T. RAM & R. IYENGAR. 2002. G protein pathways. Science **296:** 1636–1639.
15. POST, S.R., H.K. HAMMOND & P.A. INSEL. 1999. β-Adrenergic receptors and receptor signaling in heart failure. Annu. Rev. Pharmacol. Toxicol. **39:** 343–360.
16. STRAUER, B.E., M. BREHM, T. ZEUS, *et al.* 2002. Repair of infarcted myocardium by autologous intracoronary mononuclear bone marrow cell transplantation in humans. Circulation **106:** 1913–1918.

Mutations in the Human Proopiomelanocortin Gene

HEIKO KRUDE, HEIKE BIEBERMANN, AND ANNETTE GRÜTERS

Pediatric Endocrinology, Otto Heubner Center for Pediatrics, Charité-University Hospital, Humboldt-University, Berlin, Germany

ABSTRACT: The melanocortin peptides and their receptors represent one of the most complex systems in human endocrinology. Hormonal regulation includes pigmentation, weight maintenance, adrenal function, and exocrine gland secretion via endocrine, paracrine, autocrine, and neurocrine action of melanocortin peptides at five different but homologous melanocortin receptors. Genetic relevance of the melanocortin system for human physiology was initially shown by mutations in the different melanocortin receptor genes, first described in the melanocortin-2 receptor gene in 1993 as one reason for congenital hypocortisolism. Because all ligands within the melanocortin systems are derived from one single precursor hormone, proopiomelanocortin (POMC), a genetic defect in the POMC gene could have been expected to affect all functional components of the melanocortin system. Accordingly, patients with a complete defect of the POMC gene product due to homozygous or compound heterozygous loss of function mutations were shown to be affected mainly by red hair, early-onset obesity, and congenital hypocortisolism. No further obvious clinical problems were described in these patients, suggesting that no additional function of the melanocortin system has escaped recognition. However, whether partial loss of function mutations in the POMC gene might lead to more circumscribed phenotypes, especially common obesity, remains an open question.

KEYWORDS: proopiomelanocortin (POMC); adrenocorticotropin-stimulating hormone (ACTH) deficiency; obesity; pigmentation

THE POMC GENE AND HUMAN GENETICS

The finding of a central role of the proopiomelanocortin (POMC) gene in rodent body weight regulation[1,2] paved the way for the human POMC gene to attract the interest of human geneticists. Because of the tremendous increase in the prevalence of obesity over the last two decades and of epidemiologic studies that suggested a major contribution of the genome towards the maintenance of normal body weight,[3] the search for candidate genes for human obesity, including the POMC gene, was stimulated.

Address for correspondence: Heiko Krude, Pediatric Endocrinology, Otto Heubner Center for Pediatrics, Charité-University Hospital, Humboldt-University, Berlin, Germany. Voice: 0049-30-450 566251; fax: 0049-30-450566936.

heiko.krude@charite.de

Ann. N.Y. Acad. Sci. 994: 233–239 (2003). © 2003 New York Academy of Sciences.

Previously, the POMC gene had been discussed as a likely candidate gene for the very rare disease of familial isolated adrenocorticotropin-stimulating hormone (ACTH) deficiency (MIM 201400). Within families with the disorder, ACTH deficiency leads to congenital hypocortisolism, a condition that could be lethal because of severe hypoglycemia, especially in times of severe infection, or cholestasis and hepatic failure during the neonatal period. Since the introduction of substitution therapy with oral cortisone preparations, these individuals can now be treated and survive without further complications. The isolated deficiency of ACTH in the light of normal secretion of other pituitary hormones and normal morphology of the pituitary gland led some investigators to speculate about a defect in the precursor of ACTH, the POMC gene.[4] However, a mutation in the POMC gene could be excluded, leading to the idea of a defect in processing of the POMC prohormone.[4] Recently, isolated familial ACTH deficiency could have been explained by mutations in a new gene (TboxPituitary, TPIT) coding for a transcription factor, which was shown to be crucial for appropriate development of the corticotrophs, the pituitary cell line secreting ACTH.[5]

POMC-derived peptides exert their role in body weight regulation by activation of the melanocortin-3 and -4 receptors (MC3R and MC4R)[2,6] and stimulate adrenal cortisol secretion by activation of the melanocortin-2 receptor (MC2R).[7] In addition, POMC-derived peptides were shown to be processed within the skin where they bind to the melanocortin-1 receptor to influence pigmentation.[8] The fifth and most recently cloned melanocortin receptor was shown to contribute to the function of exocrine glands.[9] Moreover, pharmacologic studies with melanocortin-like substances have revealed a wide range of changes in almost every physiological system.[10] Therefore, because of the complexity of the melanocortin system, it could have been expected that inactivating mutations of the POMC gene might result in a complex disease including additional symptoms beyond obesity and ACTH deficiency, a phenotype that has not yet been described.

POMC DEFICIENCY

Based on the assumption that complete POMC deficiency might at least result in obesity, isolated ACTH deficiency, and pigmentation alterations such as red hair and pale skin, we performed a mutation screening study in two children with these clinical findings. One homozygous translation initiation mutation as well as two compound heterozygous nonsense mutations were found in these two patients. The mutations predicted a complete loss of the POMC gene product.[11]

In both children, obesity occurred in the first months of life and increased dramatically due to constant hyperphagia.[11] Since the initial description of these two children, we observed that both affected children had a continuous increase in body weight without reaching a weight-curve parallel to the normal percentiles of childhood body weight. However, both children have an accelerated growth and bone age comparable to the findings in MC4R mutant children. Therefore, the course of obesity can only be assessed by BMI-SDS values. Transforming the weight and height data into BMI-SDS values reveals that the most significant increase in obesity occurred during the first 4 to 5 years of life and that thereafter a plateau of stable BMI-SDS values was reached. In the boy, who is now 10 years old, a slight decrease in

BMI-SDS values could be observed over the last two years, suggesting some level of adaptation of body weight regulation. If these individual BMI-SDS values represent the natural course of obesity in POMC deficiency, it will be clarified by the identification of additional patients with POMC gene mutations.

A difference in phenotype expression of the two different genotypes was observed according to the time-point of ACTH deficiency manifestation. While the girl had hypoglycemic seizures and cholestasis during the neonatal period, the boy became symptomatic with hypoglycemic febrile seizures during the course of an infection only at the age of 2 years.

Consistent with the role of POMC-derived peptides in skin pigmentation, both children have red-orange hair, and their skin is pale and tends to be very vulnerable to sunlight. Thus far, no additional dermatologic symptoms have been observed.

Other symptoms in other physiologic systems are not obviously present in both children, suggesting that beyond the roles of weight regulation, adrenal function, and skin pigmentation, POMC-derived peptides might not contribute significantly to further aspects of human physiology. However, since both children are still young, no conclusions about any function of the POMC gene during later life can be drawn. For example, one might expect alterations in the timing of puberty in POMC deficiency, because in leptin deficiency retarded pubertal development was described,[12] and leptin might influence the onset of puberty by hypothalamic melanocortin peptides. In addition, recently a stimulating influence of the POMC system on sexual behavior in rodents was described,[13] which suggests a parallel role in human reproduction.

Since the description of these first two patients, we have diagnosed additional children with the characteristic clinical traits of early-onset severe obesity, ACTH deficiency, and red hair to be affected by POMC gene mutations. New nonsense mutations as well as the already described translation initiation mutation were found in homozygosity or compound heterozygosity (Krude, in press). Although these new cases confirmed the clinical spectrum of patients with a POMC gene defect, the low number of diagnosed cases demonstrated at the same time the rarity of this new syndrome.

THE POMC GENE AS A CANDIDATE FOR COMMON OBESITY

Patients with "common" non-syndromic obesity are generally not affected by hypocortisolism and red hair. Therefore, the complex phenotype of complete POMC deficiency suggested that mutations in the POMC gene might not be relevant for a genetic contribution to the occurrence of the ongoing epidemic of obesity. However, one might speculate that in patients with common obesity, partial loss of function mutations in the POMC gene might result in isolated non-syndromic obesity. Accordingly, four candidate gene mutation screening studies among a total of 601 obese patients have so far been reported.[14–17]

Besides several polymorphisms, including a variable three amino acid repeat (Ser-Ser-Cys), six missense or nonsense mutations were found to be associated or co-segregated with obesity in 7 of 601 individuals. These mutations were found in compound heterozygosity in one and heterozygosity in six patients. They affect residues located in parts of the POMC gene that have not yet been shown to be involved in the weight-regulating function of POMC.

In an Italian study, two mutations were found in the most N-terminal part of the POMC gene containing the signal peptide of the pre-prohormone.[14] The position of these two mutations implicates an alteration of appropriate trafficking or processing of the precursor. Further functional studies are obviously necessary to validate this assumption.

Interestingly, the other four mutations are located in the C-terminal part of the POMC gene coding for the β-LPH fragment, which is cleaved by the prohormones convertase-1 and convertase-2 into the functionally active peptides β-melanocyte-stimulating hormone (β-MSH) and β-endorphin. In a German mutation screen, one missense mutation was identified in the residue immediately upstream of the Lsy/Lys N-terminal cleavage side of β-MSH, changing a Glu to a Gly.[15] A second mutation was found in compound heterozygosity in the same patient, introducing a premature stop codon 9 residue upstream of the β-MSH sequence.[15] The third mutation (which was found in a British study, 2/262 tested[16]; the Italian mutation screen, 1/87 tested[14]; as well as in one subsequent French cohort, 3/182 tested[16]) changes the dibasic C-terminal cleavage side of β-LPH that separates β-MSH and β-endorphin. The same Lys residue is affected by the fourth β-LPH mutation identified in 1 of 87 obese individuals in a Danish study.[17]

While the stop mutation clearly interferes with the generation of the downstream β-MSH and β-endorphin peptides, the second German Glu-Gly mutation might have an effect on the cleavage at the N-terminal β-MSH dibasic side, which remains to be proven. The mutation of the internal β-LPH cleavage side predicts a fusion peptide containing both the β-MSH and the β-endorphin peptides. Functional studies based on cells expressing the human MC4R have shown that this fusion peptide lacks biological activity in terms of cAMP generation but still retains receptor-binding affinity.[16] This constellation led the investigators to speculate about a dominant negative effect of the mutant fusion peptide, because it would bind to the MC4R but without anorexigenic activity, which therefore would interfere with the function of the active α-MSH peptides at the receptor. Alongside this elegant hypothesis, it still seems possible that the proven loss of function of the fused β-MSH and/or β-endorphin peptide might contribute to the obesity phenotype rather than introducing a dominant negative effect on the function of α-MSH. This loss of function hypothesis is further supported by the identified stop mutation that obviously does not result in a potentially dominant negative effect. However, this alternative hypothesis would implicate an as yet unrecognized functional role of β-LPH–derived peptides in human weight regulation, which needs further investigation to be clarified.

THE POMC GENE IN GENOME SCREENS

Epidemiologic studies have implicated a strong influence of the genome on the regulation of normal human body weight.[3] Therefore, a search for a genetic component in the pathogenesis of the ongoing epidemic of obesity was begun. However, due to the stability of the genome, the increase in common obesity cannot reflect an increase in the incidence of monogenic defects. More likely, susceptibility to common obesity reflects a low capacity of the individual regulatory genetic network of weight maintenance to counteract the changes in the environment. To pinpoint the contribution of one single gene variant within the individual network of weight regulation, some genome screens were performed to identify chromosomal regions linked to obesity. In two of these studies, the POMC gene locus was potentially linked with the occurrence of obesity phenotypes.

In a French study based on 153 families with severe early-onset obesity, two loci were identified on chromosomes 10 and 5, which were not known to contain candidate genes related to weight regulation. In contrast, a third locus was found on chromosome 2 (2p21-23), which contains the POMC gene with linkage mainly to serum leptin levels.[18] In a second screen, based on a Mexican-American population of 458 individuals from 10 families, only one locus was identified with significant linkage, which was identical to the locus on chromosome 2 identified in the French study.[19] In the Mexican-American study, through the use of intragenic polymorphic markers, the LOD score was even increased up to the highly significant level of 7.46.[20] Again, within the Mexican-American group of obese individuals, linkage was more significant to the level of serum leptin compared to the individual body mass index.

These linkage data suggest that the POMC gene might contribute to obesity or obesity phenotypes (such as the level of leptin), at least in these two study populations. Subsequent screening for mutations in the POMC coding region failed to detect any alterations that are associated with the obesity phenotype.[20] Therefore, the investigators concluded that changes in the non-coding region of the POMC gene might exist that interfere with appropriate expression of the POMC gene.[20] However, the complex phenotype in POMC-deficient patients, which includes alterations of pituitary and skin POMC function, is clearly different for most patients with common obesity. A defect of gene expression of the POMC gene in common obesity, which was suggested in the genome screen studies, might affect only hypothalamic expression of the POMC gene. Those kinds of defects of tissue-specific hypothalamic gene expression further suggests the existence of a hypothalamic promoter or distant enhancer of the POMC gene that might be different from other pituitary and skin-related regulatory units of the POMC gene. Recently, it was shown that a genomic fragment containing the POMC gene locus as well as a 13-kb 5′-sequence was able to promote hypothalamic expression of a GFP reporter gene.[21,22] While expression of the POMC gene in the pituitary is mainly regulated by the basal promoter close to the transcription start site of the POMC gene,[23] deletion experiments with the 13-kb GFP fragment suggested that a hypothalamic enhancer might be located far upstream of the POMC coding region.[21] Therefore, to resolve the positive linkage data of the POMC gene locus in the light of a normal POMC coding region, it seems mandatory to identify these putative hypothalamic enhancers of the POMC gene, which might harbor mutations interfering with hypothalamic POMC gene expression.

SUMMARY OF MUTATIONS IN THE POMC GENE

To date, mutations in the human POMC gene have been investigated by three different approaches:

  (1) In a focused study based on the predicted phenotype of a complete defect of the POMC gene product, few children could be identified in clinical trials of early onset obesity, ACTH deficiency, and red hair as being affected by complete loss of function of the POMC gene. This new disease

turned out to be very rare, but it recapitulates the functional roles of POMC-derived peptides in human physiology.

(2) Several candidate gene mutation studies were performed in nonsyndromic individuals with common obesity which clearly excluded coding region mutations of the POMC gene as a major contribution to the genetics of obesity. In only 6 of 601 obese patients were mutations shown to be associated or to co-segregate with obesity.

(3) In genome screening conducted to identify genes linked to obesity phenotypes, the POMC gene in two studies was independently identified to be linked to increased levels of leptin and in one Mexican-American study to be linked to increased body fat mass. Although no mutation in the coding region could subsequently be identified, the question remains as to whether mutations in the regulatory region of the gene exist that would interfere with hypothalamic POMC gene expression. Based on the significant genome screening results, those mutations would substantially contribute to the genetic influence on common obesity.

Taken together, human genetic studies over the last few years have shown that POMC-derived peptides seem to play the same physiologic roles in humans as in rodents; however, it is still unknown whether the POMC gene contributes to the genetics of obesity.

REFERENCES

1. FAN, W. et al. 1997. Role of melanocortinergic neurons in feeding and the agouti obesity syndrome. Nature 385: 165–168.
2. HUSZAR, D. et al. 1997. Targeted disruption of the melanocortin-4 receptor results in obesity in mice. Cell 88: 131–141.
3. MAES, H. et al. 1997. Genetic and environmental factors in relative body weight and human adiposity. Behav. Genet. 27: 325–351.
4. NUSSEY, S.S. et al. 1993. Isolated congenital ACTH deficiency: a cleavage enzyme defect? Clin. Endocrinol. (Oxford) 3: 381–385.
5. LAMOLET, B. et al. 2001. A pituitary cell-restricted T box factor, Tpit, activates POMC transcription in cooperation with Pitx homeoproteins. Cell 104: 849–859.
6. CHEN, A.S. et al. 2000. Inactivation of the mouse melanocortin-3 receptor results in increased fat mass and reduced lean body mass. Nat. Genet. 26: 97–102.
7. CLARK, A.J. & A. WEBER. 1998. Adrenocorticotropin insensitivity syndromes. Endocr. Rev. 19: 828–843.
8. CONE, R. et al. 1996. The melanocortin receptors: agonists, antagonists, and the hormonal control of pigmentation. Recent Prog. Horm. Res. 51: 287–317.
9. CHEN, W. et al. 1997. Exocrine gland dysfunction in MC5-R-deficient mice: evidence for coordinated regulation of exocrine gland function by melanocortin peptides. Cell 91: 789–798.
10. OLSON, G.A., R.D. OLSON & A.J. KASTIN. 1997. Endogenous opiates: 1996. Peptides 18: 1651–1688.
11. KRUDE, H. et al. 1998. Severe early-onset obesity, adrenal insufficiency and red hair pigmentation caused by POMC mutations in humans. Nat. Genet. 19: 155–157.
12. FAROOQI, I.S. 2002. Leptin and the onset of puberty: insights from rodent and human genetics. Semin. Reprod. Med. 20: 139–144.
13. VAN DER PLOEG, L.H. et al. 2002. A role for the melanocortin 4 receptor in sexual function. Proc. Natl. Acad. Sci. USA 99: 11381–11386.
14. MIRAGLIA DEL GIUDICE, E. et al. 2001. Molecular screening of the proopiomelanocortin (POMC) gene in Italian obese children: report of three new mutations. Int. J. Obes. Relat. Metab. Disord. 25: 61–67.

15. HINNEY, A. *et al.* 1998. Systematic mutation screening of the pro-opiomelanocortin gene: identification of several genetic variants including three different insertions, one nonsense and two missense point mutations in probands of different weight extremes. J. Clin. Endocrinol. Metab. **83:** 3737–3741.
16. CHALLIS, B.G. *et al.* 2002. A missense mutation disrupting a dibasic prohormone processing site in pro-opiomelanocortin (POMC) increases susceptibility to early-onset obesity through a novel molecular mechanism. Hum. Mol. Genet. **11:** 1997–2004.
17. ECHWALD, S.M. *et al.* 1999. Mutational analysis of the proopiomelanocortin gene in Caucasians with early onset obesity. Int. J. Obes. Relat. Metab. Disord. **23:** 293–298.
18. HAGER, J. *et al.* 1998. A genome-wide scan for human obesity genes reveals a major susceptibility locus on chromosome 10. Nat. Genet. **20:** 304–308.
19. COMUZZIE, A.G. *et al.* 1997. A major quantitative trait locus determining serum leptin levels and fat mass is located on human chromosome 2. Nat. Genet. **15:** 273–276.
20. HIXSON, J.E. *et al.* 1999. Normal variation in leptin levels in associated with polymorphisms in the proopiomelanocortin gene, POMC. J. Clin. Endocrinol. Metab. **84:** 3187–3191.
21. YOUNG, J.I. *et al.* 1998. Authentic cell-specific and developmentally regulated expression of pro-opiomelanocortin genomic fragments in hypothalamic and hindbrain neurons of transgenic mice. J. Neurosci. **18:** 6631–6640.
22. COWLEY, M.A. *et al.* 2001. Leptin activates anorexigenic POMC neurons through a neural network in the arcuate nucleus. Nature **411:** 480–484.
23. LIU, B. *et al.* 1992. Identification of DNA elements cooperatively activating proopiomelanocortin gene expression in the pituitary glands of transgenic mice. Mol. Cell. Biol. **12:** 3978–3990.

Knockout Studies Defining Different Roles for Melanocortin Receptors in Energy Homeostasis

ANDREW A. BUTLER[a] AND ROGER D. CONE[b]

[a]Pennington Biomedical Research Center, Baton Rouge, Louisiana 70808, USA

[b]Vollum Institute, Oregon Health and Sciences University, Portland, Oregon 97239, USA

ABSTRACT: Proopiomelanocortin (POMC) is expressed in the arcuate nucleus of the hypothalamus (ARC) and the commissural nucleus of the solitary tract (cNTS). Post-translational processing of POMC produces two melanocortin receptor ligands, α- and γ-melanocyte–stimulating hormone (MSH). Two melanocortin receptors (MC3R, MC4R) are expressed in brain regions receiving projections of POMC fibers, most of which also receive projections from a population of ARC neurons that co-express neuropeptide Y (NPY) and the MC3R/MC4R antagonist agouti-related peptide (AgRP). MC4R haploinsufficient humans and MC4R knockout (MC4RKO) mice exhibit increased adiposity and linear growth. MC4RKO mice exhibit hyperleptinemia and hyperinsulinemia and sometimes, but not always, develop type 2 diabetes (T2D). Individually housed MC4RKO mice fed low-fat diets are not hyperphagic when food intake is corrected for lean mass, whereas hyperphagia is observed after the introduction of diets with increased fat content. POMC knockout (POMCKO) mice are similar in that the severity of hyperphagia increases with the introduction of high-fat diets. By contrast, targeted deletion of the MC3R in the mouse results in increased adiposity despite the absence of hyperphagia. MC3RKO mice also exhibit reduced linear growth and lean mass; while MC3RKO mice are hyperleptinemic and hyperinsulinemic, the development of T2D has not been reported. The MC4R, but not the MC3R, is required for the stimulation of energy expenditure in response to melanocortin agonists and voluntary hyperphagia. Evidence for altered physical activity has also been reported for both knockout models. Analysis of MC4RKO mice indicates that this receptor is involved in rapidly coordinating energy consumption with energy expenditure through diet-induced thermogenesis and activity.

KEYWORDS: diet-induced thermogenesis; activity; proopiomelanocortin; melanocortin receptor; metabolism; indirect calorimetry

INTRODUCTION

Mutagenesis of genes encoding proopiomelanocortin (POMC) or the melanocortin receptors (MC3R, MC4R) expressed in the brain is associated with the development of obesity in mice.[1–5] Analysis of the feeding behavior of the knockout models confirmed earlier studies reporting suppression of appetite by the melano-

Address for correspondence: Andrew A. Butler, Pennington Biomedical Research Center, 6400 Perkins Road, Baton Rouge, LA 70808. Voice: 225-763-2619; fax: 225-763-2525.
butleraa@pbrc.edu

Ann. N.Y. Acad. Sci. 994: 240–245 (2003). © 2003 New York Academy of Sciences.

cortin system,[6] with hyperphagia observed in mice lacking either POMC or the MC4R.[1,5] Data from studies examining the response of MC4R knockout (MC4RKO) mice to the nonspecific melanocortin agonist melanotan-II (MTII) indicate that the MC4R was mediating the anorexic and metabolic effects observed in earlier studies.[2,7] The results of these studies also indicated that activation of the MC3R, in the absence of the MC4R, does not significantly affect appetite. MC3RKO mice exhibit a normal anorexic response to MT-II,[3] although to date there have been no studies reporting the metabolic response of MC3RKO mice to MTII. The phenotype of the MC3RKO mice was interesting in that an obesity phenotype was observed with hypophagia, indicating an increase in the efficiency with which energy consumed was stored as fat.[3,4] Pair feeding of the MC4RKO mice with wild-type (WT) littermates, while significantly reducing the severity of obesity, does not completely prevent increased adiposity.[8] Both the MC3R and the MC4R, therefore, regulate adiposity independently of energy consumption through the regulation of either nutrient partitioning or energy expenditure.

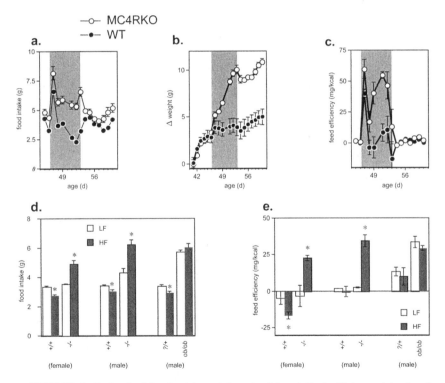

FIGURE 1. Daily food intake (**a**), weight gain (**b**) and feed efficiency (**c**) of male MC4RKO mice and WT controls fed the LF chow of HF chow (*shaded* regions). (**d**) WT mice significantly reduced food intake (in g) when given the HF chow, while MC4RKO mice increased food intake (*: $P < 0.01$ compared to LF by paired t-test). (**e**) Feed efficiency increases in male and female MC4RKO mice exposed to the HF diet (*: $P < 0.01$ compared to LF by paired t-test).

Our interest in studies examining the role of the melanocortin system in coordinating energy expenditure with energy intake was further motivated by the results of a simple experiment examining food intake in MC4RKO mice and WT littermates (FIG. 1). WT mice raised on a standard low-fat rodent chow (LF, 13% kcal from fat) exhibit a transient period of voluntary hyperphagia when exposed to a moderately high-fat diet (HF, 26% kcal from fat), a response also observed in MC3RKO, MC4RKO, and Lep^{ob}/Lep^{ob} mice (FIG. 1a).[4,9] The introduction of the HF diet was associated with a rapid gain in weight in MC4RKO mice over the comparatively short 7-day time period of the study (FIG. 1b). While the MC4RKO mice were hyperphagic on the HF chow (FIGS. 1a and d), a dramatic increase in feed efficiency (weight gain per energy intake) suggested that the increase in weight gain was not simply due to hyperphagia (FIGS. 1c and e). This result indicated other MC4R-dependent mechanisms were also involved in protecting against weight gain. Given that melanocortin agonists had already been shown to stimulate VO_2,[10] our hypothesis was that MC4RKO mice were failing to mount a metabolic response to hyperphagia associated with the introduction of the HF diet.

DIET-INDUCED THERMOGENESIS AND ENERGY HOMEOSTASIS

Maintenance of body fat content is dependent on the coordination of energy intake with energy expenditure.[11] Several mechanisms exist for adaptive thermogenesis, the process in which energy expenditure can rapidly respond to changes in environment or energy intake. For example, small mammals must rapidly increase heat production in response to a reduction in environmental temperature, a process requiring the β-adrenergic stimulation of uncoupling protein-1 (UCP1) in brown adipocyte tissue (BAT), a tissue that is replete with mitochondria.[12–14] In a cold environment, energy consumption increases to fuel the production of heat through the oxidation of fatty acids by BAT mitochondria. When energy consumption is restricted, energy expenditure is reduced as a survival mechanism. Leibel et al.[15] reported that restricting food intake in non-obese and obese volunteers to force a 10% reduction in body weight was associated with a 15% reduction in 24-h energy expenditure, corrected for body composition.

Another form of adaptive thermogenesis was proposed by Rothwell and Stock[16] over 20 years ago whereby an increase in energy consumption, in this case associated with the introduction of a "cafeteria diet," could be compensated by an increase in energy expenditure (diet-induced thermogenesis, DIT). In a recent review, the late Michael Stock indicated that DIT belongs to an evolutionarily ancient metabolic pathway, and while DIT could function to counter excessive hyperphagia associated with the discovery of a palatable diet, it could also enrich nutrient-poor diets by disposing of excess calories.[17] The underlying mechanism of DIT and whether it is involved in protecting against the development of obesity have been a matter of contention. However, a recent publication provides evidence that DIT is important for preventing obesity in mice exposed to very high-fat diets.[12] Mice lacking all three β-adrenergic receptors ("β-less" mice) exhibit a similar phenotype to that of the MC4RKO mice, gaining significantly more weight than control mice during brief periods (5 days) of exposure to a high-fat diet (58% kcal from fat). Energy consumption of the WT and β-less mice was not significantly different. In the WT mice,

TABLE 1. Percentage increase in VO_2 of male MC4RKO mice and Lep^{ob}/Lep^{ob} mice compared to WT mice after exposure to a high-fat diet[9]

	WT	MC4RKO	Lep^{ob}/Lep^{ob}
24 h – total	14.5 ± 1.9%	5.7 ± 1.7%*	11.3 ± 3.9%
24 h – basal	19.6 ± 2.1%	6.7 ± 3.1%*	29.7 ± 6.2%
48 and 72 h – basal	8.5 ± 1.4%	5.9 ± 1.6%	17.3 ± 3.4%*
48 and 72 h – total	14.7 ± 1.9%	7.8 ± 2.5%	29.5 ± 4.5%*

NOTE: VO_2 was analyzed as either the mean of all measurements taken over a 6-h period during the light phase (0600–1800 h) (*total*, 1100–1700 h), or as the three lowest values (*basal*). Each data point is represented as a mean ± SEM. *: significantly different from WT mice, $P < 0.05$.

energy expenditure (VO_2) increased by 17% after 5 days of high-fat feeding, while no increase in VO_2 was observed in β-less mice. Thus, in mice at least, DIT appears to have a significant role in maintaining body weight in response to high-fat diets.

THE MC4R IS REQUIRED FOR DIET-INDUCED THERMOGENESIS

To test the hypothesis that the MC4R regulates feed efficiency by regulating energy expenditure in response to hyperphagia, we measured metabolic rate in mice exposed to LF or HF chows using indirect calorimetry.[9] Measurements of VO_2 over a 3-day period revealed that MC4RKO mice and WT controls, when fed the standard low-fat rodent chow, had similar energy expenditure. The introduction of a HF diet resulted in a 15–20% increase in VO_2 in WT mice; however, this increase was attenuated in MC4RKO mice (TABLE 1). The MC4R is therefore involved in the acute coordination of energy expenditure with energy consumption, perhaps through the stimulation of sympathetic outflow to BAT.[8,12] MC3RKO mice exhibit a robust DIT in the same paradigm,[4] suggesting that the obesity phenotype of the MC3RKO is not due to failure to increase thermogenesis in response to hyperphagia.

While BAT has been proposed to be an important target organ of the sympathetic nervous system to stimulate DIT, two observations suggest that the defect in the MC4RKO mice might not simply represent a failure of BAT to respond to sympathetic activity. First, obese MC4RKO mice that have been back-crossed five generations into the C57BL/6J strain are not cold-sensitive and, in fact, have survived up to 2 weeks at 4°C. Second, given that obesity and insulin-resistance could affect DIT, we also examined the response of obese Lep^{ob}/Lep^{ob} mice in the same paradigm. Surprisingly, we observed a robust DIT in Lep^{ob}/Lep^{ob} mice, with VO_2 increasing 10–30% in response to the HF chow (TABLE 1). This intriguing result suggests that DIT might not be entirely dependent on the stimulation of BAT and UCP1 activity by the sympathetic nervous system. BAT of Lep^{ob}/Lep^{ob} mice is highly involuted, and since Lep^{ob}/Lep^{ob} mice exhibit cold sensitivity, it is presumably unable to rapidly respond to altered sympathetic outflow. We are currently investigating the mechanism by which Lep^{ob}/Lep^{ob} mice stimulate thermogenesis in response to HF feeding.

REGULATION OF ACTIVITY BY MELANOCORTINS

In studies using human volunteers, Leibel et al.[15] reported that an increase in adiposity associated with overfeeding led to a 16% increase in total 24-h energy expenditure. In a more recent analysis, Levine et al.[18] also observed an increase in resting metabolic rate associated with overfeeding. However, in the latter study, the amount of weight gain was shown to significantly correlate with differences in non-exercise activity thermogenesis (NEAT), while no correlation between the increase in resting-metabolic rate and weight gain was reported. To examine whether failure to increase activity-based energy expenditure is involved in the increased feed efficiency observed in MC4RKO mice, we housed young (8–9 week), comparatively lean, MC4RKO mice and WT littermates in wheel cages.[9] The wheel-running activities of the MC4RKO mice and WT controls are similar when the mice are allowed *ad libitum* access to the LF chow. However, when exposed to the HF diet, the MC4RKO mice become hypoactive due to failure to increase wheel-running activity. The MC4R might therefore protect against obesity not only by stimulating metabolic rate (DIT), but also by increasing the disposal of excess calories through activity. We have yet to examine whether the MC3R is required for the increase in activity associated with HF diets. However, it is interesting to note that two groups using different methods have reported evidence of reduced activity of the MC3RKO mice.[3,4]

CONCLUSIONS

The melanocortin system regulates energy homeostasis by affecting energy intake and energy expenditure. The rapid metabolic response that is observed in WT and Lep^{ob}/Lep^{ob} mice to the introduction of the HF chow suggests that POMC neurons, either in the ARC or in the cNTS, are rapidly receiving and responding to signals from the periphery pertaining to caloric consumption and possibly dietary fat content. The acute metabolic, behavioral, and anorexigenic effects of POMC neuronal activation in response to peripheral signals appear to predominantly involve activation of the MC4R. The MC3R was recently reported to have an autoreceptor role on POMC neurons in the ARC[19]; thus, a lean phenotype would be expected in mice lacking a functional MC3R. Although MC3RKO mice are indeed hypophagic, they exhibit an obesity phenotype possibly due to hypoactivity or altered nutrient partitioning. The additive effect of knocking out both the MC3R and the MC4R indicates that the two receptors regulate independent and nonredundant pathways.[3] The further characterization and analysis of both melanocortin receptor knockout models will undoubtedly reveal further information about the function of these receptors in regulating energy homeostasis.

REFERENCES

1. HUSZAR, D. et al. 1997. Targeted disruption of the melanocortin-4 receptor results in obesity in mice. Cell **88:** 131–141.
2. CHEN, A.S. et al. 2000. Role of the melanocortin-4 receptor in metabolic rate and food intake in mice. Transgenic Res. **9:** 145–154.

3. CHEN, A.S. *et al.* 2000. Inactivation of the mouse melanocortin-3 receptor results in increased fat mass and reduced lean body mass. Nat. Genet. **26:** 97–102.
4. BUTLER, A.A. *et al.* 2000. A unique metabolic syndrome causes obesity in the melano-cortin-3 receptor-deficient mouse. Endocrinology **141:** 3518–3521.
5. YASWEN, L. *et al.* 1999. Obesity in the mouse model of pro-opiomelanocortin deficiency responds to peripheral melanocortin. Nat. Med. **5:** 1066–1070.
6. FAN, W. *et al.* 1997. Role of melanocortinergic neurons in feeding and the agouti obesity syndrome. Nature **385:** 165–168.
7. MARSH, D.J. *et al.* 1999. Response of melanocortin-4 receptor-deficient mice to anorectic and orexigenic peptides. Nat. Genet. **21:** 119–122.
8. STE MARIE, L. *et al.* 2000. A metabolic defect promotes obesity in mice lacking melanocortin-4 receptors. Proc. Natl. Acad. Sci. USA **97:** 12339–12344.
9. BUTLER, A.A. *et al.* 2001. Melanocortin-4 receptor is required for acute homeostatic responses to increased dietary fat. Nat. Neurosci. **4:** 605–611.
10. COWLEY, M.A. *et al.* 1999. Integration of NPY, AGRP, and melanocortin signals in the hypothalamic paraventricular nucleus: evidence of a cellular basis for the adipostat. Neuron **24:** 155–163.
11. LOWELL, B.B. & B.M. SPIEGELMAN. 2000. Towards a molecular understanding of adaptive thermogenesis. Nature **404:** 652–660.
12. BACHMAN, E.S. *et al.* 2002. BetaAR signaling required for diet-induced thermogenesis and obesity resistance. Science **297:** 843–845.
13. ENERBACK, S. *et al.* 1997. Mice lacking mitochondrial uncoupling protein are cold-sensitive but not obese. Nature **387:** 90–94.
14. KOZAK, L.P. & M.E. HARPER. 2000. Mitochondrial uncoupling proteins in energy expenditure. Annu. Rev. Nutr. **20:** 339–363.
15. LEIBEL, R.L., M. ROSENBAUM & J. HIRSCH. 1995. Changes in energy expenditure resulting from altered body weight. N. Engl. J. Med. **332:** 621–628.
16. ROTHWELL, N.J. & M.J. STOCK. 1979. A role for brown adipose tissue in diet-induced thermogenesis. Nature **281:** 31–35.
17. STOCK, M.J. 1999. Gluttony and thermogenesis revisited. Int. J. Obes. Relat. Metab. Disord. **23:** 1105–1117.
18. LEVINE, J.A., N.L. EBERHARDT & M.D. JENSEN. 1999. Role of nonexercise activity thermogenesis in resistance to fat gain in humans. Science **283:** 212–214.
19. COWLEY, M.A. *et al.* 2001. Leptin activates anorexigenic POMC neurons through a neural network in the arcuate nucleus. Nature **411:** 480–484.

The Central Melanocortin System and Fever

JEFFREY B. TATRO[a,b] AND PARTHA S. SINHA[b]

[a]Division of Endocrinology, Diabetes, Metabolism, and Molecular Medicine, Department of Medicine and the Tupper Research Institute, Tufts University School of Medicine and Tufts-New England Medical Center, Boston, Massachusetts 02111, USA

[b]Department of Pharmacology and Experimental Therapeutics, Tufts University School of Medicine, Boston, Massachusetts 02111, USA

ABSTRACT: Fever is a phylogenetically ancient response that is mounted upon exposure of the host to pathogens or inflammatory agents. Melanocortin agonists act centrally to inhibit fever by acting at receptors, including the melanocortin-4 receptor, which is prominently expressed in key hypothalamic thermoregulatory centers. Furthermore, endogenous melanocortins act centrally as physiological modulators of fever, recruited during the febrile response to restrain its intensity. Functionally, these actions lie at the interface between the anti-inflammatory effects of melanocortins, which involve suppression of the synthesis and actions of proinflammatory cytokines, and the central control of thermoregulation. Considering the extensive neuroanatomic and functional overlaps between central pathways and peripheral effectors involved in thermoregulation and energy balance, it is not surprising that melanocortins have been found to influence the metabolic economy profoundly in pathological as well as normal states. For example, despite suppressing endotoxin-induced fever, endogenous melanocortins appear to mediate the associated anorexia, a classic component of the "illness syndrome" accompanying acute infections, and promote a negative energy balance. The thermoregulatory actions of melanocortins are in several respects functionally opposed, and are remarkably dependent on physiological state, indicating that responsiveness to melanocortins is a physiologically modulated variable. Elucidating the anti-inflammatory and thermoregulatory roles of central melanocortin receptors during inflammatory states may lead to novel pharmacotherapeutic targets based on selective targeting of melanocortin receptor subtypes, for clinical benefit in human disease states involving neuroinflammatory components and metabolic wasting.

KEYWORDS: antipyretic; anorexia; cachexia; hypothalamus; melanocortin receptor; thermoregulation; thermoeffector

INTRODUCTION

The central nervous system (CNS) is charged with maintaining the internal milieu by organizing the physiological responses of the body's organ systems to all manner

Address for correspondence: Dr. Jeffrey B. Tatro, Departments of Medicine and Pharmacology, Tufts-New England Medical Center, 750 Washington Street, Boston, MA 02111. Voice: 617-636-5690; fax: 617-636-2845.
jtatro@tufts-nemc.org

Ann. N.Y. Acad. Sci. 994: 246–257 (2003). © 2003 New York Academy of Sciences.

of internal metabolic demands and external stressors. Introduction of a pathogenic challenge such as infection or other proinflammatory stimulus alters the physiological needs of the host such that containing and eliminating the pathogen become important priorities. Mounting an effective host defense response entails resetting of internal environmental conditions to a state more hostile to the offending agent, a requirement that must be balanced with the need to maintain an internal milieu compatible with life in the host. The prototypical, and most familiar, host response to infection is fever, which is defined as an increase in the temperature about which body temperature is regulated (often referred to as body temperature set-point). The resulting elevation in body temperature is believed to have adaptive value, by interacting with other aspects of the host response to discourage micro-organismal growth and metabolism, and by optimizing immune responses.[1,2] Fever is only one of several physiological, biochemical, and behavioral adjustments in the stereotyped response to infection, collectively known as the acute phase response. Other components of the acute phase response include anorexia, fatigue, hyperalgesia, and a range of immunologic and hematologic changes.[3]

Studies first reported over 20 years ago demonstrated that the administration of α-melanocyte-stimulating hormone (α-MSH) or adrenocorticotropin (ACTH) can inhibit experimental fever.[4] Recent research has begun to elucidate the mechanisms involved and the physiological relevance of these antipyretic effects of melanocortins, and has also revealed important influences of melanocortins on other components of the acute phase response. The impetus for this research is not a need for new therapeutic antipyretic drugs, since effective, safe, and inexpensive antipyretic agents have been available for over a century. Rather, it is of fundamental interest to determine the intrinsic CNS pathways deployed for regulation and suppression of fever. Moreover, such studies provide insight into the mechanisms involved in at least two critical related problems: the altered metabolic economy of inflammatory states, and the central control of proinflammatory responses. These insights have already provided evidence for novel potential therapeutic opportunities. These subjects are the focus of this review.

A few technical considerations warrant brief mention (for more detailed reviews, see Kluger[1] or Tatro[5]). For example, it is worth noting the distinction between the terms *antipyretic*, which means fever-inhibiting, and *cryogenic*, which refers to an agent that lowers body temperature irrespective of the presence or absence of fever. Thermo-regulatory actions of neuropeptides and other drugs are commonly found to be highly dependent on physiological state and on other factors including dose and site of administration, ambient temperature, etc. Therefore, a given agent may act centrally, for example, to elevate body temperature in one situation, and as an antipyretic agent in others. The study of fever can also be confounded by factors such as anesthesia and by psychological stress (for example, handling, novel environment), which itself can cause fever and activate secretion of fever-altering stress hormones. Also, the character of the febrile response is highly dependent on the nature, dose, and route of exposure of the pyrogen. Therefore, the thermoregulatory effects of a given agent are highly context dependent, must be determined under precisely determined internal and environmental conditions, and should be defined and interpreted in those terms.

MECHANISMS OF ANTIPYRETIC ACTION OF MELANOCORTINS

Conventional antipyretics (aspirin, nonsteroidal anti-inflammatory drugs) act as direct inhibitors of cyclooxygenases,[6] the rate-limiting enzymes for prostanoid synthesis. It is not presumed that melanocortins act via similar mechanisms, and cyclooxygenase inhibitors do not appear to interact with central melanocortin receptors (MCRs),[7] implying that melanocortins inhibit fever by other mechanisms. The mechanisms involved in the antipyretic actions of melanocortins can be organized into three general categories: MCR subtypes, neural and humoral pathways, and thermoeffector systems.

MCR Subtypes

Studies aimed at determining the MCR subtypes involved in mediating the antipyretic effects of melanocortins have focused on the potential roles of the melanocortin-3 receptor (MC3R) and the melanocortin-4 receptor (MC4R), which are prominently distributed in autonomic sites in the hypothalamus and brain stem.[5,8] In rats, injection of α-MSH, a nonselective MCR agonist, either centrally (intracerebroventricularly, i.c.v.) or peripherally (intraperitoneally, i.p.) inhibited the fever resulting from i.p. injection of a moderately pyrogenic dose of bacterial endotoxin (*Escherichia coli* lipopolysaccharide, LPS). The minimal effective i.c.v. antipyretic dose was about 100-fold lower than the minimal effective i.p. dose, strongly suggesting a central site of action.[9,10] Intracerebroventricular coadministration of the MC3R/MC4R antagonist SHU9119 at a roughly equimolar dose blocked the antipyretic effect of exogenous α-MSH for at least 3 h, indicating a role of the MC3R and/or MC4R in mediating the antipyretic effect.[9] Furthermore, similar i.c.v. treatment with SHU9119 blocked the antipyretic effect of peripherally injected α-MSH in the same model, suggesting that peripherally injected α-MSH inhibits fever by acting at central MC3R/MC4R.[10]

Recent studies have extended these findings and indicated that selective activation of the MC4R inhibits fever in rats. Intracerebroventricular injection of a MC4R-selective agonist (MRLOB-0001) suppressed LPS-induced fever, and the antipyretic effect of the MC4R agonist was blocked by co-administration of a selective MC4R antagonist (HS014).[10A] The same dose of MC4R agonist given intravenously (i.v.) failed to suppress fever, indicating that the antipyretic effects of i.c.v. agonist treatment were mediated by MC4R within the CNS. In afebrile rats, the MC4R agonist did not lower body temperature but actually elevated it significantly. This clearly indicated that the MC4R agonist's effects in febrile animals were antipyretic rather than cryogenic, and further demonstrated the physiological state-dependent nature of central thermoregulatory responses to melanocortins. Therefore, the evidence suggests a role of the central MC4R in mediating the long-recognized antipyretic effects of centrally administered melanocortins.

Neural/Humoral Pathways

Available information concerning the neuroanatomic sites at which the antipyretic actions of melanocortins are exerted is limited. Direct injections of melanocortins into the preoptic and septal regions inhibited fever in rabbits.[11,12] These areas contain a high proportion of thermosensitive neurons, and are thought to be regions wherein critical thermoregulatory neurons reside and afferent thermosensory information is integrated.[13] The former studies[11,12] were performed at doses and injection volumes that did not provide a high degree of anatomic resolution, but the results overall suggested that neurons within the preoptic region are potential targets of the antipyretic actions of melanocortins. Consistent with this possibility, mRNAs encoding the MC4R and MC3R are prominently distributed in these regions, as are specific melanocortin binding sites, reflecting a high local density of functional MCR proteins.[8,14,15] However, the detailed localization of MC3R and MC4R proteins has yet to be determined.

The neural or humoral factors downstream of MCR signaling that mediate the antipyretic effects of melanocortins are unknown. Adrenal corticosteroids have antipyretic properties, but the antipyretic actions of melanocortins do not appear to be mediated via stimulation of adrenal glucocorticoid release, because the antipyretic effect of exogenous ACTH was preserved in adrenalectomized rabbits,[4] and because systemically administered α-MSH suppressed LPS-induced stimulation of corticosterone secretion while inhibiting the accompanying fever in rats.[10] Melanocortins are capable of suppressing the synthesis of several proinflammatory cytokines thought to act as endogenous pyrogens, including interleukin-1β, tumor necrosis factor-α, and IL-6, both peripherally and within the CNS,[16–18] but the contribution of such actions to their antipyretic effects are unknown. Systemically injected α-MSH suppressed LPS-induced elevation of circulating IL-6 levels in rats, an effect that could potentially contribute to its antipyretic effect. However, the α-MSH–induced suppression of IL-6 levels was unaffected by central administration of SHU9119, which concomitantly blocked the antipyretic effect of α-MSH, indicating that suppression of IL-6 release was not the mechanism of antipyretic action of systemically injected α-MSH.[10] The question of whether melanocortins inhibit fever primarily by suppressing pyrogenic cytokine release has been complicated by the extensive redundancy of proinflammatory cytokines in mediating fever, a feature that has also obscured the relative contributions of individual cytokines in various models of fever.[19,20]

Thermoeffectors

Several thermoeffector systems are recruited to raise core body temperature during the mounting of the febrile response, operating via increased physiological and behavioral heat conservation, and increased thermogenesis.[21] Because MCRs are localized in many autonomic centers in the CNS, and melanocortins may presumably influence thermoeffector function during fever by acting at such sites, as well as at more proximal steps in the pyrogenic signaling cascade, it is of interest to determine the effector systems involved in the antipyretic effects of melanocortins. Exogenous melanocortins act, probably at the central level, to suppress heat-loss thermoeffector responses during fever, thus inhibiting fever-associated vasoconstriction of the ear in rabbits[4] and of the tail in rats.[10A] This pathway probably contributes significantly to the febrile response,[22] but its inhibition by α-MSH cannot by itself account for the antipyretic effect of α-MSH, because a dose of α-MSH that was not sufficient to inhibit fever did prevent LPS-induced vasoconstriction in the tail, to an extent comparable to that seen with an antipyretic dose of α-MSH (Sinha et al., submitted for publication).

Melanocortins influence fever-associated suppression of spontaneous locomotor activity in some, but not all, circumstances, effects which could potentially influence fever via changes in activity-related skeletal muscle thermogenesis. For example, centrally administered antipyretic doses of α-MSH or of a selective MC4R agonist did not alter the extent of suppression of spontaneous motor activity in rats treated with a moderately pyrogenic dose of LPS (30 μg/kg)[10A] (Sinha et al., submitted for publication), but α-MSH markedly potentiated the LPS-induced behavioral suppression in rats that had been fasted overnight and received a higher dose of LPS (100 μg/kg).[23] Also, MC4R-deficient mice failed to exhibit the LPS-induced suppression of wheel-running activity seen in wild-type mice, suggesting a possible contribution of endogenous melanocortins, acting via the MC4R, to behavioral suppression by LPS in mice.[24] This influence of MC4R activation would tend to decrease LPS-induced thermogenesis, consistent with an antipyretic influence of the MC4R. However, overall febrile responses were reportedly unaffected by MC4R deficiency, and MC4R-deficient mice failed to exhibit the LPS-induced increase in metabolic rate seen in wild-type mice, despite their relatively greater locomotor activity as compared with wild-type mice.[25] Together, the latter findings indicate that metabolic thermogenesis is increased following LPS treatment to an extent that more than offsets the reduction in energy expenditure attributable to suppression of locomotor activity, and suggest a role of the MC4R in mediating these energetically opposing actions. In contrast, LPS-induced and IL-1β-induced suppression of running activity were unaffected by MC3R deficiency.[25] On the other hand, it is unknown whether the antipyretic actions of exogenous melanocortins include inhibition of metabolic pathways of thermogenesis. This is a question of some interest, since in afebrile animals, endogenous and exogenous melanocortins are thought to stimulate metabolic rate, in part by activating thermogenic effectors including uncoupling protein-1 in brown fat, potentially as mediators of leptin action.[26–30]

CONCEPT OF ENDOGENOUS ANTIPYRETIC SYSTEMS

An empirical upper limit of fever has long been recognized, and this helped lead to the concept that there are intrinsic systems in place that restrain the extent of fever.[1,31] It is not entirely clear whether febrile elevations in body temperature *per se* can be lethal, but it is unquestioned that the potent array of biological weapons employed by the host in attacking invading organisms, that is, the various components of the acute phase response, can be destructive to the host. Proinflammatory cytokines (for example, IL1-β, TNF-α) are thought to play major roles in mediating many of these effects, ranging from direct biochemical and cytotoxic actions to complex behavioral programs. Consequently, from a teleologic perspective, it is implicit that the existence of cytokine-counterregulatory pathways to mitigate the potentially destructive effects of unchecked proinflammatory cytokine actions may confer adaptive value to the host. It is now fairly well established that several endogenous antipyretic systems exist, and that these systems operate via both humoral and CNS neural pathways. For example, adrenal glucocorticoids, secretion of which is increased during fever and inflammatory states, and which are well known to exert anti-inflammatory and immunosuppressive actions in the periphery, also act centrally to inhibit fever.[32] The evidence concerning putative endogenous antipyretic systems has recently been reviewed.[5,32]

IS THERE A PHYSIOLOGICAL ANTIPYRETIC ROLE OF ENDOGENOUS MELANOCORTINS?

Studies aimed at blocking endogenous melanocortin action acutely during fever have provided evidence that endogenous melanocortins are antipyretic. Central infusions of anti-α-MSH antisera increased and prolonged experimental fevers in rabbits,[33] and central injections of the MC3R/MC4R antagonist SHU9119 increased LPS-induced fevers in rats.[9,23] The same dose of SHU9119 failed to alter fever when injected intravenously, indicating that its effects resulted from blockade of melanocortin action at MC4R and/or MC3R within the CNS.[9] These findings indicated that during inflammatory states, endogenous melanocortins, probably of neural origin, activate central MCRs to exert a counterregulatory influence on febrile responses.

Although the MC4R appears to mediate the antipyretic action of exogenous melanocortins[10A] (Sinha et al., submitted for publication), the specific roles of the MC3R and MC4R in mediating antipyretic actions of endogenous melanocortins are not known, and recent investigations of this question have introduced some seemingly conflicting results. Fevers induced by LPS[25] or by centrally or peripherally administered IL-1β were unaltered in MC4R-deficient mice vs. those in wild-type controls[34] (Huang, Sinha, and Tatro, unpublished results). LPS-induced fevers were also reportedly unaltered in MC3R-deficient mice.[25] Although these results could be taken to suggest that endogenous melanocortins may not act via MC4R and/or MC3R to modulate febrile responses in this species, the negative findings could potentially be explained by long-term compensatory or metabolic changes in the complete absence of MC4R/MC3R signaling. Alternatively, the lack of effects of MC4R deficiency on fever may be due to the fact that the MC4R exerts functionally opposing influences. Specifically, the presence of a functional MC4R was found to be essential to both LPS-induced suppression of musculoskeletal activity, and to its stimulation of metabolic thermogenesis.[24,25] These opposing influences could conceivably have contributed to the lack of net antipyretic effects of MC4R activation in the wild-type mice in those studies. In rats, central administration of a selective MC4R antagonist[10A] or a MC3R/MC4R antagonist[35] did not alter fevers induced by a mild pyrogenic dose of LPS or centrally administered IL-1β, respectively. Nevertheless, due to the fairly limited conditions tested in the latter studies, determining the potential roles and significance of endogenous melanocortins acting via the central MC3R or MC4R to modulate fever and inflammation will require further testing, using subtype-selective inhibition of MCR signaling during a wider range of proinflammatory states.

Studies using additional MCR subtype-selective antagonists, and transgenic approaches that permit conditional induction of selective MCR subtype deficiency, can be expected to shed more light on the regulatory roles of endogenous melanocortins acting via specific MCR subtypes during fever.

What physiological purpose may be served by endogenous antipyretic systems, including the central melanocortin system, remains a matter of speculation. It has been widely postulated that such systems may protect against potentially catastrophic consequences of excessive fever, but this plausible notion remains to be tested. These systems may also be involved in the fine regulation of fever. Supporting this hypothesis are the facts that during fever, blockade of endogenous melanocortin or glucocorticoid actions can increase the magnitude or duration of fevers of even moderate intensity, indicating that these systems operate even under conditions that are not life-threatening. Accordingly, the febrile state at any given moment may reflect the net result of opposing pyrogenic influences (for example, proinflammatory cytokines) and those of antipyretic systems.[5,35A] The question of whether such physiological modulation of fever by endogenous melanocortins confers any adaptive value to the host, particularly during moderate fever, remains to be addressed.

MELANOCORTIN ACTIONS IN INFLAMMATORY ANOREXIA AND CACHEXIA

From a seat of origin deep in the visceral brain, which it appears to have occupied for at least several hundred million years based on comparative studies,[36] the central melanocortin neuron system is positioned to influence an impressive range of physiological functions via neuroendocrine and descending autonomic pathways. Its influences on normal physiological control of feeding behavior and energy balance are of profound significance, as demonstrated by the consequences of disrupted central melanocortin signaling, which lead to obesity in several models.[37,38]

Interest in a potential role of melanocortins in the anorexia of inflammation and infection arose due to the profound influence of melanocortins both on proinflammatory processes and on normal energy balance.[23] On the one hand, endogenous melanocortins appear to exert a suppressive influence on fever and other central actions of proinflammatory cytokines, which would be expected to lead to suppression of the associated anorexic state. It was shown earlier that LPS-induced fever and anorexia are dissociable responses, because fever could be inhibited in rats without suppression of the associated anorexia.[39] Nevertheless, melanocortins are pleiotropic inhibitors[17] of the central actions and gene expression of cytokines that are putative mediators of inflammatory anorexia, including IL-1, TNF-α, and IL-6.[40] On this basis, one would predict a potentially suppressive influence of melanocortins on inflammation-associated anorexia. Consistent with this, one report found that the anorexic effect of centrally injected IL-1β in fasted rats was inhibited by co-administered α-MSH.[41] On the other hand, it is well established that in normal, noninflammatory states, melanocortins act centrally to suppress food intake and promote energy expenditure, which, paradoxically, would tend to promote inflammatory anorexia. To test these alternative possibilities, rats were fasted to stimulate appetite, then treated with a pyrogenic and anorexic systemic dose of LPS. Under these conditions, an antipyretic centrally administered dose of α-MSH (300 ng) potentiated the anorexic effect of LPS, even while the febrile response was suppressed. Furthermore, the sensitivity of the rats to the anorexic effect of α-MSH was markedly increased, since a 10-fold lower dose of α-MSH (30 ng), which was without effect on food intake in normal rats and was subthreshold for antipyretic effects, produced comparable suppression of the anorexic effect of LPS.[23] The results clearly indicated that the inflammatory state promotes a high degree of responsiveness to the anorexic action of exogenous melanocortins, which potentiate inflammatory anorexia despite concurrently inhibiting fever. The mechanisms underlying this apparent heightened responsiveness to melanocortin actions are unknown.

Further studies indicated that endogenous melanocortins are involved in mediating LPS-induced anorexia. The LPS-induced suppression of food intake was reversed by blockade of endogenous melanocortin action by central administration of the MC3R/MC4R antagonist SHU9119 during the 24-h period following treatments, strongly suggesting that endogenous melanocortins may contribute to LPS-induced anorexia.[23] The increase in feeding with central MCR blockade occurred despite the fact that fever was exacerbated by SHU9119 treatment, owing to blockade of the antipyretic action of centrally acting endogenous melanocortins, further demonstrating that inflammatory anorexia and fever are independently regulated CNS functions. Anorexia induced by central infusion of IL-1β was also inhibited by SHU9119 co-administration in rats, even under conditions wherein the associated fever was unaffected.[35] Similarly, in mice, central injection of another MC3R/MC4R antagonist, agouti-related peptide (AgRP)(84-132), prevented LPS-induced anorexia and body weight loss. Furthermore, MC4R-deficient mice were resistant to LPS-induced anorexia and body weight loss, strongly implicating the MC4R in mediating inflammatory anorexia and associated metabolic changes.[24] As noted above, MC4R-deficient mice also failed to exhibit another aspect of LPS-induced illness behavior, the near-complete suppression of wheel-running activity, suggesting that endogenous melanocortins acting via MC4R may be involved in mediating LPS-induced behavioral suppression.[24] In contrast with their mediating role in inflammatory anorexia, this suppressive influence of endogenous melanocortins on motor activity would tend to reduce metabolic energy expenditure during the inflammatory response. On the other hand, the MC3R seems to have a role functionally opposed to that of the MC4R. Unlike MC4R-deficient mice, MC3R-deficient mice were responsive to LPS-induced anorexia and weight loss. Further, Marks and colleagues hypothesized that activation of the MC3R, which is expressed in proopiomelanocortin (POMC)-synthesizing neurons of the arcuate nucleus, may have an inhibitory influence on inflammatory anorexia and cachexia[25] by virtue of its postulated autoinhibitory influence on POMC gene expression.[42] Consistent with this prediction, MC3R-deficient mice exhibited exaggerated anorexic responses and loss of body weight in response to central administration of IL-1β, and also in response to systemic LPS administration.[25] These findings suggest that activation of the MC3R by endogenous melanocortins exerts a suppressive influence on inflammatory anorexia, by mechanisms that are as yet unclear.

An important extension of these findings was that in addition to inflammatory anorexia, the central melanocortin system appears to play a critical role in cancer cachexia, as shown in rodent models of tumor-induced cachexia. Cachexia is a syndrome of metabolic derangements that involves both decreased food intake and increased catabolic processes, leading to wasting of lean body mass. Cachexia is notably associated with chronic diseases such as cancer, rheumatoid arthritis, and AIDS, and commonly accompanies aging in some elderly patients. Central MC3R/MC4R blockade by AgRP infusion prevented the wasting resulting from tumor growth in rats and in mice. Furthermore, tumor-induced anorexia and body weight loss were lacking in MC4R-deficient mice, strongly implicating the MC4R in mediating these effects.[24] Such a role contrasts dramatically with the normal physiological regulatory influence of endogenous melanocortins on energy balance, which is to restrain food intake and promote energy expenditure during periods of relative abundance of metabolic fuels. In normal animals, this influence of melanocortins is suppressed during periods of nutritional insufficiency (starvation) via decreased melanocortinergic tone. Thus, in animals bearing carcinomas, an inappropriate maintenance of melanocortinergic activity appears to lead to a syndrome of excessive negative energy balance.[24,43] The mechanisms involved in this phenomenon (which could presumably involve inappropriate maintenance of POMC synthesis and melanocortin neurosecretion, changes in MCR expression or sensitivity, alterations in activity of neurons expressing the endogenous MC3R/MC4R antagonist, AgRP, or changes in other neuropeptide systems that act as functional melanocortin antagonists)[43] remain to be determined.

SUMMARY AND CONCLUSIONS

Because it is organized to influence proinflammatory cytokine signaling as well as the central command of thermoregulation, energy balance, and related neuroendocrine pathways, the central melanocortin neuron system is positioned to modulate the febrile response and associated proinflammatory and metabolic responses. The antipyretic action of the central melanocortin system has been proposed as both a physiological mechanism for protecting the host against catastrophic consequences of excessive fever, and a dynamic and intrinsic regulatory component of the febrile response. Activation of the MC4R by exogenous agonists is sufficient to inhibit fever,[10A] but it is unknown whether the MC3R may participate in such pharmacological regulation of fever.

The precise roles of centrally acting endogenous melanocortins in the regulation of fever are not entirely clear. Acute MC3R/MC4R blockade or immunoneutralization of endogenous α-MSH were found to increase fever, suggesting an antipyretic role of endogenous melanocortins acting via MC3R or MC4R.[9,23,33] On the other hand, chronic MC3R or MC4R deficiency or a centrally injected MC3R/MC4R antagonist reportedly did not alter febrile responses.[25,34,35] Although the basis of these seemingly contrasting results is not certain, they are probably attributable in part to technical factors such as differences in species and genetic backgrounds, or differences in the nature, route, and dose of pyrogen and MCR ligand administration. Alternatively, they may also involve more complex physiological interactions. For example, these could involve the differential effects of acute pharmacological disruption of MCR signaling vs. chronic MCR deficiency, which may be complicated by long-term metabolic deficiencies or compensatory responses. In addition, the negative results of interfering with MCR signaling in certain cases could potentially be due to differential activation of functionally opposed MCR actions (for example, motor suppression and increased metabolic thermogenesis[24,25]) under different conditions. Consequently, further studies will be required to elucidate the nature and physiological significance of the regulation of fever by the central melanocortin system. It is likely that acute interventions, using highly selective MCR ligands and techniques for anatomic site- and cell-type-selective acute modulation of melanocortin and MCR gene expression, will be crucial to resolving the complex roles of this system *in vivo*.

The central actions of melanocortins in proinflammatory states include a number of rather dramatic examples of functional opposition. For example, in contrast with its antipyretic and other cytokine-counterregulatory central actions, the central melanocortin system appears to act via the MC4R to mediate or exacerbate inflammatory and tumor-induced anorexia, a pathophysiological role that appears to be highly maladaptive in contributing to cancer cachexia. Second, melanocortin-induced inhibition of fever,[5] and the apparent contribution of this system to LPS-induced suppression of general locomotor activity,[24] would favor reduced energy expenditure. In contrast, the fact that melanocortins mediate or exacerbate LPS-induced anorexia and wasting[23,24] reflect a predominant promotion of negative energy balance. Third, the apparent restraining effect of the MC3R on inflammatory anorexia contrasts directly with the promoting influence of the MC4R in this phenomenon.[24,25] These seemingly discordant roles of central MCR subtypes imply a highly selective functional organization of subpopulations of melanocortinergic neurons and MCR-expressing target structures in the CNS. Another critical yet largely unexplored phenomenon is the mechanism by which physiological state determines responsiveness to thermoregulatory actions of melanocortins; for example, central infusion of an MC4R agonist elevates body temperature in afebrile rats, but lowers it during fever.[10A] Therefore, to understand the neural andmolecular bases of the complex interactions of these neuronal subpopulations and their targets, an important implication for researchers is the need to utilize approaches offering increasing neuroanatomic and cell-type specificity.

Elucidating the anti-inflammatory and thermoregulatory roles of the central melanocortin system during inflammatory states will enhance our understanding of the energetic economy of coordinated host defense. Importantly, this approach may lead to the development of novel pharmacotherapeutic strategies based on selective targeting of MCR subtypes, for clinical benefit in a number of human disease states involving neuroinflammatory components.

ACKNOWLEDGMENT

This work was supported by the National Institutes of Health (Grant MH44694, awarded to J.B.T).

REFERENCES

1. KLUGER, M.J. 1991. Fever: Role of pyrogens and cryogens. Physiol. Rev. **71:** 93–127.
2. KLUGER, M.J., W. KOZAK, C.A. CONN, et al. 1996. The adaptive value of fever. Infect. Dis. Clin. North Am. **10:** 1–20.
3. SAPER, C. & C. BREDER. 1994. The neurologic basis of fever. N. Engl. J. Med. **330:** 1880–1886.
4. LIPTON, J.M., J.R. GLYN & J.A. ZIMMER. 1981. ACTH and alpha-melanotropin in central temperature control. Fed. Proc. **40:** 2760–2764.
5. TATRO, J.B. 2000. Endogenous antipyretics. Clin. Infect. Dis. **31:** S190–S201.
6. VANE, J.R. & R.M. BOTTING. 1998. Mechanism of action of nonsteroidal anti-inflammatory drugs. Am. J. Med. **104:** 2S–8S.
7. TATRO, J.B. & M.L. ENTWISTLE. 1994. Heterogeneity of brain melanocortin receptors suggested by differential ligand binding in situ. Brain Res. **635:** 148–158.

8. TATRO, J.B. 2000. Melanocortin receptor expression and function in the nervous system. *In* The Melanocortin Receptors. R.D. Cone, Ed.: 173–207. Humana Press. Totowa, NJ.

9. HUANG, Q.-H., M.L. ENTWISTLE, J.D. ALVARO, *et al*. 1997. Antipyretic role of endogenous melanocortins mediated by central melanocortin receptors during endotoxin-induced fever. J. Neurosci. **17**: 3343–3351.

10. HUANG, Q.-H., V.J. HRUBY & J.B. TATRO. 1998. Systemic α-MSH suppresses LPS fever via central melanocortin receptors independently of its suppression of corticosterone and IL-6 release. Am. J. Physiol. Regul. Integr. Comp. Physiol. **275**: R524–R530.

10A. SINHA, P.S., H.B. SCHIÖTH & J.B. TATRO. 2003. Activation of central melanocortin-4 receptor suppresses lipopolysaccharide-induced fever in rats. Am. J. Physiol. Regul. Integr. Comp. Physiol. **284**: R1595–R1603.

11. GLYN-BALLINGER, J.R., G.L. BERNARDINI & J.M. LIPTON. 1983. α-MSH injected into the septal region reduces fever in rabbits. Peptides **4**: 199–203.

12. FENG, J.D., T. DAO & J.M. LIPTON. 1987. Effects of preoptic microinjections of alpha-MSH on fever and normal temperature control in rabbits. Brain Res. **18**: 473–477.

13. BOULANT, J.A. 2000. Role of the preoptic-anterior hypothalamus in thermoregulation and fever. Clin. Infect. Dis. 31: S157–S161.

14. KISHI, T., C.J. ASCHKENASI, C.E. LEE, *et al*. 2003. Expression of melanocortin 4 receptor mRNA in the central nervous system of the rat. J. Comp. Neurol. **457**: 213–235.

15. TATRO, J.B., M.L. ENTWISTLE, V. FAIRCHILD-HUNTRESS, *et al*. 2001. MC4-R is the principal melanocortin receptor subtype of the preoptic area and lateral hypothalamus. Soc. Neurosci. Abstr. **27**: 310.2.

16. CATANIA, A. & J.M. LIPTON. 1998. Peptide modulation of fever and inflammation within the brain. Ann. N.Y. Acad. Sci. **856**: 62–68.

17. LIPTON, J.M., A. CATANIA & R. DELGADO. 1998. Peptide modulation of inflammatory processes within the brain. Neuroimmunomodulation **5**: 178–183.

18. HUANG, Q. & J.B. TATRO. 2002. α-MSH suppresses intracerebral TNF-α and IL-1β gene expression following transient cerebral ischemia in mice. Neurosci. Lett. **334**: 186–190.

19. KLUGER, M.J., W. KOZAK, L.R. LEON, *et al*. 1998. The use of knockout mice to understand the role of cytokines in fever. Clin. Exp. Pharmacol. Physiol. **25**: 141–144.

20. NETEA, M., B.J. KULLBERG & J.W.M. VAN DER MEER. 2000. Circulating cytokines as mediators of fever. Clin. Infect. Dis. **31**: S178–S184.

21. BOULANT, J.A. 1997. Thermoregulation. *In* Fever: Basic Mechanisms and Management. P.A. Mackowiak, Ed.: 35–38. Lipincott-Raven. New York.

22. GORDON, C.J. 1993. Temperature Regulation in Laboratory Rodents. Cambridge University Press. New York.

23. HUANG, Q.H., V.J. HRUBY & J.B. TATRO. 1999. Role of central melanocortins in endotoxin-induced anorexia. Am. J. Physiol. Regul. Integr. Comp. Physiol. **276**: R864–R871.

24. MARKS, D.L., N. LING & R.D. CONE. 2001. Role of the central melanocortin system in cachexia. Cancer Res. **61**: 1432–1438.

25. MARKS, D.L., A.A. BUTLER, R. TURNER, *et al*. 2003. Differential role of melanocortin receptor subtypes in cachexia. Endocrinology **144**: 1513–1523.

26. SATOH, N., Y. OGAWA, G. KATSUURA, *et al*. 1998. Satiety effect and sympathetic activation of leptin are mediated by hypothalamic melanocortin system. Neurosci. Lett. **249**: 107–110.

27. HAYNES, W.G., D.A. MORGAN, A. DJALALI, *et al*. 1999. Interactions between the melanocortin system and leptin in the control of sympathetic nerve traffic. Hypertension **33**: 542–547.

28. COWLEY, M.A., N. PRONCHUK, W. FAN, *et al*. 1999. Integration of NPY, AgRP, and melanocortin signals in the hypothalamic paraventricular nucleus: Evidence of a cellular basis for the adipostat. Neuron **24**: 155–163.

29. CHEN, A.S., J.M. METZGER, M.E. TRUMBAUER, *et al*. 2000. Role of the melanocortin-4 receptor in metabolic rate and food intake in mice. Transgenic Res. **9**: 145–154.

30. STE. MARIE, L., G.I. MIURA, D.J. MARSH, *et al*. 2000. A metabolic defect promotes obesity in mice lacking melanocortin-4 receptors. Proc. Natl. Acad. Sci. USA **97**: 12339–12344.

31. MACKOWIAK, P.A. & J.A. BOULANT. 1997. Fever's upper limit. *In* Fever: Basic Mechanisms and Management. P.A. Mackowiak, Ed.: 147–163. Lippincott-Raven. New York.

32. KLUGER, M.J., W. KOZAK, L.R. LEON, *et al.* 1998. Fever and antipyresis. Prog. Brain Res. **115:** 465–475.

33. SHIH, S.T., O. KHORRAM, J.M. LIPTON, *et al.* 1986. Central administration of alpha-MSH antiserum augments fever in the rabbit. Am. J. Physiol. **250:** 803–806.

34. TATRO, J.B., D. HUSZAR, V. FAIRCHILD-HUNTRESS, *et al.* 1999. Role of the melanocortin-4 receptor in thermoregulatory responses to central IL1β. Soc. Neurosci. Abstr. **25:** 1558.

35. LAWRENCE, C.B. & N.J. ROTHWELL. 2001. Anorexic but not pyrogenic actions of interleukin-1 are modulated by central melanocortin-3/4 receptors in the rat. J. Neuroendocrinol. **13:** 490–495.

35A. SZÉKELY, M. & A.A. ROMANOVSKY. 1998. Pyretic and antipyretic signals within and without fever: a posible interplay. Med. Hypotheses **50:** 213–218.

36. VALLARINO, M., D. TRANCHAND BUNEL & H. VAUDRY. 1993. Location and identification of α-melanocyte-stimulating hormone in the brain of the lungfish, protopterus annectens. Ann. N.Y. Acad. Sci. **680:** 634–637.

37. MARKS, D.L. & R.D. CONE. 2001. Central melanocortins and the regulation of weight during acute and chronic disease. Rec. Prog. Horm. Res. **56:** 359–375.

38. BARSH, G.S. & M.W. SCHWARTZ. 2002. Genetic approaches to studying energy balance: Perception and integration. Nat. Genet. **3:** 589–600.

39. MCCARTHY, D.O., M.J. KLUGER & A.J. VANDER. 1984. The role of fever in appetite suppression after endotoxin administration. Am. J. Clin. Nutr. **40:** 310–316.

40. PLATA-SALAMÁN, C.R. 2000. Central mechanisms contributing to the cachexia-anorexia syndrome. Nutrition **16:** 1009–1012.

41. UEHARA, Y., H. SHIMIZU, N. SATO, *et al.* 1992. Carboxy-terminal tripeptide of alpha-melanocyte-stimulating hormone antagonizes interleukin-1 induced anorexia. Eur. J. Pharmacol. **220:** 119–122.

42. COWLEY, M.A., J.L. SMART, M. RUBINSTEIN, *et al.* 2001. Leptin activates anorexigenic POMC neurons through a neural network in the arcuate nucleus. Nature **411:** 480–484.

43. LECHAN, R.M. & J.B. TATRO. 2001. Hypothalamic melanocortin signaling in cachexia [editorial]. Endocrinology **142:** 3288–3291.

The Role of the Melanocortin-3 Receptor in Cachexia

DANIEL L. MARKS[a] AND ROGER D. CONE[b]

[a]Department of Pediatric Endocrinology and
[b]The Vollum Institute, Oregon Health and Sciences University,
Portland, Oregon 97239, USA

ABSTRACT: Cachexia refers to a synergistic combination of a dramatic decrease in appetite and an increase in metabolism of fat and lean body mass. This combination is found in a number of chronic diseases and is an important determinant of mortality. In this paper, we provide evidence that in both acute and chronic disease models, blockade of the MC4-R results in a dramatic attenuation of cachexia. We have also demonstrated that blockade of the melanocortin-3 receptor (MC3-R) leads to enhanced disease-associated cachexia. Ultimately, this work may lead to investigation of drug therapy for this widespread medical problem.

KEYWORDS: cachexia; cancer; melanocortin; proopiomelanocortin; appetite

INTRODUCTION

Cachexia—from the Greek *kakos* (bad) and *hexis* (condition)—is a worldwide medical problem that has been recognized as an adverse prognostic factor for more than 2000 years. Hippocrates wrote about the relationship between dropsy (chronic heart failure) and cachexia more than 2400 years ago: "The flesh is consumed and becomes water ... the abdomen fills with water, the feet and legs swell, the shoulders, clavicles, chest and thighs melt away ... This illness is fatal."[1] Despite the long history of this disorder in medicine and the recent explosion of research into the regulation of body weight, it remains true that individuals affected with either acute or chronic diseases often lose dramatic amounts of weight when ill, even when interventions such as parenteral nutrition are used. Unlike simple starvation, cachexia implies the presence of a synergistic combination of a dramatic decrease in appetite, an increase metabolic rate, and a pathologic loss of lean body mass. This combination is found in a number of disorders, including cancer, cystic fibrosis, acquired immunodeficiency syndrome (AIDS), rheumatoid arthritis, and renal failure.[2] The ancient description of cachexia provided above implies that its severity has a dramatic impact on both quality of life and eventual mortality, a fact that is supported by numerous recent medical studies.[2,3] Indeed, in many cases, the ability of the patient to

Address for correspondence: Daniel L. Marks, Department of Pediatric Endocrinology, Oregon Health and Sciences University, Portland, OR 97239. Voice: 503-494-4667; fax: 503-494-4534.

marksd@ohsu.edu

Ann. N.Y. Acad. Sci. 994: 258–266 (2003). © 2003 New York Academy of Sciences.

maintain normal body weight has a stronger correlation with survival than does any other measure of the disease.[4] The widespread impact of cachexia on medicine has led to numerous studies of the biology of this disorder and to the proposal of numerous hypotheses regarding its etiology. At this point, most investigators suggest that cytokines released during inflammation and malignancy play a vital role in the production of a cachectic state. Although peripheral mechanisms are undoubtedly involved, it is now clear that cytokines can act on the central nervous system to alter the release and function of a number of key neurotransmitters, thereby altering both appetite and metabolic rate.[2,5–7]

THE HYPOTHALAMIC MELANOCORTIN SYSTEM

For the purposes of this review, the hypothalamic melanocortin system consists of proopiomelanocortin (POMC) neurons located in the arcuate nucleus (ARC); type 3 melanocortin receptors (MC3R) located on POMC neurons and within other areas of the hypothalamus (including the periventricular nucleus, PVN); and type 4 melanocortin receptors (MC4R), which are distributed widely throughout the brain and receive input from POMC synapses.[8–10] In addition, we discuss the endogenous antagonist of MC3/MC4 receptors, agouti-related protein (AGRP). This peptide is produced in a subset of neuropeptide Y (NPY) neurons within the ARC that project to a majority of MC4Rs expressing neurons throughout the brain.[11]

ROLE OF THE MELANOCORTIN-4 RECEPTOR IN CACHEXIA

Numerous studies have demonstrated that stimulation of the MC4R by central administration of agonists can reproduce many of the most important features of cachexia, including anorexia and an increased metabolic rate.[12] Thus, we have proposed that cytokine feedback may be mediated, in part, by POMC neurons.[13,14] To provide experimental support for this hypothesis, we administered lipopolysaccharide (LPS, a compound known to reliably produce a large release of cytokines[15–18]) to wild-type and MC4R knockout (MC4RKO) animals and determined the changes in feeding, metabolic rate, and body mass that resulted from this treatment. We demonstrated that LPS treatment of wild-type animals reliably produced anorexia, an increase in metabolic rate, and a decrease in body mass, whereas none of these effects was observed in MC4RKO mice (FIG. 1).[13] In addition, we found that these animals did not decrease their wheel-running behavior after LPS administration, consistent with the earlier findings of Tatro *et al.*,[19] who showed that MC4RKO mice resist the inhibition of locomotion produced with central interleukin (IL)-1β administration. Other studies have shown that melanocortin antagonism prevents LPS-induced anorexia in rats as well.[20]

Collectively, the existing data indicate that the central melanocortin system plays a role in transducing the signals that produce changes in food intake, metabolic rate, and movement in response to LPS. Because LPS produces a relatively short-term illness, we tested the role of the melanocortin system in chronic disease models. We initially chose to study tumor models, as there were several well-documented models that reliably produced cachexia and exhaustion of metabolic fuels without obvious

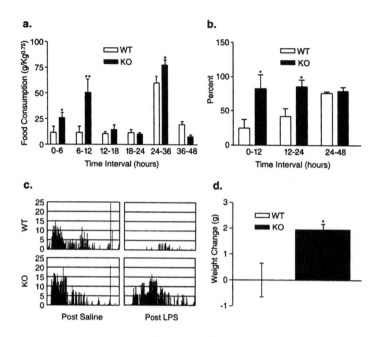

FIGURE 1. MC4 receptor knockout (MC4RKO) mice resist lipopolysaccharide (LPS)-induced cachexia and illness behavior. LPS results in a decrease in feeding for approximately 36 h in wild-type (WT), but not MC4RKO animals when expressed as (**a**) total normalized intake or (**b**) a percentage of basal (post ip saline) intake. (**c**) Normal nocturnal increase in wheel-running activity is observed in LPS-treated MC4RKO animals (results based on average turns/min in five animals, measured for 24 h, starting at 1700 h with lights out at 1900 h). (**d**) Young MC4RKO mice resist LPS-induced growth failure. (*$P < 0.05$; **$P < 0.01$ vs. wild-type controls.) (Reprinted with permission from Marks et al.[13])

complications of metastasis or infection.[21–24] In animals implanted with cachexigenic sarcomas, we found that intracerebroventricular AGRP could both prevent and reverse the anorexia and weight loss experienced during tumor development.[13] As with all the studies that we performed in tumor-bearing animals, melanocortin blockade did not affect the rate of tumor growth or final tumor mass, indicating that these receptors do not play a role in tumor growth or function. The finding that central melanocortin blockade attenuates cancer cachexia was confirmed in MC4RKO mice. Although bearing a carcinoma that produced classic cachexia in wild-type control animals, the MC4RKO animals maintained normal feeding and growth throughout the course of tumor development.

One of the most important features of cachexia is the loss of lean body mass that is greatly out of proportion to that observed in simple starvation. Unfortunately, many therapies directed at ameliorating disease-induced cachexia produce increases in body water retention and fat mass without affecting the loss of lean body mass.[25,26] Thus, in our studies, we sought to demonstrate retention of lean body mass in MC4RKO mice by serial dual-energy X-ray absorptiometry. Our scans con-

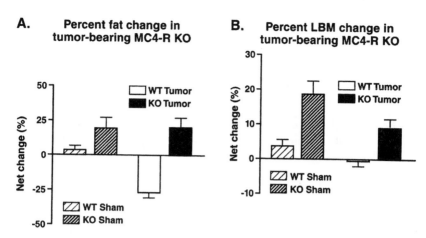

A. Percent fat change in tumor-bearing MC4-R KO

B. Percent LBM change in tumor-bearing MC4-R KO

FIGURE 2. Body composition changes in tumor-bearing MC4RKO mice. (**A**) MC4RKO mice continue to gain fat mass during tumor growth. (**B**) MC4RKO mice also accumulate lean body mass during tumor growth. (Reprinted with permission from Marks et al.[35])

firmed that the tumor-bearing MC4RKO animals continue to accumulate both lean mass and fat mass in the presence of tumor growth, whereas the wild-type animals lost both lean mass and fat mass under identical conditions (FIG. 2). Collectively, the studies to date clearly indicate that the hypothalamic MC4R plays a role in transducing cachexigenic stimuli from the periphery and that MC4R antagonists represent a logical target for weight-maintenance therapy. However, as described below, it will be important to recognize the potential pathologic consequences of MC3R blockade during disease, as loss of this receptor seems to exacerbate the process of cachexia.

While the importance of the MC4R in cachexia was becoming firmly established, the role of the melanocortin-3 receptor (MC3R) remained somewhat obscure. However, MC3R knockout (MC3RKO) mice have some features of cachexia at baseline, including reduced lean body mass, and a relative hypophagia compared to wild-type controls.[27,28] Furthermore, male MC3RKO mice exhibit an approximately 50% reduction in wheel-running behavior compared to wild-type littermates and show obvious decreases in home cage activity as well,[27,28] which is again typical of animals with acute and chronic disease. These features of the MC3RKO phenotype are consistent with the putative role for the MC3R as an inhibitory autoreceptor on POMC neurons.[29] In this capacity, deletion of the MC3R might be expected to enhance POMC neuronal function, thereby leading to a lean phenotype and enhanced susceptibility to cachexia. While these observations led to our hypothesis that the MC3RKO animal should show enhanced cachexia in disease models, the effect of MC3R deficiency on feeding and resting oxygen consumption is complex. Indeed, despite the increased adiposity in MC3RKO mice, the animals do not exhibit increased food intake or weight gain, even on a high-fat chow that is known to induce hyperphagia in MC4RKO animals.[27,30] Furthermore, it has also been impossible to demonstrate differences in resting basal or total metabolic rate in this model.[27] The

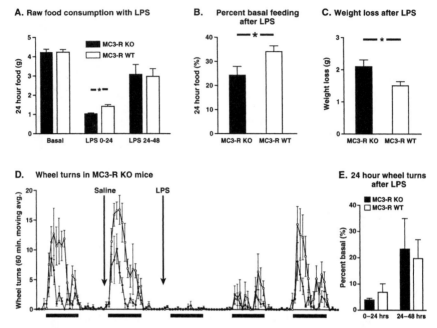

FIGURE 3. Food intake and wheel-running behavior in response to LPS treatment in MC3RKO mice. MC3RKO mice have enhanced anorexia in response to LPS (250 μg/kg) when measured both as (**A**) absolute food intake and as (**B**) a percentage of basal intake. (**C**) MC3RKO mice have increased weight loss after treatment with LPS. (**D & E**) MC3RKO mice decrease wheel-running behavior after LPS to a similar extent as do wild-type animals and show a similar recovery over the following 48 h. The dark phase is indicated by the *black bars* below the *x*-axis in **D**. Each data point represents a 60-min average for wheel turns. MC3RKO mice, *filled circles*; MC3RWT mice, *open squares*. (Reprinted with permission from Marks *et al.*[35])

anatomic data also argue that the MC3R has many more functions than providing autoinhibitory tone for POMC neurons. Within the brain, MC3R is also expressed in the ventromedial hypothalamus (VMH) and in more than 30 other brain nuclei in the rat.[31] It is also expressed in peripheral tissues, including the kidney and macrophages.[32–34] Thus, it is plausible that this receptor acts as an inhibitory autoreceptor on arcuate POMC neurons, while at the same time having an important inhibitory impact on energy storage elsewhere in the brain.

To pursue the hypothesis that MC3RKO mice should show enhanced susceptibility to cachexia, we followed an experimental paradigm analogous to that just described for the MC4RKO mice. Our first studies demonstrated that MC3RKO mice had normal metabolic and behavioral responses to simple starvation, thereby providing specificity for our studies of various models of cachexia. We then demonstrated that MC3RKO animals had significantly worse anorexia and weight loss in response to LPS relative to wild-type control animals (FIG. 3).[35] Because MC3R mice may play a role in the immune response to LPS,[34] we repeated these studies with IL-1β and again demonstrated enhanced anorexia and weight loss in MC3RKO

A.

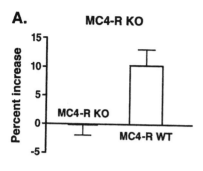

B.

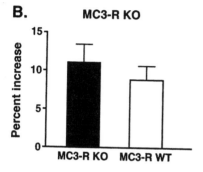

FIGURE 4. Resting energy expenditure in MCRKO mice after LPS injection. MC3RKO and wild-type mice have increased basal oxygen consumption after the injection of LPS. Data represent the percentage change in average basal VO_2 for the 8-h period after an LPS injection. Data for the LPS day are compared to those of the previous day, during which the animals received saline. (Reprinted with permission from Marks *et al.*[35])

animals relative to wild-type controls. Thus, these data argue that the MC3R is directly involved in the behavioral response to cytokines, perhaps by allowing for enhanced release of α-melanocyte–stimulating hormone (α-MSH) from POMC neurons with subsequent activation of the MC4R. An increase in metabolic rate during a time of relative anorexia is a hallmark of cachexia, and we were also able to demonstrate that the MC3RKO mouse has a normal increase in VO_2 in response to LPS, despite eating significantly less food (FIG. 4).

Perhaps the most significant finding in our studies of the role of the MC3R in cachexia is the dramatic depletion of lean body mass in tumor-bearing animals relative to that observed in wild-type controls (FIG. 5). The loss of fat mass in this model was similar to that found in tumor-bearing wild-type animals, arguing that the MC3R may play a specific role in limiting the loss of lean body mass that occurs during prolonged illness. Our data also indicate that the increased feed efficiency observed in the MC3RKO mouse at baseline is not sufficient to maintain normal lean body mass during disease.[27,28] We hypothesize that the small effect on feed efficiency is overwhelmed by the increased and unrestrained activation of the MC4R during illness. Obviously, it is also possible that peripheral factors responsible for the development of cachexia act downstream of the MC3R and therefore bypass this feature of the MC3RKO phenotype.

The hypothesis that the mechanism for enhanced cachexia in MC3RKO mice is an unrestrained release of α-MSH from POMC neurons (and thus enhanced activation of MC4Rs) implies that this phenotype should not be observed when both hypothalamic receptor subtypes are blocked. We previously showed that central

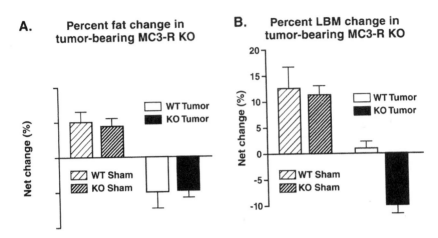

FIGURE 5. Body composition changes in tumor-bearing MC3RKO mice. (A) MC3RKO mice lose a similar amount of body fat (FM) as wild-type animals in response to tumor growth. (B) MC3RKO mice lose excess lean body mass (LBM) in response to tumor growth. (Reprinted with permission from Marks et al.[35])

injections of AGRP (which blocks both the MC3R and the MC4R) can prevent the anorexia and weight loss observed with LPS treatment or tumor growth.[13] Other investigators have demonstrated that the anorexia due to tumor growth or cytokine injection in rats can be reversed by central injections of another MC3/4 antagonist, SHU9119.[36,37] Although these studies do not demonstrate a specific conservation of lean body mass, they do provide compelling evidence that melanocortin blockade prevents cachexia via a central rather than a peripheral mechanism. Further evidence for this mechanism is derived from the observation that mice lacking both the MC3R and the MC4R, and mice overexpressing the MC3/4 antagonists agouti or AGRP show an increase in lean body mass that is indistinguishable from that observed in the MC4RKO mice rather than demonstrating the lean body mass phenotype of the MC3RKO.[28,38–40] Thus, we hypothesize that the enhanced loss of lean body mass observed in our cancer models is due to a prolonged increase in MC4R signaling brought about by loss of the autoinhibitory MC3R on POMC neurons, but further experiments will be required to solidify this conclusion.

In summary, our studies provide evidence that hypothalamic melanocortin receptors play an important role in the development of illness-induced cachexia. Specifically, our data support the hypothesis that arcuate POMC neurons are activated by peripheral signals produced during illness (including cytokines) and that the result of this activation is enhanced signaling at the MC4R throughout the brain. Numerous genetic and pharmacologic studies indicate that the outcome of enhanced signaling at the MC4R would include decreased feeding, increased metabolic rate, and loss of lean body mass, the primary defining features of cachexia. Our data also argue that the MC3R may act as an inhibitory autoreceptor on POMC neurons and that loss of this function would produce enhanced cachexia during disease. Thus, it is plausible that specific MC4R antagonists will be logical therapeutic targets for the treatment of weight loss observed in human diseases such as cancer, heart failure, Alzheimer's disease, cystic fibrosis, and AIDS.

REFERENCES

1. KATZ, A.M. & P.B. KATZ. 1962. Diseases of the heart in the works of Hippokrates. Br. Heart J. **24:** 257–264.
2. TISDALE, M.J. 1997. Biology of cachexia. J. Natl. Cancer Inst. **89:** 1763–1773.
3. LARKIN, M. 1998. Thwarting the dwindling progression of cachexia. Lancet **351:** 1336.
4. KOTLER, D.P. *et al.* 1989. Magnitude of body-cell-mass depletion and the timing of death from wasting in AIDS. Am. J. Clin. Nutr. **50:** 444–447.
5. PLATA-SALAMAN, C.R. 1989. Immunomodulators and feeding regulation: a humoral link between the immune and nervous systems. Br. Behav. Immun. **3:** 193–213.
6. PLATA-SALAMAN, C.R. 1998. Brain mechanisms in cytokine-induced anorexia. Psychoneuroendocrinology **24:** 25–41.
7. INUI, A. 1999. Feeding and body-weight regulation by hypothalamic neuropeptides—mediation of the actions of leptin. Trends Neurosci. **22:** 62–67.
8. JEGOU, S. *et al.* 2000. Melanocortin-3 receptor mRNA expression in pro-opiomelanocortin neurones of the rat arcuate nucleus. J. Neuroendocrinol. **12:** 501–505.
9. MOUNTJOY, K.G. & J.M. WILD. 1998. Melanocortin-4 receptor mRNA expression in the developing autonomic and central nervous systems. Dev. Brain. Res. **107:** 309–314.
10. XIA, Y. & J.E.S. WIKBERG. 1997. Postnatal expression of melanocortin-3 receptor in rat diencephalon and mesencephalon. Neuropharmacology **36:** 217–224.
11. HASKELL-LUEVANO, C. *et al.* 1999. Characterization of the neuroanatomical distribution of agouti-related protein (AGRP) immunoreactivity in the rhesus monkey and the rat. Endocrinology **140:** 1408–1415.
12. FAN, W. *et al.* 1997. Role of melanocortinergic neurons in feeding and the agouti obesity syndrome. Nature **385:** 165–168.
13. MARKS, D.L., N. LING & R.D. CONE. 2001. Role of the central melanocortin system in cachexia. Cancer Res. **61:** 1432–1438.
14. MARKS, D.L. & R.D. CONE. 2001. Central melanocortins and the regulation of weight during acute and chronic disease. Recent Prog. Horm. Res. **56:** 359–375.
15. HILLHOUSE, E.W. & K. MOSLEY. 1993. Peripheral endotoxin induces hypothalamic immunoreactive interleukin-1 beta in the rat. Br. J. Pharmacol. **109:** 289–290.
16. VAN DAM, A.M. *et al.* 1995. Endotoxin-induced appearance of immunoreactive interleukin-1 beta in ramified microglia in rat brain: a light and electron microscopic study. Neuroscience **65:** 815–826.
17. CAVAILLON, J.M. & N. HAEFFNER-CAVAILLON. 1990. Signals involved in interleukin-1 synthesis and release by lipopolysaccharide-stimulated monocytes/macrophages. Cytokine **2:** 313–329.
18. MOLLOY, R.G., J.A. MANNICK & M.L. RODRICK. 1993. Cytokines, sepsis and immunomodulation. Br. J. Surg. **80:** 289–297.
19. TATRO, J.B. *et al.* 1999. Role of the melanocortin-4 receptor in thermoregulatory responses to central IL-1β. Soc. Neurosci. Abstr. **25:** 1558.
20. HUANG, Q.H., V.J. HRUBY & J.B. TATRO. 1999. Role of central melanocortins in endotoxin-induced anorexia. Am. J. Physiol. **276:** R864–R871.
21. SVANINGER, G., C. DROTT & K. LUNDHOLM. 1987. Role of insulin in development of cancer cachexia in nongrowing sarcoma-bearing mice: special reference to muscle wasting. J. Natl. Cancer Inst. **78:** 943–950.
22. EMERY, P.W. 1999. Cachexia in experimental models. Nutrition **15:** 600–603.
23. EMERY, P.W., L. LOVELL & M.J. RENNIE. 1984. Protein synthesis measured in vivo in muscle and liver of cachectic tumor-bearing mice. Cancer Res. **44:** 2779–2784.
24. SVANINGER, G., J. GELIN & K. LUNDHOLM. 1989. The cause of death in non-metastasizing sarcoma bearing mice. Eur. J. Cancer Clin. Oncol. **25:** 1295–1302.

25. SIMONS, J.P. *et al.* 1998. Effects of medroxyprogesterone acetate on food intake, body composition, and resting energy expenditure in patients with advanced, nonhormone-sensitive cancer: a randomized, placebo-controlled trial. Cancer **82:** 553–560.
26. SIMONS, J.P. *et al.* 1999. Weight loss and low body cell mass in males with lung cancer: relationship with systemic inflammation, acute-phase response, resting energy expenditure, and catabolic and anabolic hormones. Clin. Sci. (London) **97:** 215–223.
27. BUTLER, A.A. *et al.* 2000. A unique metabolic syndrome causes obesity in the melano-cortin-3 receptor-deficient mouse. Endocrinology **141:** 3518–3521.
28. CHEN, A.S. *et al.* 2000. Inactivation of the mouse melanocortin-3 receptor results in increased fat mass and reduced lean body mass. Nat. Genet. **26:** 97–102.
29. BAGNOL, T. *et al.* 1999. The anatomy of an endogenous antagonist: relationship between agouti-related protein and proopiomelanocortin in brain. J. Neurosci. **19:** RC26.
30. BUTLER, A.A. *et al.* 2001. Melanocortin-4 receptor is required for adaptive thermogenesis and behavioral responses to increased dietary fat. Nat. Neurosci. **4**(6): 605–611.
31. ROSELLI-REHFUSS, L. *et al.* 1993. Identification of a receptor for γ-MSH and other proopiomelanocortin peptides in the hypothalamus and limbic system. Proc. Natl. Acad. Sci. USA **90:** 8856–8860.
32. CHHAJLANI, V. 1996. Distribution of cDNA for melanocortin receptor subtypes in human tissues. Biochem. Mol. Biol. Int. **38:** 73–80.
33. GETTING, S.J. *et al.* 1999. POMC gene-derived peptides activate melanocortin type 3 receptor on murine macrophages, suppress cytokine release, and inhibit neutrophil migration in acute experimental inflammation. J. Immunol. **162:** 7446–7453.
34. GETTING, S.J., R.J. FLOWER & M. PERRETTI. 1999. Agonism at melanocortin receptor type 3 on macrophages inhibits neutrophil influx. Inflamm. Res. **48**(Suppl. 2): S140–S141.
35. MARKS, D.L. *et al.* 2003. Differential role of melanocortin receptor subtypes in cachexia. Endocrinology **144**(4): 1513–1523.
36. LAWRENCE, C.B. & N.J. ROTHWELL. 2001. Anorexic but not pyrogenic actions of inter-leukin-1 are modulated by central melanocortin-3/4 receptors in the rat. J. Neuro-endocrinol. **13:** 490–495.
37. WISSE, B.E. *et al.* 2001. Reversal of cancer anorexia by blockade of central melanocor-tin receptors in rats. Endocrinology **142:** 3292–3301.
38. KLEBIG, M.L. *et al.* 1995. Ectopic expression of the agouti gene in transgenic mice causes obesity, features of type II diabetes, and yellow fur. Proc. Natl. Acad. Sci. USA **92:** 4728–4732.
39. OLLMANN, M.M. *et al.* 1997. Antagonism of central melanocortin receptors in vitro and in vivo by agouti-related protein. Science **278:** 135–138.
40. YEN, T.T. *et al.* 1994. Obesity, diabetes, and neoplasia in yellow A(vy)/- mice: ectopic expression of the agouti gene. FASEB J. **8:** 479–488.

Melanocortin System and Eating Disorders

ROGER A.H. ADAN, JACQUELIEN J.G. HILLEBRAND, CORINE DE RIJKE,
WOUTER NIJENHUIS, TOM VINK, KEITH M. GARNER, AND MARTIEN J.H. KAS

*Rudolf Magnus Institute of Neuroscience, Department of Pharmacology and Anatomy,
University Medical Centre Utrecht, Utrecht, the Netherlands*

ABSTRACT: The melanocortin (MC) system is involved in the regulation of
energy balance and in the development of obesity. Here we briefly review why
we became interested in investigating whether the MC system—more particularly, the increased activity of the MC system—is also involved in disorders of
negative energy balance. We provide evidence that suppression of increased
MC receptor activity by treatment with the inverse agonist agouti-related
peptide (AgRP) (83–132) rescues rats exposed to an animal model known as
activity-based anorexia. Furthermore, we found a polymorphism, Ala67Thr
AgRP, that was observed more frequently in anorexia nervosa.

KEYWORDS: activity-based anorexia; anorexia nervosa; inverse agonist; agouti-
related peptide (AgRP); stress; melanocortin (MC)

INTRODUCTION

In the last decade, the brain melanocortin (MC) system has become well known
for its involvement in the regulation of food intake and energy balance. However,
besides regulation of food-associated behaviors, the MC system has been implicated
in other aspects of behavior as well. One of the first effects observed after intracerebroventricular injection of MC is the induction of excessive grooming behavior.[1,2] Grooming behavior is often observed when a rat is exposed to a stressful
situation, such as exposure to a novel environment.

Evidence suggests that α-melanocyte–stimulating hormone (α-MSH) is the
endogenous ligand mediating grooming behavior, as observed after exposure to a
novelty. First, infusion with adrenocorticotropin-stimulating hormone (ACTH)-
antiserum into the brain ventricular system blocked novelty-induced grooming in the
rat.[3] Second, down-regulation of proopiomelanocortin (POMC) protein expression
in the hypothalamus via antisense oligonucleotides significantly reduced the grooming response to a novelty.[4] Third, a potent MC receptor antagonist (SHU9119)
reduced novelty-induced grooming in the rat (FIG. 1). Thus, the reaction of a rat to a
novelty involves activation of the MC system.

Exposing a rat to a novel environment induces stress and stimulates the
hypothalamo-pituitary-adrenal (HPA) axis. Central administration of ACTH(1–24)

Address for correspondence: Roger A.H. Adan, Rudolf Magnus Institute of Neuroscience,
Department of Pharmacology and Anatomy, University Medical Centre Utrecht, Universiteitsweg
100, 3584 CG Utrecht, the Netherlands. Voice: +31 30 2538517; fax: +31 30 2539032.
r.a.h.adan@med.uu.nl

Ann. N.Y. Acad. Sci. 994: 267–274 (2003). © 2003 New York Academy of Sciences.

activates the HPA axis independent of the direct effects of ACTH on the adrenal gland.[2] Since a novel environment or injection with MCs induces grooming behavior, which can be blocked by an MC receptor antagonist, we tested whether activation of the HPA axis by central application of MCs is also blocked by MC receptor antagonists. Indeed, this effect of ACTH(1–24) is also inhibited by coadministration of the MC4 receptor antagonists [D-Arg^8]ACTH(4–10) and SHU9119 (FIG. 2).[2] It is speculative to suggest that activation of the brain MC system is involved in activation

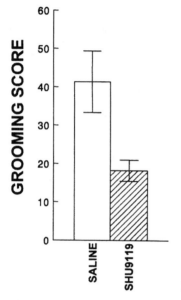

FIGURE 1. The melanocortin (MC) receptor antagonist SHU9119 reduced novelty-induced grooming in the rat. Effect of administration of saline versus 100 ng SHU9119 in rats exposed to a novel environment on spontaneous grooming behavior are presented.

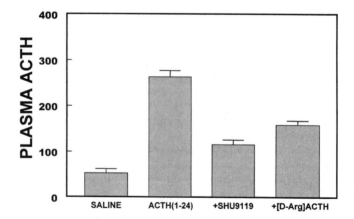

FIGURE 2. Melanocortin-4 (MC4) receptor antagonists block central MC receptor-mediated activation of the HPA axis. Plasma ACTH levels were measured 20 min after intracerebroventricular administration of 1 μg ACTH (1–24) either alone or in combination with 100 ng SHU9119 or 1 μg [D-Arg^8]ACTH(4–10).

of the HPA axis after mild emotionally stressful conditions. Therefore, whether the response of a rat to exposure to stressful situations (activation of the HPA axis) is inhibited by MC receptor antagonists needs to be determined.

Three symptoms observed in anorexia nervosa are hyperactivity, decreased food intake, and increased activity of the HPA axis. Since stimulation of MC receptors in rats results in increased motor activity, reduced food intake, and activation of the HPA axis, we investigated whether increased melanocortinergic tone contributed to the development of anorexia. We hypothesized that inadequate inhibition of MC receptors may contribute to the development of anorexia nervosa.

ACTIVITY-BASED ANOREXIA

For this purpose we introduced an animal model in the laboratory that had some similarities to the development of anorexia nervosa. When rats, in the presence of a running wheel, have access to food for only 1 h per day, they develop hyperactivity and decreased food intake. After 1 wk of exposure to the model, we found increased

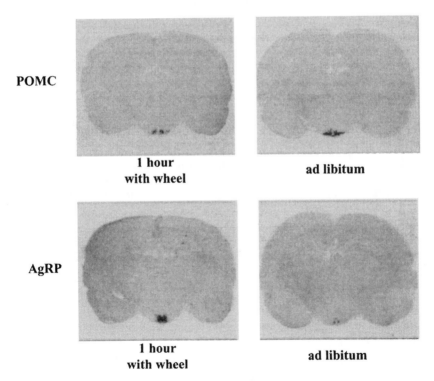

FIGURE 3. POMC mRNA levels were downregulated and AgRP mRNA levels were upregulated in activity-based anorexia. *In situ* hybridization was performed with [^{35}S]RNA probes complementary to POMC and AgRP mRNA, respectively, using brain sections (containing the arcuate nucleus) derived from control rats and rats exposed to activity-based anorexia.

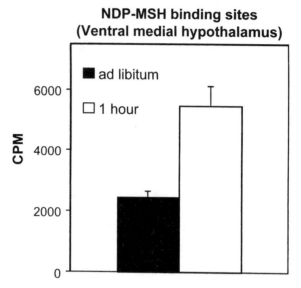

FIGURE 4. The density of melanocortin (MC) receptors increases in the ventromedial nucleus of rats exposed to activity-based anorexia. *In situ* binding was performed with [125I]NDP-MSH as radiolabel on brain sections (containing the ventromedial nucleus) derived from control rats (*ad libitum*) and rats exposed to activity-based anorexia (1 h).

activity of the HPA axis, as determined by increased plasma ACTH levels and increased adrenal gland weight. This model is known as activity-based anorexia (ABA). Using this animal model, we tested whether inadequate inhibition of MC receptor activity contributes to the development of anorexia. Therefore, we investigated whether agouti-related peptide (AgRP) was upregulated and whether POMC was downregulated in the arcuate nucleus of the hypothalamus. A comparison of the expression levels of these mRNAs among different treatment groups revealed that POMC mRNA levels were downregulated and that AgRP mRNA levels were upregulated, as expected (Fig. 3).[5] We also investigated the density of MC receptors in two brain regions, namely, the ventromedial hypothalamus and the habenular nucleus. The ventromedial hypothalamus, in contrast to the habenular nucleus, has been strongly associated with regulation of food intake. ABA is accompanied by an increase in the density of MC receptors, specifically those of the ventromedial nucleus (Fig. 4).[5] This result was in agreement with earlier work demonstrating increased density of MC4 receptors on food restriction.[6] This is paradoxical, because an increase in MC receptors will increase the sensitivity of the anorectic effect of MSH.

AgRP IS AN INVERSE AGONIST

Besides a higher sensitivity for MCs, the increase in MC receptor density also provides a signal by itself. We found that the MC4 receptor (as well as the MC3

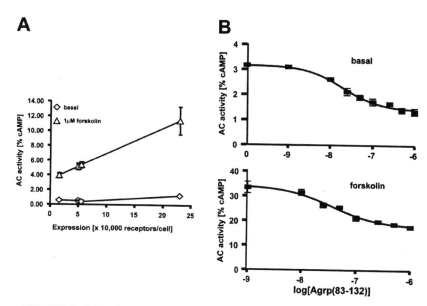

FIGURE 5. (a) Melanocortin-4 (MC4) receptors display constitutive receptor activity. B16G4F melanoma cells expressing different levels of human MC4 receptor were measured for basal and forskolin-induced adenylate cyclase activity. (b) AgRP acts as an inverse agonist. B16G4F melanoma cells expressing high levels of human MC4 receptor were incubated with AgRP (83–132), and basal- and forskolin-induced adenylate cyclase activity was determined.

receptor) displays constitutive receptor activity.[7] When clonal B16-G4F melanoma cells expressing different levels of MC4 receptors were tested for spontaneous adenylate cyclase activity, we found that the level of receptor expression correlated well with the spontaneous as well as the forskolin-induced adenylate cyclase activity (FIG. 5a).[7] Furthermore, AgRP (83–132) suppressed this spontaneous adenylate cyclase activity, thus characterizing AgRP as an inverse agonist (FIG. 5b). As expected, SHU9119 blocked this inverse agonistic effect of AgRP, demonstrating that SHU9119 acts as a competitive antagonist at the MC4 receptor. As far as we know, AgRP is the only recognized endogenous inverse agonist in the brain. Thus, the brain MC system is unique in that two neuron populations in the arcuate nucleus that produce either POMC or AgRP regulate the activity of MC receptors in cells they project on in an opposite manner. This ensures a very tightly controlled activity of MC receptors.

AgRP RESCUES RATS FROM STARVATION

To test whether self-starvation in ABA results from insufficient suppression of MC receptor activity, we treated rats exposed to ABA with AgRP (83–132). We found that central infusion with AgRP, in rats exposed to ABA, increased the surviv-

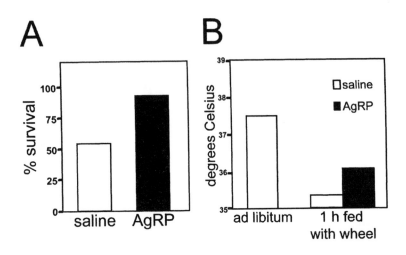

FIGURE 6. (**a**) AgRP (83–132) rescues rats from starvation. Rats exposed to activity-based anorexia (a starvation model resembling some characteristics of anorexia nervosa) were treated with saline or AgRP. Following 1 wk of exposure to the model, a considerable number of rats had to be sacrificed because they were in dramatic negative energy balance. Treatment with AgRP strongly improved health in this model. (**b**) AgRP inhibited the decrease in temperature during rest. Body temperature was determined in control rats and in rats exposed to activity-based anorexia (treated with either saline or AgRP (83–132). Body temperature was measured after a 15-min period of rest (as determined by telemetric recording of motor activity).

al rate (FIG. 6a).[5] This underscores the finding that suppression of MC receptor activity rescues these rats. Surprisingly, the first effect of AgRP in this model was not on food intake or body weight, but on body temperature in rest. We define body temperature in rest as body temperature after 15 min of rest, during which there is no activity as determined by telemetric recording of motor activity. When rats were exposed to ABA, the body temperature gradually decreased over days of exposure to ABA (FIG. 6b). Although both saline- and AgRP-treated rats had decreased body temperature in ABA, in AgRP-treated rats, body temperature was not decreased as dramatically as in saline-treated rats. Cumulative food intake was higher in AgRP-treated rats than in controls (saline-treated rats).

ASSOCIATION OF Ala67Thr AgRP WITH ANOREXIA NERVOSA

Since loss of function mutations of POMC and MC4 result in obesity, we anticipated that loss of function of AgRP would result in increased melanocortinergic tone and subsequently an inadequate response to negative energy balance. Thus, in periods of food restriction, loss of function of AgRP would result in a decreased drive to eat and inadequate adaptation to negative energy balance. Therefore, we screened the AgRP coding region for polymorphisms in DNA from patients with anorexia nervosa as compared to controls. We found an association of the (Ala67Thr)AgRP allele

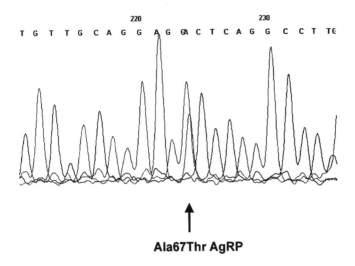

Ala67Thr AgRP

FIGURE 7. Association of the (Ala67Thr) AgRP allele with anorexia nervosa. Sequence analysis of the AgRP coding region in patients with anorexia nervosa identified a polymorphism that was significantly more frequent (almost threefold) in patients with anorexia nervosa than in controls.

with anorexia nervosa (FIG. 7).[8] This polymorphism was recently also associated with leanness and protection against obesity.[9,10]

CONCLUDING REMARKS

We found that during negative energy balance, the pharmacologic suppression of MC receptor activity protects against starvation in rats. We conclude that inverse agonists at MC receptors should be considered for the treatment of disorders of negative energy balance such as anorexia nervosa.

REFERENCES

1. ADAN, R.A., A.W. SZKLARCZYK, J. OOSTEROM, *et al.* 1999. Characterization of melano-cortin receptor ligands on cloned brain melanocortin receptors and on grooming behavior in the rat. Eur. J. Pharmacol. **378:** 249–258.
2. VON FRIJTAG, J.C., G. CROISET, W.H. GISPEN, *et al.* 1998. The role of central melano-cortin receptors in the activation of the hypothalamus-pituitary-adrenal-axis and the induction of excessive grooming. Br. J. Pharmacol. **123:** 1503–1508.
3. DUNN, A.J., E.J. GREEN & R.L. ISAACSON. 1979. Intracerebral adrenocorticotropic hor-mone mediates novelty-induced grooming in the rat. Science **203:** 281–283.
4. SPAMPINATO, S., M. CANOSSA, L. CARBONI, *et al.* 1994. Inhibition of proopiomelano-cortin expression by an oligodeoxynucleotide complementary to beta-endorphin mRNA. Proc. Natl. Acad. Sci. USA **91:** 8072–8076.
5. KAS, M.J.H., G.J. VAN DIJK, A.J.W. SCHEURINK & R.A.H. ADAN. 2003. Agouti-related protein prevents self-starvation. Mol. Psychiatry **8:** 235–240.

6. HARROLD, J.A., P.S. WIDDOWSON & G. WILLIAMS. 1999. Altered energy balance causes selective changes in melanocortin-4 (MC4-R), but not melanocortin-3 (MC3-R), receptors in specific hypothalamic regions: further evidence that activation of MC4-R is a physiological inhibitor of feeding. Diabetes **48:** 267–271.
7. NIJENHUIS, W.A., J. OOSTEROM & R.A. ADAN. 2001. AgRP(83–132) acts as an inverse agonist on the human-melanocortin-4 receptor. Mol. Endocrinol. **15:** 164–171.
8. VINK, T., A. HINNEY, A.A. VAN ELBURG, *et al.* 2001. Association between an agouti-related protein gene polymorphism and anorexia nervosa. Mol. Psychiatry **6:** 325–328.
9. ARGYROPOULOS, G., T. RANKINEN, D.R. NEUFELD, *et al.* 2002. A polymorphism in the human agouti-related protein is associated with late-onset obesity. J. Clin. Endocrinol. Metab. **87:** 4198–4202.
10. FAN, W., D.L. MARKS, A. HAQQ, *et al.* 2002. Mechanisms of melanocortin action in the CNS. Abstracts of the 5th International Melanocortin Meeting, Sunriver, August 25–28, 2002 [nr. SO11]. Ann. N.Y. Acad. Sci. 994: this volume.

Melanocortin Signaling and Anorexia in Chronic Disease States

BRENT E. WISSE,[a] MICHAEL W. SCHWARTZ,[a] AND DAVID E. CUMMINGS[b]

Division of Metabolism, Endocrinology and Nutrition, [a]Harborview Medical Center, University of Washington, Seattle, Washington 98104, USA and [b]Seattle VA Puget Sound Health Care System, Seattle, Washington 98108, USA

ABSTRACT: Data from both rodent models and humans suggest that intact neuronal melanocortin signaling is essential to prevent obesity, as mutations that decrease the melanocortin signal within the brain induce hyperphagia and excess body fat accumulation. Melanocortins are also involved in the pathogenesis of disorders at the opposite end of the spectrum of energy homeostasis, the anorexia and weight loss associated with inflammatory and neoplastic disease processes. Studies using melanocortin antagonists (SHU9119 or agouti-related peptide) or genetic approaches (melanocortin-4 receptor null mice) suggest that intact melanocortin tone is required for anorexia and weight loss induced by injected lipopolysaccharide (an inflammatory gram-negative bacterial cell wall product) or by implantation of prostate or lung cancer cells. Although the precise mechanism whereby peripheral inflammatory/neoplastic factors activate the melanocortin system remains unknown, the proinflammatory cytokines (interleukin-1, interleukin-6, and tumor necrosis factor-α) that are produced in the hypothalamus of rodents during both inflammatory and neoplastic disease processes likely play a role. The data presented in this paper summarize findings that implicate neuronal melanocortin signaling in inflammatory anorexia.

KEYWORDS: cachexia; cancer; cytokine; inflammation; definition

MELANOCORTINS AND OBESITY

The critical importance of functional melanocortin signaling in the maintenance of normal body weight is strongly supported by a large body of literature[1–5] demonstrating that genetic impairment of neuronal melanocortin signaling is strongly associated with obesity in both humans and rodents. While these findings suggest that sufficient melanocortin tone is necessary to avert weight gain, recent data also support the involvement of the melanocortin system in diseases at the opposite end of the spectrum of disturbed energy homeostasis, namely, the anorexia that results from cancer and chronic inflammatory conditions.

Address for correspondence: Brent E. Wisse, Division of Metabolism, Endocrinology, and Nutrition, Harborview Medical Center, University of Washington, 325 Ninth Avenue, Box 359757, Seattle, WA 98104-2499. Voice: 206-341-4620; fax: 206-731-8522.
bewisse@u.washington.edu

Ann. N.Y. Acad. Sci. 994: 275–281 (2003). © 2003 New York Academy of Sciences.

CANCER ANOREXIA AND CACHEXIA

Clinically, anorexia and cachexia are often joined together as the cancer anorexia-cachexia syndrome (CACS)[6] or as the "wasting syndrome" in patients with AIDS and chronic inflammatory disorders.[7] Almost all types of cancer have been reported to cause anorexia and cachexia, and more than half of cancer patients develop CACS during the course of their disease.[8] Patients with CACS have greater morbidity and mortality compared to cancer patients with equivalent tumor burden, but without CACS, and in 20% of these patients, cachexia is the main cause of death.[6] Yet, little is known about the underlying pathophysiology of CACS, and consequently, available treatment options are limited in both number and efficacy.[10]

CANCER ANOREXIA/CACHEXIA DEFINITIONS

Progress in understanding CACS requires a clear definition of the terms "anorexia" and "cachexia." For research in the area of energy homeostasis, anorexia can be defined as "a reduction in food intake caused primarily by diminished appetite," as opposed to the more inclusive literal definition, "not eating." By focusing on appetite, this definition emphasizes central mechanisms regulating food intake in the pathophysiology of cancer anorexia over other common problems, such as pain, depression, and nausea, which may cause cancer patients to consume less food. This distinction is important for clinical studies, as only a subset of cancer patients with inadequate food intake would be likely to benefit from medications that act centrally to stimulate appetite. However, this definition is not readily applicable to most animal models, where investigators must still rely on daily food intake or computerized analysis of meal patterns[11,12] to define anorexia.

Cachexia can be defined as "a metabolic disorder of energy expenditure leading to weight loss beyond that caused by reduced food intake alone." Clinically, cancer patients with significant weight loss are often diagnosed with cachexia[8] despite the fact that anorexia and "not eating" alone will reduce body weight. This distinction is important for future clinical studies in cancer patients, because treatments for anorexia and cachexia are likely to be different and targeting one without affecting the other is likely to fail. In animal models in which repeated measures of O_2 consumption and serial analysis of body composition[13] are feasible during the course of tumor progression, this definition of cachexia is more applicable. Another useful approach for animal studies is to compare weight loss and body composition in tumor-bearing animals to those in pair-fed control animals. This approach reveals the metabolic differences between restricted energy intake and cachexia.

MECHANISM OF CANCER ANOREXIA/CACHEXIA

Both central and peripheral factors are implicated in cancer anorexia and cachexia in animal models.[14–17] Much of this work has focused on proinflammatory cytokines, including tumor necrosis factor-α, interleukin-6 (IL-6), and interleukin-1β.[14] Although tumor necrosis factor-α is the original "cachectin,"[18] no single cytokine or classical central nervous system neurotransmitter[19] appears to explain the entire syndrome.[20,21] Recent evidence suggests that some tumors produce a novel cachectic factor (proteolysis-inducing factor, PIF) and that urine levels of this factor correlate with weight loss.[22]

MELANOCORTINS AND ANOREXIA

Well before central melanocortins were implicated in the regulation of energy homeostasis, it became clear that α-melanocyte–stimulating hormone (α-MSH) has potent anti-inflammatory actions in the brain.[23] Huang *et al.*[24] were the first to demonstrate that reduced central melanocortin signaling achieved by an intracerebroventricular (i.c.v.) injection of SHU9119 (a nonselective melanocortin receptor antagonist) partially reverses the anorexia caused by systemic administration of lipopolysaccharide (LPS, a potent mediator of inflammatory anorexia) in rats.[24] Increased neuronal melanocortin signaling in response to inflammation may therefore contribute to the anorexia associated with acute infection. However, LPS injection also has the potential to activate the melanocortin system indirectly by increasing circulating levels of leptin,[25] a hormone that activates hypothalamic proopiomelano-

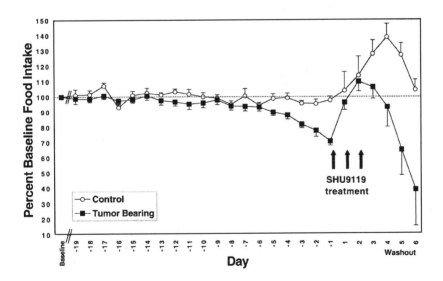

FIGURE 1. Progression of anorexia and response to SHU9119 (0.35 nmol third i.c.v.) in tumor-bearing Lobund-Wistar rats with prostate cancer. Daily food intake values, presented as a percentage of the pre-anorexic baseline, are shown for the 19 days before treatment in each group. Anorexia was defined as food intake <80% of baseline for 3 consecutive days or <75% of baseline with a steady decline 3 days or more in duration. Baseline food intake was calculated as the average daily food intake during a 10-day period following recovery from i.c.v. cannulation (that is, from day –30 to –40).). Data are means ± SEM for both curves; tumor-bearing group (■, *n* = 13) and *ad libitum* fed control animals (○, *n* = 13).

cortin (POMC) neurons, although leptin does not appear to be essential for LPS-associated anorexia.[26]

Marks *et al.*[12] confirmed that central melanocortin blockade (using agouti-related peptide) reduces LPS-induced anorexia in mice, and they extended previous findings showing that LPS-induced anorexia was diminished in melanocortin-4 receptor (MC4R) null mice, identifying the MC4R as a critical mediator of inflammatory anorexia. The same approach provided evidence that cancer anorexia depends on the melanocortin pathway, as food intake in tumor-bearing animals was partially rescued by treatment with agouti-related peptide and was attenuated in MC4R null mice. One difference between cancer anorexia and LPS-induced anorexia is that the former condition is marked by low leptin levels (due to weight loss),[27] and therefore melanocortin signaling should be reduced.

Nonetheless, we showed that central infusion of the melanocortin antagonist SHU9119 completely reversed anorexia in rats with prostate cancer during a 3-day treatment period (FIG. 1), producing weight gain comparable to that observed in non-tumor–bearing control animals treated with SHU9119.[28] Melanocortin receptor blockade also caused a much greater increase of food intake in anorexic rats than did the powerful orexigenic agents, neuropeptide Y or ghrelin.[28] Treatment with SHU9119 also reversed some of the negative metabolic consequences of anorexia as determined by serum leptin, ghrelin, insulin, and glucose values.[28]

Combined with evidence that SHU9119 can block the anorexic effects of centrally administered interleukin-1β,[29] this work suggests that inappropriate activation of neuronal melanocortin signaling may be a common mediator of anorexia in disease states involving increased cytokine signaling. Four possibilities can be postulated to explain how hypothalamic melanocortin tone might be increased by inflammatory anorexia (FIG. 2): (1) activation of hypothalamic POMC neurons; (2) enhanced processing of POMC to α-MSH and/or increased α-MSH release on target neurons;[30]

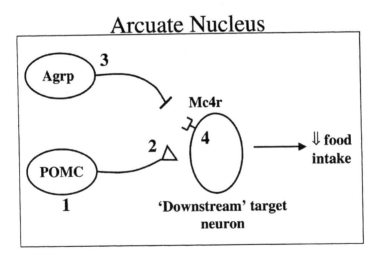

FIGURE 2. Potential mechanisms whereby cancer anorexia can increase melanocortin signaling.

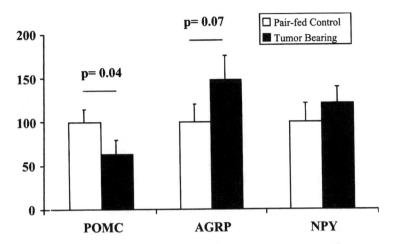

FIGURE 3. Hypothalamic neuropeptide mRNA expression in tumor-bearing and pair-fed control rats by *in situ* hybridization. Lobund-Wistar rats with prostate cancer ($n = 5$) and pair-fed controls ($n = 5$) were sacrificed when tumor-bearing animals were anorexic (as defined for Fig. 1). Brains were rapidly extracted and fresh frozen using dry ice. Then, 14-μm sections were cut on a cryostat and probed with specific antisense cRNA probes using [33P]dUTP as a label. Signal was quantified using a Cyclone Phosphoimager and Optiquant software. Data are means ± SEM expressed relative to pair-fed control animals. *P* values determined by one-way ANOVA using the LSD *post-hoc* test.

(3) decreased production or release of the endogenous melanocortin antagonist ago-uti-related peptide; or (4) increased endogenous activity of the MC4R, whereby the effect of SHU9119 and agouti-related peptide to increase food intake can be explained through inverse agonism rather than competition with a specific ligand.[31]

We recently used *in situ* hybridization techniques to evaluate two of these possibilities. Based on the assumption that POMC mRNA expression in the arcuate nucleus is reflective of POMC neuronal activation,[32] we compared the expression of POMC mRNA in the rostral arcuate nucleus of anorexic rats with prostate cancer to that of pair-fed control animals (Fig. 3). Our finding that POMC mRNA expression in tumor-bearing animals was *lower* than that in controls argues against an increase of hypothalamic melanocortin production as a mechanism to explain the anorexia. In addition, we found no evidence that cancer anorexia is characterized by reduced expression of agouti-related peptide mRNA (Fig. 3).

In an additional experiment, we analyzed the POMC mRNA expression in the rostral arcuate nucleus of third i.c.v. cannulated control and tumor-bearing rats (Fig. 4). SHU9119 treatment *increased* POMC mRNA expression in tumor-bearing animals, suggesting that the greater food intake in these animals was not due to a lowering of POMC expression in the hypothalamus. Collectively, these data argue that altered arcuate nucleus POMC or agouti-related peptide gene expression is un-likely to contribute to either the pathogenesis of cancer anorexia or the response of anorexia to treatment with a melanocortin antagonist.

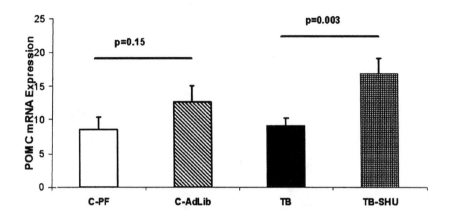

FIGURE 4. POMC mRNA expression in the rostral arcuate nucleus by *in situ* hybridization. Brains from four groups of third i.c.v. cannulated rats were prepared as described earlier. (1) Pair-fed control animals (C-PF, white bar, vehicle injected, $n = 6$); 2) *ad libitum* fed control animals (C-AdLib, *striped bar*, vehicle injected, $n = 7$); (3) anorexic tumor-bearing animals (TB, black bar, vehicle injected, $n = 9$); and (4) anorexic tumor-bearing animals injected with 0.35 nmol SHU9119 (TB-SHU, hatched bar, $n = 9$). Data are means ± SEM expressed as relative digital light units. P values determined by one-way ANOVA using the LSD *post-hoc* test.

CONCLUSIONS

Activation of the melanocortin system appears to be a key central pathway mediating inflammatory anorexia. The site(s) at which cytokine signaling or tumor products interact with the melanocortin pathway to decrease food intake remains elusive, but it does not appear to involve changes in hypothalamic POMC or agouti-related peptide gene expression. Ongoing efforts to clarify the pathophysiologic mechanisms of cancer anorexia and cachexia may identify novel therapeutic agents that may improve the treatment of cancer patients with the anorexia/cachexia syndrome.

REFERENCES

1. CONE, R.D. 1999. The central melanocortin system and energy homeostasis. Trends Endocrinol. Metab. **10:** 211–216.
2. BOSTON, B.A. 2001. Pro-opiomelanocortin and weight regulation: from mice to men. J. Pediatr. Endocrinol. Metab. **14:** 1409–1416.
3. YORK, D.A. 1999. Peripheral and central mechanisms regulating food intake and macronutrient selection. Obes. Surg. **9:** 471–479.
4. CUMMINGS, D.E. & M.W. SCHWARTZ. 2000. Melanocortins and body weight: a tale of two receptors. Nat. Genet. **26:** 8–9.
5. SCHWARTZ, M.W. *et al.* 2000. Central nervous system control of food intake. Nature **404:** 661–671.
6. BRUERA, E. 1997. ABC of palliative care. Anorexia, cachexia, and nutrition. Br. Med. J. **315:** 1219–1222.

7. GRUNFELD, C. & D.P. KOTLER. 1992. Pathophysiology of the AIDS wasting syndrome. AIDS Clin. Rev. **1992:** 191–224.
8. INUI, A. 2002. Cancer anorexia-cachexia syndrome: current issues in research and management. CA Cancer J. Clin. **52:** 72–91.
9. ALBRECHT, J.T. & T.W. CANADA. 1996. Cachexia and anorexia in malignancy. Hematol. Oncol. Clin. North Am. **10:** 791–800.
10. BRUERA, E. 1998. Pharmacological treatment of cachexia: any progress? Support Care Cancer **6:** 109–113.
11. MEGUID, M.M. *et al.* 1999. Differential feeding patterns induced by tumor growth and by TPN. Nutrition **15:** 555–562.
12. MARKS, D.L., N. LING & R.D. CONE. 2001. Role of the central melanocortin system in cachexia. Cancer Res. **61:** 1432–1438.
13. MYSTKOWSKI, P. *et al.* 2000. Validation of whole-body magnetic resonance spectroscopy as a tool to assess murine body composition. Int. J. Obes. Relat. Metab. Disord. **24:** 719–724.
14. PLATA-SALAMAN, C.R. 1996. Anorexia during acute and chronic disease. Nutrition **12:** 69–78.
15. PLATA-SALAMAN, C.R. 2000. Central nervous system mechanisms contributing to the cachexia-anorexia syndrome. Nutrition **16:** 1009–1012.
16. INUI, A. 1999. Cancer anorexia-cachexia syndrome: are neuropeptides the key? Cancer Res. **59:** 4493–4501.
17. TISDALE, M.J. 1997. Biology of cachexia. J. Natl. Cancer Inst. **89:** 1763–1773.
18. CERAMI, A. *et al.* 1987. Cachectin: a pluripotent hormone released during the host response to invasion. Recent Prog. Horm. Res. **43:** 99–112.
19. MEGUID, M.M. *et al.* 2000. Hypothalamic dopamine and serotonin in the regulation of food intake. Nutrition **16:** 843–857.
20. MALTONI, M. *et al.* 1997. Serum levels of tumour necrosis factor alpha and other cytokines do not correlate with weight loss and anorexia in cancer patients. Support Care Cancer **5:** 130–135.
21. NOGUCHI, Y. *et al.* 1996. Are cytokines possible mediators of cancer cachexia? Surg. Today **26:** 467–475.
22. CABAL-MANZANO, R. *et al.* 2001. Proteolysis-inducing factor is expressed in tumours of patients with gastrointestinal cancers and correlates with weight loss. Br. J. Cancer **84:** 1599–1601.
23. CERIANI, G. *et al.* 1994. Central neurogenic antiinflammatory action of alpha-MSH: modulation of peripheral inflammation induced by cytokines and other mediators of inflammation. Neuroendocrinology **59:** 138–143.
24. HUANG, Q.H., V.J. HRUBY & J.B. TATRO. 1999. Role of central melanocortins in endotoxin-induced anorexia. Am. J. Physiol. **276:** R864–R871.
25. GRUNFELD, C. *et al.* 1996. Endotoxin and cytokines induce expression of leptin, the ob gene product, in hamsters. J. Clin. Invest. **97:** 2152–2157.
26. FAGGIONI, R. *et al.* 1997. LPS-induced anorexia in leptin-deficient (ob/ob) and leptin receptor-deficient (db/db) mice. Am. J. Physiol. **273:** R181–R186.
27. LOPEZ-SORIANO, J. *et al.* 1999. Leptin and tumor growth in rats. Int. J. Cancer **81:** 726–729.
28. WISSE, B.E. *et al.* 2001. Reversal of cancer anorexia by blockade of central melanocortin receptors in rats. Endocrinology **142:** 3292–3301.
29. LAWRENCE, C.B. & N.J. ROTHWELL. 2001. Anorexic but not pyrogenic actions of interleukin-1 are modulated by central melanocortin-3/4 receptors in the rat. J. Neuroendocrinol. **13:** 490–495.
30. PRITCHARD, L.E., A.V. TURNBULL & A. WHITE. 2002. Pro-opiomelanocortin processing in the hypothalamus: impact on melanocortin signalling and obesity. J. Endocrinol. **172:** 411–421.
31. NIJENHUIS, W.A., J. OOSTEROM & R.A. ADAN. 2001. AgRP(83–132) acts as an inverse agonist on the human-melanocortin-4 receptor. Mol. Endocrinol. **15:** 164–171.
32. SCHWARTZ, M.W. *et al.* 1997. Leptin increases hypothalamic pro-opiomelanocortin mRNA expression in the rostral arcuate nucleus. Diabetes **46:** 2119–2123.

α-Melanocyte-Stimulating Hormone Is a Peripheral, Integrative Regulator of Glucose and Fat Metabolism

MILES B. BRENNAN,[a] JESSICA LYNN COSTA,[a] STACY FORBES,[a] PEGGY REED,[b] STEPHANIE BUI,[b] AND UTE HOCHGESCHWENDER[b]

[a]Eleanor Roosevelt Institute for Cancer Research, Denver, Colorado 80206, USA and University of Colorado—Health Sciences Center, Human Medical Genetics Program, Denver, Colorado 80262, USA

[b]Developmental Biology Program, Oklahoma Medical Research Foundation, Oklahoma City, Oklahoma 73104, USA

ABSTRACT: Melanocortins are known to affect feeding and probably insulin activity through the central nervous system. It was also recently shown that peripheral α-melanocyte-stimulating hormone (α-MSH) administration can reduce weight gain in both genetic and diet-induced obese mice. As obesity is often associated with disregulation of glucose and insulin, we investigated the nature of glucose homeostasis in the obese pro-opiomelanocortin (POMC) knockout mouse. Here we report that though they are obese, mice deficient in POMC (and, thereby, deficient in α-MSH) are euglycemic throughout their lives. While these mice are euinsulinemic, they are hypersensitive to exogenous insulin. This defect can be reversed through administration of α-MSH. We demonstrate that the actions of α-MSH in the periphery, known from our work to include lipid metabolism effects, are also involved in glucose homeostasis. These findings substantiate a pivotal role of the POMC gene products in integrating metabolism.

KEYWORDS: α-melanocyte-stimulating hormone (α-MSH); obesity; mouse; pro-opiomelanocortin (POMC); insulin; lipid; glucose; metabolism; energy homeostasis

INTRODUCTION

Lipids and glucose, in the form of glycogen, have complementary roles in the storage of energy: glycogen is a limited but quickly accessible store, while fat provides a much larger but less readily available store. The complementary roles of glucose/glycogen and free/esterified fatty acids underscore their need for integration. For example, as insulin levels rise in response to elevated blood glucose, lipid storage from free (non-esterified) fatty acids and *de novo* lipogenesis from glucose is increased.[1] Thus, insulin stimulates the accumulation of both carbohydrate and

Address for correspondence: Miles B. Brennan, Eleanor Roosevelt Institute for Cancer Research, 1899 Gaylord Street, Denver, Colorado 80206. Voice: 303-336-5600; fax: 303-333-8423.

miles@eri.uchsc.edu

Ann. N.Y. Acad. Sci. 994: 282–287 (2003). © 2003 New York Academy of Sciences.

lipid energy stores. Similarly, mobilization of lipid and carbohydrate stores depends on the perceived levels of these stores. For example, mouse *ob/ob* mutants are obese but perceive no fat stores owing to their lack of leptin.[2,3] Accordingly, these mutants are unable to mobilize energy stores for tasks such as maintaining body temperature in response to a cold challenge.

The mechanisms underlying this integration are incompletely understood. Two non-exclusive mechanisms, hormonal and neurological, have been proposed. However, both proposed mechanisms employ neuropeptides, including those encoded by the pro-opiomelanocortin (POMC) gene; thus, the POMC gene products are a possible mechanism for hormonal and neurological integrations of metabolism. For example, α-melanocyte-stimulating hormone (α-MSH) acts both as a neurotransmitter in the arcuate nucleus of the hypothalamus to regulate appetite[4] and as a peripheral hormone to modulate fat metabolism.[5,6]

The central nervous system effects of the melanocortin system on appetite, energy homeostasis, and insulin[7] have been explored, but cannot alone explain the "whole body" phenotypes of melanocortin disregulation in animals and man.[8–13] We have previously shown that peripheral α-MSH levels of *ob/ob* mice are low but rise in response to leptin,[6] which is released in proportion to adipose stores.[14] In addition, peripherally administered α-MSH decreases weight gain in genetically obese, POMC-null mice[4,5] without a corresponding decrease in food intake. Peripheral administration of α-MSH also effectively reduced weight in diet-induced obese mice.[15] The role of α-MSH in lipid usage is also seen in the improved thermoregulatory response to cold by the *ob/ob* mouse: Forbes *et al.*[6] showed that α-MSH administered peripherally to *ob/ob* mice was able to attenuate the drop in body temperature during a cold challenge. These data support the finding that α-MSH mobilizes lipid stores for metabolism, an aspect of α-MSH activity not previously reported. This hypothesis is substantiated by the finding that serum-free fatty acid levels in *ob/ob* mice treated

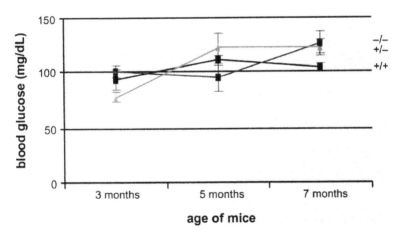

FIGURE 1. POMC-null mice are euglycemic throughout life. Blood glucose levels at different ages (3, 5, and 7 months) were tested in POMC-null mutants and heterozygous and wild-type mice. Blood glucose levels in POMC-null mutants were at no time out of the range of normal values. Blood glucose was measured using a Bayer Glucometer Elite.

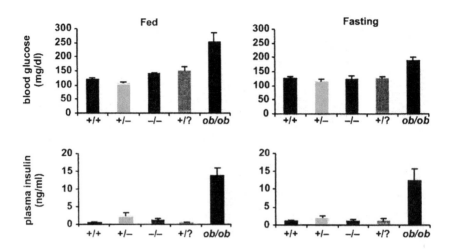

FIGURE 2. Normal glucose and insulin levels in POMC-null mutants. Blood glucose and plasma insulin levels were measured in 5-month-old females (five mice per group), either in the morning (fed state) or after an overnight fast (fasting). Mice included POMC-null mutants as well as heterozygous and wild-type littermates; controls included *ob/ob* mutants as well as wild-type or heterozygous (+/?) littermates. Blood glucose was measured using a Bayer Glucometer Elite. Plasma insulin was determined using the Linco RIA kit. The *ob/ob* mutants are hyperglycemic and hyperinsulinemic in the fed and fasted states. POMC-null mutants are indistinguishable from wild-type and heterozygous littermates.

with α-MSH are higher than in untreated *ob/ob* animals. Here we examine the effects of POMC deficiency on the glycemic levels and insulin sensitivity of the mouse as well as the effects of resupplementation of POMC hormones in an insulin challenge. The peripheral actions of α-MSH are complementary to central nervous system effects on these processes.

RESULTS

In the POMC-knockout mouse,[5] we see not only lipid disregulation resulting in obesity but also an insulin defect. That is, despite their obesity, these animals are euglycemic, but hypersensitive to insulin. FIGURE 1 shows glucose levels of wild-type, heterozygous, and homozygous POMC-null mice at 3, 5, and 7 months of age. Additionally, the POMC-null mice also have normal glucose and normal insulin levels when measured in both fed and fasted states. This is in marked contrast to the *ob/ob* mutant mice, which have elevated glucose and insulin levels (FIG. 2). Despite this euglycemia, POMC-null mutant mice lack a counter-regulatory response in an insulin-tolerance test (FIG. 3). Their plasma glucose concentrations fall to lethal levels without evidence of a counter-response. Reconstitution of POMC-null mice with exogenous α-MSH or adrenocorticotropin-stimulating hormone (ACTH) during an insulin-tolerance test (FIG. 4) show that α-MSH attenuates the severe hypoglycemia, while ACTH actually seems to exacerbate this phenomenon.

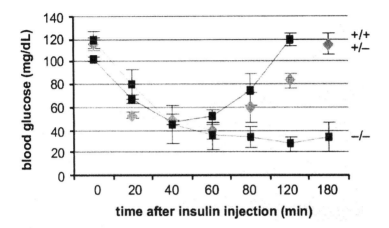

FIGURE 3. Lack of glucoregulatory response in POMC-null mutants during an insulin-tolerance test. Five-month-old females, five per group, were fasted for 6 h followed by insulin injection (human insulin, 1 U/kg, i.p.). Blood was collected at the times indicated, and glucose measured using a Bayer Elite Glucometer. Blood glucose levels were not significantly different between wild-type and heterozygous mice. Glucose levels were significantly lower in POMC-null mutants compared to wild-type littermates after the first 60 min past insulin injection (80 min: $P < 0.05$; 120 min: $P < 0.0001$; 180 min: $P < 0.0001$).

DISCUSSION

The effects of eating-mediated melanocortins in the central nervous system have been investigated in multiple rodent models[16] and macaques.[17] The peripheral immunological effects of melanocortins have also been of continued interest.[18] That α-MSH acts in the periphery as a lipolytic-enhancing factor except in the rabbit[19] was unappreciated until the analysis of our POMC-null mouse.[5] Even more unexpected was the effect of α-MSH on insulin and glucose homeostasis. These data clearly indicate that peripheral α-MSH plays a role in glucose and insulin homeostasis. As only an extremely small fraction of circulating α-MSH is able to permeate the blood-brain barrier,[20] it is highly unlikely that the central melanocortin system exercises exclusive and total control over bodily energy homeostasis.

This α-MSH regulation of lipid usage for energy and other basic body maintenance, discouragement of lipid storage, as well as α-MSH's effect on insulin sensitivity leads us to propose a pivotal role for α-MSH in integrating both glucose and fat metabolism in the periphery. α-MSH, leptin, and insulin probably cooperate to integrate fat and glucose metabolism. Our previous report, which showed that leptin can increase peripheral α-MSH concentration and that peripheral α-MSH slows weight gain when administered to genetically obese mice, has added significance in light of the data presented here, which show that the metabolism of the POMC-null mouse exhibits euglycemic obesity with high insulin sensitivity as well as a disregulation of lipid metabolism. That is, this combination of results demonstrate that α-MSH has previously been underestimated as a *peripheral*, integrative regulator of glucose and fat metabolism.

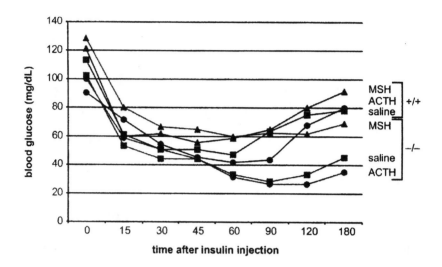

FIGURE 4. Effect of α-MSH vs. ACTH vs. saline during an insulin-tolerance test. Glucose was measured in 4–6-month-old female mice, wild-types, and POMC-null mutants (five mice per group). Mice were fasted for 5 h. Each genotype (wild-type and POMC-null mutants) received either 0.1 mL saline (i.p.), or 0.1 mL saline containing 1 μg α-MSH (i.p.), or 0.1 mL saline containing 1 μg ACTH (s.c.). One hour later, human insulin (0.5 U/g body weight) was injected intraperitoneally in all mice, and glucose was measured by using a Bayer Glucometer Elite at the indicated intervals on blood collected from a tail nick. Blood glucose levels were not significantly different between any of the wild-type groups. They also were not significantly different between the α-MSH-reconstituted POMC-null mutants and their wild-type littermates. Significant differences were found only between wild-types and either saline- or ACTH-treated POMC-null mutants. (Saline-treated wildtype vs. mutant at 90 min: $P < 0.05$; 120 min: $P < 0.01$; 180 min: $P < 0.05$. ACTH-treated wild-type versus mutants at 90 min: $P < 0.01$; 120 min: $P < 0.05$; 180 min $P < 0.05$).

REFERENCES

1. KERSTEN, S. 2001. Mechanisms of nutritional and hormonal regulation of lipogenesis. EMBO Rep. **2:** 282–286.
2. CAMPFIELD, L.A., F.J. SMITH, Y. GUISEZ, R. DEVOS & P. BURN. 1995. Recombinant mouse OB protein: Evidence for a peripheral signal linking adiposity and central neural networks. Science **269:** 546–549.
3. PELLEYMOUNTER, M.A., M.J. CULLEN, M.B. BAKER, R. HECHT, et al. 1995. Effects of the obese gene product on body weight regulation in ob/ob mice [see comments]. Science **269:** 540–543.
4. WILLIAMS, G., C. BING, X.J. CAI, J.A. HARROLD, et al. 2001. The hypothalamus and the control of energy homeostasis: Different circuits, different purposes. Physiol. Behav. **74:** 683–701.
5. YASWEN, L., N. DIEHL, M. B. BRENNAN & U. HOCHGESCHWENDER. 1999. Obesity in the mouse model of pro-opiomelanocortin deficiency responds to peripheral melanocortin [see comments]. Nat. Med. **5:** 1066–1070.
6. FORBES, S., S. BUI, B.R. ROBINSON, U. HOCHGESCHWENDER & M.B. BRENNAN. 2001. Integrated control of appetite and fat metabolism by the leptin-proopiomelanocortin pathway. Proc. Natl. Acad. Sci. USA **98:** 4233–4237.

7. FAN, W., D.M. DINULESCU, A.A. BUTLER, J. ZHOU, *et al.* 2000. The central melanocortin system can directly regulate serum insulin levels. Endocrinology **141:** 3072–3079.

8. BUTLER, A.A., R.A. KESTERSON, K. KHONG, M. J. CULLEN, *et al.* 2000. A unique metabolic syndrome causes obesity in the melanocortin-3 receptor-deficient mouse. Endocrinology **141:** 3518–3521.

9. CHEN, A.S., D.J. MARSH, M.E. TRUMBAUER, E.G. FRAZIER, *et al.* 2000. Inactivation of the mouse melanocortin-3 receptor results in increased fat mass and reduced lean body mass. Nat. Genet. **26:** 97–102.

10. VAISSE, C., K. CLEMENT, B. GUY-GRAND & P. FROGUEL. 1998. A frameshift mutation in human MC4R is associated with a dominant form of obesity [letter]. Nat. Genet. **20:** 113–114.

11. YEO, G.S., I.S. FAROOQI, S. AMINIAN, D.J. HALSALL, *et al.* 1998. A frameshift mutation in MC4R associated with dominantly inherited human obesity [letter]. Nat. Genet. **20:** 111–112.

12. KRUDE, H., H. BIEBERMANN, W. LUCK, R. HORN, *et al.* 1998. Severe early-onset obesity, adrenal insufficiency and red hair pigmentation caused by POMC mutations in humans. Nat. Genet. **19:** 155–157.

13. CHALLIS, B.G., L.E. PRITCHARD, J.W. CREEMERS, J. DELPLANQUE, *et al.* 2002. A missense mutation disrupting a dibasic prohormone processing site in pro-opiomelanocortin (POMC) increases susceptibility to early-onset obesity through a novel molecular mechanism. Hum. Mol. Genet. **11:** 1997–2004.

14. HAMILTON, B.S., D. PAGLIA, A.Y. KWAN & M. DEITEL. 1995. Increased obese mRNA expression in omental fat cells from massively obese humans. Nat. Med. **1:** 953–956.

15. PIERROZ, D.D., M. ZIOTOPOULOU, L. UNGSUNAN, S. MOSCHOS, *et al.* 2002. Effects of acute and chronic administration of the melanocortin agonist MTII in mice with diet-induced obesity. Diabetes **51:** 1337–1345.

16. MILLINGTON, G.W., Y.C. TUNG, A.K. HEWSON, S. O'RAHILLY & S.L. DICKSON. 2001. Differential effects of alpha-, beta- and gamma(2)-melanocyte-stimulating hormones on hypothalamic neuronal activation and feeding in the fasted rat. Neuroscience **108:** 437–445.

17. KOEGLER, F.H., K.L. GROVE, A. SCHIFFMACHER, M.S. SMITH & J. L. CAMERON. 2001. Central melanocortin receptors mediate changes in food intake in the rhesus macaque. Endocrinology **142:** 2586–2592.

18. LIPTON, J.M. & A. CATANIA. 1997. Anti-inflammatory actions of the neuroimmunomodulator alpha-MSH. Immunol. Today **18:** 140–145.

19. RAMACHANDRAN, J. & V. LEE. 1976. Divergent effects of adrenocortincotropin and melanocortin on isolated rat and rabbit adipocytes. Biochim. Biophys. Acta **428:** 339–346.

20. DE ROTTE, A.A., H.J. BOUMAN & T.B. VAN WIMERSMA GREIDANUS. 1980. Relationships between alpha-MSH levels in blood and in cerebrospinal fluid. Brain Res. Bull. **5:** 375–381.

Accessory Proteins for Melanocortin Signaling

Attractin and Mahogunin

LIN HE,[a] ADAM G. ELDRIDGE,[b] PETER K. JACKSON,[b] TERESA M. GUNN,[c] AND GREGORY S. BARSH[a]

[a]Departments of Pediatrics and Genetics and the Howard Hughes Medical Institute, Stanford University School of Medicine, Stanford, California 94305, USA

[b]Department of Pathology, Stanford University School of Medicine, Stanford, California 94305, USA

ABSTRACT: Switching from eumelanin to pheomelanin synthesis during hair growth is accomplished by transient synthesis of Agouti protein, an inverse agonist for the melanocortin-1 receptor (Mc1r). The coat color mutations *mahogany* and *mahoganoid* prevent hair follicle melanocytes from responding to Agouti protein. The gene mutated in *mahogany*, which is also known as *Attractin (Atrn)*, encodes a type I transmembrane protein that functions as an accessory receptor for Agouti protein. We have recently determined that the gene mutated in *mahoganoid*, which is also known as *Mahogunin (Mgrn1)*, encodes an E3 ubiquitin ligase. Like *Attractin*, *Mahogunin* is conserved in invertebrate genomes, and its absence causes a pleiotropic phenotype that includes spongiform neurodegeneration.

KEYWORDS: Agouti; Attractin; coat color; mahogany; melanocortin receptor; mouse genetics; neurodegeneration; pigmentation

INTRODUCTION

Melanocortin ligands and their receptors were discovered using physiologic and biochemical approaches, but over the last ten years genetics has come to play an increasingly prominent role in our understanding of melanocortin biology. Our laboratory first became interested in melanocortin signaling based on previous genetic studies of pigment type-switching in laboratory mice.[1,2] The ability of hair follicle melanocytes to switch between synthesizing black/brown eumelanin and red/yellow pheomelanin is responsible for a variety of coat color patterns across a broad range of mammalian species. Mutations that alter these patterns in laboratory mice have been collected and characterized over the last century, and have helped to provide the foundation for understanding Mendelian gene action and interaction.[3]

[c]Current address: Department of Biomedical Sciences, Cornell University, Ithaca, NY 14853.

Address for correspondence: Gregory S. Barsh, Departments of Pediatrics and Genetics and the Howard Hughes Medical Institute, Stanford University School of Medicine, Stanford, CA 94305. Voice: 650-723-5061; fax: 650-723-1399.

gbarsh@cmgm.stanford.edu

Ann. N.Y. Acad. Sci. 994: 288–298 (2003). © 2003 New York Academy of Sciences.

The most common coat color pattern is known as the Agouti phenotype, and consists of a subapical light or pheomelanic band on individual hairs that are otherwise dark or eumelanic. Distributed over the entire body surface, this produces an overall brushed appearance useful for camouflage.[4] As we know now, transient expression of the *Agouti* gene product (known as Agouti protein) during hair growth is directly responsible for pheomelanin synthesis. However, even before the *Agouti* gene was cloned and before the discovery of α-melanocyte-stimulating hormone (α-MSH), the Agouti phenotype was known to be regulated by at least five or six different genes, including *Agouti* itself, *extension* (now recognized as the *melanocortin-1 receptor, Mc1r*), *mahogany, mahoganoid, umbrous,* and *dorsal dark stripe.*[1,3,4]

Pigment type-switching and modification of the Agouti phenotype is a useful model system for several reasons. Mutations recognized by their effect on Agouti banding often affect specific regions of the body or have pleiotropic effects, and as such, provide insight into developmental patterning or physiologic processes outside the pigmentary system.[5] Most important, subtle changes in gene expression that might otherwise escape detection in a biochemical assay are easily recognized by a change in coat color, and a large number of available genetic reagents and techniques allow certain signaling pathways to be studied with a sophistication and depth usually associated with invertebrate model organisms.

In what follows, we summarize briefly biochemical and genetic work on melanocortin receptors and their ligands, Agouti protein and Agouti-related protein (Agrp), that provides the framework for understanding the gene products of *Attractin (Atrn)* and *Mahogunin (Mgrn1)*, the genes mutated in *mahogany* and *mahoganoid*, respectively. Recent work from our laboratory on the identification and analysis of Atrn and Mgrn1 are presented in the context of current thinking about seven-transmembrane-receptor signaling in other systems. Finally, we speculate how the Attractin-Mahogunin pathway, whose origins and functions are more ancient and pleiotropic than those of melanocortins, may have been recruited by the pigment type-switching apparatus during early vertebrate evolution.

THE AGOUTI-MELANOCORTIN RECEPTOR SYSTEM IN PIGMENTATION AND BODY WEIGHT REGULATION

Recognition that mice homozygous for the *recessive yellow (e)* allele of *extension* were resistant to the ability of exogenous α-MSH to cause darkening provided the first clue nearly forty years ago to a loss-of-function mutation in the *melanocortin-1 receptor (Mc1r).*[6] Shortly after melanocortin receptor genes were cloned, Robbins *et al.*[7] discovered molecular alterations of the *Mc1r* in several *extension* alleles, and *recessive yellow* or *e* is now known as *Mc1r^e*. Similarly, sensitivity to exogenous α-MSH in dominant gain-of-function *Agouti* mutations such as *lethal yellow (A^y)*[8,9] provided the conceptual framework for demonstrating that Agouti protein is an inhibitory ligand of the Mc1r (FIG. 1A).[10,11]

Agouti protein is normally produced by specialized dermal cells at the base of each hair follicle and has a short range of action that does not extend to adjacent hair follicles.[3,12] In *A^y* or similar dominant *Agouti* mutations, however, genomic rearrangements or retrotransposon insertions cause ectopic and ubiquitous expression of Agouti protein, which has pleiotropic effects, including a completely pheomela-

A

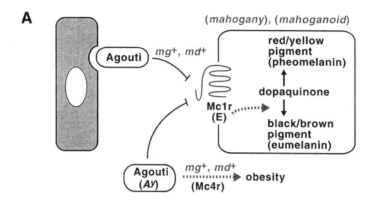

B   *In the hair follicle...*

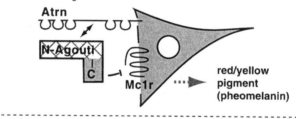

In the brain...

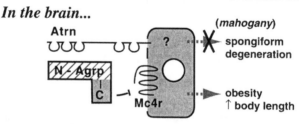

FIGURE 1. *Agouti, Mc1r, mahogany,* and *mahoganoid.* (**A**) Genetic pathway based on double-mutant studies among *Agouti, Mc1r, mahogany* (*mg*), or *mahoganoid* (*md*). As described in the text, *mahogany* and *mahoganoid* are genetically downstream of *Agouti* but genetically upstream of *Mc1r*. The *mahogany* and *mahoganoid* mutations represent a loss of function; therefore, the wild-type gene products, mg^+ and md^+, respectively, are required for Agouti signaling. The *mg* and *md* mutations suppress the effects of the gain-of-function *Agouti* allele A^y on both coat color and obesity. (**B**) *Atrn* is mutated in *mahogany.* The product of the *Atrn* gene is a large receptor-like molecule that interacts with the amino-terminal domain of Agouti protein. In the brain, deficiency of Atrn (in *mahogany* mutant mice) causes spongiform degeneration, but the normal ligand for the Mc4r, Agrp, does not interact with Atrn. The results and diagrams are based on studies described in Refs. 17, 20, and 23.

nic coat and an obesity-overgrowth syndrome.[5,13,14] In hindsight, obesity and increased body length caused by the A^y mutation are explained by the ability of ectopic Agouti protein to partially mimic the effects of Agrp, a hypothalamic neuropeptide that normally signals through the Mc3r and Mc4r (FIG. 1B).[15,16]

Besides administration of exogenous α-MSH, gain-of-function mutations that constitutively activate the Mc1r also cause hair follicle melanocytes to produce eumelanin in the presence of Agouti protein.[3,7] However, the Mc1r is normally exposed to very low levels, if any, of endogenous melanocortin ligands; instead, basal levels of Mc1r signaling are sufficient for constitutive eumelanin synthesis, and provide a background whereby production of Agouti protein inhibits Mc1r signaling to trigger pheomelanin synthesis.

Studies of animals doubly mutant for pigment type-switching mutations, or examples of epistasis analysis, have played a key role in our understanding of both Agouti-melanocortin signaling and the role of Attractin and Mgrn1. Animals carrying gain-of-function mutations at both *Mc1r* (normally black) and *Agouti* (normally yellow) exhibit a black (that is, eumelanic) coat color, while animals carrying loss-of-function mutations at both *Agouti* (normally black) and *Mc1r* (normally yellow) exhibit a yellow (that is, pheomelanic) coat color.[3] Thus, *Mc1r* is genetically downstream of *Agouti*, as expected for the relationship of a receptor to its ligand.

ATTRACTIN AND MAHOGUNIN AS ACCESSORY SIGNALING MOLECULES FOR AGOUTI PROTEIN

Studies of the *mahogany* and *mahoganoid* mutations reveal that the relationship between Agouti protein and the Mc1r is more complex than that of ligand and receptor.[17] As suggested by their names, *mahogany* and *mahoganoid* cause a similar darkened (that is, eumelanic) phenotype nearly identical to the phenotype of animals that carry *Mc1r* gain-of-function or *Agouti* loss-of-function mutations.[18,19] However, in A^y/a; $Atrn^{mg-3j}/Atrn^{mg-3j}$ or A^y/a; $Mgrn1^{md-nc}/Mgrn1^{md-nc}$ animals (the $Atrn^{mg-3j}$ and $Mgrn1^{md-nc}$ alleles are null mutations of *mahogany* and *mahoganoid*, respectively), the effects of the A^y mutation on both coat color and obesity are completely suppressed, and the double-mutant animals are black and thin (FIG. 1A). Conversely, in $Mc1r^e/Mc1r^e$; $Atrn^{mg-3j}/Atrn^{mg-3j}$ or $Mc1r^e/Mc1r^e$; $Mgrn1^{md}/Mgrn1^{md}$ animals, the pigmentary effects of the *Atrn* or *Mgrn1* alleles are completely suppressed, and the double-mutant animals are yellow. Thus, *Atrn* and *Mgrn1* lie genetically downstream of *Agouti* yet genetically upstream of the *Mc1r* (FIG. 1A).[17,20]

The original *mahoganoid* (*md*) mutation arose in the C3H/HeJ strain in 1960, and was named after its similarity to *mahogany*.[19] Another allele, initially named non-agouti curly, was first described in 1963, but was not recognized as an allele of *mahoganoid* until 1979 (when it was renamed md^{nc}).[21,22] The effects of the md^{nc} allele on coat color are more severe than that of *md*, and almost identical to those of $Atrn^{mg-3J}$. However, animals homozygous for $Mgrn1^{md-nc}$ have curly whiskers, develop premature graying in some hairs, and develop more severe behavioral abnormalities as they age than $Atrn^{mg-3J}$ mutant mice.

Atrn is a large single transmembrane-spanning protein that is widely expressed, and whose ectodomain contains several motifs found in molecules implicated in axon guidance.[23,24] Partial loss of function for *Atrn* causes the original *mahogany* mutation, and complete loss of function causes the mg^{3J} mutation. Molecular insight into the relationship between Agouti protein, Mc1r, and Atrn became apparent when we found that the amino-terminal domain of Agouti protein binds specifically to the ectodomain of Atrn (FIG. 1B).[20] Because determinants for melanocortin receptor binding were known to lie in the carboxy-terminal domain of Agouti protein, these observations suggested the possibility that Atrn might serve as an accessory receptor for Agouti protein, either by stabilizing the interaction between the carboxy-terminal domain of Agouti protein and the Mc1r, or by increasing the local concentration or stability of Agouti protein at the melanocyte cell surface. The interaction between Agouti protein and Atrn is relatively weak, but its significance is apparent from a genetic experiment: the amino-terminal domain of Agrp does not interact with the ectodomain of Atrn, and obesity induced by a ubiquitously expressed Agrp transgene is not suppressed by mutations of Atrn (FIG. 1B).[20]

In pharmacologic or *in vitro* studies, the estimated interaction between the carboxy-terminal domain of Agouti protein and Mc1r is about 500-fold stronger than the interaction between the amino-terminal domain of Agouti protein and Atrn,[25] and we do not yet know if the three molecules form a ternary complex. However, two observations suggest that Atrn may do more than simply facilitate Agouti binding to the Mc1r.[20] First, expression of Atrn is an absolute requirement for Agouti signaling *in vivo*. Several glycosaminoglycan-containing membrane proteins, including syndecan, which has been proposed as an accessory receptor for Agrp, enhance or otherwise modulate signaling by their cognate ligands.[26,27] By contrast, $Atrn^{mg-3j}$ mutant animals are *completely* resistant to the effects of Agouti, which suggests that Atrn plays a critical and integral role in Agouti signaling. Second, Mgrn1, like Atrn, is absolutely required for Agouti signaling, but Mgrn1 is an intracellular protein (see below). Because Mgrn1 does not affect levels of Atrn expression, it is likely to represent a downstream signaling component whose activity is modulated by Agouti binding.

SPONGIFORM ENCEPHALOPATHY IN *MAHOGANY* AND *MAHOGANOID* MUTANT MICE

The observation that $Atrn^{mg-3J}$ suppressed Agouti-induced obesity but not Agrp-induced obesity was somewhat perplexing, given that Agrp is normally expressed in the brain but Agouti protein is not. In addition, we found that *Atrn* is conserved in distantly related metazoan genomes, with homologues in worms and flies; by contrast, Agouti protein, Agrp, and melanocortin receptors are found only in vertebrates. Taken together with the patterns of mammalian expression—*Atrn* is expressed in many different organs while *Agouti* and the *Mc1r* are normally expressed only in the skin—we suspected that Atrn must be required for biological processes besides pigment type-switching, and carried out a comprehensive histopathologic survey of *Atrn* mutant mice. Surprisingly, we found that animals carrying a complete loss-of-function *Atrn* mutation, $Atrn^{mg-3J}$, exhibited widespread vacuolation and neuronal loss.[20] Further characterization revealed that vacuolation was accompanied by neuronal cell loss and astrocytic infiltration, and that the level of vacuolation was inversely related to the amount of normal *Atrn* mRNA present.[28,29]

We also found neurodegeneration in $Mgrn1^{md-nc}$ (but not in $Mgrn1^{md}$) mutant mice that was very similar to that observed in $Atrn^{mg-3J}$ mutant mice.[28] The onset of vacuolation in *mahoganoid* mutant mice is delayed 2–3 months compared to $Atrn^{mg-3J}$ mutant mice, but is otherwise identical, demonstrating similarities with respect to the progression, association with astrocytosis, and presence of neuronal cell loss.

Mutations of *Atrn* have also been shown to cause spongiform encephalopathy and hypomyelination in rats (*zitter*)[30] and hamsters (*black tremor*),[31] but to date, mutations of *Mgrn1* have been recognized only in laboratory mice. The pigmentary phenotypes of *Atrn* mutations in rats, mice, and hamsters are nearly identical, but the central nervous system defects in *Atrn* and *Mgrn1* mutant mice went unrecognized for several decades. In hindsight, this likely reflects the nature of the original $Atrn^{mg}$ and $Mgrn1^{md}$ alleles, which cause only a partial loss of function, and the presence of modifier genes in C3H mice that may ameliorate some of the effects on myelination.[29]

Many aspects of the spongiform encephalopathy in $Atrn^{mg-3J}$ or $Mgrn1^{md-nc}$ mutant animals are similar to prion diseases; however, tests for protease-resistant forms of prion protein in $Atrn^{mg-3J}$, $Mgrn1^{md-nc}$, and *zitter* mutant animals are negative.[28,32] Thus, in many respects, mutations of *Atrn* or *Mgrn1* cause a prion-like disease without prions.

IDENTIFICATION AND CHARACTERIZATION OF MAHOGUNIN (MGRN1)

Work from our laboratory[28] and that of Leibel and colleagues[33] has recently identified the gene mutated in *mahoganoid* as one that predicts an intracellular protein with a central C3HC4 RING domain (FIG. 2A). The RING domain has been identified in a number of transmembrane, cytoplasmic, and nuclear proteins; it is a type of "zinc finger" that contains two sites for coordinating zinc binding. Besides the RING domain, Mgrn1 contains no other protein motifs; however, homologues for Mgrn1 in the *Drosophila melanogaster* and *C. elegans* genomes are easily recognized by two blocks of sequence similarity, one in the amino-terminal region and one that contains the RING domain (FIG. 2A). *Mgrn1* has two alternative exons and gives rise to 4 mRNA isoforms that predict proteins of 532, 554, 556, or 578 amino acids (FIG. 2A and B). Differences between the isoforms affect carboxy-terminal regions of the protein; the isoforms are conserved between the human and mouse genomes, but their functional significance is not known.

Some RING-containing proteins function as transcription factors; other RING-containing proteins including Neuralized, Cbl, Hiah1, and Siah1 have been implicated in protein metabolism as E3 ubiquitin ligases, which in conjunction with a ubiquitin-activating enzyme (E1) and a ubiquitin-conjugating enzyme (E2), facilitate the transfer of ubiquitin to specific protein substrates to target them for intracellular trafficking and/or proteasomal degradation.[34,35] The RING domains of these different E3 ligases share little or no sequence similarity other than the Cys and His residues themselves, and are as different from each other as they are from the Mgrn1 RING domain (FIG. 2A). We investigated whether Mgrn1 might function as an E3 ubiquitin ligase because this activity would provide a potential explanation for the pathogenesis of neurodegeneration in *mahoganoid* mutant mice (see below).

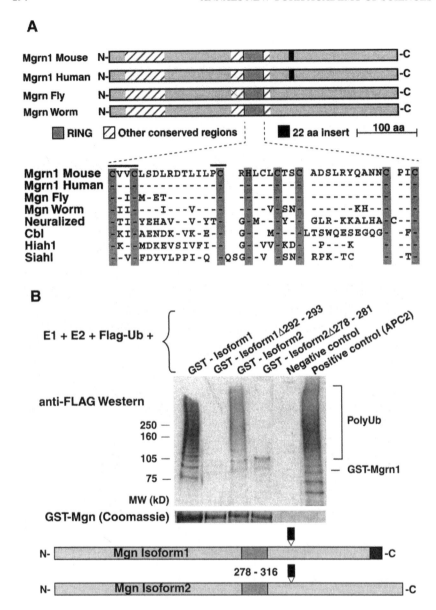

FIGURE 2. Evolutionary conservation and E3 ubiquitin ligase activity of Mahogunin. (**A**) Alignment of RING domains for mouse, human, fly (CG9941), and worm (CAA94116) Mgrn1 homologues with other RING domain-containing proteins: Neuralized (AAK84420), Cbl (A43817), Hiap1 (S68449), Siah1 (CAC35542). Dashes indicate sequence identity; residues that coordinate zinc binding are indicated in grey. Bars indicate the regions that were deleted from GST-Mahogunin expression constructs. (**B**) GST-Mahogunin fusion proteins, variants with internal deletions that disrupt the RING domain, or GST protein alone were purified and tested for E3 ubiquitin ligase activity in an *in vitro* reconstitution assay. Formation of polyubiquitinated GST-Mahogunin is ATP-dependent, requires an E1 and E2 protein, and appears similar to that of a known RING-containing E3 ubiquitin ligase, APC2. Partially adapted from Ref. 28.

In the presence of a specific E2, many E3 proteins exhibit auto-ubiquitination whether or not the endogenous target is present. To test whether or not Mgrn1 has such an activity, we developed a reconstitution assay for ubiquitination *in vitro* that contained FLAG epitope-tagged ubiquitin, an E1 protein (UBA1), and an E2 protein (UBC5) that is used frequently in the stress response to unfolded proteins. After gel electrophoresis and Western blotting of the reaction products, we probed with anti-FLAG antibody to detect polyubiquitinated derivatives of GST-Mahogunin. We tested two Mahogunin isoforms, and found that both exhibited a characteristic ladder indicative of auto-ubiquitination and E3 ligase activity (FIG. 2B). Auto-ubiquitination was not observed with variant forms of Mgrn1 in which critical Cys residues were deleted (Δ278–281 and Δ292–293) (FIG. 2B), nor was auto-ubiquitination observed with an E2 protein, UBC3, that is used frequently in cell-cycle transitions. Thus, although the endogenous substrate is not yet known, Mgrn1 is likely to function as an intracellular E3 ubiquitin ligase, and it is loss of this function that causes both the inability to respond to Agouti protein and spongiform encephalopathy.

HOW IS *ATTRACTIN* CONNECTED TO *MAHOGUNIN*?
UNRESOLVED QUESTIONS AND
FUTURE DIRECTIONS

Several observations suggest that *Atrn* and *Mgrn1* are closely linked components of a conserved biochemical and genetic pathway. Besides similar pigmentary and neurodegeneration phenotypes, both genes lie in the same epistasis group, downstream of *Agouti* but upstream of *Mc1r*, and both genes exhibit a similar pattern of evolutionary conservation. In addition, the patterns of mRNA expression of the two genes are remarkably similar, across different tissues as determined by Northern hybridization, and within the brain as determined by *in situ* hybridization.[23,28,36] An attractive hypothesis to explain these observations posits that ubiquitination of an as yet unidentified substrate by Mgrn1 is regulated by extracellular signals that act through Atrn. In the hair follicle, the extracellular signal is Agouti protein, and substrate ubiquitination would be required for Agouti-induced pheomelanogenesis. In the brain, the extracellular signal is not yet known, but substrate accumulation would represent the proximate cause of neuronal cell loss, astrocytosis, and progressive vacuolation (FIG. 3).

One prediction of this hypothesis is that *Mgrn1* lies genetically downstream of *Atrn*; thus an *Mgrn1* transgene might be able to rescue pigmentary and/or neurodegeneration phenotypes. In addition, candidate substrates for Mgrn1 might be able to induce neurodegeneration when overexpressed, and protect against neurodegeneration when ablated. Finally, if the intracellular domain of Atrn is required for its ability to activate Mgrn1, then biochemical machinery should exist that connects the intracellular domain of Atrn to Mgrn1. Some of these predictions can be tested with reagents that are currently available.

How can we explain a connection between pigmentation and neurodegeneration? *Atrn* and *Mgrn1* exhibit deeper evolutionary conservation than *Agouti* and *Mc1r*, as well as broader patterns of expression, and mutant phenotypes that are more pleiotropic. Thus, the neuronal functions of Atrn and Mgrn1 may derive from an evolutionarily conserved pathway that was recruited during vertebrate evolution by *Agouti*

In the hair follicle...

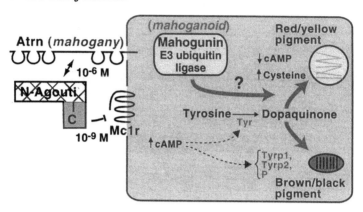

In the brain...

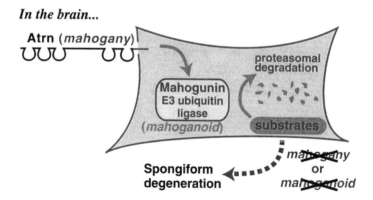

FIGURE 3. Models for the role of Atrn and Mahogunin in pigment-type switching and neurodegeneration. In the hair follicle, activity of Mahogunin may represent an unknown component of the pheomelanogenic pathway or may be required for Mc1r internalization as described in the text. In the brain, Mahogunin is likely to ubiquitinate an unknown intracellular substrate and target it for intracellular degradation. For both processes, activity of Mahogunin is dependent on, and downstream of, Atrn; thus, loss of Atrn or Mahogunin can prevent pheomelanogenesis and cause accumulation of the substrate for Mahogunin leading to spongiform degeneration.

and *Mc1r* to regulate pigment type-switching. If so, the substrate for Mgrn1 in melanocytes would likely be the same or closely related to the substrate for Mgrn1 in neurons. One potential substrate is the melanocortin receptor itself or a receptor-associated protein, because, for some seven-transmembrane-domain proteins, ubiquitination plays a role in receptor internalization, trafficking, and turnover. According to this hypothesis, Agouti protein binding to Atrn would induce ubiquitination of the Mc1r or an Mc1r-associated protein, leading to Mc1r downregulation and decreased Mc1r signaling. This scenario could explain why Atrn and Mgrn1 are genetically upstream of Mc1r yet downstream of Agouti protein.

In the brain, Mgrn1 might ubiquitinate the Mc4r or an Mc4r-associated protein in a manner that is dependent on Atrn, and that can be activated by Agouti protein but not by Agrp (FIG. 3). However, the expression of *Atrn* and *Mgrn1* brain mRNA is much broader than that of *Mc4r*, and neurodegeneration caused by mutations of *Atrn* or *Mgrn1* affects all regions of the brain, not just those that express *Mc4r*. Additional insight into these questions is likely to come from studies directed at identifying potential Mgrn1 substrates, and testing whether loss of Atrn or Mgrn1 alters melanocortin receptor internalization and/or downregulation.

Why is vacuolation milder in $Mgrn1^{md-nc}$ mutant animals than in $Atrn^{mg-3J}$ mutant animals if both represent null alleles and $Mgrn1$ is downstream of $Atrn$? The most likely explanation is cross-talk among closely related genes, since there is one additional paralogue for both genes in both the mouse and human genomes. Tentatively named *Mgrn2* and *Atrn2*, these genes show similar though not identical patterns of expression to *Mgrn1* and *Atrn*, and may play overlapping and/or redundant roles. For example, Atrn may activate both Mgrn1 and Mgrn2, with Mgrn1 playing the predominant role in melanocytes but both the *Mgrn1* and *Mgrn2* genes serving as downstream components in the brain. As is the case for Agouti-Mc1r signaling, answers to these questions are likely to come from a combination of forward and reverse genetics, and should provide additional insight into both pigment type-switching and the pathogenesis of neurodegeneration.

ACKNOWLEDGMENTS

This work was supported by grants from the National Institutes of Health (awarded to G.S.B.) and the American Heart Association (awarded to T.M.G.).

REFERENCES

1. JACKSON, I.J. 1994. Molecular and developmental genetics of mouse coat color. Ann. Rev. Genet. **28:** 189–217.
2. BARSH, G.S. 1996. The genetics of pigmentation: From fancy genes to complex traits. Trends Genet. **12:** 299–305.
3. SILVERS, W.K. 1979. The Coat Colors of Mice: A Model for Mammalian Gene Action and Interaction. Springer-Verlag. New York.
4. SEARLE, A.G. 1968. Comparative Genetics of Coat Color in Mammals. Academic Press. New York.
5. SIRACUSA, L.D. 1994. The agouti gene: Turned on to yellow. Trends Genet. **10:** 423–428.
6. TAMATE, H.B. & T. TAKEUCHI. 1964. Action of the *e* locus of mice in the response of phaeomelanic hair follicles to alpha-melanocyte stimulating hormone in vitro. Science **224:** 1241–1242.
7. ROBBINS, L.S. *et al.* 1993. Pigmentation phenotypes of variant extension locus alleles result from point mutations that alter MSH receptor function. Cell **72:** 827–834.
8. GESCHWIND, I.I. 1966. Change in hair color in mice induced by injection of α-MSH. Endocrinology **79:** 1165–1167.
9. GESCHWIND, I.I., R.A. HUSEBY & R. NISHIOKA. 1972. The effect of melanocyte-stimulating hormone on coat color in the mouse. Rec. Prog. Horm. Res. **28:** 91–130.

10. LU, D.S. *et al.* 1994. Agouti protein is an antagonist of the melanocyte-stimulating-hormone receptor. Nature **371:** 799–802.
11. OLLMANN, M.M., M.L. LAMOREUX, B.D. WILSON & G.S. BARSH. 1998. Interaction of Agouti protein with the melanocortin 1 receptor in vitro and in vivo. Genes Dev. **12:** 316–330.
12. MILLAR, S.E., M.W. MILLER, M.E. STEVENS & G.S. BARSH. 1995. Expression and transgenic studies of the mouse agouti gene provide insight into the mechanisms by which mammalian coat color patterns are generated. Development **121:** 3223–3232.
13. DUHL, D.M.J. *et al.* 1994. Pleiotropic effects of the mouse lethal yellow (A(y)) mutation explained by deletion of a maternally expressed gene and the simultaneous production of agouti fusion RNAs. Development **120:** 1695–1708.
14. MICHAUD, E.J. *et al.* 1994. A molecular model for the genetic and phenotypic characteristics of the mouse lethal yellow (Ay) mutation. Proc. Natl. Acad. Sci. USA **91:** 2562–2566.
15. OLLMANN, M.M. *et al.* 1997. Antagonism of central melanocortin receptors in vitro and in vivo by agouti-related protein. Science **278:** 135–138.
16. SHUTTER, J.R. *et al.* 1997. Hypothalamic expression of ART, a novel gene related to agouti, is up-regulated in obese and diabetic mutant mice. Genes Dev. **11:** 593–602.
17. MILLER, K.A. *et al.* 1997. Genetic studies of the mouse mutations mahogany and mahoganoid. Genetics **146:** 1407–1415.
18. LANE, P.W. & M.C. GREEN. 1960. Mahogany, a recessive color mutatioin in linkage group V of the mouse. J. Hered. **51:** 228–230.
19. Lane, P.W. 1960. New mutants. Mouse News Lett. **22:** 35.
20. HE, L. *et al.* 2001. A biochemical function for attractin in agouti-induced pigmentation and obesity. Nat. Genet. **27:** 40–47.
21. PHILLIPS, R.J.S. 1963. Non-agouti curly. Mouse News Lett. **29:** 38.
22. PHILLIPS, R.J.S. & G. FISHER. 1979. Non-agouti curly: Change of symbol. Mouse News Lett. **60:** 46–47.
23. GUNN, T.M. *et al.* 1999. The mouse mahogany locus encodes a transmembrane form of human attractin. Nature **398:** 152–156.
24. NAGLE, D.L. *et al.* 1999. The mahogany protein is a receptor involved in suppression of obesity. Nature **398:** 148–152.
25. WILLARD, D.H. *et al.* 1995. Agouti structure and function: Characterization of a potent alpha-melanocyte stimulating hormone receptor antagonist. Biochemistry **34:** 12341–12346.
26. REIZES, O. *et al.* 2001. Transgenic expression of syndecan-1 uncovers a physiological control of feeding behavior by syndecan-3. Cell **106:** 105–116.
27. BERNFIELD, M. *et al.* 1999. Functions of cell surface heparan sulfate proteoglycans. Annu. Rev. Biochem. **68:** 729–777.
28. HE, L. *et al.* 2003. Spongiform degeneration in mahoganoid mutant mice. Science **299:** 710–712.
29. GUNN, T.M. *et al.* 2001. Molecular and phenotypic analysis of Attractin mutant mice. Genetics **158:** 1683–1695.
30. KURAMOTO, T. *et al.* 2001. Attractin/Mahogany/Zitter plays a critical role in myelination of the central nervous system. Proc. Natl. Acad. Sci. USA **98:** 559–564.
31. KURAMOTO, T. *et al.* 2002. Insertional mutation of the Attractin gene in the black tremor hamster. Mamm. Genome **13:** 36–40.
32. GOMI, H. *et al.* 1994. Prion protein (PrP) is not involved in the pathogenesis of spongiform encephalopathy in zitter rats. Neurosci. Lett. **166:** 171–174.
33. PHAN, L.K., F. LIN, C.A. LEDUC, *et al.* 2002. The mouse mahoganoid coat color mutation disrupts a novel C3HC4 RING domain protein. J. Clin. Invest. **110:** 1449–1459.
34. JACKSON, P.K. *et al.* 2000. The lore of the RINGs: substrate recognition and catalysis by ubiquitin ligases. Trends Cell Biol. **10:** 429–439.
35. JOAZEIRO, C.A. & A.M. WEISSMAN. 2000. RING finger proteins: Mediators of ubiquitin ligase activity. Cell **102:** 549–552.
36. LU, X. *et al.* 1999. Distribution of Mahogany/Attractin mRNA in the rat central nervous system. FEBS Lett. **462:** 101–107.

DNA Polymorphism and Selection at the Melanocortin-1 Receptor Gene in Normally Pigmented Southern African Individuals

PREMILA R. JOHN,[a] KATERYNA MAKOVA,[b] WEN-HSIUNG LI,[b] TREFOR JENKINS,[a] AND MICHELE RAMSAY[a]

[a]*Department of Human Genetics, School of Pathology, National Health Laboratory Service and University of the Witwatersrand, Johannesburg 2000, South Africa*

[b]*Department of Ecology and Evolution, University of Chicago, Chicago, Illinois, USA*

ABSTRACT: Skin pigmentation is a polygenic multifactorial trait determined by the cumulative effects of multiple genetic variants and environmental factors. Melanocortin-1 receptor (MC1R) is one of the genes involved in pigmentation, and has been implicated in the red hair and pale skin phenotype in human Caucasoid individuals. The present study was undertaken to identify variation at the MC1R locus in normally pigmented individuals in two African populations, sub-Saharan Negroids (22 unrelated individuals) and the San (17 unrelated individuals). The study showed considerable MC1R gene sequence variation with the detection of eight synonymous and three nonsynonymous mutations. This is the first report of nonsynonymous mutations in African individuals in the MC1R gene: L99I was found in a single San individual, S47I was detected in a single Negroid individual, and F196L was detected in five Negroid individuals (5/44; 0.11). The functional significance of these mutations is not known. Three of the eight synonymous mutations found, L106L (CTG→CTA), F300F (TTC→TTT), and T314T (ACA→ACG) (also known as A942G), have been reported previously. T314T was the only variant that showed a significant difference between the Negroid and San populations (0.477 and 0.059, respectively; $P = 1.6 \times 10^{-5}$). Its low frequency in the San may be the result of random genetic drift in a population of small size, or selection. Several tests of neutrality of the MC1R coding region in these and other African populations were significant, suggesting that purifying selection (functional constraint) had occurred at this gene locus in Africans. This demonstrates that although some nonsynonymous MC1R mutations are tolerated in individuals with dark skin, this gene has likely played a significant role in the maintenance of dark pigmentation in Africans and normal pigment variation in non-African populations.

KEYWORDS: melanocortin-1 receptor (MC1R); pigmentation

Address for correspondence: Michele Ramsay, Department of Human Genetics, School of Pathology, National Health Laboratory Service (formerly the South African Institute for Medical Research) and University of the Witwatersrand, Johannesburg 2000, South Africa. Voice: 27 11 489 9214; fax: 27 11 489 9226.

michele.ramsay@nhls.ac.za

Ann. N.Y. Acad. Sci. 994: 299–306 (2003). © 2003 New York Academy of Sciences.

INTRODUCTION

One of the most visible forms of human diversity is pigmentation of the skin, hair, and eyes. Pigmentation is a trait that is determined by the synchronized interaction of various genes with environmental factors.[1] The process is stimulated by the α-melanocyte stimulating hormone (α-MSH) via a receptor on the surface of the melanocytes, the melanocyte-stimulating hormone receptor or the melanocortin-1 receptor (MC1R). The MC1R gene on chromosome 16q24.3[2-4] encodes a seven transmembrane G-protein–coupled receptor.[2,5] The hormone-receptor interaction leads to the transduction of a signal via GTP-binding protein adenyl cyclase, which results in an increase in intracellular cAMP[2,6] and affects the pigmentation pathway. In mammals, the MC1R gene, also called the *extension* locus (e), is mutated in animals with red and yellow coat colors.[7,8] Studies on the molecular basis of the red hair and pale skin phenotype in humans have shown that mutations in the MC1R gene are associated with this pigmentation type.[9-13] MC1R has been the only gene identified so far to control part of the normal pigmentation spectrum in humans.[1]

The MC1R locus is thought to have evolved more rapidly than other members of the melanocortin receptor family.[13] When one examines human populations, the level of nucleotide polymorphism (nonsynonymous mutations in particular) is high in lighter skinned individuals (non-Africans) and is low (with nonsynonymous substitutions absent) in individuals with dark skin (Africans).[13] This is opposite to the pattern observed at most other loci, where Africans are the most polymorphic,[14] and suggests that different selective pressures among individuals with dark and light skin color are shaping the genetic variation at MC1R. Evolutionarily this may be explained by functional constraints in Africans[13,15] and either by a "relaxation" of these constraints in non-Africans living in areas of lower sunshine rates[15] or by selection for MC1R mutations that lead to lighter pigmentation in northern latitudes.[13]

The aim of the present study was to search for MC1R variation in African populations, since most previous studies have focused on Caucasian populations, with the notable exception of those by Rana et al.[13] and Harding et al.,[15] who investigated various population groups, including some from Africa. We therefore investigated two groups of African individuals from southern Africa, the Negroid and San, to identify variation at the MC1R locus in these normally pigmented groups of individuals.

SUBJECTS

In this study, a total of 39 unrelated normally pigmented South African individuals, including 22 Negroid individuals and 17 unrelated San individuals, were investigated by DNA sequencing of the MC1R coding region. The mothers of the Negroid individuals were also investigated to determine the phase of the mutations when multiple variants were detected in a single individual. The Negroid samples were collected in Johannesburg from random individuals of different chiefdoms, and the San samples were obtained from Tsumkwe, in Namibia.

RESULTS

The present study detected considerably greater MC1R coding sequence variation in two African populations than had previous studies. Both synonymous and nonsynonymous mutations were detected, whereas previous studies detected only synonymous mutations in Africans.[13,15] The only variant detected by Rana *et al.*[13] in a study of 50 African MC1R genes was the synonymous mutation, T314T (ACA→ACG) (also denoted as A942G); whereas Harding *et al.*[15] detected four synonymous mutations, L106L (CTG→CTA), F300F (TTC→TTT), C273C (TGC→TGT), and T314T, in 106 African MC1R genes. We detected eight synonymous mutations (including three of the four detected by the other groups), as well as three nonsynonymous mutations in a sample of only 22 Negroid and 17 San individuals. These mutations are listed in TABLE 1. The three nonsynonymous mutations were as follows: S47I (1 in 44 Negroid chromosomes); F196L (5 in 44 Negroid chromosomes); and L99I (1 in 34 San chromosomes). The protein domains affected by the mutations are shown in TABLE 1.

The only common synonymous mutation was T314T (A942G), which occurred at similar frequencies in our study of Negroid individuals and that of Rana *et al.*,[13] 0.477 and 0.420, respectively. The frequency was higher in the African sample of Harding *et al.*,[15] at 0.576. Non-human primate data[13] suggest that the G at position 942 of the coding sequence is the ancestral base. It is, therefore, interesting to note that it is much less common in the San (0.059) and significantly different from the frequency in southern African Negroids ($P = 1.6 \times 10^{-5}$). A possible explanation may be that it has become less common in the San as a result of random genetic drift in a relatively small population or that the population went through a severe bottleneck at some time in its history. Another possibility is that the Negroid population went through a rapid population expansion. This variant has also been reported in some Caucasoid, East and Southeast Asian, and Indian individuals,[10,13] but at much lower frequencies than in the Negroid. Chi-square tests were carried out to compare the significance of variation at the MC1R locus in the Negroid and San individuals. T314T (the G allele) is significantly higher in the Negroid population when compared to the San ($P < 0.001$). F196L was found only in the Negroid sample and L50L only in the San sample (TABLE 1). Haplotypes were used to construct gene trees (networks) to illustrate the genetic backgrounds on which the mutations occurred (FIG. 1). In the Negroid and San populations, six individual point mutations had each occurred in the background of the consensus sequence (shown in black) and four mutations had occurred in the background of the root haplotype (shown in gray). Interestingly the A103A (GCC→GCT) and I168I (ATC→ATT) variants were observed only once, in the same allele, associated with the root haplotype. In the absence of further data, it is not possible to distinguish which mutation occurred first.

The differences found between the African Negroid samples in the separate studies may, in part, be explained by the geographical origins of the subjects. The earlier two studies[13,15] contained predominantly central and western African samples, whereas subjects in the present study were drawn from southern Africa only. If the differences are significant, it may suggest that the functional constraint on MC1R variation has been relaxed to some extent in southern Africa, though it is uncertain whether 2000 to 3000 years (the postulated duration of time that the Bantu-speaking peoples have been in southern Africa) would be sufficient time to accumulate this

TABLE 1. MC1R variants seen in normally pigmented Negroid and San individuals[a]

Notation	Variant	Nucleotide number	Base change	Position in protein	Number of chromosomes		P***	Reference where first described
					Negroid	San		
Non-synonymous	S471	140	AGC→ATC	First transmembrane domain	1/44	0	0.382	Present study
	L99I	295	CTC→ATC	Second transmembrane domain	0	1/34	0.252	Present study
	F196L	586	TTC→CTC	Fifth transmembrane domain	5/44	0	0.021	Present study
Synonymous	L50L	150	CTG→CTA	First transmembrane domain	0	3/34	0.044	Present study
	A103A	309	GCC→GCT	Second extracellular loop	1/44	0	0.315	Present study
	L106L	318	CTG→CTA	Second extracellular loop	3/44	0	0.078	15
	I168I	504	ATC→ATT	Fourth transmembrane domain	1/44	0	0.314	Present study
	Q233Q*	699	CAG→CAA	Third intracellular loop	—	—	—	Present study
	V265V	795	GTC→GTG	Sixth transmembrane domain	1/44	2/34	0.411	Present study
	F300F*	900	TTC→TTT	Seventh transmembrane domain	—	—	—	15
	T314T**	942	ACA→ACG	COOH terminal region	21/44	2/34	1.6×10^{-5}	10

*In the present study these variants were observed in a mother's chromosomes, in one chromosome each.

** This variant has been described as A942G in other studies.

***These are the P values from a chi-squared analysis comparing Negroid and San individuals.

[a]METHODS: MC1R was PCR amplified with the forward primer 1F 5′-AGATGAAGGAGGAGGCAGGCAT-3′ and reverse primer 1R 5′-CCGCGCTTCAA-CACTTTCAGAGATCA-3′,[10] to generate a 1238 bp product. The PCR product was amplified for 40 cycles (1 min at 95°C, 1 min at 55°C, 1 min at 72°C). The DNA was sequenced using six different primers: forward primers 2F 5′CCCCTGGCAGCACCATGAACT-3′,[10] 3F 5′-GACAATGTCATTGACGTG-3′ and 4F 5′-TCACCCTCACCATCCTGC-3′, and reverse primers 2R 5′-5′CCCCTGGCAGCACCATGAACT-3′, 3R 5′-GCGCTGCCTCTTGTGGAG-3′, and 4R 5′-ATGGAGCTGCAGGTGATC-3′.[10] Cycle sequencing was done TGCCCAGGGTCACACAGGAAC-3′, 3R 5′-GCGCTGCCTCTTGTGGAG-3′, and 4R 5′-ATGGAGCTGCAGGTGATC-3′.[10] Cycle sequencing was done using the dRhodamine or BigDye™ kits (Perkin Elmer) and the ABI Prism™ 377 sequencer. The data were collected with ABI Prism™ 377 software version 2.1.1 and the results were analyzed by the ABI Prism™ DNA sequencing analysis software version 2.1.

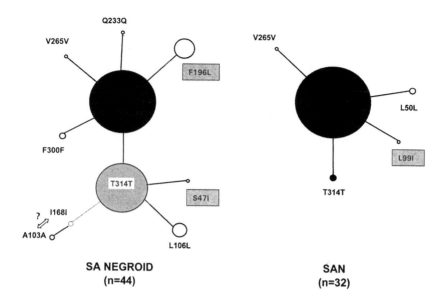

FIGURE 1. Gene trees (networks) constructed from haplotypes and showing the evolution of MC1R sequence variation in southern African Negroid and San populations. The size of the circles is proportional to the frequency of the mutations. The MC1R consensus sequence is shown as *black circles*, and the root sequence (based on the chimpanzee sequence) is shown in *gray*. The intermediate haplotype leading to the A103A/I168I haplotype was not observed and is shown as a *stippled circle*. The order of occurrence of these two mutations could not be deduced. The nucleotide changes for each variant is shown in TABLE 1. The nonsynonymous mutations are shown in *shaded boxes*. *n* = number of alleles for which haplotypes could be deduced.

extent of genetic variation in the Bantu-speakers. The San, who have significantly lighter skin colors than the Bantu-speakers,[16] appear to have been in their present areas of habitation in southern Africa for many millennia, as evidenced by their rock art in Namibia, which has been dated to roughly 30,000 years before the present, and by skeletal remains.[17–19]

To investigate whether the MC1R locus evolves under neutrality in Africans, several statistical tests were employed. Here and later in the paper, the samples from UK Africans and African Americans from the study by Harding *et al.*[15] were omitted from the analysis because of potential admixture. First, the McDonald-Kreitman test[20] was used to compare the ratio of nonsynonymous to synonymous sites between the polymorphism data in Africa and the interspecific data. There are 10 nonsynonymous and 6 synonymous human-chimpanzee differences. Thus, a segregating site has a binomial probability to generate a nonsynonymous polymorphism with a binomial probability[15] of 0.625 to generate a nonsynonymous polymorphism. In Africans from the present study (including the two variants observed only in the mothers), three nonsynonymous and eight synonymous variants were observed. The probability of observing such a ratio is $P = 0.019$. This indicates that the African

TABLE 2. Results of Tajima's D test and Fu and Li's D, F, D*, and F* tests

Test/Population	Tajima's D	Fu and Li's			
		D	F	D*	F*
San	-1.44^{**}	-0.71	-1.03	-0.64	-0.96
Negroid	-0.93	-1.87^{**}	-1.73^{*}	-1.72^{*}	-1.62^{*}
African[a]	-1.30^{*}	-1.60^{*}	-1.70^{*}	-1.51^{*}	-1.63^{*}
African[b]	-1.26^{*}	-1.82^{*}	-1.89^{**}	-1.78^{*}	-1.86^{**}

[a]This study.
[b]Rana et al.[13]; Harding et al.,[15] and this study.
*P <0.1; **P <0.05.

MC1R variation pattern is significantly different from the MC1R divergence pattern, suggesting selection acting against nonsynonymous substitutions. If the two synonymous variants that were observed only in the mothers are disregarded (as a result we have three nonsynonymous and six synonymous variants), the difference is marginally significant ($P = 0.074$).

Second, the Hudson-Kreitman-Aquade (HKA) test[21] was performed to compare the diversity and divergence over all sites by comparing the MC1R gene region (Negroids, San, all Africans from this study, and Africans from all three studies—the studies by Rana et al.[13] and Harding et al.[15] and the present study) to a 10-kb, non-coding region on chromosome 22 (results not shown).[22] Since no evidence of selection was found in the latter region, it can serve as a suitable neutral standard. In another set of HKA tests, we assumed that mutations could arise only in silent sites in Africans and compared diversity and divergence over silent sites of MC1R and a 10-kb region on chromosome 22, and again the results were not significant (results not shown). Finally, the data were also compared to variation at the β-globin locus, and all eight tests were not significant (data not shown). This set of HKA tests was not able to disprove a null hypothesis of neutrality at the MC1R locus.

Lastly, the Tajima's D test[23] as well as the Fu and Li F, D, F*, and D* tests[24] were performed for Negroids, San, all Africans in this study, and all Africans from the results of the three studies. The results given in TABLE 2 show that some of the tests were significant and most of the tests were at least marginally significant, suggesting purifying selection. However, this result could also be indicative of a rapid population expansion in Africa, as indicated by other loci.[14]

DISCUSSION

On balance, these analyses suggest that there is purifying selection at the MC1R locus in Africans. Despite this, however, the present study shows that nonsynonymous mutations are tolerated in Africa. The individuals from this study originated from southern Africa whereas those from the studies by Rana et al.[13] and Harding et al.[15] originated primarily from more northerly African populations, leading one to speculate whether the selection in southern Africa is less strong, leading to the tolerance of functional mutations in populations from this region. Although it has been shown that mutations in the MC1R gene play a significant role in the red hair and pale skin phenotype in Caucasoid individuals of European origin, the functional significance of the nonsynonymous variants observed in this study in Africans needs to be determined to fully understand the role of MC1R in normal pigment variation. Recently, Makova *et al.*[25] suggested that mutations in the MC1R promoter might also be important for normal pigmentation variation. Several other genes have been identified that are likely to play a role in normal pigment variation, either by controlling the rate and type of melanin produced, or by exerting an effect on the melanin biosynthetic pathway. These include α-MSH, ASP, the P gene, and genes producing enzymes that act in the melanin biosynthetic pathway (including TYR, TYRP1, TYRP2, and SILV[1]). Additional loci may also play significant roles in determining the pigment phenotype through their interaction with one another and the environment. Polymorphism and selection studies of such loci will be of great interest.

ACKNOWLEDGMENTS

R. Kerr and B. Dangerfield are thanked for technical support, and Drs. A. Lane and H. Soodyall for the contribution of the Negroid and San DNA samples used in this study. This study was supported by the South African Institute for Medical Research, the Foundation for Research and Development, and the University of the Witwatersrand (University of the Witwatersrand Committee for Research on Human Subjects, protocol number M970510).

REFERENCES

1. STURM, R.A., R.D. TEASDALE & N.F. BOX. 2001 Human pigmentation genes: Identification, structure and consequences of polymorphic variation. Gene **277:** 49–62.
2. MOUNTJOY, K.G., L.S. ROBBINS, M.T. MORTRUD, *et al.* 1992. The cloning of a family of genes that encode the melanocortin receptors. Science **257:** 1248–1251.
3. GANTZ, I., T. YAMADA, T. TASHIRO, *et al.* 1994. Mapping of the gene encoding the melanocortin-1 (α-melanocyte-stimulating hormone) receptor (MC1R) to human chromosome 16q24.3 by fluorescence in situ hybridization. Genomics **19:** 394–395.
4. MAGENIS, R.E., L. SMITH, J.H. NADEAU, *et al.* 1994. Mapping of the ACTH, MSH, and neural (MC3 & MC4) melanocortin receptors in the mouse and human. Mamm. Genome **5:** 503–508.
5. CHHAJLANI, V. & J.E.S. WIKBERG. 1992. Molecular cloning and expression of the human melanocyte stimulating hormone receptor cDNA. FEBS Lett. **309:** 417–420.
6. GANTZ, I., Y. KONDA, T. TASHIRO, *et al.* 1993. Molecular cloning of a new melanocortin receptor. J. Biol. Chem. **268:** 8246–8250.
7. ROBBINS, L.S., J.H. NADEAU, K.R. JOHNSON, *et al.* 1993. Pigmentation phenotypes of variant extension locus alleles result from point mutations that alter MSH receptor function. Cell **72:** 827–834.
8. VAGE, D.I., D. LU, H. KLUNGLAND, *et al.* 1997. A non-epistatic interaction of agouti and extension in the fox, *Vulpes vulpes*. Nat. Genet. **15:** 311–315.
9. VALVERDE, P., E. HEALY, I. JACKSON, *et al.* 1995. Variants of the melanocyte-stimulating hormone receptor gene are associated with red hair and fair skin in humans. Nat. Genet. **11:** 328–330
10. BOX, N.F., J.R. WYETH, L.E. O'GORMAN, *et al.* 1997. Characterization of melanocyte-stimulating hormone receptor variant alleles in twins with red hair. Hum. Mol. Genet. **6:** 1891–1897.

11. KOPPULA, S.V., L.S. ROBBINS, D. LU, *et al.* 1997. Identification of common polymorphisms in the coding sequence of the human MSH receptor (MC1R) with possible biological effects. Hum. Mutation **9:** 30–36.
12. SMITH, R., E. HEALY, S. SIDDIQUI, *et al.* 1998. Melanocortin 1 receptor variants in an Irish population. J. Invest. Dermatol. **111:** 119–122.
13. RANA, B.K., D. HEWETT-EMMETT, L. JIN, *et al.* 1999. High polymorphism at the melanocortin 1 receptor locus. Genetics **151:** 1547–1557.
14. PRZEWORSKI, M., R.R. HUDSON, & A. DI RIENZO. 2000. Adjusting the focus on human variation. Trends Genet. **16:** 296.
15. HARDING, R.M., E. HEALY, A.J. RAY, *et al.* 2000. Evidence of variable selective pressures at MC1R. Am. J. Hum. Genet. **66:** 1351–1361.
16. WEINER, J.S., G.A. HARRISON, R. SINGER, *et al.* 1964. Skin color in Southern Africa. Hum. Biol. **36:** 294–307.
17. TOBIAS, P.V. 1962. Early members of the genus *Homo* in Africa. *In* Evolution and Homonisation. G. Kruth, Ed.: 191–204. Fischer Verlag. Stuttgart.
18. NURSE, G.T., J.S. WEINER & T. JENKINS. 1985. The Peoples of Southern Africa and Their Affinities. Clarendon Press. Oxford.
19. DEACON, H.J. & J. DEACON. 1999. Human beginnings in South Africa. David Philip Publishers. Cape Town & Johannesburg.
20. MCDONALD, J.H. & M. KREITMAN. 1991. Adaptive evolution at the Adh locus in Drosophila. Nature **351:** 652–654.
21. HUDSON, R.R., M. KREITMAN & M. AQUADE. 1987. A test of neutral molecular evolution based on nucleotide data. Genetics **116:** 153–159.
22. ZHAO, Z., L. JIN, Y.X. FU, *et al.* 2000. Worldwide DNA sequence variation in a 10 kb noncoding region on human chromosome 22. Proc. Natl. Acad. Sci. USA **97:** 11354–11358.
23. TAJIMA, F. 1989. Statistical method for testing the neutral mutation hypothesis by DNA polymorphism. Genetics **123:** 585–595.
24. FU, Y.X. & W.H. LI. 1993. Statistical tests of neutrality of mutations. Genetics **133:** 693–709.
25. MAKOVA, K.D., M. RAMSAY, T. JENKINS, *et al.* 2001. Human DNA sequence variation in a 6.6-kb region containing the melanocortin 1 receptor promoter. Genetics **158:** 1253–1268.

Evolutionary Genetics of the Melanocortin-1 Receptor in Vertebrates

NICHOLAS I. MUNDY, JOANNE KELLY, EMMALIZE THERON, AND KIM HAWKINS

Department of Biological Anthropology, University of Oxford, Oxford OX2 6QS, United Kingdom

ABSTRACT: The molecular genetic basis of adaptive change in phenotype is a major outstanding issue in evolutionary biology. Evolutionary change in coat and plumage color is a promising system for making progress in this field. Most notably, recent work on the molecular genetic basis of hair and feather color has identified several genes which are candidates for involvement in evolutionary color change in mammals and birds. We have investigated the evolution of one of these candidate genes, the melanocortin-1 receptor (MC1R) gene, in relation to changes in melanin distribution among primate species, and in bananaquits (*Coereba flaveola*), which are a classic case of melanic plumage polymorphism in birds. In primates, a role of the MC1R coding region in coat color evolution can be ruled out in several cases in which closely related species have drastically different distributions of eumelanin and/or pheomelanin. However, reconstruction of MC1R sequences over primate evolution shows the presence of mutations at important functional sites in several lineages. Most notably, the lion tamarins (*Leontopithecus*) show a striking pattern of MC1R evolution, including deletions and several nonconservative amino acid changes. In the bananaquit, an E92K substitution in the MC1R is strongly associated with melanism, and this is likely to be the causative mutation. Reconstruction of the evolution of bananaquit MC1R alleles shows that melanism is a derived trait in this species. These results confirm the utility of a candidate gene approach to color evolution in vertebrates and open the way for extensive future research.

KEYWORDS: MC1R; MSH; melanin; primate; lion tamarin; bananaquit

INTRODUCTION

The genetic changes underlying adaptation and phenotypic evolution are of fundamental importance to our understanding of how evolution proceeds. However, there has been little progress in this field so far, mostly because the gene or genes responsible for particular phenotypic changes during evolution are usually obscure. The evolution of hair and feather coloration is a model system that offers many ad-

Address for correspondence: Nicholas I. Mundy, Department of Zoology, University of Cambridge, Downing Street, Cambridge CB2 3EJ, United Kingdom. Voice: 44-1223-336600; fax: 44-1223-336679.

nim21@cam.ac.uk

Ann. N.Y. Acad. Sci. 994: 307–312 (2003). © 2003 New York Academy of Sciences.

vantages for progress in this area. In particular, the key genes that regulate the melanin content of hairs and feathers have been characterized, and there is strong conservation of the function of many of these genes across mammalian and avian species.[1-3]

One such gene encodes the melanocortin-1 receptor (MC1R), which is one of the major regulators of eumelanin and pheomelanin content in hairs and feathers. In domestic mammals such as mice and cattle, gain-of-function MC1R mutations lead to darkening (increased eumelanin) in the coat, whereas loss-of-function mutations lead to a yellow or red coat, caused by pheomelanin synthesis.[4,5] MC1R mutations are also associated with hair color changes in sheep, horses, pigs, guinea pigs, foxes, dogs, humans, and chickens (see e.g., Valverde *et al.*[6] and Kijas *et al.*[7]).[8]

We were interested in the possible role of the MC1R in coat color evolution among different species of primate, and in melanic polymorphisms in birds. Five groups of primate were identified, each of which contained two or more taxa with drastic changes in eumelanin and/or pheomelanin distribution that could conceivably be attributable to MC1R effects. The five groups are the lion tamarins (*Leontopithecus*) (FIG. 1A), howler monkeys (*Alouatta*), macaques (*Macaca*), langurs (*Trachypithecus*), and ruffed lemurs (*Varecia*). None of these groups has extensive sexual dimorphism for coat color, an important potential confounding factor. For birds, we chose the bananaquit (*Coereba flaveola*), a widespread neotropical passerine. Most populations of bananaquit consist of yellow morph individuals that have a yellow breast and white eyestripe, but a fully melanic morph occurs on certain Caribbean islands, including Grenada and St. Vincent (FIG. 1B).

The full MC1R coding sequence was obtained from these species, as well as suitable outgroup species, to answer the following questions. Are evolutionary changes in pelage/plumage phenotype associated with MC1R substitutions? Do substitutions occur in the MC1R at sites of potential functional importance? If MC1R substitutions are associated with changes in phenotype, then can the evolutionary history of the color phenotype be reconstructed? A further issue that was examined was whether there was any evidence for changes in selection pressure during MC1R evolution.

MATERIALS AND METHODS

Genomic DNA was isolated from tissue and blood samples using standard methods. Primers spanning the whole single exon coding sequence of the MC1R, and internal sequencing primers, were developed separately for primates and birds.[9,10] Four populations of bananaquits were investigated: yellow morph birds from the monomorphic populations on Panama and Puerto Rico and both melanic and yellow morph birds from the polymorphic populations on Grenada and St. Vincent. MC1R sequences generally were obtained by PCR and direct sequencing using standard techniques, but for some bananaquit samples each MC1R allele was obtained separately by TA cloning of the PCR product.

Sequences were edited using Sequencher (Gene Codes, Ann Arbor, MI). Phylogenetic analysis and maximum likelihood analysis of dN/dS ratios were performed using PAUP[11] and PAML,[12] respectively.

A

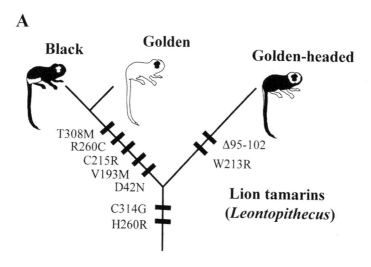

B

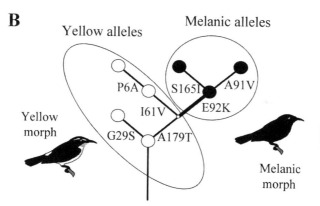

Bananaquits (*Coereba flaveola*)

FIGURE 1. MC1R evolution in lion tamarins and bananaquits (inferred amino acid mutations). (**A**) MC1R evolution among three species of lion tamarin, placed over the known phylogeny of the species.[16] Each amino acid change is represented by a *bar,* and the order of the mutations on each branch is arbitrary. *Pale areas* on the cartoons represent orange pheomelanin; *dark areas* represent black eumelanin. (**B**) Condensed network of bananaquit MC1R alleles (haplotypes) showing amino acid mutations only (for full network including silent substitutions, see Theron and colleagues[10]). The network is rooted using the MC1R of an outgroup tanager species. Each *circle* represents an MC1R allele, and each segment represents a single amino acid change. Melanic alleles (*solid circles*) are defined by the presence of the E92K mutation (shown as a *thick line*) and are evolutionarily derived from yellow alleles (*open circles*). Individual bananaquits with one or two copies of any of the melanic alleles are melanic morphs, and those with two copies of any yellow alleles have the yellow morph phenotype.

RESULTS AND DISCUSSION

Primates

To date, we have obtained MC1R sequences from 57 species and 83 individual primates (see Mundy and Kelly[9] for full details). A surprising finding is substantial length variation in the MC1R open reading frame, particularly in certain groups of New World monkeys (marmosets, Goeldi's monkey, and capuchins). This variation is mostly attributable to loss of the normal stop codon, and results in primate MC1Rs varying from 310 to 344 amino acids in length (the human MC1R is 317 amino acids long).

Among the five groups of primates in which closely related species had different patterns of eumelanin and pheomelanin, there are several examples in which the MC1R sequence is identical but the phenotype is different. One such example is provided by the golden and black lion tamarins (FIG. 1). In these cases, the MC1R coding sequence is clearly not responsible for the difference, which must instead be caused by either *cis*-induced changes in transcriptional or posttranslational regulation of MC1R and/or other loci.

Evidence from MC1R mutations in domestic animals and structure-function studies have led to the identification of numerous sites of functional importance in the MC1R. There are several examples of nonconservative mutations occurring at these sites during primate evolution. The most notable cases are (1) mutations in aspartate 117, a site implicated in ligand binding to the human MC1R,[13] in ancestral New World monkeys (D117H) and lemurs (D117G), and (2) an E94K mutation, which is associated with melanism in other species (see below), in the two subspecies of ruffed lemurs, one of which (red-ruffed lemur) has extensive pheomelanin.

The most interesting pattern of MC1R evolution was found in the lion tamarins (FIG. 1A). An 8 amino acid deletion occurred in the lineage leading to the relatively melanic golden-headed lion tamarin. This deletion probably contributed to the phenotype of this species since, remarkably, the identical deletion was recently shown to be associated with melanism in jaguarundi cats (Eizirik *et al.*[14]), suggesting that this deletion may have contributed to the relatively melanic phenotype of this species. In addition, several nonconservative mutations occurred in the ancestral lineage of the genus, and in the two major lineages leading to the present day species. This high rate of nonsynonymous substitution was not accompanied by a high rate of change at silent sites, providing evidence that the MC1R in lion tamarins has been under a different regime of selection from that in other nonhuman primates.[9] However, it is currently unclear whether this results from a relaxation of selection pressure, as appears to have occurred in humans,[15] or whether the unusual pattern results from a period of positive selection followed by a return to selective constraint. An important avenue for future research will be *in vitro* expression of lion tamarin MC1Rs to determine their functional characteristics.

Birds

There were seven variable amino acid sites in the MC1R among the 138 bananaquit chromosomes sequenced, but only one was perfectly associated with the phenotype: all individuals containing one or two copies of a lysine at position 92 were melanic, whereas all individuals homozygous for a glutamate at this position were yellow. The identical glutamate to lysine mutation causes constitutive activation of the MC1R and melanism in mice[4] and is probably responsible for melanism in chickens,[3] so it is very likely that this is also the causative mutation in bananaquits. If this is assumed to be the case, then by reconstructing the evolution of MC1R alleles (haplotypes) in bananaquits we can determine whether melanism in birds from Grenada and St. Vincent had a single origin, and what the ancestral coloration of bananaquits was. Such a reconstruction (FIG. 1B and Theron et al.[10]) shows unequivocally that the glutamate to lysine mutation occurs in a derived part of the network, and so we can conclude that melanism is a derived coloration in bananaquits. This makes geographical sense given the small, and peripheral, part of the total range of bananaquits where melanic morph birds currently are found.

CONCLUSIONS AND FUTURE DIRECTIONS

The MC1R appears to have played a key role in the evolution of melanism in bananaquits, and we currently are investigating whether it has played a similar role in other avian species such as snow geese and arctic skuas that are distant relations of passerines. In primates, the most striking mutations in the MC1R occurred during the evolution of New World monkeys (including lion tamarins) and lemurs, although the relationship of these mutations to coat color change remains to be elucidated. In examples in which coat color changes among closely related primate species are not attributable to the MC1R coding region substitutions, the likely alternatives are change in the MC1R promoters, or in other loci such as agouti. Studies of the molecular basis of the extraordinary diversity of plumage and pelage coloration are still in their infancy, but future prospects are bright.

ACKNOWLEDGMENTS

We thank the many people who provided samples for this study, especially Andrew Kitchener and Liliana Cortés-Ortiz (primates), and Eldredge Bermingham and Robert Ricklefs (bananaquits). This work was supported by the Biotechnology and Biological Sciences Research Council.

REFERENCES

1. JACKSON, I.J. 1997. Homologous pigmentation mutations in human, mouse and other model organisms. Hum. Mol. Genet. **6:** 1613–1624.
2. SEARLE, A.G. 1968. Comparative Genetics of Coat Color in Mammals. Academic Press. New York.
3. TAKEUCHI, S., H. SUZUKI, M. YABUUCHI, et al. 1996. A possible involvement of melanocortin 1-receptor in regulating feather color pigmentation in the chicken. Biochem. Biophys. Acta **1308:** 164–168.
4. ROBBINS, L.S., J.H. NADEAU, K.R. JOHNSON, et al. 1993. Pigmentation phenotypes of variant extension locus alleles result from point mutations that alter MSH receptor function. Cell **72:** 827–834.

5. KLUNGLAND, H., D.I. VÅGE, L. GOMEZRAYA, et al. 1995. The role of melanocyte-stimulating hormone (MSH) receptor in bovine coat color determination. Mamm. Genome **6:** 636–639.
6. VALVERDE, P., E. HEALY, I. JACKSON, et al. 1995. Variants of the melanocyte-stimulating hormone receptor gene are associated with red hair and fair skin in humans. Nat. Genet. **11:** 328–330.
7. KIJAS, J.M.H., R. WALES, A. TÖRNSTEN, et al.1998. Melanocortin receptor 1 (MC1R) mutations and coat color in pigs. Genetics **150:** 1177–1185.
8. NEWTON, J.M., A.L. WILKIE, L. HE, et al. 2000. Melanocortin 1 receptor variation in the domestic dog. Mamm. Genome **11:** 24–30.
9. MUNDY, N.I. & J. KELLY. 2002. Evolution of a pigmentation gene, the melanocortin-1 receptor gene, in primates. Am. J. Phys. Anthropol. **121:** 67–80.
10. THERON, E., K. HAWKINS, E. BERMINGHAM, et al. 2001. The molecular basis of an avian plumage polymorphism in the wild: a point mutation in the MC1R gene is perfectly associated with the melanic plumage morph in the bananaquit (*Coereba flaveola*). Curr. Biol. **11:** 550–557.
11. SWOFFORD, D. 1999. PAUP*: Phylogenetic analysis using parsimony and other methods. Sinauer. Sunderland, MA.
12. YANG, Z. 1997. PAML: a program package for phylogenetic analysis by maximum likelihood. CABIOS **13:** 555–556.
13. YANG, Y.-K., C. DICKINSON, C. HASKELL-LUEVANO, et al. 1997. Molecular basis of the interaction of [Nle[4], D-Phe[7]] melanocyte stimulating hormone with the human melanocortin-1 receptor (melanocyte α-MSH receptor). J. Biol. Chem. **272:** 23000–23010.
14. EIZIRIK, E., N. YUHKI, W.E. JOHNSON, et al. 2003. Molecular genetics and evolution of melanism in the cat family. Curr. Biol. **13:** 448–453.
15. HARDING, R.M., E. HEALY, A. RAY, et al. 2000. Evidence for variable selective pressures at the human pigmentation locus, MC1R. Am. J. Hum. Genet. **66:** 1351–1361.
16. MUNDY, N.I. & J. KELLY. 2001. The phylogeny of lion tamarins based on interphotoreceptor binding protein intron sequences. Am. J. Primatol. **54:** 33–40.

Melanocortin Receptor Variants with Phenotypic Effects in Horse, Pig, and Chicken

LEIF ANDERSSON

Department of Medical Biochemistry and Microbiology, Uppsala University, Uppsala, Sweden

ABSTRACT: The melanocortin system is of considerable interest in domestic animals because their energy metabolism and pigmentation have been under strong selection. This article reviews our work on *MC1R* variants in horse, pig, and chicken, as well as a study on *MC4R* polymorphism in the pig. The chestnut coat color in horses is caused by an *MC1R* missense mutation (S83F). In the pig, we have described seven *MC1R* alleles controlling four different coat color phenotypes (wild type, dominant black, black spotting, and recessive red). The most interesting allele is the one causing black spotting because it carries two causative mutations, a frameshift and a missense mutation. The frameshift mutation is somatically unstable, and the black spots reflect somatic reversion events restoring the reading frame. Classic genetics have established eight alleles at the *Extended black* locus in chicken, which is assumed to correspond to the *Extension* locus in mammals. We have analyzed the co-segregation of alleles at *MC1R* and *Extended black* using a red jungle fowl × White Leghorn intercross and provide compelling evidence that these loci are identical. A previous study indicated that a missense mutation (D298N) in pig *MC4R* has an effect on fatness, growth, and feed intake. We could not confirm this association using an intercross between the wild boar and Large White domestic pigs, but it is possible that our F_2 generation was too small to detect the rather modest effect reported for this polymorphism.

KEYWORDS: pig; chicken; MC1R; MC4R; coat color; plumage color

INTRODUCTION

Domestication of animals started approximately 10,000 years ago as part of the agricultural revolution.[1] Selective breeding has generated a rich diversity of breeds of domestic animals that are adapted to various environmental conditions and production systems. Two breeds of domestic animals may be as phenotypically divergent as different species in nature. Our domestic animals consequently provide unique resources that may be exploited to reveal the molecular basis for both monogenic and polygenic traits. Genetic analysis of phenotypic traits may be conducted by sampling existing pedigree material or by cross-breeding experiments. In some domestic species like pig and chicken, it is possible to make intercrosses involving the wild ancestor. This makes it possible to study the genetics of the domestication

Address for correspondence: Leif Andersson, Department of Medical Biochemistry and Microbiology, Uppsala University, Box 597, S-751 24 Uppsala, Sweden. Voice: 46-184714904; fax: 46-184714833.

leif.andersson@imbim.uu.se

Ann. N.Y. Acad. Sci. 994: 313–318 (2003). © 2003 New York Academy of Sciences.

process and identify variants that were selected early during the history of domestication. My group has pioneered this field, and several of the studies reviewed in this article are based on such experiments.

The melanocortin system is of considerable interest in the field of animal genetics because the energy metabolism and pigmentation have been under strong selection in domestic animals. It is obvious why there has been an interest to modify energy metabolism and food intake. For instance, in meat-producing animals there has been a strong selection for fast growth and a prerequisite for fast growth is a high appetite. Note also that in the past (before mechanization of agriculture, industry, and transport systems) there was a selection for high fat content, whereas during the last 50 years there has been a strong selection for lean meat. The genes for melanocortin receptor 3 (*MC3R*) and 4 (*MC4R*) are obvious candidate genes for controlling variation in energy metabolism and food intake.

It is less obvious why the pigmentation system has been under selection in domestic animals. A characteristic feature of domestication is that the pigmentation changes during this process. This can be explained partly by a release from purifying selection in natural populations that maintain a pigmentation pattern that is essential for camouflage and behavioral interaction (e.g., mate selection). However, it is clear that the diversification of pigmentation patterns has been promoted by selection for diversity, at least during the last 1,000 years and probably earlier. This is evident from written documentation and paintings. It is likely that the coat color (or plumage color) has been used as visual markers to distinguish improved forms with some favorable characteristic such as high-quality milk or fast growth. Thus, the colors have been used as trademarks for the breed. It is also possible that humans' fascination for fancy has contributed to the maintenance of newly arisen mutations. Maybe it was prestigious to ride a white horse or have a line of pigs that looked different from all the other pigs in the village. During the last 200 years, coat color often has been used as breed standard. This is illustrated by the fact that the color is included in the name of many breeds, for example, Large White pigs, Swedish Red and White cattle, and White Leghorn chicken. In many breeds of domestic animals, registration of the animal in the breed book requires the correct color. Because of this strong selection on pigmentation, genetic variants of the melanocortin receptor 1 (MC1R) are present in most domestic animals.

In this article, I review our work on *MC1R* polymorphism in horse, pig, and chicken and on *MC4R* polymorphism in pig.

THE CHESTNUT COLOR IN HORSES IS CAUSED BY AN *MC1R* MISSENSE MUTATION

A chestnut horse has red pigmentation on the body, mane, and tail without any black pigmentation. The shade of red varies from sorrel (light red/yellow) to dark liver chestnuts because of unknown modifying genes. The chestnut color is caused by homozygosity for a recessive allele at the *Extension* locus. Early comparative gene mapping supported the assumption that *Extension* in horses and mice are homologous loci.[2] This was later confirmed when sequence analysis showed that the chestnut color is associated with a missense mutation (S83F) in *MC1R*.[3] Extensive screening of a high number of horses in several breeds has shown a complete association between this mutation and the chestnut phenotype, strongly supporting the idea that it is the causative one. The fact that the same mutation is found in different breeds indicates that it arose before the major breeds were formed. No other *MC1R* mutation with a phenotypic effect has yet been reported in horses.

MC1R/EXTENSION VARIANTS AFFECTING COAT COLOR IN THE PIG

Classic genetic analyses have established four alleles at the *Extension* (*E*) locus in pigs.[4] These are E^+ for wild type, E^D for dominant black, E^P for black spotting, and *e* for recessive red. Subsequent genetic and sequence analyses showed that the *Extension* locus corresponds to *MC1R* as expected from previous studies in the mouse.[5] Seven *MC1R* alleles corresponding to the four phenotypically defined alleles have been sequenced[6–9] (TABLE 1); note that there is an insertion of three codons after codon 31 in the pig *MC1R* compared with *MC1R* in other mammals, and thus the numbering of codons in the pig differs from other mammals. The characterization of *MC1R* variation provided the first indication to us that pigs have been domesticated from both an Asian and a European subspecies of the wild boar and that some European breeds have a hybrid origin.[6] This was evident from the fact that we observed two *MC1R* sequences, differing by as many as five nucleotide substitutions, associated with the dominant black phenotype in European and Asian domestic pigs. Moreover, the Asian variant of dominant black was present also in the Large Black breed developed in England, suggesting introgression of Asian germplasm into this European breed. Subsequent studies of mitochondrial DNA and nuclear DNA have provided compelling evidence that pig domestication has involved both a European and an Asian subspecies of the wild boar and that introgression of Asian pigs into several European breeds of domestic pigs has occurred.[8,10]

No pharmacology has been conducted on pig variants, and therefore the following discussion on possible causative mutations is based on the nature of the amino acid substitution and comparative data from other species.

WILD TYPE (E^+)

The *MC1R* sequence from the European wild boar and the Japanese wild boar differs by a single synonymous substitution (TABLE 1). These wild-type sequences differ from the MC1R sequence in most domestic breeds, and this has been utilized for forensic analysis to distinguish wild boar meat (more highly prized) from meat from domestic pigs.

Dominant Black (E^{D1} and E^{D2})

Two variants of the Asian form of dominant black and one variant of the European form have been identified (TABLE 1). The dominant black color associated with the E^{D1} allele is most likely caused by the L102P substitution which is identical to the L99P substitution present in cattle with dominant black color. The causative mutation for E^{D2} must be the single amino acid substitution found in this allele, D124N. Interestingly, the same substitution D121N is associated with dominant black color in sheep.[11]

TABLE 1. Summary of pig MC1R/E alleles and their effect on coat color

		Codon								
Allele	Coat color	17	22	95	102	121	122	124	164	243
$MC1R*1 (E^+)$	Wild type, European wild boar	GCG Ala	CGG Arg	GTG Val	CTG Leu	AAT Asn	GTC Val	GAC Asp	GCG Ala	GCG Ala
$MC1R* (E^+)$	Wild type, Japanese wild boar	--- -	--- -	--- -	--- -	--C -	--- -	--- -	--- -	--- -
$MC1R*2 (E^{D1})$	Dominant black	--A -	--- -	A-- Met	-C- Pro	--C -	--- -	--- -	--- -	--A -
$MC1R*7 (E^{D1})$	Dominant black	--A -	--- -	A-- Met	-C- Pro	--C -	A-- Ile	--- -	--- -	--A -
$MC1R*3 (E^{D2})$	Dominant black	--- -	--- -	--- -	--- -	--- -	--- -	A-- Asn	--- -	--- -
$MC1R*6 (E^P)$	Red or white with black spots	--- -	+CC FS^a	--- -	--- -	--- -	--- -	A--	--- -	--- -
$MC1R*4 (e)$	Recessive red	--- -	--- -	--- -	--- -	--- -	--- -	--- -	-T- Val	A-- Thr

A dash (-) indicates identity to the top sequence.
^{a}Sequence out of frame after codon 22.

Black Spotting (E^P)

E^P is the most interesting *MC1R* allele in the pig, and it contains two causative mutations, an insertion of two nucleotides at codon 22 causing a frameshift and the D124N missense mutation associated with dominant black color. The frameshift is expected to cause a complete loss-of-function, and thus a solid red coat color is expected for E^P homozygotes, but in contrast it is usually associated with black spots on a red or white background. The fact that the insertion of two C nucleotides extends a stretch of six Cs to eight Cs suggested to us that the allele may be somatically unstable and that the black spots may represent somatic mutations restoring the reading frame. RT-PCR analysis using mRNA from black and red areas confirmed this hypothesis.[7] The phenotype of E^P/E^P homozygotes varies considerably and ranges from solid red in Tamworth pigs, red with black spots in Swedish Linderöd pigs, white with black spots in Pietrain pigs to almost solid black in Berkshire pigs. In fact, Sewall Wright postulated as early as 1918 that the black color of Berkshire is an extended form of black spotting.[12] Black spotting may occur on a white or red background, and this is controlled by another locus (although we cannot entirely exclude a more complicated genetic basis). The white background could either reflect that this other locus (or loci) inhibits the expression of red pheomelanin without affecting black eumelanin, or that it leads to a lack of melanocytes in the white areas. We favor the latter explanation because the black spots are consistently larger on a white background than on a red background. This phenomenon may reflect that melanocytes in the black spots are able to expand more in the absence of melanocytes in the white areas. This interpretation implies that the combined effect of this other locus and *MC1R* loss-of-function leads to the absence of melanocytes in the skin. Further studies are required to test this hypothesis.

Recessive Red (e)

The *MC1R* sequence associated with the recessive *e* allele differs from the wild-type sequence by two amino acid substitutions, A164V and A243T. We have proposed that A243T is most likely the causative one because it occurs at a highly conserved residue in the sixth transmembrane region. A164V is less likely to be causative because it is a conservative substitution and the variant amino acid (valine) is present in the functional receptor of the horse.[3]

MC1R/EXTENDED BLACK VARIANTS AFFECTING PLUMAGE COLOR IN THE CHICKEN

An allelic series of eight alleles at the *Extended black* locus in the chicken has been established by classic genetic analysis.[13] There has been some confusion whether the *Extended black* and *MC1R* loci are identical in this species. The reason for this is that the *Extended black* locus has been assigned to chromosome 1 (the largest chromosome in the chicken) by linkage analysis[14] whereas the *MC1R* locus has been assigned to a microchromosome by FISH analysis.[15] Moreover, Takeuchi *et al.*[16] determined the *MC1R* sequence in a limited number of chicken representing three different *E* alleles (*E*, *e+* and *e^y*), and the results indicated that the *E* and *MC1R* is identical. We recently have analyzed the co-segregation of *E* and *MC1R* alleles in an intercross between the red jungle fowl (*e+/e+*) and White Leghorn (*E/E*) comprising about 800 F$_2$ animals.[17] The results provided compelling evidence for that the *Extended black* and *MC1R* loci are the same. The *MC1R* allele from White Leghorn carried an E92K substitution that was present in the heterozygous or homozygous form in all black F$_2$ birds. This substitution is also associated with dominant black color in the mouse (*E^{SO-3J}*), and pharmacological studies revealed that it causes a constitutively active receptor.[5]

AN *MC4R* VARIANT IN THE PIG AND ITS POSSIBLE ASSOCIATION WITH FATNESS, GROWTH, AND FEED INTAKE

Kim *et al.*[18] used a candidate gene approach and reported that a missense mutation (D298N) in *MC4R* is associated with fatness, growth, and feed intake in pigs. The result was obtained in an extensive association study comprising samples from more than 1000 pigs representing five different commercial lines. The variant allele N298 was associated with higher growth rate, fatness, and food intake. One cannot exclude the possibility that the observed association is caused by a polymorphism in a linked gene, but a possible phenotypic effect of the D298N substitution is suggested by the fact that D298 is well conserved among all melanocortin receptors.

We decided to investigate the possible effect of the *MC4R* polymorphism using our intercross between the European wild boar and Large White domestic pig comprising 200 F$_2$ animals.[19] The two wild boar animals turned out to be homozygous for the presumed wild-type allele (D298), whereas the Large White founders were homozygous for the variant allele (N298). Thus, the polymorphism was completely informative in the pedigree. However, statistical analysis did not show any tendency for a significant effect on the growth or fatness traits recorded on the F_2 animals.[18] This result does not exclude a functional significance of the reported $MC4R$ polymorphism because the size of our F_2 population may be too small to detect the rather modest effect estimated for this polymorphism. The previous study indicated that the average backfat thickness of the two homozygotes only differs by 0.9 mm. Further studies are required to confirm or reject the functional significance of the $MC4R$ D298N substitution in the pig.

REFERENCES

1. DIAMOND, J. 2002. Evolution, consequences and future of plant and animal domestication. Nature **418:** 700–707.
2. ANDERSSON, L. & K. SANDBERG. 1982. A linkage group composed of three coat color genes and three serum protein loci in horses. J. Hered. **73:** 91–94.
3. MARKLUND, L. et al. 1996. A missense mutation in the gene for melanocyte-stimulating hormone receptor (*MC1R*) is associated with the chestnut coat color in horses. Mamm. Genome **7:** 895–899.
4. OLLIVIER, L. & P. SELLIER. 1982. Pig genetics: a review. Ann. Génét. Sél. Anim. **14:** 481–544.
5. ROBBINS, L.S. et al. 1993. Pigmentation phenotypes of variant extension locus alleles result from point mutations that alter MSH receptor function. Cell **72:** 827–834.
6. KIJAS, J.M.H. et al. 1998. Melanocortin receptor 1 (*MC1R*) mutations and coat color in pigs. Genetics **150:** 1177–1185.
7. KIJAS, J.M.H. et al. 2001. A frameshift mutation in *MC1R* and a high frequency of somatic reversions cause black spotting in pigs. Genetics **158:** 779–785.
8. GIUFFRA, E. et al. 2000. The origin of the domestic pig: independent domestication and subsequent introgression. Genetics **154:** 1785–1791.
9. GUSTAFSSON, A.C. et al. 2001. Screening and scanning single nucleotide polymorphisms in the pig melanocortin 1 receptor gene (*MC1R*) by pyrosequencing. Anim. Biotechnol. **12:** 145–153.
10. KIJAS, J.M.H. & L. ANDERSSON. 2001. A phylogenetic study of the origin of the domestic pig estimated from the near complete mtDNA genome. J. Mol. Evol. **52:** 302–308.
11. VAGE, D.I. et al. 1999. Molecular and pharmacological characterization of dominant black coat color in sheep. Mamm. Genome **10:** 39–43.
12. WRIGHT, S. 1918. Color inheritance in mammals. VIII. Swine. J. Hered. **9:** 33–38.
13. SMYTH, J.R. 1996. Genetics of plumage, skin and pigmentation in chickens. *In* Poultry Breeding and Genetics. R.D. Crawford, Ed.: 109–167. Elsevier. New York.
14. SMYTH, J.R. & A.F. PONCE DE LEON. 1992. Linkage relationship between the Pea Comb (P) and Extended black (E) loci of the chicken. Poultry Sci. **71:** 208–210.
15. SAZANOV, A. et al. 1998. Evolutionarily conserved telomeric location of BBC1 and MC1R on a microchromosome questions the identity of MC1R and pigmentation locus on chromosome 1 in chicken. Chromosome Res. **6:** 651–654.
16. TAKEUCHI, S. et al. 1996. A possible involvement of melanocortin 1-receptor in regulating feather color pigmentation in the chicken. Biochim. Biophys. Acta **1308:** 164–168.
17. KERJE, S., J. LIND, D. SCHUTZ, et al. 2003. Melanocortin 1-receptor (*MC1R*) mutations are associated with plumage colour in chicken. Anim. Genet. In press.
18. KIM, K.S. et al. 2000. A missense variant of the porcine melanocortin-4 receptor (MC4R) gene is associated with fatness, growth, and feed intake traits. Mamm. Genome **11:** 131–135.
19. PARK, H.B. et al. 2002. *Melanocortin-4 receptor* (*MC4R*) genotypes have no major effect on fatness in a Large White × Wild Boar intercross. Anim. Genet. **33:** 155–157.

Sequence Characterization of Teleost Fish Melanocortin Receptors

DARREN W. LOGAN, ROBERT J. BRYSON-RICHARDSON, MARTIN S. TAYLOR,[a] PETER CURRIE, AND IAN J. JACKSON

MRC Human Genetics Unit, Western General Hospital, Crewe Road, Edinburgh EH4 2XU, United Kingdom

ABSTRACT: Zebrafish are an excellent model system for studying the function of melanocortins in developmental and physiological processes, not least because there are a considerable number of mutant lines in which pigment patterns are affected. The behavior of fish melanophores is influenced by α-melanocyte-stimulating hormone (α-MSH) and melanin-concentrating hormone (MCH). We have used a rapid assay for α-MSH and MCH function using melanophores present on single zebrafish scales. By *in silico* analysis, we have identified the full complement of melanocortin receptors in both zebrafish and the pufferfish, *Fugu*. Mammals have five such receptors. Zebrafish have six melanocortin receptors, including two MC5R orthologues, whereas *Fugu*, lacking MC3R, has only four. We have confirmed the sequences of these 10 genes and show the comparison of the amino acid sequences of the encoded proteins with the orthologous receptor in other vertebrates.

KEYWORDS: zebrafish; *Fugu*; MC1R; MC2R; MC3R; MC4R; MC5R

INTRODUCTION

The existence of what we now call melanocortins was first indicated almost 90 years ago with the demonstration that the pituitary played a role in regulation of pigmentation of frogs and tadpoles.[1,2] Much of the early work on melanocyte-stimulating hormone (MSH) was conducted on cold-blooded vertebrates (poikilotherms) and the regulation of pigmentation in amphibia and fish in particular.[3]

In recent years, with the identification by molecular cloning of the family of five melanocortin receptors in mammals, attention has turned to mammalian systems. The realization that melanocortins regulate a wide range of physiological processes, from feeding behavior to sexual function, has greatly increased interest and scientific activity in the area. Poikilotherm melanocortin function has not received a corresponding increase in study. However, the zebrafish, *Danio rerio*, is an excellent model organism for which several powerful genetic tools have been developed that allow the

Address for correspondence: Ian J. Jackson, MRC Human Genetics Unit, Western General Hospital, Crewe Road, Edinburgh, EH4 2XU, United Kingdom. Voice: 44-0-131-567-8409; fax: 44-0-131-343-2620.

ian.jackson@hgu.mrc.ac.uk

[a]Present address: Wellcome Trust Centre for Human Genetics, Oxford University, Oxford, United Kingdom.

Ann. N.Y. Acad. Sci. 994: 319–330 (2003). © 2003 New York Academy of Sciences.

identification and study of genes and mutations affecting a range of developmental and physiological processes.[4,5] There is clearly an opportunity to study the genetics of melanocortin function in fish using these resources. What role, if any, melanocortins may play in the behavior of zebrafish is largely unknown. What is known, however, is that the primary effect of MSH on pigment cells, melanophores, is different from that seen in mammals and birds. Genetic and cell biology studies on a range of mammals show that α-MSH acts through the melanocortin receptor-1 (MC1R) on melanocytes and causes the cells to produce dark eumelanin. Lack of signaling through MC1R results in the synthesis of pheomelanin, which is a red or yellow pigment depending on the species and genetic background (reviewed in Jackson[6]). In contrast, action of MSH on the melanophores of fish and amphibia causes a rapid redistribution of melanin granules in the cell, so that intracellular pigment changes from being concentrated around the nucleus to being dispersed throughout the cell.[7] This movement gives the appearance of the cell itself rapidly changing shape and expanding. In poikilotherms, yellow and red pigments are generated in separate cell types (termed xanthophores and erythrophores). There is some evidence that α-MSH can promote a similar cellular redistribution of these pigments in some species, although these are more difficult to study than melanophores. It will be interesting to elucidate the basis for the difference in response of these homologous cell types in cold-blooded versus warm-blooded vertebrates.

Availability of large-scale DNA sequence information from two fish species, zebrafish and the pufferfish *Fugu rubripes*, has facilitated gene identification in these animals.[8] We have used these sequence resources to identify what we believe to be the entire repertoire of melanocortin receptors in the two species.

A SINGLE CELL ASSAY FOR MSH FUNCTION IN ZEBRAFISH

The dark melanophores of zebrafish scales can be readily visualized under brightfield optics. The effect of addition of factors to the medium on melanin dispersion and aggregation can be seen by the apparent shape change of the cells. Sequential collection of digital images over a period of several minutes can generate a movie of the pigment movement within the cell. Use of appropriate image intensity thresholds allows the area occupied by the pigment granules to be rapidly calculated for each time point. Addition of α-MSH, or analogues such as NDP-MSH, to the medium causes a rapid dispersion of pigment (in FIG. 1, 50% maximal dispersion is seen in 215 s). The reverse effect is seen by addition of the peptide hormone melanin concentrating hormone (MCH) (in this example, 50% maximal aggregation in 165 s). Although MCH is a functional antagonist of MSH, it probably signals through a different receptor and inhibits MSH action downstream.[20] The regulation of pigment distribution by MSH/MCH is augmented in some poikilotherms by other control mechanisms. For example, melatonin, a hormone produced in the pineal gland in response to photoperiod, can also aggregate pigment in zebrafish melanophores as can the neurotransmitter noradrenaline.

FIGURE 1 illustrates a single cell assay for NDP-MSH and MCH. More reliable data can be obtained by averaging the intensity across the whole field. This is of course analogous to the frog melanophore assay that has been widely used to assay MSH receptor agonist and antagonist activity but one in which individual cell behavior can be observed and recorded.[9,10]

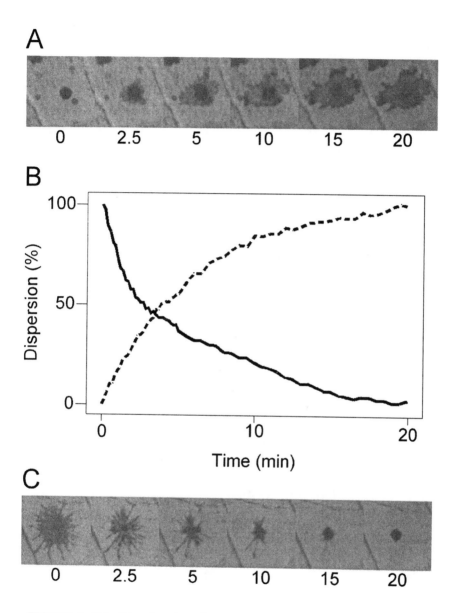

FIGURE 1. Zebrafish melanophore dispersion assay. (**A**) A single dorsal fin scale melanophore after addition of 10 μM NDP-MSH. Noradrenaline (10 μM) was added before the assay to aggregate the melanophores. Images were collected at the times indicated in minutes. (**B**) Percentage dispersion or aggregation over time after addition of 10 μM NDP-MSH (*broken line*) or 10 μM MCH (*solid line*), as in **A** and **C**. Sixty-one images of a field containing several melanophores were collected automatically at 10-s intervals for the first 5 min and 30-s intervals thereafter. Image intensity threshold was set to permit automatic measurement of the area occupied by melanin granules. (**C**) A single dorsal fin scale melanophore after addition of 10 μM MCH. Images were collected at the times indicated in minutes.

IDENTIFICATION OF THE FISH MELANOCORTIN RECEPTOR GENE FAMILY

Extensive DNA sequence data are available for zebrafish and for two pufferfish species, *Fugu rubripes* and *Tetraodon nigroviridis.*[8] (Resources at: www.sanger.ac.uk/ Projects/D_rerio/, http://genome.jgi-psf.org/fugu6/fugu6.home.html and www.geno-scope.cns.fr/externe/tetraodon/). We used TBLASTN analysis of whole genome shotgun databases of each species using the chicken MC1R amino acid sequence (accession number D78272). Positive sequence reads were assembled using Phred/ Phrap[11] and then extended by iterated BLAST searching and assembly. The assembled DNA sequences were confirmed by PCR and resequencing. The process was facilitated by the fact that all previously described melanocortin receptor genes lacked introns in the coding region, and indeed we found that most of the fish orthologues also had only a single coding exon. However, we identified two exceptions. The pufferfish (but not the zebrafish) orthologues of *MC2R* and *MC5R* contained one and three introns, respectively. The single intron of *MC2R* is in the same position (same codon and phase) as the second intron of *MC5R* .[20]

Orthologues of each of the mammalian melanocortin receptor gene family members were clearly identified in fish. We found that MC1R, MC2R, and MC4R have single orthologues in both zebrafish and pufferfish. We identified an orthologue of MC3R in zebrafish but could not find MC3R in pufferfish.

The *Fugu* genome sequence is not yet complete. However, the coverage afforded by the available sequence data is estimated to be 5.6-fold,[8] giving a greater than 95% chance of a read containing the desired sequence. Furthermore, an extensive data set consisting of 6-fold coverage of DNA sequence from the closely related pufferfish *Tetraodon nigroviridis* also lacks sequence corresponding to an *MC3R* orthologue. The probability that these combined databases lack the *MC3R* gene by chance is less than 0.002, and it is highly likely that MC3R is absent from the pufferfish lineage. The pufferfish genome sequence databases contain a single *MC5R* gene. In contrast, and as has been reported previously, zebrafish have two MC5R orthologues.[12]

PROTEIN SEQUENCE COMPARISONS

FIGURES 2 to 6 show the protein sequences of the five melanocortin receptor family members from human, mouse, chicken, *Fugu*, and zebrafish. We examined the sequences for significant differences between the fish receptor sequences and those of other vertebrates.

A good deal of information is known about the effects of both natural and site-directed mutations in MC1R.[13–16] Almost all the amino acid residues known to be important from these mutations are conserved in the fish MC1R. One major difference is in the third intracellular loop. Replacement of certain residues in this loop, between amino acids 226 and 238, by alanine result in loss of coupling of the human

---

FIGURE 2. Alignment of the amino acid sequences of MC1R from human (Hs), mouse (Mm), chicken (Gg), *Fugu* (Fr), and zebrafish (Dr). Amino acids identical in all sequences are shaded *black*, those different in one species are *dark gray*, and those different in two species are *light gray*.

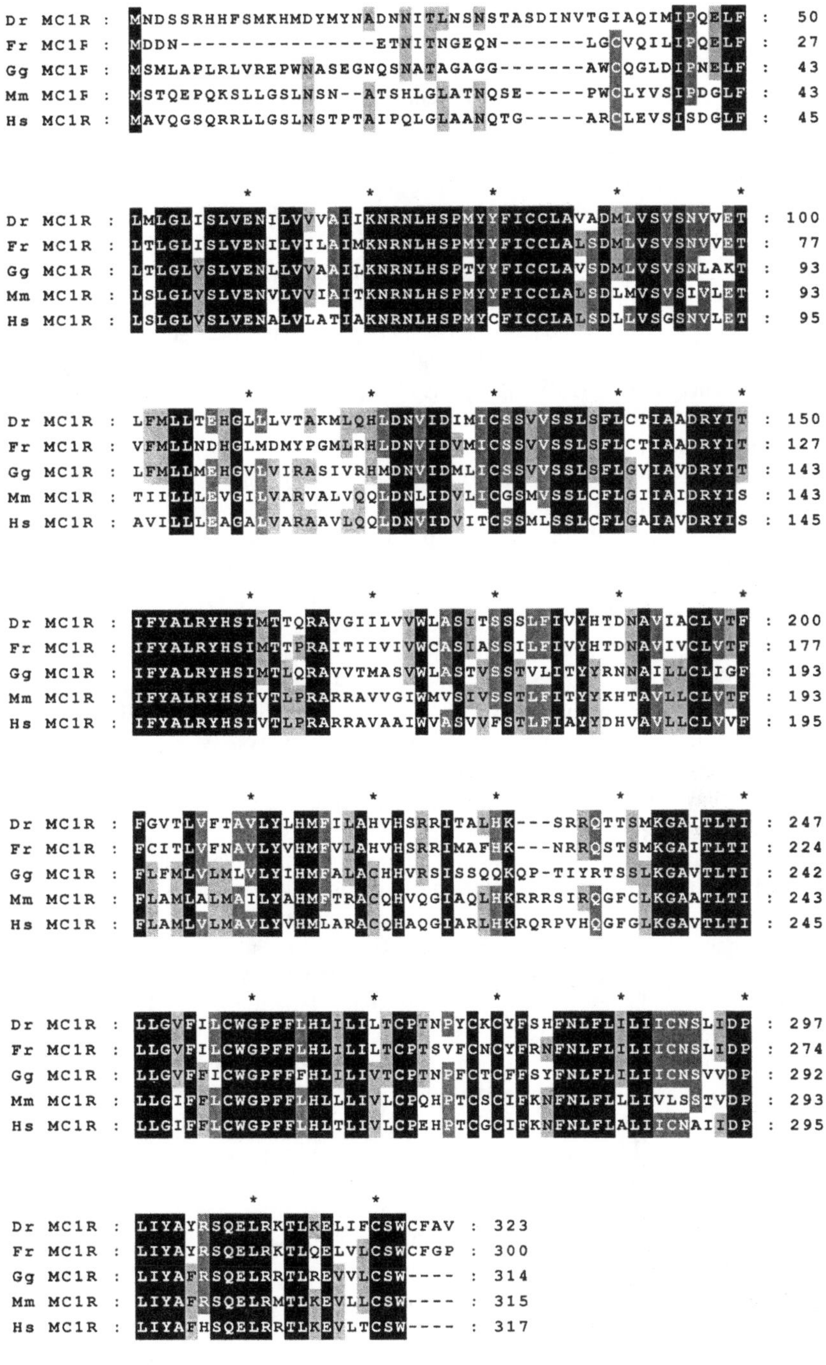

```
Dr MC1R :  MNDSSRHHFSMKHMDYMYNADNNITLNSNSTASDINVTGIAQIMIPQELF :  50
Fr MC1F :  MDDN---------------ETNITNGEQN-------LGCVQILIPQELF :  27
Gg MC1F :  MSMLAPLRLVREPWNASEGNQSNATAGAGG-------AWCQGLDIPNELF :  43
Mm MC1F :  MSTQEPQKSLLGSLNSN--ATSHLGLATNQSE-----PWCLYVSIPDGLF :  43
Hs MC1R :  MAVQGSQRRLLGSLNSTPTAIPQLGLAANQTG-----ARCLEVSISDGLF :  45

                 *          *          *          *          *
Dr MC1R :  LMLGLISLVENILVVVAIIKNRNLHSPMYYFICCLAVADMLVSVSNVVET : 100
Fr MC1R :  LTLGLISLVENILVILAIMKNRNLHSPMYYFICCLALSDMLVSVSNVVET :  77
Gg MC1R :  LTLGLVSLVENLLVVAAILKNRNLHSPTYYFICCLAVSDMLVSVSNLAKT :  93
Mm MC1R :  LSLGLVSLVENVLVVIAITKNRNLHSPMYYFICCLALSDLMVSVSIVLET :  93
Hs MC1R :  LSLGLVSLVENALVLATIAKNRNLHSPMYCFICCLALSDLLVSGSNVLET :  95

                 *          *          *          *          *
Dr MC1R :  LFMLLTEHGLLLVTAKMLQHLDNVIDIMICSSVVSSLSFLCTIAADRYIT : 150
Fr MC1R :  VFMLLNDHGLMDMYPGMLRHLDNVIDVMICSSVVSSLSFLCTIAADRYIT : 127
Gg MC1R :  LFMLLMEHGVLVIRASIVRHMDNVIDMLICSSVVSSLSFLGVIAVDRYIT : 143
Mm MC1R :  TIILLLEVGILVARVALVQQLDNLIDVLICGSMVSSLCFLGIIAIDRYIS : 143
Hs MC1R :  AVILLLEACALVARAAVLQQLDNVIDVITCSSMLSSLCFLGAIAVDRYIS : 145

                 *          *          *          *          *
Dr MC1R :  IFYALRYHSIMTTQRAVGIILVVWLASITSSSLFIVYHTDNAVIACLVTF : 200
Fr MC1R :  IFYALRYHSIMTTPRAITIIVIVWCASIASSILEIVYHTDNAVIVCLVTF : 177
Gg MC1R :  IFYALRYHSIMTLQRAVVTMASVWLASTVSSTVLITYYRNNAILLCLIGF : 193
Mm MC1R :  IFYALRYHSIVTLPRARRAVVGIWMVSIVSSTLFITYYKHTAVLLCLVTF : 193
Hs MC1R :  IFYALRYHSIVTLPRARRAVAAIWVASVVFSTLFIAYYDHVAVLLCLVVF : 195

                 *          *          *          *          *
Dr MC1R :  FGVTLVFTAVLYLHMFILAHVHSRRITALHK---SRRQTTSMKGAITLTI : 247
Fr MC1R :  FCITLVFNAVLYVHMFVLAHVHSRRIMAFHK---NRRQSTSMKGAITLTI : 224
Gg MC1R :  FLFMLVLMLVLYIHMFALACHHVRSISSQQKQP-TIYRTSSLKGAVTLTI : 242
Mm MC1R :  FLAMLALMAILYAHMFTRACQHVQGIAQLHKRRRSIRQGFCLKGAATLTI : 243
Hs MC1R :  FLAMLVLMAVLYVHMLARACQHAQGIARLHKRQRPVHQGFGLKGAVTLTI : 245

                 *          *          *          *          *
Dr MC1R :  LLGVFILCWGPFFLHLILILTCPTNPYCKCYFSHFNLFLILIICNSLIDP : 297
Fr MC1R :  LLGVFILCWGPFFLHLILILTCPTSVFCNCYFRNFNLFLILIICNSLIDP : 274
Gg MC1R :  LLGVFFICWGPFFFHLILIVTCPTNPFCTCFFSYFNLFLILIICNSVVDP : 292
Mm MC1R :  LLGIFFLCWGPFFLHLLLIVLCPQHPTCSCIFKNFNLFLLLIVLSSTVDP : 293
Hs MC1R :  LLGIFFLCWGPFFLHLTLIVLCPEHPTCGCIFKNFNLFLALIICNAIIDP : 295

                 *          *
Dr MC1R :  LIYAYRSQELRKTLKELIFCSWCFAV : 323
Fr MC1R :  LIYAYRSQELRKTLQELVLCSWCFGP : 300
Gg MC1R :  LIYAFRSQELRRTLREVVLCSW---- : 314
Mm MC1R :  LIYAFRSQELRMTLKEVLLCSW---- : 315
Hs MC1R :  LIYAFHSQELRRTLKEVLTCSW---- : 317
```

FIGURE 2. *See previous page for legend.*

receptor to G proteins.[16] The region containing these important residues is not well conserved in fish. It is three amino acids shorter than the mammalian receptor, and although K226 and K238 flanking the domain are conserved in the fish MC1R, the intervening amino acids are quite different. The sequence is, however, very basic in charge, similar to the region in humans (and the poorly conserved mouse region). It will be informative to determine whether specific amino acid residues are required for coupling, or whether it is the basic nature that is important.

Another difference between the fish and most other MC1R proteins is at the C terminus. Most G protein–coupled receptors have a cysteine residue in the intracellular, C-terminal tail. This has been shown to be modified by palmitoylation and to be essential for function in some receptors.[17] Previously described MC1R proteins contain the C-terminal tripeptide motif CSW, and truncation of the C-terminal 12 amino acids of MC1R appears to cause loss of receptor function.[18] Although both fish MC1R proteins have the CSW motif, it is not directly at the end of the protein. Both have an additional four residues, including another cysteine residue, which may be an alternative target for palmitoylation.

Human MC3R and MC4R have differential affinity for α–MSH, MC3R having approximately 30-fold higher affinity. Domain swapping between the two receptors indicates that the third extracellular loop is responsible for the difference.[19] Mutation of individual amino acids shows that residues 267 and 268 of MC4R are particularly important, especially Y268. Where MC4R has Phe-Tyr at these positions, MC3R has Leu-Ile. Substitution of the MC4R residues with those found in MC3R at these positions makes the affinity for α-MSH of the resulting receptor similar to that of MC3R. The equivalent positions in fish MC4R are Leu-Met, much more like MC3R. In particular, the bulky Tyr268 is thought to impair binding of α-MSH. In fish MC4R, this is a more compact Met. We suggest based on these comparisons that fish MC4R has a higher affinity for α-MSH than their mammalian and bird counterparts. This may have removed selective pressure to retain MC3R and thus may have allowed loss of *Fugu MC3R*.

The predicted protein sequences of the receptors can be placed in a phylogenetic tree that indicates their likely evolutionary relationship to each other. FIGURE 7 is such a tree in which the branch lengths are an indication of the divergence. Almost all the branch points have a high degree of statistical support, based on bootstrap replicates from 1,000 repetitions. The two exceptions are the branch points where MC1R splits from the rest of the gene family and where MC5R and MC3R split after separating from MC4R. If these nodes are discounted, then it appears that the original melanocortin receptor gene may have split into three members initially, causing the ancestors of MC1R, MC2R, and a single ancestor for the rest of the family. This single ancestor may then have split into three, which became MC3R, MC4R, and MC5R. However, all five family members must have been established before the teleost fish lineage separated from the rest of the vertebrate lineage. Because the melanocortin receptor family is not found in any invertebrate genome, including the urochordate *Ciona intestinalis* (data not shown), it is likely that the family emerged soon after vertebrate evolution and there were several rounds of gene duplication and divergence early on in vertebrate evolution.

The tree indicates that the pair of *MC5R* genes of zebrafish probably duplicated before the separation of the zebrafish and *Fugu* lineages, and the second *MC5R* was subsequently lost from *Fugu*. It is possible, however, that the duplication occurred

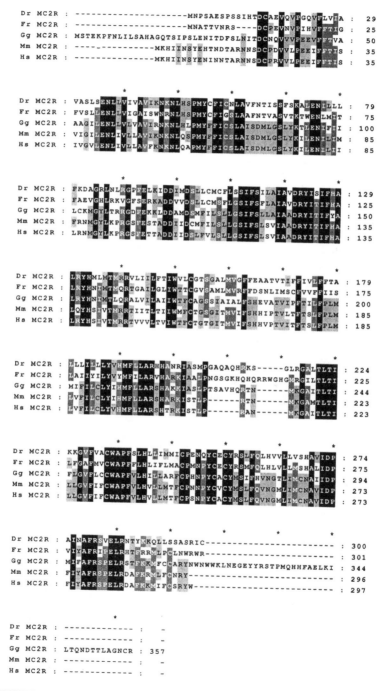

FIGURE 3. Alignment of the amino acid sequences of MC2R. Species and shading are as in FIGURE 2.

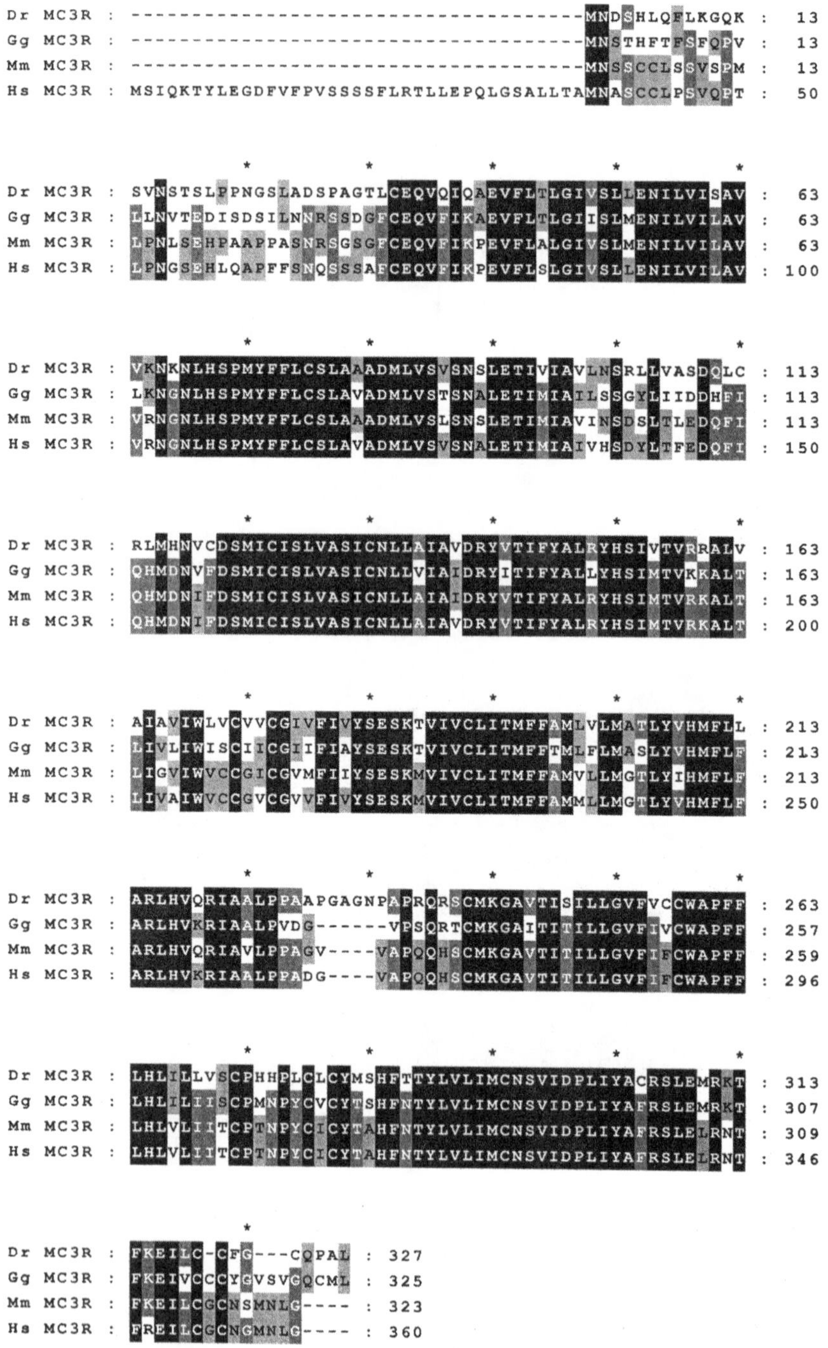

```
Dr MC3R : ------------------------------------MNDSHLQFLKGQK :  13
Gg MC3R : ------------------------------------MNSTHFTFSFQPV :  13
Mm MC3R : ------------------------------------MNSSCCLSSVSPM :  13
Hs MC3R : MSIQKTYLEGDFVFPVSSSSFLRTLLEPQLGSALLTAMNASCCLPSVQPT :  50


Dr MC3R : SVNSTSLPPNGSLADSPAGTLCEQVQIQAEVFLTLGIVSLLENILVISAV :  63
Gg MC3R : LLNVTEDISDSILNNRSSDGFCEQVFIKAEVFLTLGIISLMENILVILAV :  63
Mm MC3R : LPNLSEHPAAPPASNRSGSGFCEQVFIKPEVFLALGIVSLMENILVILAV :  63
Hs MC3R : LPNGSEHLQAPFFSNQSSSAFCEQVFIKPEVFLSLGIVSLLENILVILAV : 100


Dr MC3R : VKNKNLHSPMYFFLCSLAAADMLVSVSNSLETIVIAVLNSRLLVASDQLC : 113
Gg MC3R : LKNGNLHSPMYFFLCSLAVADMLVSTSNALETIMIAIISSGYLIIDDHFI : 113
Mm MC3R : VRNGNLHSPMYFFLCSLAAADMLVSLSNSLETIMIAVINSDSLTLEDQFI : 113
Hs MC3R : VRNGNLHSPMYFFLCSLAVADMLVSVSNALETIMIAIVHSDYLTFEDQFI : 150


Dr MC3R : RLMHNVCDSMICISLVASICNLLAIAVDRYVTIFYALRYHSIVTVRRALV : 163
Gg MC3R : QHMDNVFDSMICISLVASICNLLVIAIDRYITIFYALLYHSIMTVKKALT : 163
Mm MC3R : QHMDNIFDSMICISLVASICNLLAIAIDRYVTIFYALRYHSIMTVRKALT : 163
Hs MC3R : QHMDNIFDSMICISLVASICNLLAIAVDRYVTIFYALRYHSIMTVRKALT : 200


Dr MC3R : AIAVIWLVCVVCGIVFIVYSESKTVIVCLITMFFAMLVLMATLYVHMFLL : 213
Gg MC3R : LIVLIWISCIICGIIFIAYSESKTVIVCLITMFFTMLFLMASLYVHMFLF : 213
Mm MC3R : LIGVIWVCCGICGVMFIIYSESKMVIVCLITMFFAMVLLMGTLYIHMFLF : 213
Hs MC3R : LIVAIWVCCGVCGVVFIVYSESKMVIVCLITMFFAMMLLMGTLYVHMFLF : 250


Dr MC3R : ARLHVQRIAALPPAAPGAGNPAPRQRSCMKGAVTISILLGVFVCCWAPFF : 263
Gg MC3R : ARLHVKRIAALPVDG------VPSQRTCMKGAITITILLGVFIVCWAPFF : 257
Mm MC3R : ARLHVQRIAVLPPAGV----VAPQQHSCMKGAVTITILLGVFIFCWAPFF : 259
Hs MC3R : ARLHVKRIAALPPADG----VAPQQHSCMKGAVTITILLGVFIFCWAPFF : 296


Dr MC3R : LHLILLVSCPHHPLCLCYMSHFTTYLVLIMCNSVIDPLIYACRSLEMRKT : 313
Gg MC3R : LHLFLIISCPMNPYCVCYTSHFNTYLVLIMCNSVIDPLIYAFRSLEMRKT : 307
Mm MC3R : LHLVLIITCPTNPYCICYTAHFNTYLVLIMCNSVIDPLIYAFRSLERNT  : 309
Hs MC3R : LHLVLIITCPTNPYCICYTAHFNTYLVLIMCNSVIDPLIYAFRSLERNT  : 346


Dr MC3R : FKEILC-CFG---CQPAL     : 327
Gg MC3R : FKEIVCCCYGVSVGQCML     : 325
Mm MC3R : FKEILCGCNSMNLG----     : 323
Hs MC3R : FREILCGCNGMNLG----     : 360
```

FIGURE 4. Alignment of the amino acid sequences of MC3R. Species and shading are as in FIGURE 2, except that *Fugu* is absent.

```
Dr MC4R : MTSHHHGLHH--SFRNHSQGALPVGKPSHGDRCSASG-CYEQLLISTEV : 47
Fr MC4R : NATDPPGRVQ--DFSNGSQ--TPETDFPNEEKESSTG-CYEQMLISTEV : 45
Gg MC4R : NFTQHRGTLQPLHFWNQS-NGLHRGASEPSAKCHSSGGCYEQLFVSPEV : 49
Mm MC4R : NSTHHHGMYTSLHLWNRSSYGLHSNASESLGKGHPDGGCYEQLFVSPEV : 50
Hs MC4R : VNSTHRGMHTSLHLWNRSSYRLHSNASESLGKGYSDGGCYEQLFVSPEV : 50


                 *         *         *         *         *
Dr MC4R : FLTLGLVSLLENILVIAAIVKNKNLHSPMYFFICSLAVADLLVSVSNASE : 97
Fr MC4R : FLTLGIISLLENILVVAAIVKNKNLHSPMYFFICSLAVADMLVSVSNASE : 95
Gg MC4R : FVTLGIISLLENVLVIVAIAKNKNLHSPMYFFICSLAVADMLVSVSNGSE : 99
Mm MC4R : FVTLGVISLLENILVIVAIAKNKNLHSPMYFFICSLAVADMLVSVSNGSE : 100
Hs MC4R : FVTLGVISLLENILVIVAIAKNKNLHSPMYFFICSLAVADMLVSVSNGSE : 100


                 *         *         *         *         *
Dr MC4R : TVVMALITGGNLTNRESIIKNMDNVFDSMICSSLLASIWSLLAIAVDRYI : 147
Fr MC4R : TIVIALINSGTLTIPATLIKSMDNVFDSMICSSLLASICSLLAIAVDRYI : 145
Gg MC4R : TIVITLLNN-TDTDAQSFTINIDNVIDSVICSSLLASICSLLSIAVDRYF : 148
Mm MC4R : TIVITLLNS-TDTDAQSFTVNIDNVIDSVICSSLLASICSLLSIAVDRYF : 149
Hs MC4R : TIIITLLNS-TDTDAQSFTVNIDNVIDSVICSSLLASICSLLSIAVDRYF : 149


                 *         *         *         *         *
Dr MC4R : TIFYALRYHNIMTQRRAGTIITCIWTFCTVSGVLFIVYSESTTVLICLIS : 197
Fr MC4R : TIFYALRYHNIVTLRRASLVISSIWTCCTVSGVLFIVYSESTTVLICLIT : 195
Gg MC4R : TIFYALQYHNIMTVKRVGVIITCIWAACTVSGILFIIYSDSSVVVICLIS : 198
Mm MC4R : TIFYALQYHNIMTVRRVGIIISCIWAACTVSGVLFIIYSDSSAVLICLIS : 199
Hs MC4R : TIFYALQYHNIMTVKRVGIIISCIWAACTVSGILFIIYSDSSAVLICLIT : 199


                 *         *         *         *         *
Dr MC4R : MFFTMLALMASLYVHMFLLARLHMKRIAALPGNGPIWQAANMKGAITITI : 247
Fr MC4R : MFFTMLVLMASLYVHMFLLARLHMKRIAAMPGNAPIHQRANLKGAITLTI : 245
Gg MC4R : MFFTMLILMASLYVHMFMMARMHIKKIAVLPGTGPIRQGANMKGAITLTI : 248
Mm MC4R : MFFTMLVLMASLYVHMFLMARLHIKRIAVLPGTGTIRQGTNMKGAITLTI : 249
Hs MC4R : MFFTMLALMASLYVHMFLMARLHIKRIAVLPGTGAIRQGANMKGAITLTI : 249


                 *         *         *         *         *
Dr MC4R : LLGVFVVCWAPFFLHLILMISCPRNPYCVCFMSHFNMYLILIMCNSVIDP : 297
Fr MC4R : LLGVFVVCWAPFFLHLILMITCPKNPYCTCFMSHFNMYLILIMCNSVIDP : 295
Gg MC4R : LIGVFVVCWAPFFLHLIFYISCPYNPYCVCFMSHFNFYLILIMCNSIIDP : 298
Mm MC4R : LIGVFVVCWAPFFLHLLFYISCPQNPYCVCFMSHFNLYLILIMCNAVIDP : 299
Hs MC4R : LIGVFVVCWAPFFLHLIFYISCPQNPYCVCFMSHFNLYLILIMCNSIIDP : 299


                 *         *         *
Dr MC4R : LIYALRSQEMRKTFKEICCWYG--LASLCV-- : 326
Fr MC4R : IIYALRSQEMRKTFKEIFCCSQM--LVCM---- : 322
Gg MC4R : LIYALRSQELRKTFKEIICCNLRGLCDLPGKY : 331
Mm MC4R : LIYALRSQELRKTFKEIICFYPLGGICELSSRY : 332
Hs MC4R : LIYALRSQELRKTFKEIICCYPLGGLCDLSSRY : 332
```

FIGURE 5. Alignment of the amino acid sequences of MC4R. Species and shading are as in FIGURE 2.

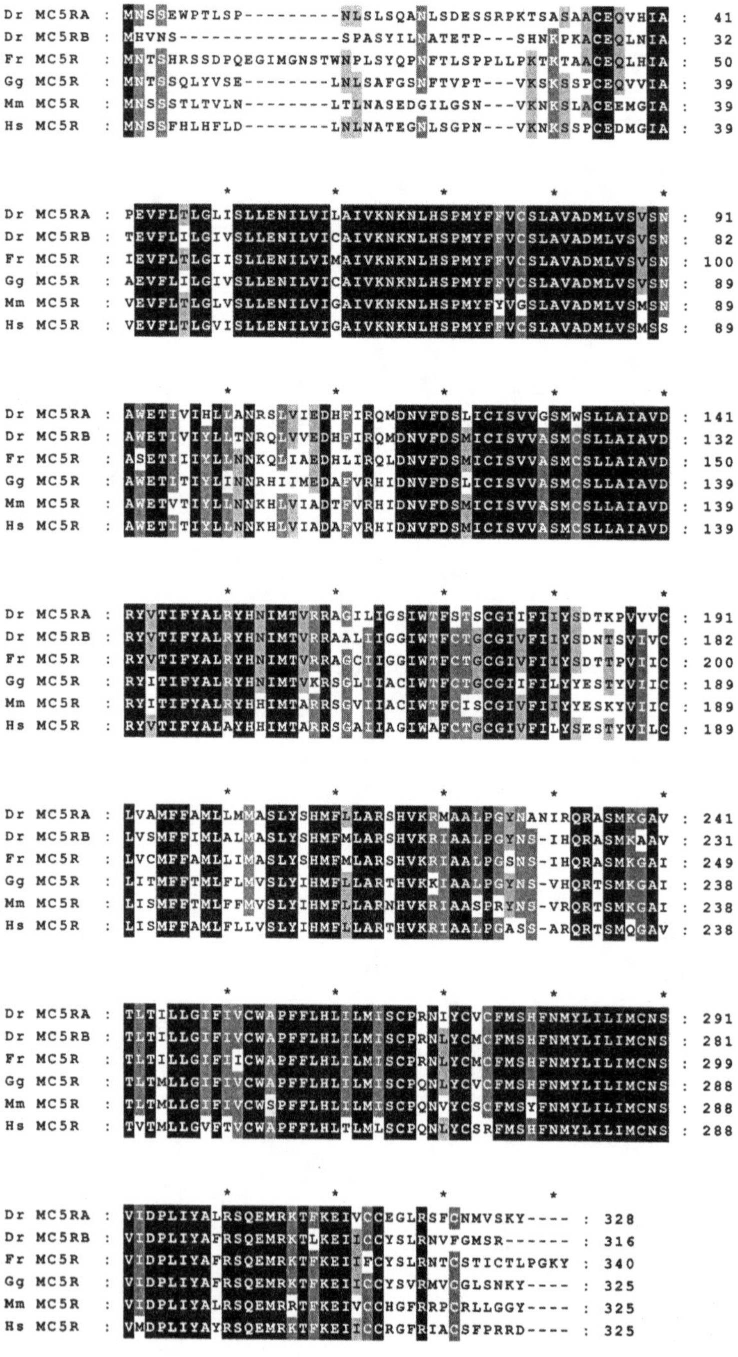

FIGURE 6. Alignment of the amino acid sequences of MC5R. Species and shading are as in FIGURE 2. Note that zebrafish has two MC5R orthologues.

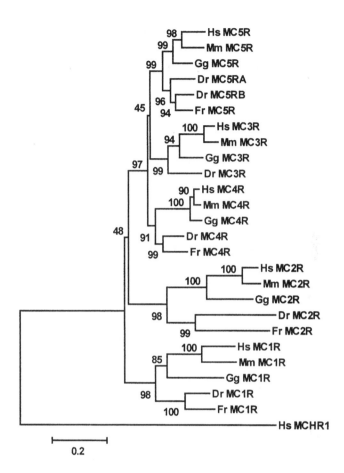

FIGURE 7. Neighbor-joining tree of melanocortin receptor protein sequences as in FIGURES 2 to 6. The tree uses a Poisson correction distance between two sequences where the proportion of amino acid sites at which the two differ is corrected for multiple substitutions at the same site. Thus, the distance is equal to $-\ln(1-p)$, where p is the proportion of sites that differ. The number at each node is the percentage of bootstrap replicates of the node from 1,000 repetitions.

specifically in the zebrafish lineage, and that *MC5RA* has been freed of evolutionary constraint and thus has evolved more rapidly.

SUMMARY

In summary, the extensive sequence data now available for zebrafish and *Fugu* have allowed us to identify *in silico* the entire family of melanocortin receptors for these species. We have confirmed the sequences of these 10 genes. These are a good foundation for the investigation of melanocortin function in cold-blooded vertebrates.

REFERENCES

1. ALLEN, B.M. 1916. Extirpation of the hypophysis and thyroid glands of *Rana pipiens*. Science **44:** 755–757.
2. SMITH, P.E. 1916. Experimental ablation of the hypophysis in the frog embryo. Science **44:** 280–282.
3. BAGNARA, J.T. 1998. Comparative anatomy and physiology of pigment cells in non-mammalian tissues. *In* The Pigmentary System: Physiology and Pathophysiology. J.J. Nordlund, R.E. Boissy, V.J. Hearing, R.A. King, and J-P. Ortonne, Eds.: 9–40. Oxford University Press. New York.
4. KELSH, R.N., M. BRAND, Y.J. JIANG, *et al.* 1996. Zebrafish pigmentation mutations and the processes of neural crest development. Development **123:** 369–389.
5. ODENTHAL, J., K. ROSSNAGEL, P. HAFFTER, *et al.* 1996. Mutations affecting xanthophore pigmentation in the zebrafish, *Danio rerio*. Development **123:** 391–398.
6. JACKSON, I.J. 1997. Homologous pigmentation mutations in human, mouse and other model organisms. Hum. Mol. Genet. **6:** 1613–1624.
7. FUJII, R. 2000. The regulation of motile activity in fish chromatophores. Pigm. Cell Res. **13:** 300–319.
8. APARICIO, S., J. CHAPMAN, E. STUPKA, *et al.* 2002. Whole-genome shotgun assembly and analysis of the genome of *Fugu rubripes*. Science **297:** 1301–1310.
9. POTENZA, M.N. & M.R. LERNER. 1992. A rapid quantitative bioassay for evaluating the effects of ligands upon receptors that modulate cAMP levels in a melanophore cell line. Pigm. Cell Res. **5:** 372–378.
10. JAYAWICKREME, C.K., J.M. QUILLAN, G.F. GRAMINSKI, *et al.* 1994. Discovery and structure-function analysis of alpha-melanocyte-stimulating hormone antagonists. J. Biol. Chem. **269:** 29846–29854.
11. EWING, B. & P. GREEN. 1998. Base-calling of automated sequencer traces using phred. II. Error probabilities. Genome Res. **8:** 186–194.
12. RINGHOLM, A., R. FREDRIKSSON, N. POLIAKOVA, *et al.* 2002. One melanocortin 4 and two melanocortin 5 receptors from zebrafish show remarkable conservation in structure and pharmacology. J. Neurochem. **82:** 6–18.
13. VALVERDE, P., E. HEALY, I.J. JACKSON, *et al.* 1995. Variants of the melanocyte-stimulating hormone receptor gene are associated with red hair and fair skin in humans. Nat. Genet. **11:** 328–330.
14. STURM R.A., R.D. TEASDALE & N.F. BOX. 2001. Human pigmentation genes: identification, structure and consequences of polymorphic variation. Gene **277:** 49–62.
15. HEALY, E., S.A. JORDAN, P.S. BUDD, *et al.* 2001. Functional variation of MC1R alleles from red-haired individuals. Hum. Mol. Genet. **10:** 2397–2402.
16. FRANDBERG, P.A., M. DOUFEXIS, S. KAPAS, *et al.* 1998. Amino acid residues in third intracellular loop of melanocortin 1 receptor are involved in G-protein coupling. Biochem. Mol. Biol. Int. **46:** 913–922.
17. MORELLO, J.P. & M. BOUVIER. 1996. Palmitoylation: a post-translational modification that regulates signalling from G-protein coupled receptors. Biochem. Cell Biol. **74:** 449–457.
18. NEWTON, J.E., A.L. WILKIE, L. HE, *et al.* 2000. Melanocortin 1 receptor variation in the domestic dog. Mamm. Genome **11:** 24–30.
19. OOSTEROM, J., W.A. NIJENHUIS, W.M. SCHAAPER, *et al.* 1999. Conformation of the core sequence in melanocortin peptides directs selectivity for the melanocortin MC3 and MC4 receptors J. Biol. Chem. **274:** 16853–16860.
20. LOGAN, D.W., R.J. BRYSON-RICHARDSON, K.E. PAGAN, *et al.* 2003. The structure and evolution of the melanocortin and MCH receptors in fish and mammals. Genomics **81:** 184–191.

Pigmentary Switches in Domestic Animal Species

H. KLUNGLAND[a] AND D.I. VÅGE[b]

[a]Department of Laboratory Medicine, Children's and Women's Diseases,
Faculty of Medicine, Norwegian University of Science and Technology,
N-7006 Trondheim, Norway

[b]Department of Animal Science, Agricultural University of Norway, Ås, Norway

ABSTRACT: Although homogeneous pigmentation usually is observed in wild animals, most domestic animal species display a wide variety of coat colors. In fur animals, the coat color is an important production trait, and in other species such as cattle and sheep, the coat color is a major breed characteristic. Variability in coat color is seen both within and between breeds, and makes domesticated species unique for studying gene function and gene regulation of loci affecting pigmentation. In several species, mutations in the MC1-R gene have been shown to cause the dominant expression of black pigment. In fox, alleles of both the agouti and the MC1-R gene could cause eumelanin synthesis. In addition, a nonepistatic interaction between MC1-R and agouti has been observed, resulting in several different coat color phenotypes expressing a mixture of red and black pigmentation. Also in cattle and sheep, amino acid substitutions within the MC1-R explain the dominant inheritance of black pigmentation. Unlike the constitutively activated MC1-R found in the Alaska silver fox, dominant variants of the MC1-R found in cattle and sheep seem to be completely dominant with no antagonizing effect of agouti. MC1-R variants with premature stop codons are widespread in several cattle populations, indicating that this well-conserved gene has no other fundamental function beside pigmentation. Other well-established breed characteristics include distinct coat color patterns in which the distribution of melanocytes, partly regulated by the c-*kit* gene, seems to be involved.

KEYWORDS: MC1-R; melanocortin receptor; melanocyte-stimulating hormone receptor agouti; melanocytes; pigment; fox; cattle; sheep; evolution; coat color

INTRODUCTION

In mammalian species, the distribution of pheomelanin (red or yellow pigment) and eumelanin (brown or black pigment) is mainly determined by the melanocortin receptor 1 (MC1-R) and its antagonist agouti.[1–4] Although limited variation in coat coloration can be seen in wild living animals, a fascinating variation is present

Address for correspondence: Helge Klungland, Department of Laboratory Medicine Children's and Women's Diseases, Faculty of Medicine, Norwegian University of Science and Technology, N-7006 Trondheim, Norway. Voice: 47-73598776; fax: 47-73867322.
helge.klungland@medisin.ntnu.no

Ann. N.Y. Acad. Sci. 994: 331–338 (2003). © 2003 New York Academy of Sciences.

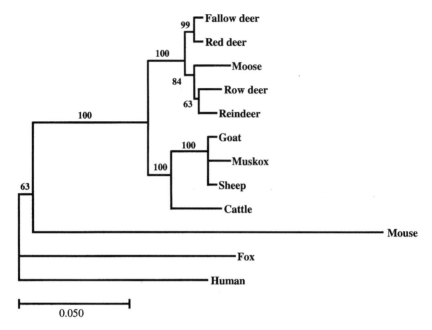

FIGURE 1. Rooted Neighbour Joining phylogram utilizing the complete 954-bp amino acid coding region of the MC1-R gene from Norwegian ruminants (domesticated as well as wild species), human, mouse, and fox.[5]

among domesticated animals, both between and within species. Domestic animal species therefore do represent important model organisms for basic studies of pigmentation, and unique correlations between genotype and phenotype have been revealed.

The syntheses of pigment do have several functions. Coat color is of importance for mating, for protection against predators, and as an adaptation to the environment. As a consequence of its central function in pigmentation, the MC1-R gene is well conserved, and has been shown to be suitable for evolutionary studies[5] (FIG. 1). Although the gene is well conserved, several domestic species have nonfunctional variants of the MC1-R, as well as for the agouti gene.[6,7] A change in pigmentation is apparently the only effect seen in these naturally occurring knockouts. Coat color mutations are associated with reduced fitness in wild animal species but have been amplified in domestic animals in which breeds with distinct phenotypes have been established by selective breeding (FIG. 2). Breeds with limited variation in coat color (often a unique coat color combined with a unique coat color pattern) showed reduced genetic variation for loci that are of importance for the pigment production.[8] This association was established in cattle for both the MC1-R gene where the alleles E^+, E^D, and e were examined and the c-*kit* gene for a nonfunctional polymorphism. These breeds are now utilized for evaluating genomics responses of modern breeding programs.

To understand the basic molecular mechanisms regulating pigmentation in domestic animal species, we have used a comparative approach based on previous stud-

A.

B.

FIGURE 2. Selection that is based on coat color and horn has produced breeds with phenotypical characteristics. Whereas some breeds express a single unique coat color phenotype, as exemplified in **A** and **B**, others express a wide variety of phenotypes. (**A**) Individuals of Østlandsk raukolle are red and polled and has no spotting. (**B**) All individuals of the Telemark cattle breed are red-sided with horns. Similar color-sided breeds exist that are black and polled (black-sided Trønder and Nordlands cattle).

ies in mouse.[3,9] By applying this strategy, we have identified major coat color genes in fox, as well as in cattle and sheep. Several of the genes that contain important coat color mutations have been studied previously by classic genetic experiments,[10–12] and these results have provided valuable information for the molecular analysis of the same mutations.

The silver fox may appear in two different genotypes that both are caused by a mutation in the red fox. Although the Alaska silver phenotype is caused by a mutation in MC1-R (Cys→Arg in position 125; C125R) that shows complete linkage to the actual phenotype, the standard silver fox phenotype is caused by a larger deletion in the *agouti* gene which gives a nonfunctional antagonist.[7] In the absence of agouti protein, α-MSH may bind to MC1-R in a nonrestricted manner, and synthesis of eumelanin will be induced. Pharmacological analysis showed that the C125R mutation turned the MC1-R receptor into a constitutive active state (FIGS. 3 and 4). The interaction between MC1-R and the agouti protein also may be responsible for the expression of red hairs that is observed in individuals of the Alaska silver fox. As for the MC1-R, many dominant, recessive, or less modified agouti alleles are known in mice.[13–15] On the other hand, there are no recognized effects of agouti on human pigmentation.[13]

Contrary to the red fox (*Vulpes vulpes*), the closely related species polar fox (*Alopex lagopus*) change color during the winter season. As for the red fox, several phenotypic variants exist of the polar fox. The major variant used for fur farming, the blue fox, does not change color during the winter season. We recently have identified mutations that could explain the blue fox phenotype (unpublished results). The dominant blue fox mutation(s) also suppresses the seasonal change of coat color found in winter-white polar foxes. Our results indicate that the molecular interaction

A.

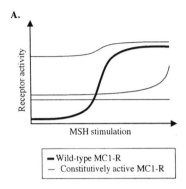

MSH stimulation

— Wild-type MC1-R
— Constitutively active MC1-R

B.

FIGURE 3. Pharmacological studies have shown that the constitutively active forms of MC1-R in fox and sheep do not respond upon α-MSH stimulation. (**A**) Typically, constitutively active forms of the MC1-R are active in absence of hormone and show little response upon stimulation.[7,9,17] (**B**) The dominant acting allele E^D produces a constitutively active receptor resulting in black pigment synthesis for one of the twin lambs.

between the G protein–coupled receptor MC1-R and the agouti protein is involved in the development of the winter-white coat color. These studies will now be followed up with pharmacological studies.

Following the characterization of dominant black coat color in mice,[3] the bovine MC1-R gene was sequenced in animals expressing different colors such as red, brown, or black.[6] Two mutations were observed, a base substitution giving an L99P amino acid substitution, and a single base deletion producing a frame shift (FIGS. 4 and 5). Both the mutated E^D and e alleles are widespread and are found in Nordic cattle breeds and Holstein, as well as in African breeds.[8] The truncated bovine e allele is also found within a closely related species, the Egyptian buffalo, showing that this allele probably did exist before the speciation of cattle and buffalo.[8] The bovine E^D allele is almost identical to the mouse E^{so} allele (sombre, L98P) that is known to produce a constitutively active MC1-R.[3] The wide variety of phenotypes expressed in E^+ animals, including red, brown, and black animals, could be explained by two different hypotheses. Because several agouti alleles are proposed in cattle,[16] it is likely to assume that the agouti gene does regulate the wild-type receptor produced by E^+ animals. Alternatively, the variety of coat colors expressed by E^+ animals also could be explained by additional mutations within the MC1-R gene that have not yet been discovered.

Another species in the Bovidae family, the sheep, have a similar dominant acting MC1-R as found in cattle and other species.[17] Two mutations, M73K and D121N, are identified that show complete cosegregation to the dominant black coat color. Although the M73K mutation alone showed constitutive activation in pharmacological studies, it is difficult to conclude that this is the mutation causing black pigment syn-

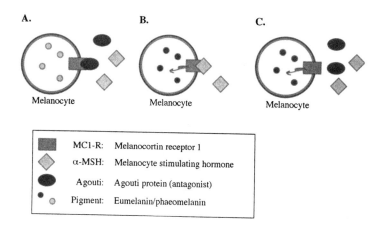

FIGURE 4. The following models illustrate our findings for the regulation of pheo-melanin or eumelanin synthesis in fox, cattle, and sheep. (**A**) MC1-R is blocked by the antagonist agouti and pheomelanin is synthesized (red or yellow pigment; gray in this illustration). The wild-type phenotype of the red fox is caused by the antagonizing effect of agouti, whereas the constitutively activated MC1-R produce cross-fox phenotypes when heterozygous and Alaska silver fox while homozygous.[7] Similar models may explain the phenotype of red cattle and white sheep. Recessive black individuals exist for both species, but so far a linkage to agouti has not been proved. The same effect is obtained with nonfunctional receptors, in our study represented by the cattle *e* allele producing a prematurely terminated receptor.[6] (**B**) In the absence of agouti α-MSH binds MC1-R and eumelanin is produced (black or brown pigment). The standard silver fox lack functional agouti protein, resulting in a black phenotype in homozygous animals.[7] (**C**) A constitutively active variant of the MC1-R produce black pigment. This simple model explains dominant black variants of cattle and sheep.[6,17] In the fox, a nonepistatic interaction occurs between MC1-R and agouti, and only animals homozygous for either or both loci have the silver fox phenotype.[7]

thesis. From other studies, it is known that the D121 position is required for ligand binding. Illustrations of animals possessing different alleles at the extension locus are given in FIGURE 3B.

Based on the thoroughly characterized MC1-R regulatory system, three distinct classes of the receptor could be outlined. Constitutively active MC1-R will increase the intracellular level of cAMP and subsequently produce eumelanin independent of the α-MSH or agouti interactions. This altered mode of action could be explained either by the lack of an agouti binding site or simply by structural changes in the MC1-R which then becomes active even without α-MSH binding.[9] On the other side, loss-of-function variants of the MC1-R that are not stimulated by α-MSH will produce pheomelanin.[3] In the wild type, or in individuals with less severe mutations in the MC1-R gene, the receptor function could either be modified by agouti or α-MSH binding or alter the MC1-R response to α-MSH or agouti. Contradictory to the two previously described regulatory changes, these modifications do not exclude regulation of the pigment synthesis that is caused by interactions between the MC1-R and the agouti protein.

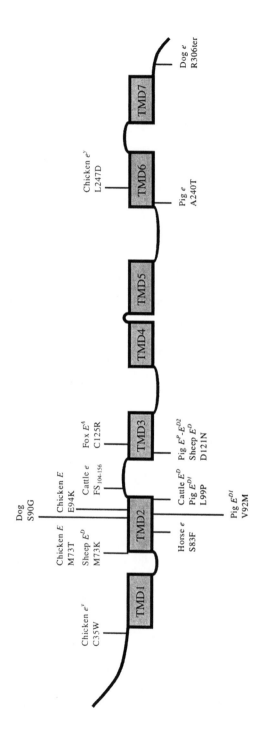

FIGURE 5. The melanocortin receptor 1 (MC1-R) has seven transmembrane domains, TMD1-7. Dominant mutations (*in capitals*) are localized to TMD2-3.[18] Altered amino acids are shown.

Similar to what is found in other species, including mouse, only point mutations and exchange of specific amino acids do produce a constitutively active receptor in domestic animals.[18] Dominant MC1-R mutations that induce the synthesis of eumelanin are mainly located in the second and third transmembrane domain of the MC1-R (FIG. 5). It therefore is likely that this region is crucial for generating a constitutively active receptor. Similar results also have been published in mice.[3,9] In addition to the fox E^A allele, and cattle and sheep E^D alleles, this region also contains mutations that are associated with a black phenotype in chicken, pig, and dog.[19–21] This is also in agreement with pharmacological studies conducted with the mouse MC1-R.[9] It is also interesting to observe the existence of identical mutations in sheep and pig (D121N), as well as in cattle and pig (L99P), although these species are relatively distantly related.[6,17,19] Because several mutations that are correlated with a constitutive active receptor and eumelanin synthesis affect identical amino acids,[3,6,9,17,19] there are probably a limited number of substitutions that have the potential of producing a constitutive active receptor. Moreover, other amino acids may be associated with reduced fitness, thereby reducing the possibility for preserving that mutation.

REFERENCES

1. BULTMAN, S.J., E.J. MICHAUD & R.P. WOYCHIK. 1992. Molecular characterization of the mouse agouti locus. Cell **71:** 1195–1204.
2. CONE, R.D., D. LU, S. KOPPULA, et al. 1996. The melanocortin receptors: agonists, antagonists, and the hormonal control of pigmentation. Recent Prog. Horm. Res. **51:** 287–318.
3. ROBBINS, L.S., J.H. NADEAU, K.R. JOHNSON, et al. 1993. Pigmentation phenotypes of variant extension locus alleles result from point mutations that alter MSH receptor function. Cell **72:** 827–834.
4. WILSON, B.D., M.M. OLLMANN, L. KANG, et al. 1995. Structure and function of ASP, the human homolog of the mouse agouti gene. Hum. Mol. Genet. **4:** 223–230.
5. KLUNGLAND, H., K. RØED, C.L. NESBØ, et al. 1999. Melanocyte stimulating hormone receptor (MC1-R) gene as a tool for evolutionary studies in artiodactyles. Hereditas **131:** 39–46.
6. KLUNGLAND, H., D.I. VÅGE, L.G. RAYA, et al. 1995. The role of melanocyte-stimulating hormone (MSH) receptor in bovine coat color determination. Mamm. Genome **6:** 636–639.
7. VÅGE, D.I., D. LU, H. KLUNGLAND, et al. 1997. A non-epistatic interaction of agouti and extension in the fox, *Vulpes vulpes*. Nat. Genet. **15:** 311–315.
8. KLUNGLAND, H., H.G. OLSEN, M.S. HASSANANE, et al. 2000. Coat colour genes in diversity studies. J. Animal Breed. Genet. **117:** 217–224.
9. LU, D., D.I. VÅGE & R.D. CONE. 1998. A ligand-mimetic model for constitutive activation of the melanocortin-1 receptor. Mol. Endocrinol. **12:** 592–604.
10. ADALSTEINSSON, S. 1977. Albinism in Icelandic sheep. J. Hered. **68:** 347–349.
11. ADALSTEINSSON, S., P. HERSTEINSSON & E. GUNNARSSON. 1987. Fox colors in relation to colors in mice and sheep. J. Hered. **78:** 235–237.
12. SILVERS, W.K. 1979. The Coat Colors of Mice, A Model for Mammalian Gene Action and Interaction. Springer-Verlag. New York.
13. DINULESCO, D. & R.D. CONE. 2000. Agouti and agouti-related protein: analogies and contrasts. J. Biol. Chem. **275:** 6695–6698.
14. HUSTAD, C.M., W.L. PERRY, L.D. SIRACUSA, et al. 1995. Molecular genetic characterization of six recessive viable alleles of the mouse agouti locus. Genetics **140:** 255–265.
15. SIRACUSA, L.D. 1994. The agouti gene: turned on to yellow. Trends Genet. **10:** 423–428.
16. ADALSTEINSSON, S., S. BJARNADOTTIR, D.I. VÅGE & J.V. JONMUNDSSON. 1995. Brown coat color in Icelandic cattle produced by the loci homolExtension and agouti. J. Hered. **86:** 395–398.

17. Våge, D.I., H. Klungland, D. Lu & R.D. Cone. 1999. Molecular and pharmacological characterization of dominant black coat colour in sheep. Mamm. Genome **10:** 39–43.
18. Klungland, H. & D.I. Våge. 2000. Molecular genetics of pigmentation in domestic animals. Curr. Genomics **1:** 223–242.
19. Kijas, J.M., R. Wales, A. Tornsten, *et al.* 1998. Melanocortin receptor 1 (MC1R) mutations and coat color in pigs. Genetics **150:** 1177–1185.
20. Newton, J.M., A.L. Wilkie, L. He, *et al.* 2000. Melanocortin 1 receptor variation in the domestic dog. Mamm. Genome **11:** 24–30.
21. Takeuchi, S., H. Suzuki, M. Yabuuchi & S. Takahashi. 1996. A possible involvement of melanocortin 1-receptor in regulating feather color pigmentation in the chicken. Biochim. Biophys. Acta **1308:** 164–168.

Defining the Quantitative Contribution of the Melanocortin 1 Receptor (MC1R) to Variation in Pigmentary Phenotype

T. HA, L. NAYSMITH, K. WATERSTON, C. OH, R. WELLER, AND J.L. REES

Systems Group, Dermatology, University of Edinburgh, Edinburgh EH3 9YW, United Kingdom

ABSTRACT: The melanocortin 1 receptor (MC1R) is a key determinant of pigmentary phenotype. Several sequence variants of the MC1R have been described, many of which are associated with red hair and cutaneous sensitivity to ultraviolet radiation even in the absence of red hair. Red hair approximates to an autosomal recessive trait, and most people with red hair are compound heterozygote or homozygous for limited numbers of mutations that show impaired function in *in vitro* assays. There is a clear heterozygote effect on sun sensitivity (even in those without red hair) and with susceptibility to the most common forms of skin cancer.

KEYWORDS: pigmentation; skin color; hair color; melanocortin 1 receptor; melanin; pheomelanin; eumelanin

A better quantitative understanding of the relation between genotype and phenotype requires improved methods of phenotypic assessment. Hair color can be measured using tristimulus colorimetry and by measures of eumelanin and pheomelanin using chemical degradation methods. Skin pigmentation can be assessed using colorimetry and in addition measurement of melanins. Sensitivity to ultraviolet radiation (UVR) can be assessed using erythema as a convenient end point, and the use of noradrenaline iontophoresis allows manipulation of blood flow such that skin color and blood flux can be measured on the same area.

Following the identification of the *extension* locus as the melanocortin 1 receptor (MC1R) in mouse (reviewed in Cone *et al.*[1]), human studies quickly showed that there was an association between sequence variation of the coding region of the MCR1 locus and hair color.[2–5] Initially, exploration of this association was made difficult by the (apparently) high degree of sequence polymorphism at this locus in humans.[6,7] Subsequent work has shown that there are strong associations between certain sequence variants in the coding region of the MC1R and the pigmentary phenotype, for both hair color and skin type (skin type is defined as the propensity of the skin to burn and tan in response to UVR).[2,3,8] Here, we review selectively the

Address for correspondence: J.L. Rees, Systems Group, Dermatology, University of Edinburgh, First Floor, Lauriston Buildings, Lauriston Place, Edinburgh EH3 9YW, United Kingdom. Voice: 44-131-5362041; fax: 44-131-2298769.

jonathan.rees@ed.ac.uk

Ann. N.Y. Acad. Sci. 994: 339–347 (2003). © 2003 New York Academy of Sciences.

published work on humans and outline a case for why new methods of quantification are required to elucidate the relationship between MC1R genotype and pigmentary phenotype. A broader case is also worth considering. Pigmentation is genetically determined, is polygenic, and has widely been used as a experimental system for understanding gene action in mouse and now zebrafish. Despite considerable advances in experimental tractability of these model systems, application to humans remains limited by our poor ability to define human phenotypes with sufficient precision. To date, the MC1R is the only locus that explains "normal" variation in pigmentary phenotype (with the possible exception of agouti; see Kanetsky et al.[9] and Voisey et al.[10]). This is likely to reflect bias because of the ease of studying the MC1R because it is small (<1 kb), and many of the functional changes are in the coding rather than control regions.

Studies in mouse showed that the MC1R was a key control point in determining the amounts of eumelanin and pheomelanin produced in hair.[11] Eumelanin is brown or black whereas pheomelanin is red or yellow. The ratio of these two types of melanin discriminates well between different hair colors. In the mouse, several homozygous loss of function mutations or mutations leading to diminished signaling through the MC1R produce yellow hair, whereas dominant gain of function lead to black hair (with a high eumelanin/pheomelanin ratio).[1,11] In humans, more than 30 single nucleotide polymorphisms have been detected within the MCR1, some of which show a striking association with red hair and skin type. No gain of function mutations has been described in humans.

Studies by the Davenports and others early in the 20th century suggested that red hair approximated to an autosomal recessive trait, although some studies were compatible with other modes of inheritance.[12–14] Studies at the MC1R in families show that based on known and functionally defined sequence variants at the MC1R[15] an autosomal recessive model fits the data well: based on analysis of particular single nucleotide polymorphisms within families hair color could be predicted correctly in almost 90% of cases.[3] That this relationship is not perfect can be accounted for in part by the difficulties associated with phenotyping, including changes in hair color with age (which are largely unexplained), and site variation in hair color which, although well known in humans, is also unexplained at the molecular level (at least in humans).

Early studies showed that three particular variants of the MC1R—the R151C, the R160W, and the D294H—were strongly associated with red hair.[2,4] Perhaps 80% of individuals defined as having red hair by using L'Oreal color charts are homozygous or compound heterozygous for either these mutations or one of a lower number of highly penetrant mutations including some frameshifts. Apart from these highly penetrant variants, other sequence variants also appear to be associated with red hair such as the V60L. It would appear that the functional status of the various alleles associated with red hair is not equivalent. The V60L allele in humans is a case in point. Data based on transfection of COS cells and assay of cAMP show that the V60L allele shows diminished function but that signaling is less impaired than that seen with the R151C, R160W, or D294H alleles.[15] Human studies also show that the V60L allele, with a frequency of near 0.1, one of the most common sequence variants, is a low-penetrance "red-hair allele."[3] Further support for the non-equivalence of different variants is provided by work using BAC rescue of null mice in which the R151C and the R160W are shown not to be equivalent to the D294H allele in rescuing the null phenotype.[16]

The need for improved phenotype measures is, in addition to the above, illustrated by the fact that the particular sequence variants of the MC1R have a clear effect in individuals without red hair. Studies show a clear heterozygote effect on tumor risk (melanoma and nonmelanoma skin cancer) and also on an individual's skin type.[4,8,17,18] Skin type is a widely used, but unsatisfactory, measure of an individual's cutaneous response to UVR.[19,20] According to an ordinal scoring system from 1 to 4 in Northern Europeans, persons who tend to burn but not tan in response to UVR are classified as skin type 1, whereas persons who tan very well but never burn will be scored 4. Persons who are intermediate in phenotype are scored 2 or 3 as appropriate.

A study conduced in a UK population showed not only that a clear phenotypic effect of the homozygous or compound heterozygous sequence variants associated with red hair on skin type, but also that individuals who were heterozygous for one of these sequence variants also were more likely to have a lower skin type.[8] Therefore, whereas the initial phenotype associated with changes of the MC1R was based on hair, it is clear that the whole pigmentary phenotype is influenced by this gene. This is of course not too surprising because there is a striking correlation between skin type and hair color. This relation is not, however, absolute: most individuals with red hair tend to be sun sensitive and therefore have a low skin type, but there are individuals who also appear to have striking black hair and yet from a cutaneous point alone also resemble individuals with red hair. As yet, beyond saying that such individuals are more likely to be heterozygote, we do not understand what other molecular controls produce this black hair/pale skin phenotype. It also seems evident that the MC1R phenotype is influenced by genetic background; to us the red hair phenotype differs between Scandinavia, the United Kingdom, and southern Europe.

METHODS OF DEFINING PHENOTYPE

Defining an individual's pigmentary phenotype (at minimum) will include an objective measure of constitutive skin color and pigmentation in response to UVR, and of hair color and analyses of eumelanin and pheomelanin.

Cutaneous pigmentation is either constitutive or facultative. Constitutive refers to "resting" pigmentation in the absence of UVR exposure, whereas facultative pigmentation encompasses the increase in pigmentation that is part of the physiological response to exposure to UVR. There is an association between these two factors in that individuals with low constitutive pigmentation tend not to tan well, whereas those with greater basal pigmentation appear more able to increase their pigment in response to UVR (i.e., they show a greater adaptive response). That being said, there are still unexplained differences between individuals, at least at the anecdotal level which are nonetheless convincing. For instance, many individuals with red hair have pale skin and do not tan in response to UVR. In contrast, some Scandinavians have light-colored hair with a relative absence of both eumelanin and pheomelanin but pale skin which in response to UVR appears to tan well. Systematic study of these differences is required.

The approach we have followed is to use simple color charts (provided by L'Oreal) and conventional degradation assays of eumelanin and pheomelanin to as-

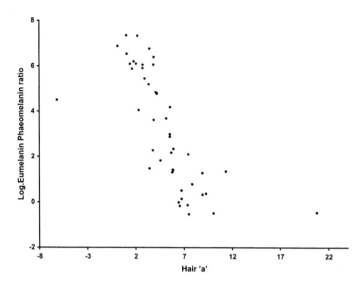

FIGURE 1. Decreasing eumelanin/pheomelanin ratio of hair is observed in hair with increasing degree of redness (increasing a* on the L*ab score) as assessed using colorimetry.

sess hair phenotype. Hair color can be usefully represented using the $L^*a^*b^*$ tristimulus system. In this system, defined by international standards, color is represented in three-dimensional space using three axes. The L^* is a lightness axis, the a^*, a red/green axis, and the b^* a yellow/blue axes. Such measurements can be performed using portable machines, are rapid and reproducible, and independent of ambient lighting. The values obtained will of course be altered by anything that appears to alter the color of hair including hair dyes or conditioners, or some cosmetics applied to the hair. The advantages of such a system is that whereas the human eye is extremely sensitive at comparing or spotting differences between two similar hair colors, ambient lighting effects influence human interpretation, and it is difficult to obtain reproducibility between different observers or scale the results appropriately. Furthermore, a problem that affects skin typing as well as hair color is the tendency of individuals to normalize the data to a particular population. In other words, what is perceived as dark in a population that is largely blonde may well be different from a population in which a larger proportion is dark haired.

The disadvantage of using an objective color system, rather than subject recall at age 21, is that hair color changes with age. The molecular basis for these changes is not understood. Generally, hair color is lighter in early childhood, darkens with increasing maturity, and then color is lost as an individual ages. For those with red hair as a young adult, they may be lighter at birth and strikingly red in early childhood but then turn white in older life. Conversely, however, some individuals who are clearly red in childhood appear to darken in early adulthood. Color therefore is age dependent, and ideally studies need to either adjust for this or perhaps better still stratify results by age.

The second method of assessing hair color relies on the chemical degradation assays developed by Ito or Prota and their colleagues for measuring eumelanin and pheomelanin.[21] Recent modifications to this assay[22] better distinguish pheomelanin from other products, and, as would be expected, such assays show a good relation with color as judged by colorimetry (FIG. 1). Hair melanin assays are, of course, like colorimetry, not independent of changes in melanin content with age, and hair dying may lead to other high-performance liquid chromatography peaks that interfere with interpretation of the results.

MEASURING SKIN COLOR AND THE CUTANEOUS RESPONSE TO ULTRAVIOLET RADIATION

Skin color can be assessed with the same method used to measure hair color using tristimulus colorimetry. Note that the color of skin reflects not just melanin pigment, but also blood (chiefly the effect of hemoglobin) and, to a more limited degree, some other pigments including carotenoids. Under pathological conditions, other constituents also many contribute to skin color, but in healthy subjects the effects of these can be ignored. The practical corollary is that measurement of color is only valid as a marker of pigmentary status in the absence of changes in blood flow. For measures of basal skin color at a particular site, this might seem reasonable, assuming that changes in blood flow between subjects of different skin colors are random. This assumption breaks down when attempts are made to assess pigmentation after UVR, because UVR, depending on dose and time scale, will increase cutaneous blood flow and usually results in increased pigmentation. It is this methodological issue that previously has limited studies of facultative pigmentation.

Skin color and the response to UVR vary at different body sites.[23,24] Some standardization of body sites examined therefore is required. Ideally, an area that is infrequently exposed to UVR would be the best for assessment of constitutive pigmentation. In many studies, the inner upper arm is used, but this is less than ideal because there are seasonal differences in color as this site is exposed to ultraviolet in many populations.[25] The buttock is a better site although there may well still be seasonal changes due to slight transmission of UVR through clothing.[25]

An alternative instrument to the tristimulus colorimeter is to use a reflectance machine, such as the Diastron, which is more commonly used to assess erythema.[26] This machine has been used less than the tristimulus colorimeter but, subject to the caveats mentioned above about changes in blood flow, performs well.

Finally, more invasive methods could be used to assess melanin within the epidermis directly. For instance, suction blistering or biopsy and subsequent splitting of the epidermis and dermis may provide enough material to be processed for eumelanin and pheomelanin measurement.[27] Such studies have been rarely performed.

DYNAMIC STUDY OF PIGMENTATION

It is an everyday observation that people with pale skin and red hair tend to burn in the sun, whereas individuals with olive skin seem to fair better in response to sunshine. What is often not appreciated is that the acute response to erythema may be

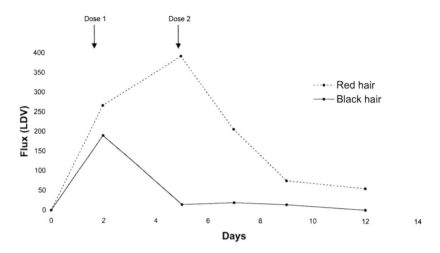

FIGURE 2. The difference in skin erythemal response to two serial doses of UVB demonstrates more rapid photoadaptation in the black-haired Asian individual compared with the red-haired individual.

more similar in individuals with different hair color and skin colors than is expected.[26] For example, results from two individuals are shown in FIGURE 2. One is of an individual with light red hair with very pale skin and the other from an Asian subject with skin which is classified as skin type 5. The acute erythemal responses to UVR of these two subjects, while not identical, are similar. The individual with pale skin does indeed go more red than the other individual but the differences fall well within those seen in many whites. What is striking, however, is that when the irradiation is repeated the two phenotypes separate. The Asian individual clearly switches on photoadaptive mechanisms, presumably pigmentation accounting for the difference in this case, whereas the individual with pale skin appears unable to do this. Therefore, exposing phenotype differences in pigmentation may be facilitated by stressing the system or dynamic challenge to the system. The difficulty to date in using such approaches has been the limitation imposed by technology in how to measure responses, because, as outlined above, changes in blood flow and the presence of changes in pigmentation vitiate objective measure of each. Alternative approaches therefore are required.

One alternative we have used is to measure blood flow by measuring red cell flux directly. Laser Doppler instruments, working on the Doppler principle, are available to do this.[23] They are not so easily portable as reflectance instruments but allow correct assessment of blood flow, irrespective of pigmentary status (at least for certain machines). It therefore is possible with these instruments to record the response to UVR objectively in the presence of changes in pigmentary status.

There are two broad classes of instruments in use. The first, and oldest, is a point laser Doppler that is attached to the skin and has the advantage of great sensitivity but suffers from possible artifact because the skin has been touched and a limited

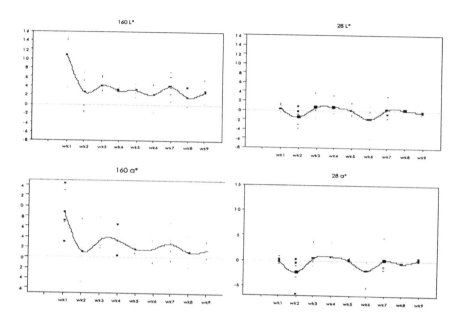

FIGURE 3. The concurrent use of noradrenaline-induced vasoconstriction and colorimetry demonstrates UVB dose-related increase in facultative skin pigmentation (*y* axis shows Facultative minus Constitutive Readings of Chromameter L* and Chromameter a*). The increase in pigmentation (increasing L* and increasing a*) is demonstrated in the 160-mJ UVB site but absent from the 28-mJ UVB site.

area is sampled. A more recent alternative is based on a scanning Doppler which can be adjusted to scan a larger area of skin, and then the results are integrated or assessed visually as pseudocolor images. The advantage of the scanning instrument is that a larger sample is taken and the machine does not need to have direct contact with the skin. The disadvantage is that the sensitivity is less than that with the contact instrument.

The converse of this problem is how to measure changes in pigmentation in the presence of changes in blood flow. One approach would be to measure blood flow and then biopsy the skin and conduct studies of epidermal melanin. This is adequate, but obviously invasive, and there therefore are practical limitations in the number of samples that can be obtained. An alternative strategy which we have developed relies on measuring color in the absence of cutaneous blood flow after pharmacological alteration in the cutaneous microcirculation. Iontophoresis of noradrenaline, a potent vasoconstrictor, allows measurements of skin color which, in the absence of blood flow, reflects melanin pigmentation. Use of the laser Doppler allows the absence of red cell flux to be confirmed. Such an assay therefore allows measurement of pigment in the presence of changes in blood flow. An example of this approach for two different doses of irradiation is shown in FIGURE 3.

REFERENCES

1. CONE, R.D., D. LU, S. KOPPULA, et al. 1996. The melanocortin receptors: agonists, antagonists, and the hormonal control of pigmentation [review]. Rec. Prog. Horm. Res. **51:** 287–317.
2. BOX, N.F., J.R. WYETH, L.E. O'GORMAN, et al. 1997. Characterization of melanocyte stimulating hormone receptor variant alleles in twins with red hair. Hum. Mol. Genet. **6:** 1891–1897.
3. FLANAGAN, N., E. HEALY, A. RAY, et al. 2000. Pleiotropic effects of the melanocortin 1 receptor (MC1R) gene on human pigmentation. Hum. Mol. Genet. **9:** 2531–2537.
4. SMITH, R., E. HEALY, S. SIDDIQUI, et al. 1998. Melanocortin 1 receptor variants in an Irish population. J. Invest. Dermatol. **111:** 119–122.
5. VALVERDE, P., E. HEALY, I. JACKSON, et al. 1995. Variants of the melanocyte-stimulating hormone receptor gene are associated with red hair and fair skin in humans. Nat. Genet. **11:** 328–330.
6. HARDING, R.M., E. HEALY, A.J. RAY, et al. 2000. Evidence for variable selective pressures at the human pigmentation locus, MC1R. Am. J. Hum. Genet. **66:** 1351–1361.
7. MAKOVA, K.D., M. RAMSAY, T. JENKINS, et al. 2001. Human DNA sequence variation in a 6.6-kb region containing the melanocortin 1 receptor promoter. Genetics **158:** 1253–1268.
8. HEALY, E., N. FLANNAGAN, A. RAY, et al. 2000. Melanocortin-1-receptor gene and sun sensitivity in individuals without red hair. Lancet **355:** 1072–1073.
9. KANETSKY, P.A., J. SWOYER, S. PANOSSIAN, et al. 2002. A polymorphism in the agouti signaling protein gene is associated with human pigmentation. Am. J. Hum. Genet. **70:** 770–775.
10. VOISEY, J., N.F. BOX & A. VAN DAAL. 2001. A polymorphism study of the human agouti gene and its association with MC1R. Pigm. Cell Res. **14:** 264–267.
11. ROBBINS, L.S., J.H. NADEAU, K.R. JOHNSON, et al. 1993. Pigmentation phenotypes of variant extension locus alleles result from point mutations that alter MSH receptor function. Cell **72:** 827–834.
12. DAVENPORT, C.C. & C.B. DAVENPORT. 1909. Heredity of hair color in man. Am. Nat. **43:** 193.
13. NEEL, J.V. 1943. Concerning the inheritance of red hair. J. Hered. **34:** 93–96.
14. REED, T.E. 1952. Red hair colour as a genetical character. Ann. Eugen. **17:** 115–139.
15. SCHIOTH, H.B., S. PHILLIPS, R. RUDZISH, et al. 1999. Loss of function mutations of the human melanocortin 1 receptor are common and associated with red hair. Biochem. Biophys. Res. Commun. **260:** 488–491.
16. HEALY, E., S.A. JORDAN, P.S. BUDD, et al. 2001. Functional variation of MC1R alleles from red-haired individuals. Hum. Mol. Genet. **10:** 2397–2402.
17. BOX, N.F., D.L. DUFFY, R. IRVING, et al. 2001. Melanocortin 1 receptor genotype is a risk factor for basal and squamous cell carcinoma. J. Invest. Dermatol. **116:** 224–229.
18. PALMER, J.S., D.L. DUFFY, N.F. BOX, et al. 2000. Melanocortin-1 receptor polymorphisms and risk of melanoma: is the association explained solely by pigmentation phenotype? Am. J. Hum. Genet. **66:** 176–186.
19. FITZPATRICK, T.B. 1988. The validity and practicality of sun-reactive skin types I through VI. Arch. Dermatol. **124:** 869–871.
20. RAMPEN, F.H., B.A. FLEUREN, T.M. DE BOO, et al. 1988. Unreliability of self-reported burning tendency and tanning ability. Arch. Dermatol. **124:** 885–888.
21. WAKAMATSU, K. & S. ITO. 2002. Advanced chemical methods in melanin determination. Pigm. Cell Res. **15:** 174–183.
22. WAKAMATSU, K., S. ITO & J.L. REES. 2002. Usefulness of 4-amino-3-hydroxyphenylalanine as a specific marker of pheomelanin. Pigm. Cell Res. **15:** 225–232.
23. FARR, P.M. & B.L. DIFFEY. 1984. Quantitative studies on cutaneous erythema induced by ultraviolet radiation. Br. J. Dermatol. **111:** 673–682.
24. RHODES, L.E. & P.S. FRIEDMANN. 1992. A comparison of the ultraviolet B-induced erythemal response of back and buttock skin. Photodermatol. Photoimmunol. Photomed. **9:** 48–51.

25. LOCK-ANDERSEN, J. & H.C. WULF. 1997. Seasonal variation of skin pigmentation. Acta Derm. Venereol. (Stockh.) **77:** 219–221.
26. FLANAGAN, N., A.J. RAY, C. TODD, *et al.* 2001. The relation between melanocortin 1 receptor genotype and experimentally assessed ultraviolet radiation sensitivity. J. Invest. Dermatol. **117:** 1314–1317.
27. THODY, A.J., E.M. HIGGINS, K. WAKAMATSU, *et al.* 1991. Pheomelanin as well as eumelanin is present in human epidermis. J. Invest. Dermatol. **97:** 340–344.

Genetic Association and Cellular Function of MC1R Variant Alleles in Human Pigmentation

R.A. STURM,[a] D.L. DUFFY,[b] N.F. BOX,[a] R.A. NEWTON,[a] A.G. SHEPHERD,[a]
W. CHEN,[a] L.H. MARKS,[a,b] J.H. LEONARD,[b] AND N.G. MARTIN[b]

[a]*Institute for Molecular Bioscience, University of Queensland, Brisbane,
Queensland 4072, Australia*

[b]*Queensland Institute of Medical Research, Brisbane, Queensland 4029, Australia*

ABSTRACT: We have examined MC1R variant allele frequencies in the general
population of South East Queensland and in a collection of adolescent dizygotic
and monozygotic twins and family members to define statistical associations
with hair and skin color, freckling, and mole count. Results of these studies are
consistent with a linear recessive allelic model with multiplicative penetrance
in the inheritance of red hair. Four alleles, D84E, R151C, R160W, and D294H,
are strongly associated with red hair and fair skin with multinomial regression
analysis showing odds ratios of 63, 118, 50, and 94, respectively. An additional
three low-penetrance alleles V60L, V92M, and R163Q have odds ratios 6, 5,
and 2 relative to the wild-type allele. To address the cellular effects of MC1R
variant alleles in signal transduction, we expressed these receptors in permanently transfected HEK293 cells. Measurement of receptor activity via induction of a cAMP-responsive luciferase reporter gene found that the R151C and
R160W receptors were active in the presence of NDP-MSH ligand, but at much
reduced levels compared with that seen with the wild-type receptor. The ability
to stimulate phosphorylation of the cAMP response element binding protein
(CREB) transcription factor was also apparent in all stimulated MC1R variant
allele–expressing HEK293 cell extracts as assessed by immunoblotting. In contrast, human melanoma cell lines showed wide variation in the their ability to
undergo cAMP-mediated CREB phosphorylation. Culture of human melanocytes of known MC1R genotype may provide the best experimental approach
to examine the functional consequences for each MC1R variant allele. With
this objective, we have established more than 300 melanocyte cell strains of
defined MC1R genotype.

KEYWORDS: hair color; skin color; cAMP; melanocyte; melanoma

The human melanocortin–1 receptor (MC1R) is specifically expressed on the surface of melanocytes and is a key regulator of intracellular signaling to the melanin
biosynthetic pathway governing pigment formation. The MC1R locus is highly polymorphic in human populations, with variant forms of the receptor underlying the diverse range of human pigmentation phenotypes and skin phototypes.[1] Several of the

Address for correspondence: R.A. Sturm, Institute for Molecular Bioscience, University of
Queensland, Brisbane, Queensland, Australia. Voice: 61-7-3346-2038; fax: 61-7-3346-2101.
r.sturm@imb.uq.edu.au

Ann. N.Y. Acad. Sci. 994: 348–358 (2003). © 2003 New York Academy of Sciences.

MC1R variant alleles have been associated with the red hair and fair skin (denoted RHC; red hair color) phenotype, a condition that is caused by the synthesis of a high level of pheomelanin and that can place individuals at higher risk of skin cancer (reviewed in Sturm[2]). To begin to quantify the relative contribution of each independent MC1R variant allele to human pigmentary phenotypes, we have examined statistical associations of some common MC1R gene variants with red hair, skin color, degree of freckling, and nevus counts in a large collection of adolescent twins, parents, and siblings. Findings from these genetic studies also must be compared and contrasted with those found using functional approaches to probe changes in cellular physiological responses initiated by variant MC1R receptor proteins. Ultimately, this will give a mechanistic understanding of these pigmentary effects in melanocytes, and such approaches also have been initiated in this study.

MC1R VARIANT ALLELE PENETRANCE IN RED HAIR AND SKIN PIGMENTARY TRAITS

Previous studies examining variant MC1R alleles in relation to hair color all have been consistent in finding the RHC phenotype to be a recessive trait, although the reported influence of some of variant alleles has varied between studies (reviewed in Sturm[2]). Homozygote or compound heterozygote variant MC1R genotype carriers are generally red haired, but because this is not always the case it is likely that other loci are involved in the expressivity of the trait.[3] In addition, red hair occurs in a significant proportion of heterozygote consensus MC1R allele carriers, some alleles displaying greater influence than others. It is unlikely that each of these variant alleles represent complete loss of receptor function or that they have a simple Mendelian recessive mode of inheritance; rather, variant alleles more likely represent a linear series of differential strength alleles.

There were 2,331 family members in 645 pedigrees for whom some phenotypic data were recorded, and DNA was available for genotyping for 1,779 individuals within 460 of these pedigrees.[4] Excluding one member of each genotyped MZ twin pair, there were 1,569 individuals with complete MC1R genotype, hair color, eye color, and gender recorded. The allele frequencies and odds ratios for nine polymorphic amino acid sites within the MC1R coding region are summarized in TABLE 1. RHC variant alleles R151C, R160W, D294H are common in the Queensland population and are responsible for most of the red hair color in our community, with at least one of these three alleles found in 93% of those with red hair. With the exception of the R142H and I155T variants, which occur at relatively low frequencies, all variants could be found in at least one red-haired individual, most commonly paired with one of the other RHC alleles as a compound heterozygote. The consensus allele in combination with a variant allele was found in only 11% of redheads, and no consensus homozygotes with red hair were found.

Penetrance for red hair of each MC1R variant allele was modeled in a logistic regression analysis.[4] This indicated that the RHC alleles D84E, R151C, R160W, and D294H were highly associated with red hair and fair skin, showing odds ratios (ORs) of 63, 118, 50, and 94 relative to the consensus MC1R allele compared with low-strength alleles V60L and V92M with ORs 6 and 5. The weakest allele was R163Q, and this gave only a twofold increase in association with red hair. From these anal-

TABLE 1. MC1R variant allele frequency in South East Queensland and odds ratio for red hair and fair skin

Variant allele	Frequency[a] (%)	Red hair OR (95% CI)[b]	Fair/pale skin OR (95% CI)[b]
Consensus (+)	50.4	1	1
V60L	12.2	6.4 (2.8–14.9)	1.7 (1.3–2.1)
D84E	1.2	62.8 (17.6–223.7)	12.5 (4.8–42.8)
V92M	9.7	5.3 (2.2–12.9)	2.3 (1.8–3.1)
R142H	0.4	0.0 (0.0–∞)[c]	1.8 (0.6–5.7)
R151C	11.0	118.3 (51.5–271.7)	4.4 (3.3–5.7)
I155T	0.9	0.0 (0.0–∞)[c]	2.1 (1.0–4.3)
R160W	7.0	50.5 (22.0–115.8)	3.2 (2.4–4.4)
R163Q	4.7	2.4 (0.5–11.3)	2.0 (1.4–2.9)
D294H	2.7	94.1 (33.7–263.1)	7.5 (4.4–13.7)
r	26.7	5.1 (2.5–11.3)	1.9 (1.6–2.3)
R	21.8	63.3 (31.9–139.6)	4.2 (3.4–5.2)

[a]The total number of individuals ranged from 1,747 to 1,787 scored at the different variants.
[b]Odds ratio relative to consensus genotype (95% Confidence Interval).
[c]Insufficient numbers to test for statistical association.
r = V60L, V92M, R163Q.
R = D84E, R151C, R160W, D294H.

TABLE 2. MC1R genotype and penetrance of red hair phenotype

Allele	+	r	R
	0[a]	0.2	2.5
+	**0**		
	0:425[b]		
		1.0	11.8
r	**1.0**	**0.9**	
	4:408	1:110	
			62.1
R	**1.5**	**10.8**	**67.1**
	5:339	22:181	49:24

[a]*Top triangle matrix* is the expected frequency (%) of red-haired individuals, using a linear multiplicative inheritance model for each genotype (allele 1OR * allele 2OR from TABLE 1).

[b]*Bottom triangle matrix* gives the penetrance (in *boldface*) and below that the numbers of red/non-red–haired individuals (RHC:NRHC) of each MC1R genotype [penetrance = 100*RHC/(RHC + NRHC)].

yses, it can be concluded that the D84E, R151C, R160W, and D294H variants can be considered strong RHC alleles, which we designate "*R*." The V60L,V92M, and R163Q variants are relatively weak RHC alleles and are designated "*r*." The penetrances of the six genotypes formed upon combining these alleles grouped as weak and strong, together with the consensus "+" allele for red–nonred hair color are shown in TABLE 2 as a matrix of genotypes. This shows that significant numbers of red-haired individuals are seen only in *R/r* and *R/R* genotypes with 10.8% and 67.1%, respectively. The frequency of red hair in those with a heterozygous *R/+* genotype is only 1.5%, and less than 1% in those with an *r/r* genotype or *r/+* genotype. Using a regression model in which these alleles act multiplicatively on expression of red hair, we compared the predicted penetrance for the various compound heterozygote and homozygote genotypes (top triangle, TABLE 2) to that observed in our sample (bottom triangle). The concordance between the observed and expected frequencies of red hair for each grouped genotype supports a multiplicative penetrance model with the RHC alleles acting in a linear recessive fashion.

Subjects with a consensus MC1R genotype showed the darkest induced skin color, with mean skin reflectances of 60.5% for the inner arm and 49.8% for the back of hand. To address the quantitative relationship of variant MC1R alleles with skin reflectance, we calculated the increase in the mean skin reflectance measurement per variant allele relative to the consensus genotype. In general, alleles acted in an additive manner to increase skin reflectance, *R* alleles showing a greater effect (+1.9%) than *r* alleles (+0.9%) on inner arm reflectance.[4] Red-haired subjects had the greatest number of freckles and least number of moles, and this phenotypic association was further tested for association with MC1R genotype. No consensus homozygote (+/+) had a severe freckling score, and there was little effect of the *r* allele upon this correlation, but the *R/R* genotype had the lowest number of moles and greatest number of freckles, consistent with the phenotypic association found in redheads. In the heterozygous state, the *R/+* genotype displays an initial positive association with moliness until severe freckling is reached, after which there is a significant decrease in the number of moles. The attributable fraction due to carrying at least one variant MC1R allele increased linearly with the degree of freckling from 23.4% for mild to 100% for severely freckled subjects.

APPROACHES TO FUNCTIONAL ASSAY OF MC1R VARIANT ALLELES

Several approaches have been used for functional analysis of variant MC1R receptors including expression in MC1R-null transgenic mice to determine their effects on coat color and in the MC1R-deficient mouse melanoma cell line, B16G4F, to monitor cellular and physiological response to ligand. The resultant coat color pigmentation phenotypes suggest that these alleles do not result in a complete loss of activity nor are they functionally equivalent.[5] However, only the wild-type receptor was responsive in B16G4F cell line growth inhibition experiments.[6]

Variant MC1R receptors have been tested through heterologous expression in nonmelanocytic cells for ligand binding and cAMP dose-response coupling. Amino acid substitutions can result in altered receptor activity either by reduced affinity for hormone binding[7] or through deficient coupling to intracellular cAMP activation without significantly affecting ligand binding.[8–10] Heterologous expression in nonmelanocytic cells overcomes the problem of endogenous genomic MC1R-allele expression and allows for testing in a range of cell lines with potentially greater sensitivity through higher levels of receptor expression. However, this must be offset against potential artifacts seen with levels of receptor produced in a nonphysiological cellular environment. The advantages of functional testing in homologous cell systems such as melanoma lines are that melanocytic pigmentary functions, cell growth, and adhesion can be assayed in a cellular system partially equating to a normal cellular milieu. Considering these limitations, primary human melanocyte strains may represent the best system for MC1R functional analysis because they are a normal cell, and in them the pigmentation pathway may give a true response to physiological melanogenic stimulators. Unfortunately, there are several impediments to their use because they are slow-growing cells with limited growth span that must be freshly established and genotyped. With these factors in mind, we have used both homologous and heterologous test systems for RHC variant MC1R receptors.

FUNCTIONAL TESTING OF MC1R VARIANT ALLELES IN HEK293 CELL LINES

Functional testing of the wild-type and R151C and R160W receptors was first approached by stable expression in HEK293 cells by using the pcDNA3.1 plasmid vector. Total membrane extracts[11] from several initial cell isolates were tested for binding specificity and affinity of radiolabeled MC1R superagonist ligand ^{125}I-NDP-MSH and demonstrated specific binding of each receptor-expressing cell line compared with the parental HEK293 cells, with relative assay levels 10- to 20-fold above background.

The MC1R receptor is known to couple to second messenger systems that result in the increase of intracellular cAMP levels[12]; therefore, the PathDetect in Vivo Signal Transduction Pathway cis-Reporting System for cyclic-AMP responsiveness (Stratagene) was used to measure agonist response in MC1R-expressing HEK cells transfected with the pCRE-Luc reporter plasmid vector as a surrogate measurement of cAMP activation.[13,14] The HEK-pcDNA3.1 parental control vector–transfected cells did not respond to ligand administration but demonstrated several hundredfold activation after cotransfection with pFC-PKA encoding the protein kinase A catalytic subunit (FIG. 1A). This level of fold-activation over background in response to PKA was also seen in each of the MC1R-transfected HEK cells, demonstrating the utility of this approach to normalize activation relative to this positive control. In the presence of 10 nM NDP-MSH, cells expressing the wild-type receptor had an equivalent level of activation to that seen with pFC-PKA, but with the R151C and R160W variant receptors, activation was reduced to approximately half that level. In an attempt to competitively inhibit agonist activation, we also added a synthetic agouti peptide derived from the agouti protein, known to be an antagonist at the MC1R receptor.[15] No inhibition was seen. This system though may be dependent on agonist/antagonist ratios which would need to be determined using a range of NDP-MSH and agouti peptide concentrations.

To further test the function of wild-type and variant receptors, we assayed for phosphorylation of the CREB using Western blots on MC1R-expressing HEK293 cell line extracts to measure the level of the 43-kDa activated form of CREB using a phosphoserine-133–specific antibody (Cell Signaling Technology). Assay of CREB

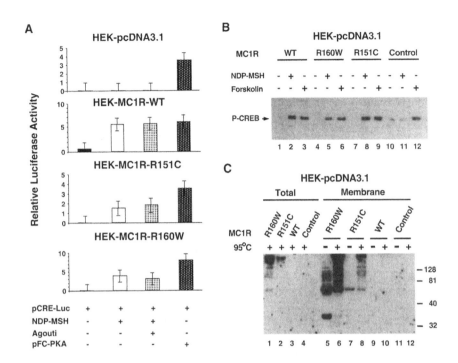

FIGURE 1. (A) Relative luciferase activities detected in HEK293 cells in response to NDP-MSH. HEK293 cells permanently transfected with the pcDNA3.1 vector (Invitrogen) containing MC1R receptors as indicated were harvested and plated into 12-well tissue culture plates at 80% confluence and allowed to attach for 24 h. The next day cells were transfected with 100 ng of pCRE-Luc reporter plasmid per well using Lipofectomine reagent (LifeTechnologies, Gaithersburg, MD), with control activation seen through cotransfection of 50 ng of pFC-PKA encoding the protein kinase A catalytic subunit. At 5 h after transfection, the cells were stimulated with 10 nM NDP-MSH ligand included in the DMEM media plus 10% fetal calf serum and left for another 6 h. The assay medium was aspirated and cells were washed twice with PBS and frozen dry at -20°C overnight. These cells were harvested in 100 mL of DMEM media without phenol red indicator and used to assay relative luciferase levels (Roche Luciferase Reporter Gene Assay, Constant Light Signal). **(B)** CREB phosphorylation in MC1R-expressing HEK293 cell lines. HEK293 cells stably transfected with cDNAs for MC1R receptor variants **(lanes 1–9)** or vector alone **(lanes 10–12)** were stimulated for 30 min with either NDP-MSH (10 nM) or forskolin (10 μM). Whole-cell lysates were analyzed by Western immunoblot using a specific anti–phospho-CREB antibody (Cell Signaling Technology). **(C)** MC1R levels in HEK293 cell lines. Whole-cell lysates or membrane preparations from stably transfected HEK293 cell lines were analyzed by Western immunoblot using an antibody raised against the amino terminus of the human MC1R (N19 Santa Cruz Biotechnology, Santa Cruz, CA). Whole-cell lysates were denatured in SDS-PAGE sample buffer (0.025M Tris-HCl, pH 6.8; 10% glycerol; 2% SDS; 5% 2-mercaptoethanol) for 2 min at 95°C prior to separation using 10% SDS-PAGE gels. Membrane preparations either were heat denatured in a manner identical to that for the lysates or were simply resuspended in sample buffer and incubated at room temperature for 10 min. Similar levels of protein were loaded for each cell line.

phosphorylation after 30-min stimulation with either NDP-MSH or the adenylate-cyclase activator forskolin induced similar levels of CREB phosphorylation in HEK293-expressing wild-type, R151C, and R160W receptors, with the HEK-pcDNA3.1 parental control again nonresponsive to ligand administration (FIG. 1B). The similar pattern of CREB-phosphorylation activity levels seen with each of these receptors contrasts with the lower activity of reporter gene transcriptional activity directed by the RHC variant alleles, suggesting that there may be differences in the temporal activation of adenylate-cyclase during extended periods of NDP-MSH administration, which were 30 min for kinase activity but 6 h for the luciferase reporter assay.

In the process of selecting MC1R-expressing HEK293 cell lines, we initially screened for receptor activation by assaying for luciferase reporter activity. Although this approach confirmed that RHC variant MC1R receptors do indeed have the ability to respond to ligand activation, indicating that these receptors are still partially functional, this may have biased the selection process toward cell lines with higher than normal physiological levels of the receptor. To investigate this possibility, we assayed total protein and membrane fractions from cell extracts for MC1R levels by immunoblotting using an N-terminal MC1R antibody (FIG. 1C). Only high molecular weight MC1R cross-reactive aggregates were detected in heat-denatured samples, as has been reported previously.[16] However, using non–heat denaturing conditions,[17,18] it was possible to resolve this aggregate into several bands, an expected monomeric species slightly above the 32-kDa marker, together with several other distinct higher molecular weight complexes with a predominant 66-kDa band. Considerably greater levels of receptor expression were detectable in the R151C and R160W variant cell lines than was seen with wild-type MC1R-expressing cells. No cross-reactivity was seen in pcDNA3.1 control–transfected cells. Therefore, although the wild-type and variant receptors have some functional activity, the excess levels of the variant receptors suggest that qualitative differences in receptor coupling are apparent. We have now established additional HEK293 cell lines expressing comparable levels of the MC1R receptor to address this issue. In addition, cell lines expressing V60L, V92M, and D294H MC1R receptor variants have been established for functional testing.

FUNCTIONAL TESTING OF MC1R VARIANT ALLELES IN MELANOCYTIC CELLS

Human melanoma cells are known to express from 400 to 1,600 specific receptors for MSH ligands per cell.[19,20] These have been utilized to study effects on hormone induction of melanogenesis and proliferation/growth inhibition and have suggested new modalities for melanoma treatment using toxic analogues. There have been conflicting reports on the actions of hormones on melanoma cell lines which reflect differences in the cell lines and culture conditions used. Moreover, the MC1R genotype of each cell line must be considered in assessment of any such response to ligand. To begin to address the functional properties of MC1R receptors in melanoma cells, we genotyped 56 melanoma cell lines. To attribute function solely to an individual variant receptor, we sought melanoma lines that were homozygote for each allele, thus restricting the capacity of these cells to expression of a single MC1R receptor isoform. Heterozygous and compound heterozygous cell lines having the potential to produce two forms of receptor cannot be used to assign function to a specific allele. We found several consensus, compound heterozygote and homozygote MC1R variant–carrying lines.

Selected homozygote consensus and variant genotype lines were assessed for their ability to respond to either NDP-MSH or forskolin via coupling to CREB phosphorylation. These cell lines included two wild-type, homozygote V60L, R151C, R160W, and one D294H compound heterozygote (FIG. 2A) and were compared with the response observed in HEK293 wild-type MC1R-expressing cells. Direct stimulation of adenylate-cyclase by forskolin showed a variable pattern with apparent stimulation in some melanoma cells such as MM200, MM96L, ME10538, and A06MLC with nonresponsiveness in MM418 and CJM; however, no cell line showed an increase equivalent to that obtained in the HEK293 wild-type MC1R control. In contrast and irrespective of genotype, none of the cell lines increased the phosphorylation of CREB after NDP-MSH treatment. Variability in response to MSH stimulation of melanoma cell lines has been reported, but increases in cAMP above the basal level also have been described,[21–23] and this lack of CREB responsiveness may reflect innate diversity in MC1R signal transduction pathways between homologous and heterologous cell lines (see below). Receptor levels on each of the melanoma cell lines used in this assay must also be determined before definitive conclusions can be reached.

Given the variability and genetic instability that are characteristic of melanoma cell lines, possibly reflecting disrupted signaling pathways, we next turned to primary melanocytic cell strains as a suitable system to assess MC1R receptor function. Cultured human melanocytes have been found to express high-affinity receptors for the α-MSH ligand on their surface with approximately 700 binding sites per cell.[24–26] We have performed a systematic genotype screen for several of the common MC1R polymorphisms by using melanocytes cultured from individual foreskins to ascertain clonal cell strains. Over 1,000 foreskin samples were processed. The haplotype frequencies for the three R151C, R160W, and D294H RHC alleles were 8.3%, 7.3%, and 1.6%, respectively, which is comparable to the haplotype frequency range seen in the South East Queensland population (TABLE 1). It was our intention to isolate homozygote strains for each of these alleles to allow the specific functional characterization of individual forms of the receptor.

More than 300 clonal genotyped melanocyte cell strains were successfully established in culture from this initial screen with many wild-type +/+ and heterozygote +/− strains, including five R151C−/− and one R160W−/− homozygotes. However, no D294H−/− homozygotes were identified. When MC1R genotype was considered, all consensus +/+ allele strains had very dark pellets, being scored as black or dark brown by eye. When strains with a variant genotype were grouped, homozygous R151C−/− and R160W−/− cell pellets were considerably less pigmented, whereas heterozygous R151C+/− and R160W+/− cell pellets varied from black to white but were generally of intermediate pigmentation between the consensus and homozygotes. Upon culture of these cells with known second messenger inducers of pigmentation including IBMX and dbcAMP, cell growth was stimulated as seen by larger cell pellets, but there was no effect on color of the R151C−/− and R160W−/− cell pellets. However, when these cells were incubated with the melanogenic precursor substrate DOPA, we were able to demonstrate melanin production in each of

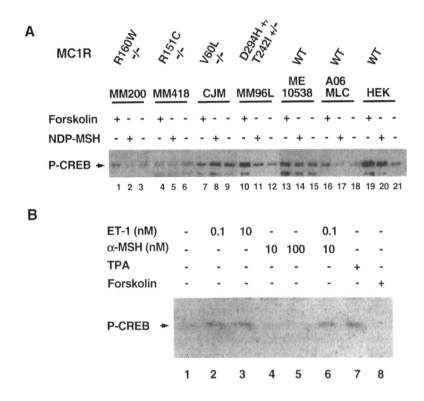

FIGURE 2. (**A**) CREB phosphorylation in melanoma cell lines. Cells were stimulated for 30 min with either NDP-MSH (10 nM) or forskolin (10 μM). Whole-cell lysates were analyzed by Western immunoblot using a specific anti–phospho-CREB antibody. Analysis of HEK293 cells expressing wild-type MC1R (**lanes 19–21**) is included for comparison. (**B**) CREB phosphorylation in primary human melanocytes. Cells were stimulated for 15 min with ET-1, α-MSH, TPA (16 nM), forskolin (10 μM), or combinations of each as indicated above the panel before whole-cell lysates were prepared for immunoblot.

these cell strains as compared with the nonmelanocytic HeLa cell line. Under the growth conditions used, a wild-type MC1R receptor appeared to be required for full melanization, and the cAMP second messenger stimulators could not mimic the pigmentary effects in variant MC1R homozygote melanocytes.

Melanocytes respond to a variety of growth factors, hormones, and ultraviolet light to increase pigmentation,[27–29] and attention increasingly is being focused on the interplay between the various signaling pathways. Recently, a synergistic action between endothelin, α-MSH, and bFGF was demonstrated for the phosphorylation of CREB.[30] Our initial studies have utilized a consensus melanocyte strain to investigate CREB phosphorylation in response to various mitogens. Both endothelin and TPA were able to couple to CREB phosphorylation. However, as previously seen for melanoma cells, α-MSH was not able to elicit such phosphorylation (FIG. 2B). Co-stimulation of melanocytes with endothelin and α-MSH did not reveal any obvious latent synergy between these factors. Lack of α-MSH–induced CREB phosphorylation has also been described previously.[30]

Note that α-MSH–mediated cAMP induction does occur in melanocytes,[31] but these measurements are yet to be performed in our melanocyte strains. A direct loss of receptor-activated cAMP intracellular signaling recently has been reported in other studies of primary human melanocytes of defined MC1R genotype,[32] although kinase signaling pathways were not examined. In these studies, R160W homozygote and R151C/D294H, R160W/D294H compound heterozygote melanocyte strains demonstrated an impairment of tyrosinase activation in response to α-MSH stimulation and pronounced sensitivity to ultraviolet radiation treatment. These results confirm the utility of using primary melanocytes for MC1R functional analysis, but pose new questions about the signaling pathways that function in this cell which form a new tissue-specific paradigm.

ACKNOWLEDGMENTS

This work was supported by grants from the Queensland Cancer Fund and National Health and Medical Research Council of Australia (950998, 981339, 142988) and the National Cancer Institute (CA88363).

REFERENCES

1. STURM, R.A. *et al.* 2001. Human pigmentation genes: identification, structure and consequences of polymorphic variation. Gene **277:** 49–62.
2. STURM, R.A. 2002. Skin color and skin cancer –MC1R, the genetic link. Melanoma Res. **12:** 405–416.
3. BOX, N.F. *et al.* 1997. Characterization of melanocyte stimulating hormone variant alleles in twins with red hair. Hum. Mol. Genet. **6:** 1891–1897.
4. DUFFY, D.L. *et al.* Interactive effects of MC1R and OCA2 on melanoma risk phenotypes. Submitted.
5. HEALY, E. *et al.* 2001. Functional variation of MC1R alleles from red-haired individuals. Hum. Mol. Genet. **10:** 2397–2402.
6. ROBINSON, S.J. & E. HEALY. 2002. Human melanocortin-1 receptor (MC1R) gene variants alter melanoma cell growth and adhesion to extracellular matrix. Oncogene **21:** 8037–8046.
7. JIMENEZ-CERVANTES, C. *et al.* 2001. Thr40 and Met122 are new partial loss-of-function natural mutations of the human melanocortin 1 receptor. FEBS Lett. **508:** 44–48.
8. KOPPULA, S.V. *et al.* 1997. Identification of common polymorphisms in the coding sequence of the human MSH receptor (MC1R) with possible biological effects. Hum. Mutat. **9:** 30–36.
9. FRANDBERG, P.A. *et al.* 1998. Human pigmentation phenotype: a point mutation generates nonfunctional MSH receptor. Biochem. Biophys. Res. Commun. **245:** 490–492.
10. SCHIOTH, H.B. *et al.* 1999. Loss of function mutations of the human melanocortin 1 receptor are common and are associated with red hair. Biochem. Biophys. Res. Commun. **260:** 488–491.
11. BRAND, S.H. *et al.* 1996. Role of myristoylation in membrane attachment and function of G alpha i- 3 on Golgi membranes. Am. J. Physiol. **270:** C1362–C1369.
12. BUSCA, R. & R. BALLOTTI. 2000. Cyclic AMP a key messenger in the regulation of skin pigmentation. Pigm. Cell Res. **13:** 60–69.
13. CHEN, W. *et al.* 1995. A colorimetric assay for measuring activation of Gs- and Gq-coupled signaling pathways. Anal. Biochem. **226:** 349–354.

14. FITZGERALD, L.R. *et al.* 1999. Measurement of responses from Gi-, Gs-, or Gq-coupled receptors by a multiple response element/cAMP response element-directed reporter assay. Anal. Biochem. **275:** 54–61.
15. LU, D. *et al.* 1994. Agouti protein is an antagonist of the melanocyte-stimulating-hormone receptor. Nature **371:** 799–802.
16. EBERLE, A.N. 1988. The Melanotropins: Chemistry, Physiology and Mechanisms of Action. Basel. Kargel, Switzerland.
17. SCHIOTH, H.B. *et al.* 1996. Expression of functional melanocortin 1 receptors in insect cells. Biochem. Biophys. Res. Commun. **221:** 807–814.
18. SALAZAR-ONFRAY, F. *et al.* 2002. Tissue distribution and differential expression of melanocortin 1 receptor, a malignant melanoma marker. Br. J. Cancer. **87:** 414–422.
19. SIEGRIST, W. *et al.* 1989. Characterization of receptors for alpha-melanocyte-stimulating hormone on human melanoma cells. Cancer Res. **49:** 6352–6358.
20. EBERLE, A.N. *et al.* 1993. Receptors for melanocyte-stimulating hormone on melanoma cells. Ann. N. Y. Acad. Sci. **680:** 320–341.
21. FULLER, B.B. & F.L. MEYSKENS, JR., 1981. Endocrine responsiveness in human melanocytes and melanoma cells in culture. J. Natl. Cancer Inst. **66:** 799–802.
22. SIEGRIST, W. *et al.* 1994. Homologous and heterologous regulation of alpha-melanocyte-stimulating hormone receptors in human and mouse melanoma cell lines. Cancer Res. **54:** 2604–2610.
23. MAS, J.S. *et al.* 2002. Loss-of-function variants of the human melanocortin-1 receptor gene in melanoma cells define structural determinants of receptor function. Eur. J. Biochem. **269:** 6133–6141.
24. DONATIEN, P.D. *et al.* 1992. The expression of functional MSH receptors on cultured human melanocytes. Arch. Dermatol. Res. **284:** 424–426.
25. DE LUCA, M. *et al.* 1993. Alpha melanocyte stimulating hormone (alpha MSH) stimulates normal human melanocyte growth by binding to high-affinity receptors. J. Cell Sci. **105:** 1079–1084.
26. THODY, A.J. *et al.* 1993. Human melanocytes express functional melanocyte-stimulating hormone receptors. Ann. N. Y. Acad. Sci. **680:** 381–390.
27. BOHM, M. *et al.* 1995. Identification of p90RSK as the probable CREB-Ser133 kinase in human melanocytes. Cell Growth Differ. **6:** 291–302.
28. IM, S. *et al.* 1998. Activation of the cyclic AMP pathway by alpha-melanotropin mediates the response of human melanocytes to ultraviolet B radiation. Cancer Res. **58:** 47–54.
29. TADA, A. *et al.* 1998. Endothelin-1 is a paracrine growth factor that modulates melanogenesis of human melanocytes and participates in their responses to ultraviolet radiation. Cell Growth Differ. **9:** 575–584.
30. TADA, A. *et al.* 2002. Mitogen- and ultraviolet-B-induced signaling pathways in normal human melanocytes. J. Invest. Dermatol. **118:** 316–322.
31. SUZUKI, I. *et al.* 1996. Binding of melanotropic hormones to the melanocortin receptor MC1R on human melanocytes stimulates proliferation and melanogenesis. Endocrinology **137:** 1627–1633.
32. SCOTT, M.C. *et al.* 2002. Human melanocortin 1 receptor variants, receptor function and melanocyte response to ultraviolet radiation. J. Cell Sci. **115:** 2349–2355.

Significance of the Melanocortin 1 Receptor in Regulating Human Melanocyte Pigmentation, Proliferation, and Survival

ANA LUISA KADEKARO, HIROMI KANTO, RENNY KAVANAGH, AND
ZALFA A. ABDEL-MALEK

Department of Dermatology, University of Cincinnati, Cincinnati, Ohio 45267, USA

ABSTRACT: The characterization of the melanocortin 1 receptor (MC1R) expressed on human melanocytes and the findings that certain mutations in the
POMC gene or the *MC1R* gene result in red hair phenotype underscore the significance of melanocortins and MC1R in regulating human pigmentation. We
demonstrated that human melanocytes respond to α-melanocortin (α-MSH) or
ACTH with increased proliferation and melanogenesis, and to agouti signaling
protein by abrogation of these effects. α-MSH and ACTH were equipotent and
more potent than β-MSH, and γ-MSH was the least potent in activating the
MC1R and stimulating melanogenesis and proliferation of human melanocytes. We characterized the *MC1R* genotype in a panel of human melanocyte
cultures and identified three cultures that were homozygous for Arg160Trp,
heterozygous for Arg151Cys and Asp294His, and heterozygous for Arg160Trp
and Asp294His substitutions, respectively. Those cultures failed to respond to
α-MSH with increase in cAMP levels, tyrosinase activity, or proliferation and
had an exaggerated response to the cytotoxic effect of ultraviolet (UV) radiation. These loss-of-function mutations have been associated with red hair phenotype and increased risk for skin cancer. Melanocytes homozygous for
Val29Met substitution in MC1R responded normally to α-MSH and UVB, suggesting that this variant is a polymorphism. We observed that α-MSH promotes human melanocyte survival by inhibiting the UV-induced apoptosis
independently of melanin synthesis. This effect was absent in human melanocytes with loss of function *MC1R* mutations. We predict that the survival effect
of α-MSH is caused by reduction of UV-induced DNA damage and contributes
to the prevention of melanoma.

KEYWORDS: pigmentation; melanocortin 1 receptor; melanocytes; melanocortins; ultraviolet radiation; skin cancer; melanoma

ROLE OF MELANOCORTINS AND MC1R IN HUMAN PIGMENTATION

α-Melanocyte-stimulating hormone (α-MSH; α-melanocortin) was named according to its first and best-described function as a physiological regulator of integ-
umental pigmentation of many vertebrate species. It is common knowledge that α-
MSH induces rapid color change in fish, amphibians, and reptiles via stimulating

Address for correspondence: Zalfa A. Abdel-Malek, Department of Dermatology, University
of Cincinnati, Cincinnati, OH 45267-0592. Voice: 513-558-6242; fax: 513-558-0198.
abdelmza@email.uc.edu.

Ann. N.Y. Acad. Sci. 994: 359–365 (2003). © 2003 New York Academy of Sciences.

melanosome dispersion, and increases pigmentation in various mammals by increasing the synthesis of melanin by melanocytes.[1-3] However, a role for melanocortins in regulating human pigmentation remained questionable for many years. It was assumed that α-MSH functions as an endocrine hormone, yet serum levels of the hormone were too low to produce a physiological effect, and a direct effect of α-MSH on cultured human melanocytes could not be demonstrated.[4-6] Decades ago, injection of human subjects with melanocortins was found to increase skin pigmentation.[7] More recently, similar effects were obtained after injection of human volunteers with the potent synthetic analogue of α-MSH, [Nle[4], D-Phe[7]]-α-MSH (NDP-α-MSH).[8] However, these results could not lead to the conclusion that melanocortins have a direct effect on human melanocytes. The cloning of the human *MC1R* gene from human melanocytes, the demonstration that melanocortins are synthesized in the skin and thus function as paracrine and autocrine factors, and that cultured human melanocytes respond avidly to α-MSH put an end to a long-lasting controversy about the significance of melanocortins in human pigmentation.[9-15]

In humans, skin and hair color is determined by the relative amounts of eumelanin, the brown-black pigment, and pheomelanin, the red-yellow pigment, by epidermal and follicular melanocytes, respectively.[16] Results of genetic studies on mouse coat color indicated that in follicular melanocytes the synthesis of eumelanin is regulated by the *extension* locus, which codes for the melanocortin 1 receptor (MC1R), whereas the synthesis of pheomelanin is under the regulation of *agouti* gene, which codes for the agouti signaling protein, the physiological antagonist of α-MSH.[17,18] In human skin, epidermal melanocytes synthesize both eumelanin and pheomelanin, and synthesis of eumelanin can be stimulated by treatment of cultured human melanocytes with the potent α-MSH analogue NDP-α-MSH, or with α-MSH.[19,20] It is known that human melanocytes express functional MC1R.[15,21,22] We have demonstrated that human melanocytes respond to α-MSH with dose-dependent increase in proliferation and melanogenesis, as measured by stimulation of the activity of tyrosinase, the rate-limiting enzyme in the melanin synthetic pathway.[14,15] We also found that ACTH mimics the effects of α-MSH, and that the effects of both hormones on melanocytes are mediated by binding to the MC1R and activating adenylate cyclase. We reported that treatment of human melanocytes with α-MSH increases the protein levels of the melanogenic enzymes tyrosinase and tyrosinase-related proteins 1 and 2, changes known to be associated with eumelanin synthesis.[14] Upon comparison of the relative potencies of the melanocortins α-MSH, β-MSH, γ-MSH, and ACTH on human melanocytes, we found that α-MSH and ACTH were equipotent and more potent than β-MSH and that γ-MSH was the least potent.[15]

Compelling evidence for the importance of the MC1R in mediating the melanogenic response of melanocytes to melanocortins came from the observation that expression of the human *MC1R* in the amelanotic G4F B16 mouse melanoma cells that lack the expression of *MC1R* resulted in eumelanin synthesis.[23] In addition, it was shown recently that expression of the wild-type human *MC1R* in recessive yellow mice (e/e mice) restored the wild-type coat color.[24] The best evidence for the significance of the proopiomelanocortin (POMC) in human pigmentation came from the observations that mutations that disrupt the expression of *POMC* result in red hair (because of lack of eumelanin synthesis), in addition to metabolic abnormalities that include adrenal insufficiency and obesity.[25] These findings strongly suggest that the MC1R and the melanocortins play a similar role in human and mouse melanocytes.

AGOUTI SIGNALING PROTEIN, ANOTHER MC1R LIGAND

Another ligand for the MC1R is agouti signaling protein (ASP). In the mouse, temporal expression of the *agouti* gene results in the switch from the synthesis of eumelanin to pheomelanin.[17,18] The effect of ASP is attributed to its ability to compete with α-MSH for binding to the MC1R, and by acting as an inverse agonist of the MC1R.[17] We demonstrated that human melanocytes respond to treatment with ASP by complete abrogation of the melanogenic and mitogenic effects of α-MSH.[26] We found that in human melanocytes, ASP functions as a competitive inhibitor of α-MSH binding to the MC1R. Treatment of human melanocytes with ASP reduced the protein levels of tyrosinase, and tyrosinase-related proteins 1 and 2, indicative of pheomelanin synthesis. Indeed, we found that treatment with ASP increased the level of pheomelanin in human melanocytes by 25% (unpublished results). We also reported that the MC1R is the principal mediator of ASP function, because mouse follicular melanocytes harboring loss-of-function mutations in *MC1R* fail to respond to ASP.[27]

The human *MC1R* is subject to regulation by its own ligands and by various growth factors and hormones known to affect melanogenesis. We have reported that α-MSH and ACTH increase the level of *MC1R* mRNA.[15] A similar increase in the *MC1R* transcript was observed in melanocytes treated with β-estradiol, a hormone implicated in hyperpigmentation that occurs during pregnancy and in melasma.[28] Treatment of human melanocytes with the potent keratinocyte-derived mitogen endothelin-1 markedly increased *MC1R* mRNA levels.[28] These effects are expected to sustain and possibly increase the ability of human melanocytes to respond to melanocortins. On the other hand, treatment of human melanocytes with ASP drastically reduced the level of *MC1R* transcript.[28] This effect is expected to contribute to the inhibition of eumelanin synthesis by ASP.

MC1R and Skin Cancer Risk

Interest in the regulation of human pigmentation stems mainly from the important role of melanin in photoprotection. Eumelanin is thought to be more photoprotective than pheomelanin because it is more resistant to degradation by exposure to UV and is more efficient at scavenging reactive oxygen radicals that are produced due to exposure to UV.[29] Ultrastructural studies revealed that melanosomes in dark skin, which are expected to be enriched with eumelanin, remain intact throughout the epidermal layers, whereas in fair skin, no intact melanosomes are evident in the suprabasal layers of the epidermis.[30] Epidemiological data clearly show that the risk for skin cancer is highest in individuals with fair skin and red hair, who have a poor tanning ability.[31,32] Human pigmentation is regulated by approximately 70 genes. Among those, the *MC1R* gene seems to be the most polymorphic, suggesting its significance in the wide variation in human pigmentation.[33] To date, approximately 35 allelic variants of the *MC1R* have been identified in various populations.[34,35] Most of those variants are the result of a single amino acid substitution in the MC1R, and some lead to either disruption of binding or signaling of the receptor. Some variants, namely, Arg151Cys, Arg160Trp, and Asp294His substitutions, are highly associated with red hair phenotype and poor tanning ability in Northern European and Australian populations.[34–36] Individuals homozygous for any of these alleles or compound heterozygous for at least two of these alleles most likely have red hair.

The significance of constitutive pigmentation in photoprotection and the correlation between fair skin and the high risk for skin cancer have sparked interest in elucidating the role of the *MC1R* gene in determining the susceptibility to melanoma and nonmelanoma skin cancers.[31,32,37–40] The *MC1R* gene is considered a candidate skin cancer susceptibility gene. Epidemiological studies found that some *MC1R* allelic variants, particularly Arg160Trp, Arg151Cys, and Asp294His, known to be associated with poor tanning ability, are also associated with increased risk for melanoma and nonmelanoma skin cancers, independently of skin or hair color.[37–40] This strongly implies that skin cancer susceptibility cannot be accurately and objectively assessed merely based on pigmentary phenotype.

MC1R Variants, MC1R Function, and Response of Melanocytes to UV

We have been interested in elucidating the biological consequences of the naturally expressed *MC1R* alleles on human melanocytes, the physiological target for melanocortins, and a main target for UV. We used a panel of primary human melanocyte cultures to compare their dose responses to α-MSH, their eumelanin and pheomelanin contents, *MC1R* genotype, and their survival and proliferation after UV exposure.[41] Sequencing the coding region of the *MC1R* gene in thirteen different melanocyte cultures showed that three cultures were wild-type, six were heterozygous for known *MC1R* alleles, and one was homozygous for Val92Met substitution. Those 10 cultures responded to α-MSH with dose-dependent increase in cAMP formation, proliferation, and tyrosinase activity. Three of the thirteen cultures were refractory to α-MSH, as determined by lack of stimulation of tyrosinase activity, proliferation, or cAMP formation by increasing doses of α-MSH. Those three cultures were homozygous for Arg160Trp, compound heterozygous for Arg160Trp and Asp294His, or compound heterozygous for Arg151Cys and Asp294His, respectively. As discussed above, these mutations are strongly associated with skin cancer risk.[37,38,40] Interestingly, the third culture had a high eumelanin to pheomelanin ratio, thus confirming the important observation that the impact of *MC1R* variants on the response of melanocytes to UV is independent of constitutive pigmentation. Importantly, these cultures showed increased sensitivity to killing by UV. Therefore, Arg160Trp, Arg151Cys, and Asp294His, but not Val92Met, substitutions result in loss of function of the MC1R and sensitize melanocytes to the apoptotic effect of UV.

The latter results led us to investigate the role of α-MSH in the survival of melanocytes after UV irradiation. We found that α-MSH counteracts the apoptotic effect of UV and promotes melanocyte survival. This effect was independent of the ability of α-MSH to increase melanin synthesis, because it was evident in tyrosinase-negative albino melanocytes. The survival effect of α-MSH on melanocytes may extent to various cell types that express different MC receptors and may be elicited by other melanocortins that activate those receptors. The survival effect of α-MSH was absent from melanocytes expressing loss-of-function MC1R. Given that loss-of-function mutations in the *MC1R* gene increase the risk for skin cancer, we predict that α-MSH reduces the extent of UV-induced DNA damage and thus prevents mutations and carcinogenesis in cells expressing functional MC1R. We attribute the increased apoptosis of melanocytes with nonfunctional MC1R in response to UV exposure to their inability to cope efficiently with DNA damage. Enhancement of DNA repair by survival factors already has been demonstrated for interleukin 12 and insulin-like growth factor–1 that rescue human keratinocytes from UV-induced apoptosis by enhancing the repair of DNA photoproducts.[42,43] Demonstrating a link between α-MSH and the mechanisms of DNA repair will provide a valuable strategy for skin cancer prevention. Determining the *MC1R* genotype will accurately identify individuals with a high risk for melanoma, and designing α-MSH–like analogues or agents that mimic the signaling pathway of α-MSH that can be delivered transdermally might enhance the protection against UV-induced DNA damage and mutagenesis.

ACKNOWLEDGMENTS

This article is dedicated to Mac Hadley, Ph.D., who mentored the corresponding author and inspired her to investigate the role of melanocortins in human pigmentation. We thank the previous laboratory staff and collaborators who contributed significantly to this research project. This work was supported by National Institutes of Health Grant R01 ES 09110, by a grant from the Ohio Cancer Research Associates, and by POLA Chemical Company (to Z.A.A.-M.).

REFERENCES

1. SAWYER, T.K. *et al*. 1983. α-Melanocyte stimulating hormone: chemical nature and mechanism of action. Am. Zool. **23:** 529–540.
2. SHERBROOKE, W.C. *et al*. 1988. Melanotropic peptides and receptors: an evolutionary perspective in vertebrate physiologic color change. *In* Melanotropic Peptides. Vol. II. M.E. Hadley, Ed.: 175–190. CRC Press. Boca Raton, FL.
3. GESCHWIND, I.I. *et al*. 1972. The effect of melanocyte-stimulating hormone on coat color in the mouse. Rec. Prog. Horm. Res. **28:** 91–130.
4. THODY, A.J. *et al*. 1983. MSH peptides are present in mammalian skin. Peptides **4:** 813–816.
5. FRIEDMAN, P.S. *et al*. 1990. α-MSH causes a small rise in cAMP but has no effect on basal or ultraviolet-stimulated melanogenesis in human melanocytes. Br. J. Dermatol. **123:** 145–151.
6. HALABAN, R. *et al*. 1993. Pigmentation and proliferation of human melanocytes and the effects of melanocyte-stimulating hormone and ultraviolet B light. Ann. N. Y. Acad. Sci. **680:** 290–301.
7. LERNER, A.B. & J.S. MCGUIRE. 1961. Effect of alpha- and beta-melanocyte stimulating hormones on the skin colour of man. Nature **189:** 176–179.
8. LEVINE, N. *et al*. 1991. Induction of skin tanning by the subcutaneous administration of a potent synthetic melanotropin. JAMA **266:** 2730–2736.
9. MOUNTJOY, K.G. *et al*. 1992. The cloning of a family of genes that encode the melanocortin receptors. Science **257:** 1248–1251.
10. CHHAJLANI, V. *et al*. 1993. Molecular cloning of a novel human melanocortin receptor. Biochem. Biophys. Res. Commun. **195:** 866–873.
11. CHAKRABORTY, A.K. *et al*. 1996. Production and release of proopiomelanocortin (POMC) derived peptides by human melanocytes and keratinocytes in culture: regulation by ultraviolet B. Biochim. Biophys. Acta **1313:** 130–138.
12. WAKAMATSU, K. *et al*. 1997. Characterization of ACTH peptides in human skin and their activation of the melanocortin-1 receptor. Pigm. Cell Res. **10:** 288–297.
13. HUNT, G. *et al*. 1994. α-Melanocyte stimulating hormone and its analogue Nle⁴DPhe⁷α-MSH affect morphology, tyrosinase activity and melanogenesis in cultured human melanocytes. J. Cell Sci. **107:** 205–211.

14. ABDEL-MALEK, Z. *et al.* 1995. Mitogenic and melanogenic stimulation of normal human melanocytes by melanotropic peptides. Proc. Natl. Acad. Sci. USA **92:** 1789–1793.
15. SUZUKI, I. *et al.* 1996. Binding capacity and activation of the MC1 receptors by melanotropic hormones correlate directly with their mitogenic and melanogenic effects on human melanocytes. Endocrinology **137:** 1627–1633.
16. PATHAK, M.A. *et al.* 1980. Photobiology of pigment cells. *In* Phenotypic Expression in Pigment Cells. M. Seiji, Ed.: 655–670. University of Tokyo Press. Tokyo.
17. YEN, T.T. *et al.* 1994. Obesity, diabetes, and neoplasia in yellow A^{vy}/- mice: ectopic expression of the *agouti* gene. FASEB J. **8:** 479–488.
18. BARSH, G.S. 1996. The genetics of pigmentation: from fancy genes to complex traits. Trends Genet. **12:** 299–305.
19. THODY, A.J. *et al.* 1991. Pheomelanin as well as eumelanin is present in human epidermis. J. Invest. Dermatol. **97:** 340–344.
20. HUNT, G. *et al.* 1995. Nle⁴DPhe⁷ α-melanocyte-stimulating hormone increases the eumelanin:phaeomelanin ratio in cultured human melanocytes. J. Invest. Dermatol. **104:** 83–85.
21. DONATIEN, P.D. *et al.* 1992. The expression of functional MSH receptors on cultured human melanocytes. Arch. Dermatol. Res. **284:** 424–426.
22. DE LUCA, M. *et al.* 1993. α-Melanocyte stimulating hormone (αMSH) stimulates normal human melanocyte growth by binding to high-affinity receptors. J. Cell Sci. **105:** 1079–1084.
23. TAPIA, J.C. *et al.* 1996. Induction of constitutive melanogenesis in amelanotic mouse melanoma cells by transfection of the human melanocortin-1 receptor gene. J. Cell Sci. **109:** 2023–2030.
24. HEALY, E. *et al.* 2001. Functional variation of MC1R alleles from red-haired individuals. Hum. Mol. Genet. **10:** 2397–2402.
25. KRUDE, H. *et al.* 1998. Severe early-onset obesity, adrenal insufficiency and red hair pigmentation caused by *POMC* mutations in humans. Nat. Genet. **19:** 155–157.
26. SUZUKI, I. *et al.* 1997. Agouti signaling protein inhibits melanogenesis and the response of human melanocytes to α-melanotropin. J. Invest. Dermatol. **108:** 838–842.
27. ABDEL-MALEK, Z.A. *et al.* 2001. The melanocortin 1 receptor is the principal mediator of the effects of agouti signaling protein on mammalian melanocytes. J. Cell Sci. **114:** 1019–1024.
28. SCOTT, M.C. *et al.* 2002. Regulation of the human melanocortin 1 receptor expression in epidermal melanocytes by paracrine and endocrine factors, and by UV radiation. Pigm. Cell Res. **15:** 433–439.
29. MENON, A. *et al.* 1983. Effects of ultraviolet-visible radiation in the presence of melanin isolated from human black or red hair upon Ehrlich ascites carcinoma cells. Cancer Res. **43:** 3165–3169.
30. PATHAK, M.A. *et al.* 1971. The photobiology of melanin pigmentation in human skin. *In* Biology of Normal and Abnormal Melanocytes. T. Kawamura *et al.*, Eds.: 149–169. University Park Press. Baltimore.
31. EPSTEIN, J.H. 1983. Photocarcinogenesis, skin cancer and aging. J. Am. Acad. Dermatol. **9:** 487–502.
32. SOBER, A.J. *et al.* 1991. Epidemiology of cutaneous melanoma. An update. Dermatol. Clin. **9:** 617–629.
33. STURM, R.A. *et al.* 1998. Human pigmentation genetics: the difference is only skin deep. Bioessays **20:** 712–721.
34. SMITH, R. *et al.* 1998. Melanocortin 1 receptor variants in Irish population. J. Invest. Dermatol. **111:** 119–122.
35. BOX, N.F. *et al.* 1997. Characterization of melanocyte stimulating hormone receptor variant alleles in twins with red hair. Hum. Mol. Genet. **6:** 1891–1897.
36. FLANAGAN, N. *et al.* 2000. Pleiotropic effects of the melanocortin 1 receptor (*MC1R*) gene on human pigmentation. Hum. Mol. Genet. **9:** 2531–2537.
37. PALMER, J.S. *et al.* 2000. Melanocortin-1 receptor polymorphisms and risk of melanoma: is the association explained solely by pigmentation phenotype? Am. J. Hum. Genet. **66:** 176–186.

38. KENNEDY, C. *et al.* 2001. Melanocortin 1 receptor (*MC1R*) gene variants are associated with an increased risk for cutaneous melanoma which is largely independent of skin type and hair color. J. Invest. Dermatol. **117:** 294–300.
39. BOX, N.F. *et al.* 2001. Melanocortin-1 receptor genotype is a risk factor for basal and squamous cell carcinoma. J. Invest. Dermatol. **116:** 224–229.
40. BASTIAENS, M.T. *et al.* 2001. Melanocortin-1 receptor gene variants determine the risk of non-melanoma skin cancer independent of fair skin type and red hair. Am. J. Hum. Genet. **68:** 884–894.
41. SCOTT, M.C. *et al.* 2002. Human *melanocortin 1 receptor* variants, receptor function and melanocyte response to ultraviolet radiation. J. Cell Sci. **115:** 2349–2355.
42. SCHWARZ, A. *et al.* 2002. Interleukin-12 suppresses ultraviolet radiation-induced apoptosis by inducing DNA repair. Nat. Cell Biol. **4:** 26–31.
43. DECRAENE, D. *et al.* 2002. Insulin-like growth factor-1-mediated AKT activation postpones the onset of ultraviolet B-induced apoptosis, providing more time for cyclobutane thymine dimer removal in primary human keratinocytes. J. Biol. Chem. **277:** 32587–32595.

Avian Melanocortin System: α-MSH May Act as an Autocrine/Paracrine Hormone

A Minireview

SAKAE TAKEUCHI,[a] SUMIO TAKAHASHI,[a] RONALD OKIMOTO,[b]
HELGI B. SCHIÖTH,[c] AND TIMOTHY BOSWELL[d]

[a]Department of Biology, Faculty of Science, Okayama University,
Okayama 700-8530, Japan

[b]Poultry Science Department, University of Arkansas, Fayetteville, Arkansas 72701, USA

[c]Department of Neuroscience, Uppsala University, 751 24 Uppsala, Sweden

[d]Division of Integrative Biology, Roslin Institute (Edinburgh),
Midlothian, EH25 9PS, United Kingdom

ABSTRACT: The interest in the physiological role of α-MSH in birds has been limited because they lack the intermediate lobe of the pituitary, the main source of circulating α-MSH in most vertebrates. Recent studies have improved our understanding of the avian melanocortin system. We have cloned and characterized all five MC-R subtypes, POMC, and AGRP in chicken. Analyses of the tissue distribution of expression of these genes revealed widespread expression throughout the body, corresponding to the situation in mammals in which α-MSH exerts a multiplicity of effects in different tissues by acting as a local mediator. We showed that the *extended black* locus controlling feather pigmentation in the chicken encodes MC1-R. Moreover, black chickens carrying the dominant allele, the *extended black*, express the MC1-R with ligand-independent activity as the *somber-3J* black mice. α-MSH and AGRP were expressed in the infundibular nucleus of POMC and NPY neurons, respectively, in the brain of Japanese quail. Furthermore, fasting stimulated AGRP expression and lowered POMC expression. These data indicate that at least two of the major melanocortin systems reported in mammals, that is, regulation of pigmentation and energy homeostasis, was developed in a common ancestor to chicken and mammals at least 300 million years ago. Furthermore, α-MSH peptide was identified in developing chicken eye, suggesting a possible involvement of α-MSH in regulation of ocular development. Collectively, the data reviewed here indicate that α-MSH is produced locally and acts as an autocrine/paracrine hormone in birds.

KEYWORDS: α-MSH; agouti-related protein; melanocortin receptor; feather pigmentation; energy homeostasis; growth hormone; chicken; Japanese quail; review

Address for correspondence: Sakae Takeuchi, Department of Biology, Faculty of Science, Okayama University, Okayama 700-8530, Japan. Voice: 81-86-251-7868; fax: 81-86-251-7876.
stakeuch@cc.okayama-u.ac.jp

Ann. N.Y. Acad. Sci. 994: 366–372 (2003). © 2003 New York Academy of Sciences.

INTRODUCTION

α-Melanocyte-stimulating hormone (α-MSH) was discovered approximately 90 years ago as a potent stimulator of pigment cells, modulating color change in amphibia. Surgical ablation of the intermediate lobe of the pituitary results in frogs turning pale, providing the idea that the gland plays a key role in α-MSH function.[1] The absence of this region in the avian pituitary has led to the general thought that α-MSH is a hormone of negligible interest in birds, and, up until now, neither the occurrence nor a physiological role for α-MSH has been established. To obtain insight into the evolution of the α-MSH endocrine system, it is necessary to clarify whether α-MSH is produced and having physiological roles in birds.

CLONING AND TISSUE DISTRIBUTION OF EXPRESSION OF MELANOCORTIN COMPONENT GENES

We cloned genomic and complementary DNAs of the chicken proopiomelanocortin (POMC),[2] agouti-related protein (AGRP)[3], and melanocortin receptors (MC-Rs).[4–7] Our analyses of MC-R genes showed that the chicken genome contains all five MC-R subtypes and they are distributed in various tissues.[2,3,5–7] We found that the chicken POMC contains nine proteolytic cleavage sites,[2] suggesting that α-MSH and other melanocortins could be produced by proteolytic cleavage of POMC. In addition, the POMC mRNA was found to be widely expressed throughout the body,[2] corresponding to the situation in mammals in which α-MSH exerts a multiplicity of effects in different tissues by acting as a local mediator. These results imply the possibility that the physiological roles of α-MSH have been conserved during evolution so that it acts as a paracrine and/or autocrine hormone in birds as it does in mammals.

ENERGY HOMEOSTASIS SYSTEM

It is now well known that in mammals the melanocortin system within the hypothalamus regulates appetite and energy balance.[8] α-MSH and AGRP are the peptides implicated most strongly in the regulation of energy balance. They are expressed by two distinct populations of neurons within the arcuate nucleus of the hypothalamus, POMC neurons, and neurons expressing neuropeptide Y (NPY), respectively. Both α-MSH and AGRP released by these neurons exert effects on energy balance by signaling through MC3-R and MC4-R that induce an inhibitory influence on appetite and body weight. α-MSH acts as an agonist to the receptors and suppresses feeding, and AGRP conversely stimulates food intake by antagonizing the α-MSH signaling. Both the AGRP and POMC genes are sensitive to nutritional signals as demonstrated by their respective up- and down-regulation during fasting.

Although it has been demonstrated that human AGRP and α-MSH exert comparable effects on feeding to those seen in mammals when they are administrated into the brain of chicks,[9,10] the existence of the hypothalamic melanocortin system regulating food intake remains to be established in birds. To obtain evidence for the central melanocortin system regulating appetite in the Japanese quail, we performed a series of *in situ* hybridization as well as immunocytochemistry. Dual labeling of

AGRP mRNA and NPY mRNA showed that they are coexpressed in the hypothalamic infundibular nucleus,[11] the avian equivalent structure to the arcuate nucleus, reflecting the situation in mammals and suggesting that the NPY/AGRP neuronal cell type has been conserved during evolution. A combination of *in situ* hybridization and immunocytochemistry detected the expression of POMC mRNA in the infundibular nucleus and production of α-MSH in the POMC neurons.[12] Again, the presence of α-MSH–expressing neurons in the region is similar to that seen in mammals and suggests evolutionary conservation of these neurons. *In situ* hybridization demonstrated the MC4-R mRNA expression in the paraventricular nucleus and ventromedial nucleus known to be important brain regions for the regulation of food intake (T. Boswell, personal communication). Furthermore, 24-h fasting increased AGRP mRNA and decreased POMC mRNA in the infundibular nucleus of Japanese quail.[12] Collectively, these results suggest that the melanocortin system plays a physiologically relevant role in the control of appetite in the Japanese quail, comparable to its effects in mammals. The avian hypothalamic melanocortin system regulating appetite is presented in FIGURE 1.

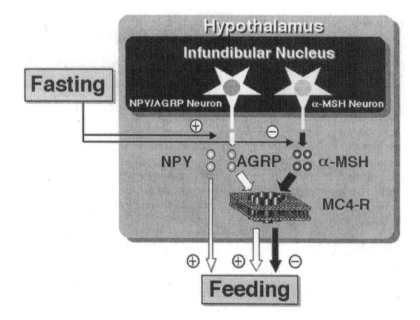

FIGURE 1. Schematic representation of the hypothalamic melanocortin system regulating appetite in the Japanese quail. The α-MSH–expressing neurons and AGRP/NPY-expressing neurons have been neuroanatomically and functionally conserved during long periods of vertebrate evolution.

PIGMENTARY SYSTEM

The *extension* locus, together with the *agouti* locus, regulates the relative amount of black pigment (eumelanin) and red/yellow pigment (pheomelanin) in mammals. The *extension* locus encodes MC1-R, and the *agouti* locus encodes agouti-signaling protein (ASP). α-MSH acts as an agonist by binding to MC1-R and elevating intracellular cAMP to induce eumelanin synthesis in the follicular melanocytes. ASP produced by the hair follicle acts in trans on the follicular melanocytes to inhibit eumelanin synthesis, thereby resulting in production of pheomelanin, the default product of melanin biosynthetic pathway. In mouse, allelic MC1-R variants have been well characterized. The dominant *somber-3J* allele, for example, encodes MC1-R with Glu92Lys mutation responsible for a constitutively active MC1-R and resultant black coat color in mice, and the *recessive yellow* allele encodes a truncated nonfunctional MC1-R.[13]

Genetic locus similar to the murine *extension* locus has been reported in the chicken, the *extended black* locus on chromosome 1 (GGA1).[14] As is the case of the *extension* locus in mammals, alleles that make more eumelanin are dominant over those that make less.[15] By restriction fragment length polymorphism[16] and PAMSA (PCR amplification of multiple specific alleles)[17] analyses, we demonstrated that the *extended black* locus encodes MC1-R. Typing of MC1R polymorphisms on the East Lansing reference population revealed that the MC1-R gene cosegregates with ADL33, but not with MSU34, ADL19, or ADL234.[17] The ADL33 is located on the E30 linkage group that is probably the GGA11 microchromosome and other markers are on the GGA1 where the extended black locus has been thought to be. FISH analysis localized the chicken MC1-R gene on microchromosome.[18,19] Based on these results, we conclude that the *extended black* locus encoding the chicken MC1-R gene is located on microchromosome, but not on the GGA1.

The allelic variants of the chicken MC1-R that we identified so far are shown in FIGURE 2. Those variants were expressed in mammalian cells and pharmacologically characterized.[20] The *wild-type* (e^+) MC1-R was found to bind α-MSH and increase intracellular cAMP level in a dose of ligand-dependent manner. The skin of the chicken does express POMC mRNA,[3] suggesting that α-MSH produced in the skin regulates feather color pigmentation. The *extended black* (*E*), the most dominant and conferring a uniformly black pigmentation, was shown to encode MC1-R with ligand-independent activity caused by the same mutation as that of the murine *somber-3J*,[16] suggesting that constitutively active MC1-R variant confers black pigmentation in the chicken as in the case of the mouse. Notably, the same mutation, Glu92Lys, was reported in the bananaquit, *Coereba flaveola*, associated with melanic plumage.[21] Our pharmacological results, however, could not completely explain the association of MC1-R variants with the corresponding phenotypes; the *recessive wheaten* (e^y) allele exhibiting yellow-red pigmentation in chicks and females, for example, encodes normally functioning MC1-R. Furthermore, it remains unclear why sexual difference is observed in influence of the *extended black* locus (FIG. 2). Further analyses such as evaluation of promoter activity of the MC1-R gene in each *extended black* locus allele as well as cloning of the ASP gene thus are required to clarify molecular mechanisms for all the different feather pigmentation patterns in the chicken. Nevertheless, our experiments revealed that mutations in the MC1-R are capable of causing different pigmentation phenotypes in the chicken, just as they do

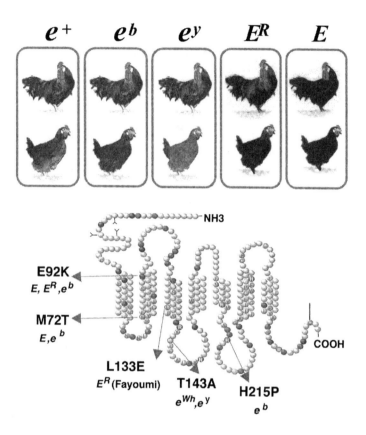

FIGURE 2. Cartoon representation of the chicken *extended black* locus. The figure shows the adult *extended black* locus color patterns, providing that all other feather color genes are wild type (*top*). The allelic variants of the MC1-R are shown below.

in mammals. This suggests that the melanocortin system regulating pigmentation has been conserved during evolution.

IDENTIFICATION OF AVIAN α-MSH IN THE EYE

Many lines of evidence have shown that administration of α-MSH affects numerous functions of the eye; however, the source of the α-MSH acting in the eye remains unclear.[1] By a combination of RT-PCR, *in situ* hybridization, and immunocytochemistry, we demonstrated that retinal pigment epithelial (RPE) cells in chick embryo express α-MSH in a developmentally regulated pattern.[22] The MC-R subtypes MC1-R, MC4-R, and MC5-R are expressed in the layers of the choroid and the neural retina,[22] adjacent to RPE cells, suggesting that RPE cells secrete α-MSH to regulate ocular development in a paracrine fashion. Although the exact physiological role of α-MSH during ocular development remains to be elucidated, this was the first evidence for the expression of α-MSH peptide in peripheral tissues in birds. Recently, we demonstrated that a novel growth hormone isoform (s-GH) is expressed in the chicken eye, and that its binding to RPE cells is temporally regulated and correlates well with the production of α-MSH in RPE cells during embryonic development,[23] implying a possible involvement of s-GH in the regulation of ocular development by acting on the intraocular melanocortin system in the chicken.

CONCLUSION

Great progress has been made over the years in the understanding of the molecular mechanisms of melanocortin actions in mammals. In contrast, little is known about avian melanocortin system. Yet, progress has been made in recent years. The data reviewed here lead to the idea that α–MSH is a local mediator exerting a multiplicity of effects in different tissues in birds as it does in mammals. Besides birds, some mammalian species, including humans, also lack the intermediate lobe of the pituitary. The gland may have developed as the primary secretory gland of circulating α-MSH involved in physiological color change in lower vertebrates, and the main physiological roles of the gland might have altered during evolution toward animals bearing feather or hair. α-MSH may be essentially an autocrine and/or paracrine hormone.

ACKNOWLEDGMENTS

This work was supported by Japan Society for the Promotion of Science Grant-in-Aid for Scientific Research (C) (S.T.), a Biotechnology and Biological Sciences Research Council Advanced Fellowship (T.B.), and The Swedish Research Council and Melacure Therapeutics (H.B.S.).

REFERENCES

1. EBERLE, A.N. 1988. The Melanotropins: Chemistry, and Mechanisms of Action. Karger Publishers. Basel.
2. TAKEUCHI, S. *et al.* 1999. Molecular cloning and characterization of the chicken proopiomelanocortin (POMC) gene. Biochim. Biophys. Acta **1450:** 452–459.
3. TAKEUCHI, S. *et al.* 2000. Widespread expression of agouti-related protein (AGRP) in the chicken: a possible involvement of AGRP in regulating peripheral melanocortin systems in the chicken. Biochim. Biophys. Acta **1496:** 261–269.
4. TAKEUCHI, S. *et al.* 1996. Molecular cloning and sequence analysis of the chick melanocortin 1-receptor gene. Biochim. Biophys. Acta **1306:** 122–126.
5. TAKEUCHI, S. *et al.* 1998. Molecular cloning of the chicken melanocortin 2 (ACTH)-receptor gene. Biochim. Biophys. Acta **1403:** 102–108.
6. TAKEUCHI, S. & S. TAKAHASHI. 1999. A possible involvement of melanocortin 3 receptor in the regulation of adrenal gland function in the chicken. Biochim. Biophys. Acta **1448:** 512–518.
7. TAKEUCHI, S. & S. TAKAHASHI. 1998. Melanocortin receptor genes in the chicken—tissue distributions. Gen. Comp. Endocrinol. **112:** 220–231.
8. CONE, R.D. 1999. The central melanocortin system and energy homeostasis. Trends Endocrinol. Metab. **10:** 211–216.

9. KAWAKAMI, S-I. *et al.* 2000. Central administration of α-melanocyte stimulating hormone inhibits fasting- and neuropeptide Y-induced feeding in neonatal chicks. Eur. J. Pharmacol. **398:** 361–364.

10. TACHIBANA, T. *et al.* 2001. Intracerebroventicular injection of agouti-related protein attenuates the anorexigenic effect of alpha-melanocyte stimulating hormone in neonatal chicks. Neurosci. Lett. **305:** 131–134.

11. BOSWELL, T. *et al.* 2002. Neurons expressing neuropeptide Y mRNA in the infundibular hypothalamus of Japanese quail are activated by fasting and co-express agouti-related protein mRNA. Mol. Brain Res. **100:** 31–42.

12. PHILLIPS-SINGH, D. *et al.* 2003. Fasting differentially regulates expression of agouti-related peptide, pro-opiomelanocortin, prepro-orexin and vasoactive intestinal polypeptide mRNAs in the hypothalamus of Japanese quail. Cell Tissue Res. In press.

13. ROBBINS, L.S. *et al.* 1993. Pigmentation phenotypes of variant extension locus alleles result from point mutations that alter MSH receptor function. Cell **72:** 827–834.

14. CAREFOOT, W.C. 1993. Further studies of linkage and mapping of the loci of genes in group 3 on chromosome 1 of the domestic fowl. Br. Poult. Sci. **34:** 205–209.

15. BRUMBAUGH, J.A. & W.F. HOLLANDER. 1965. A further study of the E pattern locus in the fowl. Iowa State J. Sci. **40:** 51–64.

16. TAKEUCHI, S. *et al.* 1996. A possible involvement of melanocortin 1-receptor in regulating feather color pigmentation in the chicken. Biochim. Biophys. Acta **1308:** 164–168.

17. OKIMOTO, R. *et al.* 2000. Melanocortin 1-receptor (MC1-R) gene polymorphisms associated with the chicken E locus alleles [abstract]. Poult. Sci. **79:** 9.

18. SAZANOV, A. *et al.* 1998. Evolutionarily conserved telomeric localization of BBC1 and MC1R on a microchromosome questions the identity of MC1R and a pigmentation locus on chromosome 1 in chicken. Chromosome Res. **6:** 651–654.

19. SCHIÖTH, H.B. *et al.* 2003. Remarkable synteny conservation of melanocortin receptors (MC1-5R) in chicken, human, and other vertebrates. Genomics **81:** 504–509.

20. LING, M.K. *et al.* 2003. Association of feather colour with constitutively active melanocortin 1 receptors in chicken. Eur. J. Biochem. **270:** 1441–1449.

21. THERON, E. *et al.* 2001. The molecular basis of avian plumage polymorphism in the wild: a melanocortin-1-receptor point mutation is perfectly associated with the melanic plumage morph of the bananaquit, *Coerebra flaveola.* Curr. Biol. **11:** 1–20.

22. TESHIGAWARA, K. *et al.* 2001. Identification of avian α-melanocyte-stimulating hormone in the eye: temporal and spatial regulation of expression in the developing chicken. J. Endocrinol. **168:** 527–537.

23. TAKEUCHI, S. *et al.* 2001. Identification of a novel GH isoform: a possible link between GH and melanocortin systems in the developing chicken eye. Endocrinology **142:** 5158–5166.

P-Locus Is a Target for the Melanogenic Effects of MC-1R Signaling

A Possible Control Point for Facultative Pigmentation

JANIS ANCANS,[a] NIAMH FLANAGAN,[b] MARTIN J. HOOGDUIJN,[a] AND ANTHONY J. THODY[a]

[a]Department of Biomedical Sciences, University of Bradford, Bradford, United Kingdom

[b]Department of Dermatology, Addenbrooke's NHS Trust, Cambridge, United Kingdom

ABSTRACT: Melanocortin receptor type 1 (MC-1R) is an important control point for ultraviolet ray (UVR)–induced tanning response in the skin. In this study, we show that p-locus is a downstream target for MC-1R signaling. The expression of p-locus was up-regulated by α-MSH as well as db-cAMP, a synthetic analogue of cAMP that mimics activation of MC-1R. Furthermore, p-locus transcript abundance was significantly increased in epidermal melanocytes of white skin with facultative (UVR-induced) pigmentation. Because p-locus product is essential for pigmentation and also has been shown to be highly polymorphic in human population, we propose that the pigmentary response to the melanocortin peptides/UVR would be affected not only by MC-1R mutations but also by the functionality of p-locus product. These factors together could account for the many different levels of tanning ability seen in the white population.

KEYWORDS: melanocortin peptides/receptor; pigmentation; p-locus

The melanocortin peptides, MSH and ACTH, are the best-characterized hormones that regulate human pigmentation. *In vitro* these peptides increase melanocyte dendricity, proliferation rate, and melanogenesis (reviewed by Tsatmali *et al.*[1]). The ability of melanocortin peptides to induce pigmentation of human skin *in vivo* has been demonstrated on several occasions.[2,3] More recently, a non-sense mutation in the POMC gene which causes a complete absence of ACTH/α-MSH was shown to result in red hair and pale skin.[4]

Melanocytes express the melanocortin receptor type 1 (MC-1R), which has a high affinity for the melanocortin peptides α-MSH and ACTH.[5] The MC-1R couples to cAMP second messenger pathway, and this in turn activates protein kinase A and CREB, leading to an up-regulation of the transcription factors and genes involved in

Address for correspondence: Janis Ancans, Department of Biomedical Sciences, University of Bradford, United Kingdom. Voice: 371-6494422; fax: 371-7323403.

j.ancans@lycos.co.uk

Ann. N.Y. Acad. Sci. 994: 373–377 (2003). © 2003 New York Academy of Sciences.

melanogenesis.[6] In recent years, there has been considerable interest in MC-1R gene mutations, and more than 30 allelic variants have now been identified. Four of these, Arg142His, Arg151Cys, Arg160Trp, and Asp294His, have an association with red-hair/ pale skin phenotype (reviewed by Rees[7]). These alleles were shown to be loss-of-function mutants with disruption of receptor coupling to the cAMP pathway.[8] MC-1R allele Val60Leu has been correlated with blond hair color and shown to produce a partial loss-of-function receptor. However, loss-of-function MC-1R alleles are not the only determinants for hair color. Thus, homozygous and compound heterozygous mutations of the MC-1R have been found in individuals who do not have red hair, and, conversely, individuals with red hair can have functional MC-1R alleles.[9–13] This was illustrated in a study in which five pairs of siblings were found to have identical loss-of-function MC-1R alleles, but did not share red hair color.[9]

There is less correlation between MC-1R alleles and skin color/phototype. Although individuals with two loss-of-function MC-1R alleles are likely to have pale skin and poor tanning ability (skin types I/ II), many white individuals with pale skin and poor tanning ability have been identified as having no MC-1R mutations.[9–13] Furthermore, the role of MC-1R in populations other than white seems to be minimal. The fact that there are no known hypopigmentary characteristics associated with MC-1R mutations in Asian or African populations supports the view that in these darker skin types the inherited high level of melanogenesis is independent of MC-1R signaling. This is in agreement with the finding that despite Val60Leu and Arg160Trp mutations in the MC-1R alleles a melanocyte culture from a black donor showed a high level of melanogenesis comparable to that seen in other black donor melanocytes (Dr. N.P. Smit, University of Leiden, personal communication). This is not consistent with a recent proposal that loss-of-function MC-1R alleles do not exist in the black African population.[14] It seems that more studies involving larger and more diverse donor populations should be conducted to draw conclusions regarding the status of MC-1R gene in Africa.

The evidence therefore would suggest that, whereas loss-of-function MC-1R alleles are associated with the "red/blond hair, pale skin/no tan" phenotype in whites, the wide range of hair/skin color and tanning abilities seen in the rest of the white

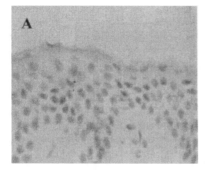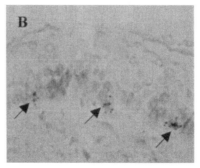

FIGURE 1. Detection of p-locus transcripts in the white donor's skin. Samples represent nonexposed area (scalp; **A**) and tanned exposed area (cheek; **B**) with visible facultative pigmentation. *In situ* hybridization was conducted as described by Suzuki *et al.*[22]; *arrows* indicate presence of p-locus transcript in melanocytes.

population cannot be accounted for on the basis of MC-1R functionality alone. A rational target for further investigation on natural pigmentary variations would be the polymorphism of genes acting downstream of the MC-1R.

Polymorphism of genes directly involved in the enzymatic processes and essential for melanin production, such as tyrosinase and TRP-1, has been found to be low, and apart from loss-of-function mutations which cause albinism there is no correlation with pigmentary phenotype (reviewed by Sturm *et al.*[15]). We recently have turned our attention to the posttranslational mechanisms that regulate tyrosinase activity. The latter is dependent on the pH of the melanosomes,[16] the melanin-producing organelles. It was shown that neutralization of melanosomal pH in white melanocytes increased tyrosinase activity, the eumelanin/pheomelanin ratio and melanosome maturation rate;[17] such increases are characteristic for the tanning response. Melanosomal pH therefore is a critical factor for melanogenesis, and its control could be an inherited feature that is vital for pigmentation.[17,18] If, as suspected, melanosomal pH is regulated by the p-locus product,[19,20] then the latter could serve as a rate-limiting gene for tyrosinase activity with the potential to determine a range of melanin production levels. It is known that expression of p-locus is highly specific to melanocytes, and defects of this gene cause several hypopigmentary conditions (reviewed by Brilliant[21]). Because reduced activity of p-locus product (P-protein) can "recapitulate" less pigmented skin types, it has been suggested that P-protein might be a physiological determinant of human skin color.[15,21]

In a recent study, increases in p-locus transcript abundance were found in Asian skin in response to ultraviolet B irradiation.[22] The time course of the increase suggested an involvement of p-locus in the tanning process. Indeed, similar experiments conducted by us indicated that the presence of p-locus transcripts was virtually undetectable in nonexposed skin of white donors (see FIG. 1A), and this was consistent with

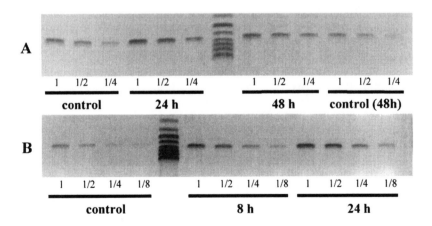

FIGURE 2. Regulation of p-locus expression via MC-1R/cAMP signaling system. Effects of 10 nM α-MSH (**A**) and 0.5 mM db-cAMP (**B**) on p-locus transcript levels were estimated by semiquantitative PCR. Each time point is represented by two (**A**) or three (**B**) serial double dilutions of a cDNA sample made from 1.0 μg of total RNA.

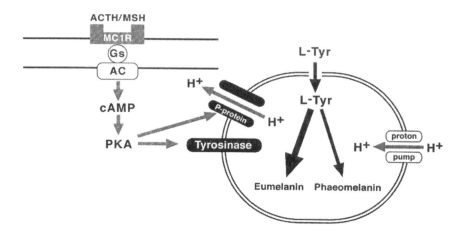

FIGURE 3. The relationship between melanocortin signaling system, p-locus, and melanosomal pH. Activation of MC-1R induces up-regulation of p-locus expression which enhances activity of tyrosinase and eumelanin production via neutralization of melanosomal pH. The abundance and polymorphism of MC-1R and P-protein could serve as key control points for pigmentation.

the finding in Asian skin prior to ultraviolet irradiation.[22] However, a good signal for p-locus transcript was seen in exposed white skin that had facultative pigmentation (see FIG. 1B).

Semiquantitative PCR was conducted to investigate whether α-MSH, a key mediator of tanning response, enhances expression of p-locus. Thus, α-MSH increased p-locus abundance in B16 melanoma cells by more that twofold, and expression remained elevated for the 48-h duration of the experiment (FIG. 2A). Dibutyryl-cAMP (db-cAMP) had a similar effect, indicating that the effect of α-MSH was mediated via the PKA signaling pathway (FIG. 2B).

We conclude that because P-protein expression is a downstream target for the melanocortin/MC-1R signaling system it could provide an important control point for the facultative pigmentation (FIG. 3). It would follow that pigmentary responses to UVR/melanocortin peptides are affected not only by MC-1R mutations but also by the functionality of p-locus alleles. These two factors together could account for the wide range of tanning ability and skin color that is seen in the white population. A high polymorphism of p-locus in human population has been already indicated.[23]

REFERENCES

1. TSATMALI, M. *et al.* 2002. Melanocyte function and its control by melanocortin peptides. J. Histochem. Cytochem. **50:** 125–133.
2. LERNER, A.B. & J.S. MCGUIRE. 1964. Melanocyte-stimulating hormone and adrenocorticotropic hormone. Their relation to pigmentation. N. Engl. J. Med. **270:** 539–546.
3. LEVINE, N. *et al.* 1991. Induction of skin tanning by subcutaneous administration of a potent synthetic melanotropin. JAMA **266:** 2730–2736.

4. KRUDE, H. *et al.* 1998. Severe early-onset obesity, adrenal insufficiency and red hair pigmentation caused by POMC mutations in humans. Nat. Genet. **19:** 155–157.
5. TSATMALI, M. *et al.* 1999. ACTH1-17 is a more potent agonist at the human MC1 receptor than alpha-MSH. Cell. Mol. Biol. (Noisy-le-grand) **45:** 1029–1034.
6. BUSCA, R. & R. BALLOTTI. 2000. Cyclic AMP a key messenger in the regulation of skin pigmentation. Pigm. Cell Res. **13:** 60–69.
7. REES, J.L. 2000. The melanocortin 1 receptor (MC1R): more than just red hair. Pigm. Cell Res. **13:** 135–140.
8. SCHIOTH, H.B. *et al.* 1999. Loss of function mutations of the human melanocortin 1 receptor are common and are associated with red hair. Biochem. Biophys. Res. Commun. **260:** 488–491.
9. BOX, N.F. *et al.* 1997. Characterization of melanocyte stimulating hormone receptor variant alleles in twins with red hair. Hum. Mol. Genet. **6:** 1891–1897.
10. SMITH, R. *et al.* 1998. Melanocortin 1 receptor variants in an Irish population. J. Invest. Dermatol. **111:** 119–122.
11. PALMER, J.S. *et al.* 2000. Melanocortin-1 receptor polymorphisms and risk of melanoma: is the association explained solely by pigmentation phenotype? Am. J. Hum. Genet. **66:** 176–186.
12. BASTIAENS, M.T. *et al.* 2001. Melanocortin-1 receptor gene variants determine the risk of nonmelanoma skin cancer independently of fair skin and red hair. Am. J. Hum. Genet. **68:** 884–894.
13. KENNEDY, C. *et al.* 2001. Melanocortin 1 receptor (MC1R) gene variants are associated with an increased risk for cutaneous melanoma which is largely independent of skin type and hair color. J. Invest. Dermatol. **117:** 294–300.
14. HARDING, R.M. 2000. *et al.* Evidence for variable selective pressures at MC1R. Am. J. Hum. Genet. **66:** 1351–1361.
15. STURM, R.A. *et al.* 2001. Human pigmentation genes: identification, structure and consequences of polymorphic variation. Gene **277:** 49–62.
16. ANCANS, J. & A.J. THODY. 2000. Activation of melanogenesis by vacuolar type H(+)-ATPase inhibitors in amelanotic, tyrosinase positive human and mouse melanoma cells. FEBS Lett. **478:** 57–60.
17. ANCANS, J. *et al.* 2001. Melanosomal pH controls rate of melanogenesis, eumelanin/phaeomelanin ratio and melanosome maturation in melanocytes and melanoma cells. Exp. Cell Res. **268:** 26–35.
18. FULLER, B.B. *et al.* 2001. Regulation of the catalytic activity of pre-existing tyrosinase in black and Caucasian human melanocyte cell cultures. Exp. Cell Res. **262:** 197–208.
19. ANCANS, J. *et al.* 2001. Melanosomal pH, pink locus protein and their roles in melanogenesis. J. Invest. Dermatol. **117:** 158–159.
20. MANGA, P. & S.J. ORLOW. 2001. Inverse correlation between pink-eyed dilution protein expression and induction of melanogenesis by bafilomycin A1. Pigm. Cell Res. **14:** 362–367.
21. BRILLIANT, M.H. 2001. The mouse p (pink-eyed dilution) and human P genes, oculocutaneous albinism type 2 (OCA2), and melanosomal pH. Pigm. Cell Res. **14:** 86–93.
22. SUZUKI, I. *et al.* 2002. Increase of pro-opiomelanocortin mRNA prior to tyrosinase, tyrosinase-related protein 1, dopachrome tautomerase, Pmel-17/gp100, and P-protein mRNA in human skin after ultraviolet B irradiation. J. Invest. Dermatol. **118:** 73–78.
23. LEE, S.T. *et al.* 1995. Organization and sequence of the human P gene and identification of a new family of transport proteins. Genomics **26:** 354–363.

DOTA α-Melanocyte-Stimulating Hormone Analogues for Imaging Metastatic Melanoma Lesions

SYLVIE FROIDEVAUX, MARTINE CALAME-CHRISTE, LAZAR SUMANOVSKI, HEIDI TANNER, AND ALEX N. EBERLE

Laboratory of Endocrinology, Department of Research, University Hospital and University Children's Hospital, CH-4031, Basel, Switzerland

ABSTRACT: Scintigraphic imaging of metastatic melanoma lesions requires highly tumor-specific radiopharmaceuticals. Because both melanotic and amelanotic melanomas overexpress melanocortin-1 receptors (MC1R), radio-labeled analogues of α-melanocyte-stimulating hormone (α-MSH) are potential candidates for melanoma diagnosis. Here, we report the *in vivo* performance of a newly designed octapeptide analogue, [βAla^3, Nle^4, Asp^5, D-Phe^7, Lys^{10}]-α-MSH$_{3-10}$ (MSH$_{OCT}$), which was conjugated through its N-terminal amino group to the metal chelator 1,4,7,10-tetraazacyclododecane-1,4,7,10-tetraacetic acid (DOTA) to enable incorporation of radiometals (e.g., indium-111) into the peptide. DOTA-MSH$_{OCT}$ displayed high *in vitro* MC1R affinity (IC$_{50}$ 9.21 nM). *In vivo* [^{111}In]DOTA-MSH$_{OCT}$ exhibited a favorable biodistribution profile after injection in B16-F1 tumorbearing mice. The radi-opeptide was rapidly cleared from blood through the kidneys and, most importantly, accumulated preferentially in the melanoma lesions. Lung and liver melanoma metastases could be clearly imaged on tissue section autoradi-ographs 4 h after injection of [^{111}In]DOTA-MSH$_{OCT}$. A comparative study of [^{111}In]DOTA-MSH$_{OCT}$ with [^{111}In]DOTA-[Nle^4, D-Phe^7]-α-MSH ([^{111}In]-DOTA-NDP-MSH) demonstrated the superiority of the DOTA-MSH$_{OCT}$ peptide, particularly for the amount of radioactivity taken up by nonmalignant organs, including bone, the most radiosensitive tissue. These results demonstrate that [^{111}In]DOTA-MSH$_{OCT}$ is a promising melanoma imaging agent.

Keywords: melanoma imaging; MC1R; α-MSH; DOTA; receptor autoradiog-raphy; scintigraphy; tumor targeting

INTRODUCTION

Cutaneous melanoma incidence and mortality rates are increasing in the United States, as they are in most countries, and this skin cancer is currently the most common malignancy among young adults.[1,2] Unless tumors are detected early and adequate surgery can be performed, the prognosis of this disease is poor. This has

Address for correspondence: Sylvie Froidevaux, Laboratory of Endocrinology, Department of Research, University Hospital and University Children's Hospital, CH-4031, Basel, Switzerland. Voice: 41-61-265-2361; fax 41-61-265-2350.
sylvie.froidevaux@unibas.ch

Ann. N.Y. Acad. Sci. 994: 378–383 (2003). © 2003 New York Academy of Sciences.

motivated several investigations aimed at developing melanoma-specific radiopharmaceuticals for imaging and staging, but also with the intention of using these tools for internal radiotherapy. Thus, radiolabeled monoclonal antibodies or antibody fragments directed against specific melanoma cell epitopes[3–6] have been studied, but their clinical application has been restrained because of the occurrence of individual tumor variants. The finding that isolated melanoma cells and melanoma tissues overexpress high melanocortin-1 receptors (MC1R)[7–13] has opened the perspective of using radiolabeled α-MSH analogues as imaging tools. As compared with the use of antibodies or antibody fragments, the use of small peptides as vectors offers several advantages such as negligible immunogenicity, better tumor penetration, and faster blood clearance. The suitability of peptides as diagnostic tools is very well documented by the successful development of the peptide chelator conjugate OctreoScan which contains an N-terminal diethylenetriamine pentaacetic acid (DTPA) chelator for indium-111 attached to octreotide ($\rightarrow$ [^{111}In]DTPA-octreotide) for the visualization of somatostatin-receptor–positive malignancies in the clinic. In a similar approach, DTPA–α-MSH analogues have been synthesized and were shown to accumulate in melanoma.[14–16] However, their nonspecific accumulation in tissues known to be common sites of distant melanoma metastases such as liver[17] prevented further clinical development. Very recently, a technetium-labeled α-MSH analogue, incorporating the radiometal directly into the peptide molecule, was reported to exhibit favorable melanoma-targeting properties in a mouse model.[18,19]

A chelator with the capability to strongly incorporate a multitude 2+ or 3+ charged radiometals such as DOTA (1,4,7,10-tetraazacyclododecane-1,4,7,10-tetraacetic acid)[20,21] would greatly increase the range of applications for diagnosis and therapy of melanoma. Furthermore, it has been demonstrated that DOTA compares favorably with DTPA in studies of tumor-targeting with radiopeptides, both in animal models[22,23] and in patients.[24] Here, we report the development and biological characteristics of a novel DOTA-MSH$_{OCT}$ analogue suitable for melanoma targeting.

RESULTS AND DISCUSSION

[βAla^3, Nle^4, Asp^5, D-Phe^7, Lys^{10}]-α-MSH$_{3-10}$ (MSH$_{OCT}$) and the superpotent α-MSH analogue [Nle^4, D-Phe^7]-α-MSH (NDP-MSH)[25] were synthesized in our laboratory using the continuous flow technology and Fmoc (9-fluorenylmethoxycarbonyl) strategy and then conjugated to DOTA as previously described.[26] The resulting DOTA-peptides were found to retain good MC1R affinity as demonstrated by *in vitro* competition binding experiments (see Siegrist *et al.*[7] for methodology) using B16-F1 mouse melanoma cells and [^{125}I]-NDP-MSH as radioligand: IC$_{50}$ = 9.21 nM for DOTA-MSH$_{OCT}$ and 0.25 nM for DOTA-NDP-MSH. These data prompted us to perform tissue distribution experiments in a melanoma mouse model consisting of female B6D2F1 mice (C57Bl/6 × DBA/2 F1 hybrids, breeding pairs obtained from IFFA-CREDO, L'Arbresle, France) implanted subcutaneously with 0.5 million B16-F1 cells. All animal experiments were performed in compliance with the Swiss regulation for animal treatment. Incorporation of ^{111}In into DOTA-peptides was performed at 95°C for 25 min as previously described[26]; the resulting radioligands were found to be of high purity as demonstrated by reversed-phase high-performance liquid chromatography, and the specific activity was always of greater than 40 GBq/

TABLE 1. Tumor to non-target organ uptake ratios of [^{111}In]DOTA-NDP-MSH and [^{111}In]DOTA-MSH$_{OCT}$ 4 h after injection

	[^{111}In]DOTA-NDP-MSH	[^{111}In]DOTA-MSH$_{OCT}$
Blood	38.2 ± 5.28	165.0 ± 31.4
Muscle	7.25 ± 0.61	202.8 ± 47.3
Liver	14.0 ± 2.03	10.2 ± 0.70
Kidney	0.80 ± 0.07	0.3 ± 0.04
Spleen	10.9 ± 1.37	31.3 ± 1.30
Lung	30.2 ± 5.68	45.5 ± 3.50
Small intestine	22.67 ± 2.22	61.6 ± 7.90
Heart	101.67 ± 9.63	156.3 ± 12.2
Bone	4.96 ± 0.33	54.1 ± 5.90
Pancreas	42.6 ± 0.33	162.1 ± 10.2
Skin	6.95 ± 0.81	51.0 ± 6.80

[^{111}In]DOTA-MSH$_{OCT}$ or [^{111}In]DOTA-NDP-MSH was injected to melanoma-bearing mice, and the tissue-associated radioactivity was measured 4, 24, and 48 h after injection. Results are expressed as ratios of tumor uptake to normal organ uptake (each in percentage of injected dose per gram of tissue, % I.D./g), mean ± SEM, n = 4–8. The 4-h tumor uptake values were 6.32 ± 0.16% I.D./g for [^{111}In]DOTA-NDP-MSH and 4.31 0.30% I.D./g for [^{111}In]DOTA-MSH$_{OCT}$.

μmol. The radioligands (5 μCi), diluted in NaCl, 0.1% BSA, pH 7.5, were injected intravenously in the lateral tail vein in 200 μL 1 week after implantation of the tumor. The mice were killed 4 to 48 h later. Organs and tissues of interest were dissected and rinsed of excess blood, weighed, and assayed for radioactivity in a gamma-counter. Percentage of injected dose per gram (I.D./g) was calculated for each tissue and expressed as mean ± SEM (n = 4–8). Both DOTA peptides were found to accumulate in the tumor and to reach a maximum of 4.31 ± 0.30% I.D./g for [^{111}In]-DOTA-MSH$_{OCT}$ and 6.32 ± 0.16% for [^{111}In]DOTA-NDP-MSH 4 h after injection. Co-injection of an excess of α-MSH (50 μg) to block MC1R reduced the 4-h tumor uptake by an average of 86%, indicating that the radioligands are taken up in the melanoma by a receptor-mediated process. The blood clearance of [^{111}In]DOTA-MSH$_{OCT}$ was very rapid, and, 4 h after injection, the blood-associated radioactivity was as little as 0.03 ± 0.00% I.D./g. This was associated with a fast elimination of radioactivity from MC1R-negative tissues including those that are involved in the excretion process such as the liver and the small intestine. The only exception was the kidney which showed persistent retention of radioactivity, starting from 13.5 ± 1.12% I.D./g at 4 h after injection and still represented 6.56 ± 0.77% I.D./g 20 h later. Blockage of MC1R with α–MSH could not reduce the kidney uptake that demonstrates that this uptake is MC1R independent (not shown). The ratios of radioactivity in the melanoma to that in nontarget tissues 4 h after injection all were at least superior to 10 and often greater than 100 with, again, the exception of the kidney that exhibited a ratio equal to 0.3 (TABLE 1). In contrast, [^{111}In]DOTA-NDP-MSH accumulated to a much larger extent in nonmalignant organs (with the exception of

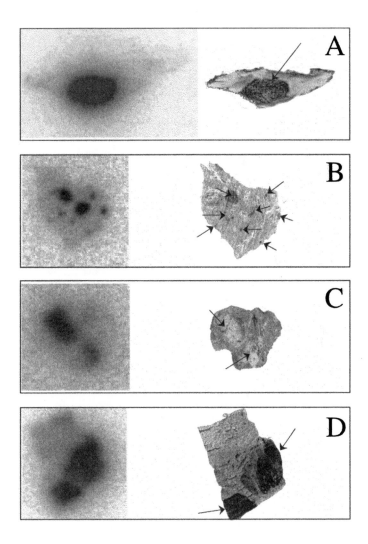

FIGURE 1. Autoradiographs of tissue sections from melanoma-bearing mice. $[^{111}In]DOTA\text{-}MSH_{OCT}$ was injected in mice inoculated with B16F1 cells, and tissues were collected 4 h after injection. (**A–D**) Autoradiographs (*left*) and scanner images (*right*) of a primary melanoma with surrounding skin tissue (**A**), lung with melanotic melanoma metastases (**B**), lung with amelanotic melanoma metastases (**C**), and liver with melanotic melanoma metastases (**D**). *Arrows* indicate the melanoma lesions.

the kidneys), leading to higher tumor/nontarget tissue ratios (TABLE 1). The suitability of $[^{111}In]$DOTA-MSH$_{OCT}$ to image melanoma lesions was demonstrated by tissue section autoradiography of a variety of tissues (skin, liver, lung) containing melanoma lesions collected from B6D2F1 mice previously injected with B16-F1 cells (FIG. 1): radioactivity indeed was found only in melanotic or amelanotic melanoma lesions, and surrounding tissue contained only background activity.

CONCLUSION

The data presented here demonstrate that the newly designed $[^{111}In]$DOTA-MSH$_{OCT}$ exhibits a very favorable performance in a melanoma tumor mouse model which makes this radioligand a promising imaging agent for the clinic. Moreover, because DOTA is able to form stable complexes with various radiometals[20,21,27,28] including those exhibiting suitable physical characteristics for consideration as potential therapeutic radionuclides (e.g., Y-90, Ga-67), $[^{111}In]$DOTA-MSH$_{OCT}$ is a good lead candidate for radiotherapy if administered in combination with pharmaceuticals known to lower nonspecific renal uptake of radiopeptides.[22,29]

ACKNOWLEDGMENTS

We thank Dr. J. Baumann for her critical review of the manuscript and A. M. Meier for her excellent technical assistance. This work was supported by the Swiss Cancer League and the Roche Research Foundation.

REFERENCES

1. DENNIS, L.K. 1999. Analysis of the melanoma epidemic, both apparent and real: data from the 1973 through 1994 surveillance, epidemiology, and end results program registry. Arch. Dermatol. **135:** 275–280.
2. VOSMIK, F. 1996. Malignant melanoma of the skin. Epidemiology, risk factors, clinical diagnosis. Cas. Lek. Cesk. **135:** 405–408.
3. MARCHITTO, K.S., W.R. KINDSVOGEL, P.L. BEAUMIER, et al. 1989. Characterization of a human-mouse chimeric antibody reactive with a human melanoma associated antigen. Prog. Clin. Biol. Res. **288:** 101–105.
4. BEAUMIER, P.L., K.A. KROHN, J.A. CARRASQUILLO, et al. 1985. Melanoma localization in nude mice with monoclonal Fab against p97. J. Nucl. Med. **26:** 1172–1179.
5. SALK, D. 1988. Technetium-labeled monoclonal antibodies for imaging metastatic melanoma: results of a multicenter clinical study. Semin. Oncol. **15:** 608–618.
6. TAYLOR, A., JR., W. MILTON, H.P. EYRE, et al. 1988. Radioimmunodetection of human melanoma with indium-111-labeled monoclonal antibody. J. Nucl. Med. **29:** 329–337.
7. SIEGRIST, W., F. SOLCA, S. STUTZ, et al. 1989. Characterization of receptors for alpha-melanocyte-stimulating hormone on human melanoma cells. Cancer Res. **49:** 6352–6358.
8. BAGUTTI, C., M. OESTREICHER, W. SIEGRIST, et al. 1995. alpha-MSH receptor autoradiography on mouse and human melanoma tissue sections and biopsies. J. Recept. Signal Transduct. Res. **15:** 427–442.
9. SIEGRIST, W., C. BAGUTTI, F. SOLCA, et al. 1992. MSH receptors on mouse and human melanoma cells: receptor identification, analysis and quantification. Prog. Histochem. Cytochem. **26:** 110–118.

10. SIEGRIST, W., J. GIRARD & A.N. EBERLE. 1991. Quantification of MSH receptors on mouse melanoma tissue by receptor autoradiography. J. Recept. Res. **11:** 323–331.
11. TATRO, J.B., M.L. ENTWISTLE, B.R. LESTER, *et al.* 1990. Melanotropin receptors of murine melanoma characterized in cultured cells and demonstrated in experimental tumors in situ. Cancer Res. **50:** 1237–1242.
12. LOIR, B., C. PEREZ SANCHEZ, G. GHANEM, *et al.* 1999. Expression of the MC1 receptor gene in normal and malignant human melanocytes. A semiquantitative RT-PCR study. Cell. Mol. Biol. **45:** 1083–1092.
13. JIANG, J., S.D. SHARMA, J.L. FINK, *et al.* 1996. Melanotropic peptide receptors: membrane markers of human melanoma cells. Exp. Dermatol. **5:** 325–333.
14. BARD, D.R., C.G. KNIGHT & D.P. PAGE-THOMAS. 1990. Targeting of a chelating derivative of a short-chain analogue of alpha-melanocyte stimulating hormone to Cloudman S91 melanomas. Biochem. Soc. Trans. **18:** 882–883.
15. BAGUTTI, C., B. STOLZ, R. ALBERT, *et al.* 1994. [^{111}In]-DTPA-labeled analogues of alpha-melanocyte-stimulating hormone for melanoma targeting: receptor binding in vitro and in vivo. Int. J. Cancer **58:** 749–755.
16. WRAIGHT, E.P., D.R. BARD, T.S. MAUGHAN, *et al.* 1992. The use of a chelating derivative of alpha melanocyte stimulating hormone for the clinical imaging of malignant melanoma. Br. J. Radiol. **65:** 112–118.
17. BALCH, C.M., S.J. SOONG, T.M. MURAD, *et al.* 1983. A multifactorial analysis of melanoma. IV. Prognostic factors in 200 melanoma patients with distant metastases (stage III). J. Clin. Oncol. **1:** 126–134.
18. CHEN, J., Z. CHENG, T.J. HOFFMAN, *et al.* 2000. Melanoma-targeting properties of ^{99m}technetium-labeled cyclic alpha-melanocyte-stimulating hormone peptide analogues. Cancer Res. **60:** 5649–5658.
19. GIBLIN, M.F., N. WANG, T.J. HOFFMAN, *et al.* 1998. Design and characterization of alpha-melanotropin peptide analogs cyclized through rhenium and technetium metal coordination. Proc. Natl. Acad. Sci. USA **95:** 12814–12818.
20. CAMERA, L., S. KINUYA, K. GARMESTANI, *et al.* 1994. Evaluation of the serum stability and in vivo biodistribution of CHX-DTPA and other ligands for yttrium labeling of monoclonal antibodies. J. Nucl. Med. **35:** 882–889.
21. KOZAK, R.W., A. RAUBITSCHEK, S. MIRZADEH, *et al.* 1989. Nature of the bifunctional chelating agent used for radioimmunotherapy with yttrium-90 monoclonal antibodies: critical factors in determining in vivo survival and organ toxicity. Cancer Res. **49:** 2639–2644.
22. DE JONG, M., W.H. BAKKER, W.A. BREEMAN, *et al.* 1998. Pre-clinical comparison of [DTPA^0] octreotide, [DTPA^0,Tyr^3] octreotide and [DOTA^0,Tyr^3] octreotide as carriers for somatostatin receptor-targeted scintigraphy and radionuclide therapy. Int. J. Cancer **75:** 406–411.
23. FROIDEVAUX, S., A. HEPPELER, A.N. EBERLE, *et al.* 2000. Preclinical comparison in AR4-2J tumor bearing mice of four radiolabeled DOTA-somatostatin analogs for tumor diagnosis and internal radiotherapy. Endocrinology **141:** 3304–3312.
24. OTTE, A., R. HERRMANN, A. HEPPELER, *et al.* 1999. Yttrium-90 DOTATOC: first clinical results. Eur. J. Nucl. Med. **26:** 1439–1447.
25. SAWYER, T.K., P.J. SANFILIPPO, V.J. HRUBY, *et al.* 1980. 4-Norleucine, 7-D-phenylalanine-alpha-melanocyte-stimulating hormone: a highly potent alpha-melanotropin with ultralong biological activity. Proc. Natl. Acad. Sci. USA **77:** 5754–5758.
26. HEPPELER, A., S. FROIDEVAUX, H.R. MÄCKE, *et al.* 1999. Radiometal labelled macrocyclic chelator derivatised somatostatin analogue with superb tumour targeting properties and potential for receptor mediated internal radiotherapy. Chem. Eur. J. **5:** 1016–1023.
27. DESHPANDE, S.V., S.J. DENARDO, D.L. KUBIS, *et al.* 1990. Yttrium-90-labeled antibody for therapy: labeling by a new macrocyclic bifunctional chelating agent. J. Nucl. Med. **31:** 473–479.
28. CHINOL, M., G. PAGANELLI, F. SUDATI, *et al.* 1997. Biodistribution in tumour-bearing mice of two ^{90}Y-labelled biotins using three-step tumour targeting. Nucl. Med. Commun. **18:** 176–182.
29. HAMMOND, P.J., A.F. WADE, M.E. GWILLIAM, *et al.* 1993. Amino acid infusion blocks renal tubular uptake of an indium-labelled somatostatin analogue. Br. J. Cancer **67:** 1437–1439.

Index of Contributors